U.S. Customary unit		Times conversion factor		Equals SI unit	
		Accurate	Practical		
Moment of inertia (area)					
inch to fourth power	in.^{4}	416,231	416,000	millimeter to fourth power	mm^4
inch to fourth power	in.^{4}	0.416231×10^{-6}	0.416×10^{-6}	meter to fourth power	m^4
Moment of inertia (mass)					
slug foot squared	slug-ft^2	1.35582	1.36	kilogram meter squared	kg·m^2
Power					
foot-pound per second	ft-lb/s	1.35582	1.36	watt (J/s or N·m/s)	W
foot-pound per minute	ft-lb/min	0.0225970	0.0226	watt	W
horsepower (550 ft-lb/s)	hp	745.701	746	watt	W
Pressure; stress					
pound per square foot	psf	47.8803	47.9	pascal (N/m^2)	Pa
pound per square inch	psi	6894.76	6890	pascal	Pa
kip per square foot	ksf	47.8803	47.9	kilopascal	kPa
kip per square inch	ksi	6.89476	6.89	megapascal	MPa
Section modulus					
inch to third power	in.^{3}	16,387.1	16,400	millimeter to third power	mm^3
inch to third power	in.^{3}	16.3871×10^{-6}	16.4×10^{-6}	meter to third power	m^3
Velocity (linear)					
foot per second	ft/s	0.3048*	0.305	meter per second	m/s
inch per second	in./s	0.0254*	0.0254	meter per second	m/s
mile per hour	mph	0.44704*	0.447	meter per second	m/s
mile per hour	mph	1.609344*	1.61	kilometer per hour	km/h
Volume					
cubic foot	ft^3	0.0283168	0.0283	cubic meter	m^3
cubic inch	in.^{3}	16.3871×10^{-6}	16.4×10^{-6}	cubic meter	m^3
cubic inch	in.^{3}	16.3871	16.4	cubic centimeter (cc)	cm^3
gallon (231 in.^{3})	gal.	3.78541	3.79	liter	L
gallon (231 in.^{3})	gal.	0.00378541	0.00379	cubic meter	m^3

*An asterisk denotes an *exact* conversion factor

Note: To convert from SI units to USCS units, *divide* by the conversion factor

Temperature Conversion Formulas

$$T(°C) = \frac{5}{9}[T(°F) - 32] = T(K) - 273.15$$

$$T(K) = \frac{5}{9}[T(°F) - 32] + 273.15 = T(°C) + 273.15$$

$$T(°F) = \frac{9}{5}T(°C) + 32 = \frac{9}{5}T(K) - 459.67$$

INTRODUCTION TO ENVIRONMENTAL ENGINEERING

THIRD EDITION, SI

P. AARNE VESILIND *Bucknell University*

SUSAN M. MORGAN *Southern Illinois University Edwardsville*

LAUREN G. HEINE *Clean Production Action*

SI EDITION *Prepared by*

TARUN GUPTA *Indian Institute of Technology Kanpur*

CENGAGE
Learning™

Australia · Brazil · Japan · Korea · Mexico · Singapore · Spain · United Kingdom · United States

CENGAGE Learning™

Introduction to Environmental Engineering, Third Edition, SI

P. Aarne Vesilind, Susan M. Morgan, and Lauren G. Heine

SI Edition prepared by Tarun Gupta

Publisher: Christopher M. Shortt

Senior Developmental Editor:
Hilda Gowans

Acquisitions Editor, SI Edition:
Swati Meherishi

Editorial Assistant: Tanya Altieri

Team Assistant: Carly Rizzo

Marketing Manager:
Lauren Betsos

Content Project Manager: Diane Bowdler

Production Service: RPK Editorial
Services, Inc.

Copyeditor: Fred Dahl

Proofreader: Erin Wagner

Indexer: Shelly Gerger-Knechtl

Compositor: Integra

Senior Art Director:
Michelle Kunkler

Internal Designer: Carmela Pereira

Cover Designer: Andrew Adams

Cover Image: © Carlosthecat/
Dreamstime.com

Text and Photo Permissions Researcher:
Kristiina Paul

Senior First Print Buyer:
Arethea L. Thomas

For product information and technology assistance, contact us at
**Cengage Learning Customer & Sales Support,
1-800-354-9706.**

For permission to use material from this text or product, submit all requests online at
www.cengage.com/permissions.
Further permissions questions can be emailed to
permissionrequest@cengage.com.

Library of Congress Control Number: 2009940223

ISBN-13: 978-0-495-29585-3
ISBN-10: 0-495-29585-X

Cengage Learning
200 First Stamford Place, Suite 400
Stamford, CT 06902
USA

Cengage Learning is a leading provider of customized learning solutions with office locations around the globe, including Singapore, the United Kingdom, Australia, Mexico, Brazil, and Japan. Locate your local office at: **international.cengage.com/region.**

Cengage Learning products are represented in Canada by Nelson Education Ltd.

For your course and learning solutions, visit
www.cengage.com/engineering.

Purchase any of our products at your local college store or at our preferred online store **www.ichapters.com.**

Printed in the United States of America
1 2 3 4 5 6 7 12 11 10 09

This book is dedicated, with gratitude, to
the late Edward E. Lewis—
publisher, curmudgeon, and friend.

P. Aarne Vesilind

And to Steven J. Hanna, PhD, PE—
who introduced me to the profession of
engineering and answered all those
questions the first few years on the job.

Susan M. Morgan

And to Abbie, Orion, Aragorn, and Scout
and to friends of all species who share the
love that powers our lives.

Lauren G. Heine

CONTENTS

P A R T   T W O

CHAPTER TEN
Water Supply and Treatment   281

CHAPTER ELEVEN
Wastewater Treatment   342

PREFACE TO THE SI EDITION

This edition of **Introduction to Environmental Engineering** has been adapted to incorporate the International System of Units (*Le Système International d'Unités or SI*) throughout the book.

LE SYSTÈME INTERNATIONAL D'UNITÉS

The United States Customary System (USCS) of units uses FPS (foot-pound-second) units (also called English or Imperial units). SI units are primarily the units of the MKS (meter-kilogram-second) system. However, CGS (centimeter-gram-second) units are often accepted as SI units, especially in textbooks.

USING SI UNITS IN THIS BOOK

In this book, we have used both MKS and CGS units. USCS units or FPS units in the US Edition of the book have been converted to SI units throughout the text and problems. However, in case of data sourced from handbooks, government standards, and product manuals, it is not only extremely difficult to convert all values to SI, it also encroaches upon the intellectual property of the source. Also, some quantities such as the ASTM grain size number and Jominy distances are generally computed in FPS units and would lose their relevance if converted to SI. Some data in figures, tables, examples, and references, therefore, remain in FPS units. For readers unfamiliar with the relationship between the FPS and the SI systems, conversion tables have been provided inside the front and back covers of the book.

To solve problems that require the use of sourced data, the sourced values can be converted from FPS units to SI units just before they are to be used in a calculation. To obtain standardized quantities and manufacturers' data in SI units, the readers may contact the appropriate government agencies or authorities in their countries/regions.

INSTRUCTOR RESOURCES

A printed Instructor's Solution Manual in SI units is available on request. An electronic version of the Instructor's Solutions Manual and PowerPoint slides of the figures from the SI text are available through www.cengage.com/engineering.

The readers' feedback on this SI Edition will be highly appreciated and will help us improve subsequent editions.

The Publishers

PREFACE

The third edition of *Introduction to Environmental Engineering* continues to have two unifying themes: material balances and environmental ethics. It also adds more information regarding sustainability, an increasingly important component of engineering practice and intricately intertwined with the two unifying themes.

ORGANIZATION

Part One opens with a brief look at the history and future of environmental engineering and then provides examples of the complex issues that surround identifying and solving environmental problems. It continues by introducing various tools engineers use in making decisions, including technology, benefit/cost, risk, and ethics. The discussion on ethics is brief, serving merely as an introduction to some of the value-laden problems engineers face and providing some background for the discussions throughout the text, but it suggests that ethical decision making is just as important in engineering as technical decision making.

Part Two opens with a review of and introduction to basic concepts: dimensions and units, density, concentration, flow rate, retention time, and approximations. This material leads directly to the introduction of material balances, a theme used throughout the book. The discussion of reactions is similar to what would be covered in an introductory physical chemistry or biochemical engineering course and is followed by ideal reactor theory, similar to material found in a chemical engineering unit operations course but at a level readily understood by freshmen engineering students. The mass balance approach is then applied to energy flow. This part concludes with the recognition that some of the most fascinating reactions occur in ecosystems, and ecosystems are described using the mass balance approach and the reaction kinetics introduced earlier.

Part Three applies the fundamental concepts covered in Part Two to the major areas of environmental engineering. It begins with the quest for clean water. When is water clean enough, and how do we measure water quality characteristics? This discussion is followed by an introduction to water supply and treatment and then wastewater treatment. Because meteorology determines the motion of air pollutants, this topic introduces the section on air pollution. The types and sources of air pollutants of concern are discussed, concluding with the evolution of air quality standards, treatment of emissions, and the dispersion of pollutants. The management of municipal solid waste and hazardous waste are discussed with an emphasis on the need to prevent their generation. Noise pollution is then introduced. The book concludes with an evaluation of the motivations for pursuing green engineering.

REQUIREMENTS AND PRESENTATION

The material on ethics is at a basic level, so it is readily understandable by any engineering instructor or student. No formal preparation in ethics is required. The technical material is at a level that a freshman engineer or a BS environmental science student can readily digest. Calculus is used in the text, and it is assumed that the student has had at least one course in differential and integral calculus. A college chemistry course is useful but not mandatory. Fluid mechanics is not used in the textbook, and hence the material is readily applicable to freshman courses in environmental engineering or environmental engineering courses for science students.

Experience in sequentially teaching the core material of Part Two (balances, reactions, reactors) by Dr. Vesilind has shown that it can be covered in four to six weeks. He recommends, however, that the instructor take lateral excursions into areas of environmental ethics during this time as well as embellish the lectures with personal "war stories" to maintain student interest. Introducing complementary material throughout the course is an effective teaching technique, grounded in modern learning theory. The material in Part Three may be used in any sequence deemed best without losing its value or meaning.

The third edition introduces a feature called "Focus On"—a collection of vignettes and case studies illustrating approaches to solving technical, ethical, and sustainability-oriented problems. They present opportunities for the instructor and students to delve into material not often covered in undergraduate education, such as the aesthetics and potential social implications of design. The ethical materials introduced throughout the text similarly require that both the student and the instructor pause and discuss technical problems in a different light, thus reinforcing the material learned. Because ethics and taste are such personal issues, the discussion of technical matters from these perspectives tends to internalize the subject and create, in effect, a learning experience similar to what would be achieved by a field trip or a narrative of a real-world experience.

ACKNOWLEDGMENTS

Thank you to those who took the time to send comments on the second edition. We have incorporated them as appropriate in this edition. Continued thanks go to my family, who patiently put up with me during the revisions, and to colleagues who willingly shared their expertise answering out-of-the-blue questions.

Susan M. Morgan, PE
2008

Thank you to Susan and Aarne for inviting me to participate in this revision and to add some examples and perspective from the sustainability field. I hope it will be of value and inspiration to students. It has been a pleasure to work with you.

I am very grateful to my mentors and friends Michael Braungart, P. Aarne Vesilind, and Paul Anastas whose brilliance, humor, and vision informed and inspired me deeply. I am also grateful for contributions to this book by Margaret Whittaker of ToxServices LLC, Erin Kanoa of UrbanMarmot, Dan Bihn, Namara Brede, and James Hill. Finally, I am grateful to my dear husband Carl for his patience and support.

Lauren G. Heine, PhD
2008

ABOUT THE AUTHORS

P. Aarne Vesilind, PE

Vesilind was born in Estonia and emigrated to the United States in 1949. He grew up in Beaver, Pennsylvania, a small town downriver from Pittsburgh. Following his undergraduate degree in civil engineering from Lehigh University, he received his PhD in environmental engineering from the University of North Carolina. He spent a postdoctoral year with the Norwegian Institute for Water Research in Oslo and a year as a research engineer with Bird Machine Company. He joined the faculty at Duke University in 1970. In 1999 he was appointed to the R. L. Rooke Chair of the Historical and Societal Context of Engineering at Bucknell University. He served in this capacity until his retirement in 2006.

His research has resulted in the authorship of over 175 professional articles. He has also authored over 20 books on environmental engineering and professional ethics, one of the latest being *Socially Responsible Engineering* (John Wiley & Sons), which considers the role of justice in engineering decisions. His wildly popular book *Estonian Jokes* was published in 2009 by Punkt&Koma in Tallinn.

Since June 2006 he has lived in New London, New Hampshire, and has recently been appointed visiting scholar by the Ethics Institute at Dartmouth College. In retirement, he has continued his lifelong interest in music and is playing euphonium in several local bands, including The Exit 13 Tuba Quartet. In 2007 he was asked to take over the job of conductor of the Kearsarge Community Band.

Susan M. Morgan, PE

Professor and Chair
Department of Civil Engineering
Southern Illinois University Edwardsville

Morgan received her BS in civil engineering from southern Illinois University Carbondale. A recipient of a National Science Foundtion Fellowship, she earned her PhD in environmental engineering from Clemson University. She joined the faculty in the Department of Civil Engineering at Southern Illinois University Edwardsville in 1996. From 1999 to 2007, she served as the graduate program director for the department. Currently, she is a tenured professor and department chair. She is a licensed professional engineer in Illinois.

Dr. Morgan has been active on the Environmental Technical Committee of the St. Louis Section of the American Society of Civil Engineers and in the St. Clair Chapter of the Illinois Society of Professional Engineers. She has received multiple awards, including the National Society of Professional Engineers' Young Engineer of the Year Award in 2001. She is a member of several honor societies, including Chi Epsilon and Tau Beta Pi, as well as other engineering organizations. She has conducted research in a variety of areas. Currently, her focus is on stormwater management, particularly through the use of green roofs.

Lauren G. Heine, PhD

Senior Science Advisor, Clean Production Action
Principal, Lauren Heine Group LLC

Heine earned her doctorate in civil and environmental engineering from Duke University. She is one of America's leading experts in applying green chemistry, green engineering, and design for the environment for sustainable business practices. As senior science advisor for Clean Production Action and as principal for the Lauren Heine Group, she guides organizations seeking to integrate green chemistry and engineering into their product and process design and development activities—eliminating toxics and the concept of waste and moving toward economic, environmental and community sustainability. Specific areas of expertise include the development of technical tools and strategies for identifying greener chemicals, materials, and products and for the facilitation of multistakeholder initiatives, particularly those that are technically based.

From 2003 to 2007, Dr. Heine served as director of applied science at Green Blue Institute, a nonprofit organization founded by architect William McDonough and German chemist Michael Braungart to focus on sustainable product design. She initiated and directed the development of CleanGredients™, a unique, web-based information platform, developed in partnership with the USEPA's Design for the Environment Program that provides information on key human and environmental health, safety, and sustainability attributes of chemical raw materials to help with cleaning product formulation. She also led the development of the Sustainability Assessment Standard for Contract Furnishing Fabrics, an American National Standards Institute standard, in collaboration with the Association for Contract Textiles and NSF International.

Dr. Heine served on the California Green Chemistry Initiative Science Advisory Panel and is cochairing the tool development subcommittee for Wal-Mart's Sustainability Network for Chemical Intensive Products. She publishes on green chemistry metrics, alternatives assessment, and multistakeholder process. She cofounded the Oregon-based Zero Waste Alliance and was a fellow with the American Association for the Advancement of Science in the Green Chemistry Program of the Industrial Chemicals Branch of the USEPA in Washington, D.C. Prior to that, she taught organic chemistry labs at Bowdoin College in Brunswick, Maine, where she helped to develop the Microscale Organic Lab program. Dr. Heine currently lives in Bellingham, Washington.

ENVIRONMENTAL ENGINEERING

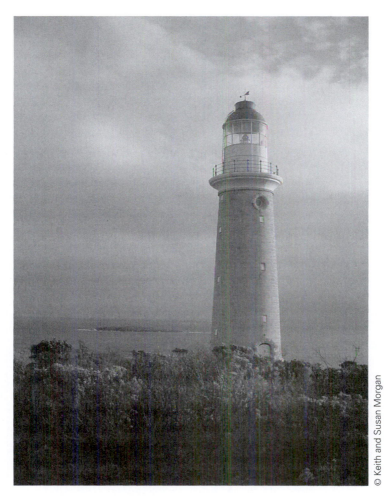

Cape du Couedic Lighthouse, Australia

Environmental engineers provide not only warnings of danger but light to lead the way towards a sustainable standard of living to protect human health and the environment.

CHAPTER ONE

Identifying and Solving Environmental Problems

Bald eagle

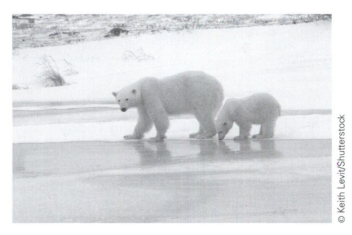

Polar bears

Environmental engineers need to be aware of the lessons of the past—how problems came about and how scientists, engineers, policy makers, and others worked together to solve them. We then need to apply those lessons as appropriate to solve current problems and prevent similar mistakes in the future.

1.1 WHAT IS ENVIRONMENTAL ENGINEERING?

Environmental engineering has a long history, although the phrase "environmental engineering" is relatively new. It is useful to review briefly that history and look at what the future holds before delving into specific examples and the nitty-gritty of concepts and calculations.

1.1.1 The Origins of Environmental Engineering

The roots of environmental engineering reach back to the beginning of civilization. Providing clean water and managing wastes became necessary whenever people congregated in organized settlements. For ancient cities, the availability of a dependable water source often meant the difference between survival and destruction, and a water supply became a defensive necessity. The builders of wells and aqueducts were the same people who were called on to build the city walls and moats, as well as the catapults and other engines of war. These men became the engineers of antiquity. It was not until the mid-1700s that engineers who built facilities for the civilian population began to distinguish themselves from the engineers primarily engaged in matters of warfare, and the term "civil engineering" was born. In the formative years of the United States, engineers were mostly self-educated or were trained at the newly formed United States Military Academy. Civil engineers—the builders of roads, bridges, buildings, and railroads—were called on to design and construct water supplies for the cities, and to provide adequate systems for the management of waterborne wastes and storm water.

The advent of industrialization brought with it unbelievably unsanitary conditions in the cities because of the lack of water and waste management. There was no public outcry, however, until it became evident that water could carry disease. From that time on, civil engineers had to more than just provide an adequate supply of water; they now had to make sure the water would not be a vector for disease transmission. Public health became an integral concern of the civil engineers entrusted with providing water supplies to the population centers, and the elimination of waterborne disease became the major objective in the late 19th century. The civil engineers entrusted with the drainage of cities and the provision of clean water supplies became public health engineers (in Britain) and sanitary engineers (in the United States).

1.1.2 Environmental Engineering Today

Sanitary engineers have achieved remarkable reductions in the transmission of acute disease by contaminated air or water. In the United States, the acute effects of pollution are for all intents and purposes eliminated. These acute concerns have been replaced, however, by more complex and chronic problems such as climate change; depleting aquifers; indoor air pollution; global transport of persistent, bioaccumulating and toxic chemicals; synergistic impacts of complex mixtures of human-made chemicals from household products and pharmaceuticals in wastewater effluents, rivers and streams; endocrine-disrupting chemicals; and a lack of information on the effect on human and environmental health and safety of rapidly emerging new materials, such as nanoparticles. Challenges to individual environmental media such as air and water can no longer be considered and managed within individual compartments. They must be managed at the ecosystem level to avoid shifting pollution concerns from one environmental medium to another. To address these chronic problems before they become acute, scientists and engineers are seeking to understand the environment, cities, and industry as interacting systems (i.e., as interconnected ecosystems, social systems, and industrial systems) and to think proactively and preemptively so that we can avoid unintended consequences rather than having to manage them reactively.

In most developed countries today, public opinion has evolved to where the direct and immediate health effects of environmental contamination are no longer the sole concern. The cleanliness of streams, for the benefit of the stream itself, has become a driving force, and legislation has been passed addressing our desire for a clean environment. The protection of wildlife habitat, the preservation of species, and the health of ecosystems have become valid objectives for the spending of resources. Such a sense of mission, often referred to as an *environmental ethic*, is a major driving force behind modern environmental engineering and is demanded by the public as a public value. In the 20th century, an environmental ethic was often pitted against the desires of those who wished to exploit natural resources for human gain. Common thinking assumed that a trade-off had to be made: One had to choose between the economy *or* the environment.

1.1.3 Environmental Engineering on the Horizon

In the 21st century it is apparent that ecosystems and the natural capital on our planet are not inexhaustible. Preserving and maintaining the health, economic, and social well-being of people depends on preserving and maintaining the integrity of ecosystems and the ecosystem services they provide. The solution is not a trade-off. The solution is the well-being of the economy *and* the environment. A common goal has emerged throughout the world—the goal of sustainable development—defined by a 1987 United Nations commission in the Brundtland Report as "development that meets the needs of the present without compromising the ability of future generations to meet their own needs." Sustainable development (sometimes referred to as "sustainability") means different things to different people and communities depending on the nature of their activities and the cultural, geographic, economic, and environmental contexts in which they operate.

In the past, environmental engineering was a reactive profession, reacting to the problems created by the growth of world populations and the increase in our standard of living. The future problems that environmental engineers will address can be extrapolated. We know, for example, the pollution problems that only a few decades ago were local problems are now global in scale. We also know that the continued use and discharge of new and old chemicals into the environment will have unpredictable and sometimes synergistic effects. And we know that focusing on energy efficiency and other efficiency improvements alone will not adequately address emerging resource constraints as populations grow and the quality of life in developing countries improves. Environmental engineers, now and in the future, will play a key role in realizing sustainable development.

Needless to say, classical sanitary engineering education based on applied hydraulics, public health, and chemical engineering processes is inadequate to deal with these complex matters. The new breed of environmental engineers entrusted with the protection of human health and the environment will embrace the natural sciences and will deal on a cutting-edge level with the application of biological and chemical sciences, including nanotechnology, biotechnology, information technology, chemical fate and toxicity, and life cycle impact considerations. Environmental engineers will also gain greater understanding of industrial processes, including product design and development. And finally, because these problems are all in the public domain, they will learn to apply the social sciences such as public policy, communications, and economics and will learn to work with diverse stakeholders to solve problems. Since all of these topics are unlikely to be fully addressed in current civil or environmental engineering curricula, students may need to seek exposure to them in sister academic programs, such as business management, environmental science and management, and public policy.

Environmental engineers will use their design training to be proactive and preemptive in the development of solutions. As they move away from a focus on end-of-pipe treatment and even beyond pollution prevention through engineering controls, they will move toward the use of design to prevent problems from the start. Emerging fields such as green chemistry, green engineering, and design for the environment, discussed at various points in this textbook, will help environmental and other engineers develop sustainable products and foster sustainable materials management.

As the problems faced by environmental engineers continue to grow more complex, there is a growing need for principles and frameworks to guide the development of solutions. In 2003, approximately 65 engineers and scientists convened in Sandestin, Florida to develop a set of principles for green engineering. The U.S. Environmental Protection Agency (EPA) defines green engineering as "the design, commercialization, and use of processes and products [that] are feasible and economical while minimizing 1) generation of pollution at the source and 2) risk to human health and the environment." The group conceived a set of principles that went beyond what was typically seen as the scope of green engineering to address social elements. As such, they became known as the Sandestin Principles for Sustainable Engineering (Sandestin Principles).[1,2] Other sets of principles have been developed to support the design and development of products and processes with benefits for human health and the environment. These include the 12 Principles of Green Chemistry and the 12 Principles of Green Engineering.[3] The USEPA has adopted the nine Sandestin Principles as the Principles of Green Engineering. Together, these principles provide a guide and a broad framework for all engineers who seek to help solve the problems of the 21st century. They characterize well the expanding role of the environmental engineer (Table 1.1).

Sustainable engineering transforms existing engineering disciplines and practices into those that promote sustainability. Sustainable engineering incorporates the development and implementation of technologically and economically viable products, processes, and systems that promote human welfare, while protecting human health and elevating the protection of the biosphere as a criterion in engineering solutions.

These principles can serve as a guide for environmental engineers in all types of employment—whether in government, consulting, academia, or private industry. The public recognizes and appreciates the work of the environmental engineer and is prepared to use societal resources to achieve sustainability. The professional environmental engineering community must prepare now to meet this challenge, ensuring that environmental engineers continue to achieve the depth of technical expertise typically expected of environmental engineers and complemented with the ability to understand problems at the system level and to collaborate productively with experts and nonexperts from other disciplines and sectors.

Table 1.1 Sandestin Principles for Sustainable Engineering

1. Engineer processes and products holistically, use systems analysis, and integrate environmental impact assessment tools.
2. Conserve and improve natural ecosystems while protecting human health and well-being.
3. Use life-cycle thinking in all engineering activities.
4. Ensure that all material and energy inputs and outputs are as inherently safe and benign as possible.
5. Minimize depletion of natural resources.
6. Strive to prevent waste.
7. Develop and apply engineering solutions, while being cognizant of local geography, aspirations, and cultures.
8. Create engineering solutions beyond current or dominant technologies; improve, innovate, and invent (technologies) to achieve sustainability.
9. Actively engage communities and stakeholders in development of engineering solutions.

1.2 CASE STUDIES

Following are examples of environmental problems that have been identified and solved. They illustrate some of the principles and controversies inherent in the field of environmental engineering.

1.2.1 The Holy Cross College Hepatitis Outbreak

Following the Dartmouth game, members of the 1969 Holy Cross College football team got sick.[4] They had high fever, nausea, abdominal pain, and were becoming jaundiced— all characteristics of infectious hepatitis. During the next few days over 87 members of the football program—players, coaches, trainers, and other personnel—became ill. The college cancelled the remainder of the football season and became the focus of an epidemiological mystery. How could an entire football team have contracted infectious hepatitis?

The disease is thought to be transmitted mostly from person to person, but epidemics can also occur, often due to contaminated seafood or water supplies. There are several types of hepatitis virus, with widely ranging effects on humans. The least deadly is Hepatitis A virus, which results in several weeks of feeling poorly and seldom has lasting effects; Hepatitis B and C, however, can result in severe problems, especially liver damage, and can last for many years. At the time of the Holy Cross epidemic the hepatitis virus had not been isolated and little was known of its etiology or effects.

When the college became aware of the seriousness of the epidemic, it asked for and received help from state and federal agencies, which sent epidemiologists to Worcester. The epidemiologists' first job was to amass as much information as possible about the members of the football team—who they had been with, where they had gone, what they had eaten, and what they had drunk. The objective was then to deduce from the clues how the epidemic had occurred. Some of the information they knew or found out was as follows:

- Since the incubation period of hepatitis is about 25 days, the infection had to have occurred sometime before 29 August or thereabouts.
- Football players who left the team before 29 August were not infected.
- Varsity players who arrived late, after 29 August, similarly were not infected.
- Freshman football players arrived on 3 September, and none of them got the disease.
- Both the freshmen and varsity players used the same dining facilities, and since none of the freshmen became infected, it was unlikely that the dining facilities were to blame.
- A trainer who developed hepatitis did not eat in the dining room.
- There was no common thread of the players having eaten at restaurants where contaminated shellfish might have been the source of the virus.
- The kitchen prepared a concoction of sugar, honey, ice, and water for the team (the Holy Cross version of Gatorade), but since the kitchen staff sampled this drink before and after going to the practice field and subsequently none of the kitchen staff developed hepatitis, it could not have been the drink.

The absence of alternatives forced the epidemiologists to focus on the water supply. The college receives its water from the city of Worcester, and a buried line provides water from Dutton Street, a dead-end street, to the football practice field, where a drinking fountain is used during practices. Samples of water taken from that fountain showed no contamination. The absence of contamination during the sampling did not, however, rule out the possibility of disease transmission through this water line. The line ran to the practice field through a meter pit and a series of sunken sprinkler boxes used for watering the field (Figure 1.1).

Two other bits of information turned out to be crucial. One of the houses on Dutton Street was found to have kids who had hepatitis. The kids played near the sprinkler boxes during the summer evenings and often opened them, splashing around in the small ponds created in the pits. But how did the water in the play ponds, if the children had contaminated it, get into the water line with the line always under positive pressure?

The final piece of the puzzle fell into place when the epidemiologists found that a large fire had occurred in Worcester during the evening of 28 August lasting well into the early hours of the next day. The demand for water for this fire was so great that the residences on Dutton Street found themselves without any water pressure at all. That is, the pumpers putting out the fire were pumping at such a high rate that the pressure in the

Figure 1.1 The pipeline that carried water to the Holy Cross practice field came from Dutton Street and went through several sprinkler boxes.

water line became negative. If, then, the children had left some of the valves in the sprinkler boxes open and if they had contaminated the water around the box, the hepatitis virus must have entered the drinking water line. The next morning, as pressure was resumed in the water lines, the contaminated water was pushed to the far end of the line—the drinking fountain on the football field—and all those players, coaches, and others who drank from the drinking fountain were infected with hepatitis.

This case illustrates a classical *cross connection*, or the physical contact between treated drinking water and contaminated water and the potentially serious consequences of such connections. One of the objectives of environmental engineering is to design systems that protect public health. In the case of piping, engineers need to design systems in such a way as to avoid even the possibility of cross connections being created, although as the Holy Cross College incident shows, it is unlikely that all possibilities can be anticipated.

Discussion Questions

1. The next time you take a drink from a drinking fountain or buy a bottle of water, what would be your expectations about the safety of the water? Who exactly would be responsible for fulfilling these expectations? (Careful with that last question. Remember that you (fortunately) live in a democracy.)
2. Given what we now know about hepatitis, how would the investigation by the epidemiologists have been different if the incident had occurred last year? You will have to do a little investigation here. Most universities have excellent online information of hepatitis and other communicable diseases.
3. Pretend you are a personal injury lawyer who is hired by the family of one of the football players. How would you establish fault? Who should be sued?

1.2.2   The Disposal of Wastewater Sludge

The famous American linguist and writer H. L. Mencken, in his treatise *The American Language: An Inquiry into the Development of English in the United States* (Alfred A. Knopf Inc. 1977), observed that many of the newer words in our language have been formed as a combination of sounds that in themselves convey a picture or a meaning. For example, "crud" started out as C.R.U.D., chronic recurring unspecified dermatitis, a medical diagnosis for American soldiers stationed in the Philippines in the early 1900s. The word has a picture without a definition. Try smiling, and in your sweetest, friendliest way, say "crud." It just can't be done. It always sounds . . . well . . . cruddy.

A combination of consonants that Mencken points out as being particularly ugly is the "sl" sound. Scanning the dictionary for words starting with "sl" produces slimy, slither, slovenly, slug, slut, slum, and, of course, sludge. The very sound of the word is ugly, so the stuff must be something else!

And it is. Sludge is produced in a wastewater treatment plant as the residue of wastewater treatment. Wastewater treatment plants waste energy because humans are inefficient users of the chemical energy they ingest. And like the human body, the metabolism of the wastewater treatment plant is inefficient. While these plants produce clear water that is then disposed of into the nearest watercourse, the plants also produce a byproduct that still has substantial chemical energy. This residue cannot be simply disposed of because it would easily overwhelm an aquatic ecosystem or cause nuisance problems or even be hazardous to human health. The treatment and disposal of wastewater sludge is one of environmental engineering's most pressing problems. (To reduce the negative public opinion of sludge disposal, one water quality association suggested that the stuff leaving the treatment plant be called "biosolids" instead of "sludge." Of course, "a rose by any other name. . . .")

Because sludge disposal is so difficult and because improper disposal can cause human health problems, governmental regulations are needed. In the setting of environmental regulations by governmental agencies, human health or well-being is often weighed against economic considerations. That is, how much are we willing to pay to have a healthier environment? The assumption, or at least hope, is that the regulating agency has the information necessary to determine just what effect certain regulations will have on human health. Unfortunately, this is seldom the case, and regulatory agencies are forced to make decisions based on scarce or unavailable scientific information. The regulator must balance competing interests and diverse constituents. (For example, in Iceland the presence of elves has been taken seriously, and roads have been rerouted to prevent damage to the suspected homes of the little people!) In a democracy the regulator represents the interests of the public. If the regulations are deemed unacceptable, the public can change the regulations (or can change the regulator!).

An example of an unpopular regulation in the United States was the 90 km/h speed limit on interstate highways, a regulation that was commonly ignored and eventually repealed. The U.S. Department of Transportation regulation makers misjudged the willingness of the public to slow down on highways. The two benefits—reduction in gasoline use and saving of lives—were admirable goals, but the regulation was rejected because it asked too much of the public. In the case of the speed limit, transportation engineers were able to state unequivocally that the lowering of the speed limit from 105 km/h to 90 km/h would save about 20,000 lives annually, but this benefit did not sway public opinion. The public was not willing to pay the price of lower highway speeds.

Environmental regulations similarly seek admirable and morally justifiable goals, usually the enhancement of public health (or dealing with the negative, the prevention of disease or premature death). Environmental regulations require the regulator to weigh the benefits accrued by the regulations against the costs of the regulations. Often the value of human health protection is balanced against the imposition of regulatory actions that may entail economic costs and restraints on freedom by curtailing polluting behaviors. That is, the regulator, by setting environmental regulations that enhance the health of the public, takes away freedom from those who would discharge pollutants into the environment. The regulator balances the good of public health against the loss of freedom or wealth—in effect reducing liberty and taking wealth. Setting severe limits on discharges from municipal wastewater treatment plants requires that public taxes be raised to pay for the additional treatment. Prohibiting the discharge of a heavy-metal-laden industrial sludge requires companies to install expensive pollution-prevention systems and prevents them from discharging these wastes by least-cost means. Setting strict drinking water standards similarly results in greater expenditure of disposable wealth in building better water treatment plants. In every case the regulator, when setting environmental regulations, balances the moral value of public health against the moral value of taking wealth. Thou shalt not hurt versus thou shalt not steal. This is a moral dilemma, and this is exactly what the regulator faces in setting environmental regulations.

Earle Phelps was the first to recognize that most environmental regulatory decisions are made by using what he called the *principle of expediency*.[5] A sanitary engineer known for his work with stream sanitation and the development of the Streeter–Phelps dissolved oxygen sag curve equation (Chapter 8), Phelps described expediency as "the attempt to reduce the numerical measure of probable harm, or the logical measure of existing hazard, to the lowest level that is practicable and feasible within the limitations of financial resources and engineering skill." He recognized that "the optimal or ideal condition is seldom obtainable in practice, and that it is wasteful and therefore inexpedient to require a nearer approach to it than is readily obtainable under current engineering practices and at justifiable costs." Most importantly for today's standard setters, who often have difficulty defending their decisions, he advised that "the principle of expediency is the logical basis for administrative standards and should be frankly stated in their defense." Phelps saw nothing wrong with the use of standards as a kind of speed limit on pollution affecting human health. He also understood the laws of diminishing returns and a lag time for technical feasibility. Yet he always pushed toward reducing environmental hazards to the lowest expedient levels. (Note that there is a competing philosophy called the precautionary principle. This philosophy sanctions erring on the side of caution in the face of uncertainty to avoid the problems we have repeatedly created by assuming we had adequate information when we did not—for example, disposing of hazardous waste in open, unlined lagoons.)

The responsibility of the regulator is to incorporate the best available science into regulatory decision making. But problems arise when only limited scientific information is available. The complexity of the environmental effect of sludge on human health leads to scientific uncertainty and makes sludge disposal difficult. The problem in developing sludge disposal regulations is that wastewater sludge has unknown and dynamic properties and behaves differently in different environmental media. Regulators must determine when the presence of sludge is problematic and what can and should be done about it.

In the face of such complexity, in the mid-1980s the USEPA initiated a program to develop health-based sludge disposal regulations. The agency waited as long as it could, even though they were mandated by the 1972 Clean Water Act to set such regulations. The task was daunting, and they knew it. They set about it in a logical way, first specifying all the means by which the constituents of sludge could harm humans and then defining the worst-case scenarios. For example, for sludge incineration they assumed that a person lives for 70 years immediately downwind of a sludge incinerator and breathes the emissions 24 hours per day. The person never moves, the wind never shifts, and the incinerator keeps emitting the contaminants for 70 years. Of most concern would be volatile metals, such as mercury. Using epidemiological evidence, such as from the Minemata tragedy in Japan, and extrapolating several orders of magnitude, the USEPA estimated the total allowable emissions of mercury from a sludge incinerator.

By constructing such worst-case but totally unrealistic scenarios, the USEPA developed a series of draft sludge disposal regulations and published them for public comment. The response was immediate and overwhelming. They received over 600 official responses, almost all of them criticizing the process, the assumptions, and the conclusions. Many of the commentaries pointed out that there are presently no known epidemiological data to show that proper sludge disposal is in any way harmful to the public. In the absence of such information, the setting of strict standards seemed unwarranted.

Buffeted by such adverse reaction, the USEPA abandoned the health-based approach and adopted Phelps-type expediency standards that define two types of sludge, one (Class B) that has been treated by such means as anaerobic digestion and the other (Class A) that has been disinfected. Class A sludge can be disposed of on all farmland, but Class B sludge has restrictions, such as having to wait 30 days before cattle could be reintroduced to a pasture on which sludge had been sprayed. Sludge that has not been treated (presumably Class C, although this is not so designated) is not to be disposed of into the environment. This regulation is expedient because all wastewater treatment plants in the United States now have some type of sludge stabilization, such as anaerobic digestion, and a regulation that most of the treatment plants are already complying with is a popular regulation.

The absence of useful epidemiological information on the effect of sludge constituents on human health forced the USEPA, in developing their worst-case scenarios, to err so much on the conservative side that the regulations became unrealistic and would not have been accepted by the public. The downfall of the health-based regulations was that the regulators could not say how many people would be harmed by sludge disposal that did not meet the proposed criteria. In the absence of such information, the public decided that it simply did not want to be saddled with what they perceived as unnecessary regulations. The USEPA would have been taking too much from them (money) and giving back an undefined and apparently minor benefit (health). So the USEPA decided to do what was expedient—to have the wastewater treatment plants do what they can (such as anaerobic digestion in some cases or disinfection by heat in others), knowing that these regulations would still be better than none at all. As our skill at treatment improves and as we decide to spend more money on wastewater treatment, the standards can be tightened because this would then be ethically expedient.

Regulatory decision making, such as setting sludge disposal regulations, has ethical ramifications because it involves distributing costs and benefits among affected citizens. The principle of expediency is an ethical model that calls for a regulator to optimize the benefits of health protection while minimizing costs within the constraints of technical feasibility. Phelps' expediency principle, proposed over 50 years ago, is still a useful application of ethics using scientific knowledge to set dynamic and yet enforceable environmental regulations. In the case of sludge disposal the USEPA made an ethical decision based on the principle of expediency, weighing the moral good of human health protection against the moral harm of taking wealth by requiring costly wastewater sludge treatment and disposal.

Discussion Questions

1. Discuss your driving habits from the standpoint of Phelps' principle of expediency.
2. A gubernatorial candidate in the state of New Hampshire once ran on a single issue: to stop disposing of wastewater sludge on land in New Hampshire. Suppose you had the opportunity to ask him three questions during a public panel discussion. What would they be, and what do you think his answers might have been?
3. People who live in Japan, a country with a strong sense of public health and cleanliness, were found to have more severe and more frequent colds than people who live in other countries. Why might this be true?

1.2.3   The Donora Episode

It was a typical Western Pennsylvania fall day in 1948, cloudy and still.[6] The residents of Donora, a small mill town on the banks of the Monongahela River, did not pay much attention to what appeared to be a particularly smoggy atmosphere. They had seen worse. Some even remembered days when the air was so thick that streamers of carbon would actually be visible, hanging in the air like black icicles. So the children's Halloween parade went on as scheduled, as did the high school football game Saturday afternoon, although the coach of the opposing team vowed to protest the game. He claimed that the Donora coach had contrived to have a pall of smog stand over the field so that, if a forward pass were thrown, the ball would completely disappear from view and the receivers would not know where it would reappear.

But this was different from the usual smoggy day. By that night 11 people were dead, and ten more were to die in the next few hours. The smog was so thick that the doctors treating patients would get lost going from house to house. By Monday almost half the people in the small town of 14,000 were either in hospitals or sick in their own homes with severe headaches, vomiting, and cramps. Pets suffered most, with all the canaries and most of the dogs and cats dead or dying. Even houseplants were not immune to the effects of the smog.

There were not enough emergency vehicles or hospitals able to assist in a catastrophe of this magnitude, and many people died for lack of immediate care. Firefighters were sent out with tanks of oxygen to do what they could to assist the most gravely ill. They did not have enough oxygen for everyone, so they gave people a few breaths of oxygen and went on to assist others.

When the atmosphere finally cleared on 31 October, six days of intense toxic smog had taken its toll, and the full scope of the episode (as these air quality catastrophies came to be known) became evident. The publicity surrounding Donora ushered in a new awareness and commitment to control air quality in our communities. Health workers speculated that, if the smog had continued for one more night, almost 10,000 people might have died.

What is so special about Donora that made this episode possible? First, Donora was a classical steel belt mill town. Three large industrial plants were on the river—a steel plant, a wire mill, and a zinc plant for galvanizing the wire—the three together producing galvanized wire. The Monongahela River provided the transport to world markets, and the availability of raw materials and dependable labor (often imported from eastern Europe) made this a most profitable venture. During the weekend when the air quality situation in town became critical, the plants did not slow down production. Apparently, the plant managers did not sense that they were in any way responsible for the condition of the citizens of Donora. Only Sunday night, when the full extent of the tragedy became known, did they shut down the furnaces.

Second, Donora sits on a bend in the Monongahela River, with high cliffs to the outside of the bend, creating a bowl with Donora in the middle (Figure 1.2 on the next page). On the evening of 25 October, 1948, an inversion condition settled into the valley. This meteorological condition, having itself nothing to do with pollution, simply limited the upward movement of air and created a sort of lid on the valley. Pollutants emitted from the steel plants thus could not escape and were trapped under this lid, producing a steadily increasing level of contaminant concentrations.

Ⓐ

Figure 1.2   Donora was a typical steel town along the Monongahela River, south of Pittsburgh, with (A) high cliffs creating a bowl and (B) three steel mills producing the pollutants.

The steel companies insisted that they were not at fault, and indeed there never was any fault implied by the special inquiry into the incident. The companies were operating within the law and were not coercing any of the workers to work in their plants or anyone to live in Donora. In the absence of legislation, the companies felt no obligation to pay for air pollution equipment or to change processes to reduce air pollution. They believed that, if only their companies were required to pay for and install air pollution control equipment, they would be at a competitive disadvantage and would eventually go out of business.

The tragedy forced the State of Pennsylvania and eventually the U.S. government to act and was the single greatest impetus to the passage of the Clean Air Act of 1955, although it wasn't until 1972 that effective federal legislation was passed. In Donora and nearby Pittsburgh, however, there was a sense of denial. Smoke and poor air quality constituted a kind of macho condition that meant jobs and prosperity. The Pittsburgh press gave the news of the Donora tragedy equal billing to a prison breakout. Even in the early 1950s

Figure 1.2   Continued.

there was a fear that, if people protested about pollution, the plants would close down and the jobs would disappear. And indeed, the zinc plant (thought to be the main culprit in the formation of the toxic smog) shut down in 1957, and the other two mills closed a decade later. Donora, however, lives on as the location of the single most significant episode that put into motion our present commitment to clean air.

Discussion Questions

1. Some years after the Donora episode, the local paper lamented that "The best we can hope is that people will soon forget about the Donora episode." Why did the editors of the paper feel that way? Why did they not want people to remember the episode?
2. The ages of the people who died ranged from 52 to 85. Old people. Most of them were already cardiovascular cripples, having difficulty breathing. Why worry about them? They would have eventually died anyway, after all.
3. The fact that pets suffered greatly has been almost ignored in the accounts of the Donora episode. Why? Why do we concentrate on the 21 people who died, and not on the hundreds and hundreds of pets who perished in the smog? Are they not important also? Why are people more important to us than pets?

1.2.4 Jersey City Chromium

Jersey City, in Hudson County, New Jersey, was once the chromium processing capital of America, and over the years, 18 million tonnes of chromite ore processing residue were sold or given away as fill.[7] There were many contaminated sites, including ball fields and basements underlying both homes and businesses. It was not uncommon for brightly colored chromium compounds to crystallize on damp basement walls and to bloom on soil surfaces where soil moisture evaporates, creating something like an orange hoar frost of hexavalent chromium—Cr(VI). A broken water main in the wintertime resulted in the formation of bright green ice due to the presence of trivalent chromium—Cr(III).

The companies that created the chromium waste problem no longer exist, but three conglomerates inherited the liability through a series of takeovers. In 1991, Florence Trum, a local resident, successfully sued Maxus Energy, a subsidiary of one of the conglomerates, for the death of her husband, who loaded trucks in a warehouse built directly over a chromium waste disposal site. He developed a hole in the roof of his mouth and cancer of the thorax; it was determined by autopsy that chromium poisoning caused his death. While the subsidiary company did not produce the chromium contamination, the judge ruled that they knew about the hazards of chromium.

The State of New Jersey initially spent $30 million to locate, excavate, and remove some of the contaminated soil. But the extent of the problem was overwhelming, so they stopped these efforts. The director of toxic waste cleanup for New Jersey admitted that, even if the risks of living or working near chromium were known, the state did not have the money to remove it. Initial estimates for site remediation were well over $1 billion.

Citizens of Hudson County were angry and afraid. Those sick with cancer wondered if it could have been prevented. Mrs. Trum perceived the perpetrators as well dressed business people who were willing to take chances with other peoples' lives. "Big business can do this to the little man," she said.[7]

The contamination in Jersey City was from industries that used chromium in their processes, including metal plating, leather tanning, and textile manufacturing. The deposition of this chrome in dumps resulted in chromium-contaminated water, soils, and sludge. Chromium is particularly difficult to regulate because of the complexity of its chemical behavior and toxicity, which translates into scientific uncertainty. Uncertainty exacerbates the tendency of regulatory agencies to make conservative and protective assumptions, the tendency of the regulated to question the scientific basis for regulations, and the tendency of potentially exposed citizens to fear potential risk.

Chromium exists in nature primarily in one of two oxidation states—Cr(III) and Cr(VI). In the reduced form of chromium, Cr(III), there is a tendency to form hydroxides that are relatively insoluble in water at neutral pH values. Cr(III) does not appear to be carcinogenic in animal and bioassays. Organically complexed Cr(III) became one of the more popular dietary supplements in the United States and can be purchased commercially as chromium picolinate or with trade names like Chromalene to help with proper glucose metabolism, weight loss, and muscle tone.

When oxidized as Cr(VI), however, chromium is highly toxic. It is implicated in the development of lung cancer and skin lesions in industrial workers. In contrast to Cr(III), nearly all Cr(VI) compounds have been shown to be potent mutagens. The USEPA has classified chromium as a human carcinogen by inhalation based on evidence that Cr(VI) causes lung cancer. However, chromium has not been shown to be carcinogenic by ingestion.

What complicates chromium chemistry is that, under certain environmental conditions, Cr(III) and Cr(VI) can interconvert. In soils containing manganese, Cr(III) can be oxidized to Cr(VI). While organic matter may serve to reduce Cr(VI), it may also complex Cr(III) and make it more soluble, facilitating its transport in ground water and increasing the likelihood of encountering oxidized manganese present in the soil. Given the heterogeneous nature of soils, these redox reactions can occur simultaneously.

Cleanup limits for chromium were originally based on contact dermatitis, which was controversial. While some perceive contact dermatitis as a legitimate claim to harm, others jokingly suggested regulatory limits for poison ivy, which also causes contact dermatitis. The methodology by which dermatitis-based soil limits were determined came under attack by those who questioned the validity of skin patch tests and the inferences by which patch test results translate into soil Cr(VI) levels.

Through the controversy, there evolved some useful technologies to aid in resolution of the disputes. For example, analytical tests to measure and distinguish between Cr(III) and Cr(VI) in soils were developed. Earlier in the history of New Jersey's chromium problem, these assays were unreliable and would have necessitated remediating soil based on total chromium. Other technical/scientific advances included remediation strategies designed to chemically reduce Cr(VI) to Cr(III) to reduce risk without excavation and removal of soil designated as hazardous waste.

The frustration with slow cleanup and what the citizens perceived as double-talk by scientists finally culminated in the unusual step of amending the state constitution so as to provide funds for hazardous waste cleanups. State environmentalists depicted the constitutional amendment as a referendum on Gov. Christine Todd Whitman's (R) environmental record, which relaxed enforcement and reduced cleanups. (Whitman was subsequently President George W. Bush's administrator of the USEPA.)

Chromium is also the culprit in the highly successful film *Erin Brockovich*, starring Julia Roberts and Albert Finney (Figure 1.3). Erin Brockovich (Julia Roberts) was a dedicated and enthusiastic public advocate, unsophisticated in legal niceties, who helped

© Everett collection

Figure 1.3   A scene from *Erin Brockovich*, starring Julia Roberts and Albert Finney.

win a significant settlement in the pollution of groundwater with chromium around an industrial site.

Discussion Questions

1. Given what you now know about chromium, what qualms might you have in taking a chrome supplement along with your vitamins?
2. Suppose you are a resident of Jersey City. What three research questions would you want answered? Make sure these are reasonable questions for which answers can be found through chemical, biological, or epidemiological research.
3. Is it possible for something to be beneficial to human health at low doses but detrimental at high doses? Name at least three chemicals that might be good and bad, depending on the dose. Can something be good at high doses and detrimental at low doses?

1.2.5   The Discovery of Biological Wastewater Treatment

Before 1890, chemical precipitation with land farming was the standard method of wastewater treatment in England.[8] The most popular option was to first allow the waste to go anaerobic in what we today call septic tanks. Such putrefaction was thought to be a purely chemical process because the physical nature of the waste obviously changed. The effluent from the septic tanks was then chemically precipitated, and the sludge was applied to farmland or transported by special sludge ships to the ocean. The partially treated effluent was discharged to streams, where it usually created severe odor problems.

At the time London's services were provided by the London Metropolitan Board of Works, which, among its other responsibilities, was charged with cleaning up the River Thames. The chief engineer for this organization was Joseph Bazalgette, who approached

Figure 1.4   Early sewer construction.

the Thames water quality problem from what was at the time a perfectly rational engineering perspective. If the problem was bad odor in London, why not build long interceptor sewers along both banks of the Thames and discharge the wastewater far downstream (Figure 1.4)? Although expensive, this solution was adopted and the city spent large sums of money to export the wastewater to Barking Creek on the north bank and Crossness Point on the south bank. The idea was to collect the sewage at these central locations and then treat it to produce a useful product, such as fertilizer. None of the recycling schemes came to fruition and the wastewater was discharged untreated from the outfalls into the lower Thames. Because at the location of the outfalls the Thames is a tidal estuary, the initial plan was to discharge wastewater only during the outgoing tide. Unfortunately, the wastewater had to be discharged continuously, and the incoming tide carried the evil-smelling stuff back up to the city and put great pressure on the politicians to do something.

Several solutions were considered. One was simply to continue the interceptor sewers and extend them all the way to the North Sea, but this proved to be prohibitively expensive. Another solution was to spray the wastewater on land, but the amount of land to be purchased far outstripped the budget of the Board. The problem required a new approach, one which was to come from the emerging science of microbiology.

The chief chemist working for the Board at that time was William Joseph Dibdin (Figure 1.5). Dibdin, a self-educated son of a portrait painter, began work with the Board in 1877, rising to chief chemist in 1882 but with the responsibilities of the chief engineer. In seeking a solution to the wastewater disposal problem at the Barking Creek and Crossness outfalls, he initiated a series of experiments using various flocculating chemicals—such as alum, lime, and ferric chloride—to precipitate the solids before discharging to the river.

© Royal Society of Civil Engineers

Figure 1.5   William Dibdin (1850–1925).

This was not new, of course, but Dibdin discovered that using only a little alum and lime was just as effective as using a lot, a conclusion that appealed to the cost-conscious Metropolitan Board.

Dibdin recognized that the precipitation process did not remove the demand for oxygen, and he had apparently been convinced by one of his staff, a chemist named August Dupré, that it was necessary to maintain positive oxygen levels in the water to prevent odors. Dibdin decided to add permanganate of soda (sodium permanganate) to the water to replenish the oxygen levels. Because Dibdin's recommendations were considerably less expensive than the alternatives, the Board went along with his scheme.

Dibdin's plan was adopted, and in 1885 construction of the sewage treatment works at the Barking outfall commenced. Given the level of misunderstanding at the time, there were a great many who doubted that Dibdin's scheme would work, so he had to continually defend his project. He again argued that the presence of the addition of the permanganate of soda was necessary to keep the odor down, and he began to explain this by suggesting that it was necessary to keep the aerobic microorganisms healthy. Christopher Hamlin, a historian at the University of Notre Dame, has written widely on Victorian sanitation and believes that this was a rationalization on Dibdin's part and that he did not yet have an insight into biological treatment. The more Dibdin was challenged by his detractors, however, the more he apparently became an advocate of beneficial aerobic microbiological activity in the water because this was his one truly unique contribution that could not be refuted.

When Dibdin started to conduct his experiments at the outfall, Dupré was conducting experiments with aerobic microorganisms. Dupré, a German émigré and a public health chemist, argued that minute aerobic microorganisms cleansed rivers, and, therefore, these same microorganisms might be used for treating wastewater. Dupré tried to convert Dibdin to the understanding of microbial action. In one letter to him Dupré wrote, "The destruction of organic matter discharged into the river in the sewage is practically wholly accomplished by minute organisms. These organisms, however, can only do their work in the presence of oxygen, and the more of that you supply the more rapid the destruction."[8] Later, in an 1888 address to the Royal Society of Arts, Dupré suggested that "our treatment should be such as to avoid the killing of these organisms or even hampering them in their actions, but rather to do everything to favor them in their beneficial work."

But Dibdin and Dupré were not totally successful in convincing the Board that their ideas were right. Many scientists, still believing in the evils of the microbial world, argued that odor control could be achieved only by killing the microorganisms. These scientists managed in 1887 to wrest control of the treatment works from Dibdin and initiated a summer deodorization control suggested by a college professor that involved antiseptic treatment with sulfuric acid and chloride of lime. This process failed; Dibdin was vindicated, and biological wastewater treatment became the standard for all large municipal sewage facilities.

Discussion Questions

1. How was human civilization saved in the book *The War of the Worlds* by H. G. Wells (written in 1898)? Why was it such a sensation?
2. The River Thames during the 19th century was the single recipient of all of London's wastewater. There were no wastewater treatment plants. Human waste was collected

in cesspools and transported by carts to farms. Often these cesspools leaked or were surreptitiously connected to storm sewers that emptied directly into the river. The stench from the Thames was so bad that the House of Commons, meeting in the Parliament building next to the river, had to stuff rags soaked with chloride of lime [calcium hypochlorite, $Ca(ClO)_2 \cdot 4H_2O$] into the cracks in the shutters to try to keep out the awful smell. Gentlemen used to carry pomegranates stuffed with cloves to help mask the odors. Waste from trade people was simply thrown in the streets, where it would be washed into the sewers by rain. Shambles was the street where the butchers sold their wares and where they left their wastes to rot. Eventually the street name became a common word for any big mess. On one pretty Sunday a private party was socializing on a barge on the Thames when the barge overturned and dumped everyone into the water. Nobody drowned but almost everyone came down with cholera as a result of swimming in the contaminated water. Most of the smaller streams feeding the River Thames were lined with outdoor privies overhanging into the river. In short, the conditions were abominable. Why is it, then, that we seldom if ever read about these conditions in novels and stories written during these times? Nobody, for example, goes to the toilet in any of Jane Austen's novels, and nobody steps in poo on the sidewalk in any Charles Dickens story. How come?

3. Edwin Chadwick launched in the 1840s the great sanitary awakening, arguing that filth was detrimental and that a healthy populace would be of higher value to England than a sick one. He had many schemes for cleaning up the city, one of which was to construct small-diameter sanitary sewers to carry away wastewater, a suggestion that did not endear him to the engineers. A damaging confrontation between Chadwick, a lawyer, and the engineers ensued, with the engineers insisting that their hydraulic calculations were correct and that Chadwick's sewers would plug up, collapse, or otherwise be inadequate. The engineers wanted to build large-diameter egg-shaped brick sewers that allowed human access. These were, however, three times as expensive as Chadwick's vitrified clay conduits. Who eventually won out, and why?

1.2.6   The Garbage Barge

Awareness of municipal solid waste problems was greatly heightened by the saga of the garbage barge.[9] The year was 1987, and the barge named *Mobro* had been loaded in New York with municipal solid waste and found itself with nowhere to discharge the load. Because disposal into the ocean is illegal, the barge was towed from port to port, with six states and three countries rejecting the captain's pleas to offload its unwanted cargo (Figure 1.6 on the next page).

The media picked up on this unfortunate incident and trumpeted the so-called garbage crisis to anyone who would listen. Reporters honed their finest hyperbole, claiming that the barge could not unload because all our landfills were full and that the United States would soon be covered by solid waste from coast to coast. Unless we did something soon, they claimed, we could all be strangled in garbage.

The story of the hapless *Mobro* is actually a story of an entrepreneurial enterprise gone sour. An Alabama businessman, Lowell Harrelson, wanted to construct a facility for converting municipal refuse to methane gas and recognized that baled refuse would be

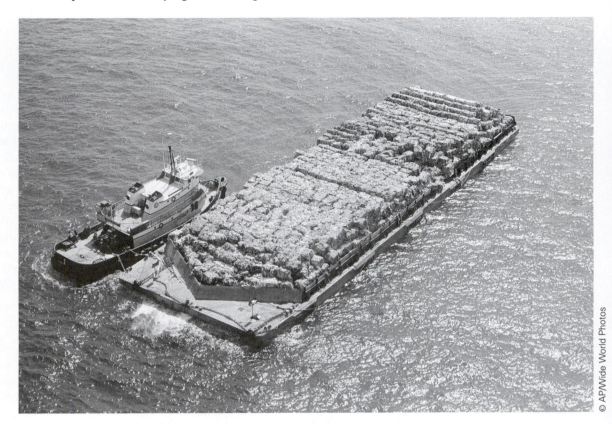

© AP/Wide World Photos

Figure 1.6 The garbage barge.

the best form of refuse for that purpose. He purchased the bales of municipal solid waste from New York City and was going to find a landfill somewhere on the east coast or in the Caribbean where he could deposit the bales and start making methane. Unfortunately, he did not get the proper permits for bringing refuse into various municipalities, and the barge was refused permission to offload its cargo. As the journey continued, the press coverage grew, and no local politicians would agree to allow the garbage to enter their ports. Poor Harrelson finally had to burn his investment in a Brooklyn incinerator.

The garbage crisis never developed, of course. Large waste disposal corporations constructed huge landfills in remote areas and began competing for the municipal solid waste from the eastern seaboard. However, entrepreneurs continue to seek ways to extract value from garbage.

Discussion Questions

1. What is the fate of the municipal solid waste (or garbage in the vernacular) in your community or home town?
2. Is municipal solid waste (garbage) a hazardous material? What constituents of garbage might make it a hazardous material? What types of wastes should be prevented from mixing with normal garbage, and how would this be done?

3. Harrelson, the owner of the garbage barge, had good intentions. He wanted to use the garbage to make methane, a useful product. Should the various governments of the United States and other countries have been more helpful to him since his intentions were admirable? What might be the proper and right governmental responses to plights of private citizens, such as Harrelson?

1.3   SUSTAINABILITY AND CRADLE-TO-CRADLE DESIGN

The prosperity of the Western world can be considered to be largely a product of the Industrial Revolution. While the industrialization of the past two centuries produced enormous benefits, it also left us with a legacy of unintended and negative consequences: vast quantities of waste, the depletion of natural resources, and the contamination of people and ecosystems with toxic substances dispersed throughout the planet.

The Industrial Revolution is partially based on a cradle-to-grave model with a pattern of "take, make, and waste." Products go from being raw materials to products and then to waste in very short order. While a product itself may be intended for rapid consumption, it may generate waste that stays around for hundreds of years. For example, some food items need to stay fresh for just a few days, but they are packaged in materials that could take hundreds of years to decompose by natural processes in the environment. In the past few decades, scientists, engineers, and policymakers have begun to address the deficiencies of this unsustainable model and to define and experiment with what would be desirable.

Many people have looked to nature for inspiration and models of more sustainable ways to create and manage chemicals, materials, and products in society. Sustainable materials management (SMM) has emerged as "an approach to promote sustainable materials use, integrating actions targeted at reducing negative environmental impacts and preserving natural capital throughout the life-cycle of materials, taking into account economic efficiency and social equity."[10]

1.3.1   Framework for Sustainability

In 1989, a framework for sustainability called The Natural Step emerged from Sweden through the efforts of Dr. Karl-Henrik Robèrt, a leading Swedish oncologist.[11] This framework provides a set of four system conditions that define a sustainable society based on the laws of thermodynamics and natural cycles. The Natural Step System Conditions consider the Earth as a closed system for materials and as an open system for energy that sustains life through a complex interactive network of material cycles that uses solar energy to counteract the tendency of materials to dissipate and otherwise increase in entropy.

Therefore, for a society to be sustainable, nature must not be subjected to the following systematically increasing processes:[11]

1. *Extracting concentrations of substances from the Earth's crust.* This condition refers to the extraction of minerals and fossil fuels. Substances that are scarce in nature should be substituted with those that are more abundant. Mined materials should be used efficiently and recycled, and dependence on fossil fuels should be systematically reduced.

Figure 1.7 Ecological aspects of The Natural Step System Conditions. (Illustrated by Larry Chalfan and Lauren Heine)

2. *Building up concentrations of human-made compounds in nature*. This condition refers to the manufacture of persistent and unnatural compounds. Persistent and unnatural compounds should be replaced with those that are normally abundant and or that break down completely and easily in nature. All substances produced by society should be used efficiently.

3. *Utilizing renewable resources at rates faster than they are regenerated and reducing the productive capacity of nature*. This condition refers to the use of natural resources. Resources should be drawn only from well managed ecosystems, systematically pursuing the most productive and sustainable uses both of those resources and land, and exercising caution in all kinds of modification of nature.

And in that society:

4. *People are able to meet their needs worldwide*. This condition means using all of our resources efficiently, effectively, fairly and responsibly so that the needs of all people, including the future needs of people who are not yet born, stand the best chance of being met.

The ecological aspects of The Natural Step System Conditions are illustrated in Figure 1.7. Materials flow in a closed system comprised of two loops. The outer loop represents the cycling of materials within earth's ecosystems. The inner loop represents cycling within the industrial/economic system. Arrow 1 represents the extraction of natural resources for use in the industrial/economic system. In a sustainable society, the rate of natural resource extraction equals the rate of regeneration. Arrows 2 and 3 represent the extraction and resettling of materials from the earth's crust, primarily fossil fuel and mined materials. In a sustainable society, material extraction from the earth's crust will be displaced by the use of recycled and recyclable materials. Arrows 4 and 5 represent substances that flow from the industrial/economic system to the greater ecosystem. Substances that assimilate quickly without harm are represented by Arrow 4. Arrow 5 represents substances that are toxic, persistent, bioaccumulative, or otherwise cause harm to humans or the environment. In a sustainable society, Arrow 5 will disappear.

1.3.2 Cradle-to-Cradle Design

One of the leading practical strategies for achieving sustainable materials management is through positive industrial activity, called "cradle-to-cradle" design. The term "cradle-to-cradle was coined in the 1970s by Walter Stahel and Michael Braungart. The key principles of cradle-to-cradle design were first systematically outlined as the Intelligent Product System (IPS) by Braugart et al. in 1992 and further developed and articulated by Michael Braungart and William McDonough in 2002 in their book *Cradle to Cradle: Remaking the Way We Make Things.*[12] Just as in natural systems where one organism's waste becomes food for another, cradle-to-cradle design applies the same concept to the design of human industry. Cradle-to-cradle design defines two metabolisms within which materials are conceived as nutrients circulating benignly and productively through metabolisms. Biological nutrients cycle within biological metabolisms, and technical nutrients cycle within technical metabolisms.

Biological metabolism is the system of natural processes that supports life. Biological processes are cyclical, ultimately fueled by the energy of the sun, and include the biodegradation (and possibly other forms of degradation) of organic materials and their incorporation into organisms. Materials that contribute to the productivity of biological metabolisms are *biological nutrients*. They are renewable, degradable, and ecologically benign. Products of industry made from biological nutrients can be integrated into natural or engineered biological metabolisms, including water treatment processes and organic processing systems such as composting or anaerobic digestion. The output of biological metabolisms can be resources that engender new biological nutrients, such as beneficial soil amendments. Products that are intended for release to the environment should designed as biological nutrients that are benign for their intended functional use.

Industry can also mimic natural processes by creating *technical metabolisms* that circulate *technical nutrients*. Technical nutrients are typically nonrenewable and they are valuable for their performance qualities. Examples include metals such as copper or aluminum. When designed in cradle-to-cradle systems, technical nutrients can be recovered and recycled over and over—without degrading their quality and without harm to handlers—into similar or dissimilar products. Some companies view the materials in their products as so valuable that they even engage in leasing programs whereby products are essentially leased from the manufacturer to the customer until they are no longer wanted. Then the manufacturer will take them back for remanufacturing of the valuable materials and components into new products. Technical nutrients can be designed for reuse within a company or between companies in similar or dissimilar industries, depending on the material. Products made from technical nutrients should be designed to facilitate material recovery at its highest value with minimal expenditure of energy and cost.

Cradle-to-cradle design, as described by McDonough and Braungart, uses a model of human industry based on three design principles derived from natural systems.

1. *Use current solar income*. With very few exceptions, life on earth is ultimately fueled by energy from the sun. We are only beginning to expand our capacity to harness solar energy, directly and indirectly, for human purposes.
2. *Celebrate diversity*. Natural systems thrive on richness and diversity. Likewise, industry should promote the development of diverse products that are fitting for different preferences, cultures, geographies, and ecosystems.

3. *Waste equals food.* There is no waste in nature. The product of one organism is food or structure for another. Human systems can also be designed to circulate materials productively, eliminating the concept of waste.

Some people call for a strategy of *eco-efficiency*: to reduce the amount of resources used and to generate less waste in industrial activities. But eco-efficiency alone is not a strong enough strategy for sustainability. Improvements in eco-efficiency are often quickly overwhelmed by increases in demand. For example, improvement in automobile fuel efficiency has been offset by an increase in the number of cars on the roads and the number of miles driven. Think of eco-efficiency as a way to slow the loss of resources, similar to helping a wound bleed slowly and not hemorrhage. Cradle-to-cradle design calls for *eco-effectiveness,* which is analogous to healing the wound and supporting the health of the whole body. Eco-effectiveness is concerned with increasing cyclical material flows (hence "effectiveness"), so waste equals food.

By redesigning industry based on nature's models, economic activity can reinforce, rather than compromise, social and environmental prosperity. While this may sound a bit like pie in the sky, there are some unintended benefits of using cradle-to-cradle as a design strategy. First of all, it drives innovation. Businesses are always looking for new ideas and ways to distinguish their businesses, to add more value to their products, and, of course, to make more profit. Cradle-to-cradle design stimulates fresh thinking and creativity. According to Roger McFadden, now chief science officer and vice president for product science and technology for Corporate Express A Staples Company, "Chemical product manufacturers should recognize the opportunity that the sustainability movement is creating for innovation. The movement is exciting in part because it is driving real product innovation and development of new raw materials for formulation after a long period of incremental change."[13]

What are the benefits of products designed for cradle-to-cradle systems? For one, the public is in an increasingly green mood! Given the opportunity to buy two products with similar performance and similar price, who would not prefer to buy the one with greener chemicals and innovative packaging that eliminates waste? Cradle-to-cradle design also drives creativity around new business models. For example, products designed as technical nutrients can be viewed as products of service. If you think about it, do people really need to own computers or televisions? (Now you are sure we are dreaming, right?) What people really want is to be able to afford and select the highest-quality equipment possible for their unrestricted enjoyment in the privacy of their homes (or offices). If when you were ready to change your model, you knew that you could conveniently return the product to the manufacturer and that its components would be reused or remade, would you mind? We predict not, especially if its return brought you value, such as refunds or discounts toward the next item of service that you desire.

Cradle-to-cradle thinking may come as second nature to environmental engineers, who are charged with protecting both human health and the environment and who learn to design material flow systems for materials such as water and wastewater. Rather than treating the waste at the end of the pipe, environmental engineers can use the tools of the trade to design sustainable systems up front that reuse valuable resources without degrading their quality over time. While we cannot deny the current presence of unsustainable infrastructure, it should not deter us from improving designs and applying cradle-to-cradle

FOCUS ON

Designing Products for Technical Cycles: Herman Miller Mirra Office Chair [14,15]

The Herman Miller Mirra office chair was the first office chair designed to flow in technical cycles according to the principles of cradle-to-cradle design. The chair was the result of collaboration between Herman Miller, McDonough Braungart Design Chemistry, Studio 7.5, and EPEA Internationale Umweltforschung GmbH. The goal was to synchronize product development with cradle-to-cradle design.

The team created a Design for the Environment (DfE) product assessment tool to evaluate the environmental performance of new product designs in three key areas: material chemistry, disassembly, and recyclability. Evaluating material chemistry involves three core steps:

1. Identifying all the chemicals in a material used to manufacture a product—such as the steel shaft in a chair—down to 100 ppm
2. Evaluating the hazards posed by the chemicals in the material
3. Assigning the material a score of green, yellow, orange, or red based on

> green being little to no hazard
> yellow being low to moderate hazard
> orange being incomplete data
> red being high hazard

Disassembly refers to the ease of breaking a final product—such as the office chair—down into its constituent parts for recycling or reuse. Evaluating recyclability considers whether a part contains recycled material and, more importantly, whether that part can be recycled at the end of the product's useful life.

The Mirra chair was designed for technical cycles as a product of service. Just like nutrients in the natural world, materials in technical cycles are nutrients in the industrial world and can be used multiple times. The materials used in the Mirra chair were analyzed by EPEA in Hamburg, Germany. Materials that are toxic or unable to flow in a technical cycle were eliminated from the design. Of the chemicals and materials selected, 69% of them are considered green. With respect to recycling, the chair contains 42% by weight pre- and postconsumer recycled content, and 96% by weight of the chair is recyclable. With respect to disassembly, 93% of the product by weight can be readily disassembled, a necessity to facilitate recycling. See Figure 1.8.

Figure 1.8   The Herman Miller Mirra office chair. Left: recyclable parts, 96% by weight. Right: non-recyclable parts (4% by weight)—mixed plastic armpads (white parts), seat pan, and leaf springs (black parts). © Courtesy Hermann Miller Canada, Inc.

FOCUS ON

Designing Products for Biological Cycles—Baypure®:
Eco-Effective Chemistry[16]

Baypure® is a line of commercial chemicals developed to have high environmental performance while meeting or exceeding the performance of conventional chemicals. It consists of chemicals used in industrial processes that were designed to have improved biodegradability, good ecotoxicological characteristics, and production processes with closed-loop material flows. The Baypure® line is made up primarily of two chemicals: Baypure® DS100 is a dispersing agent and Baypure CX100 is a complexing agent. Dispersing agents help dissolve solids in liquids and are used, for example, to disperse solid pigments into liquids during the manufacture of chemical dyes. A complexing agent binds metals in liquid solutions, keeping them from interfering negatively with product performance. They are used, for example, in detergents and in paper production.

EPEA Internationale Umweltforschung GmbH compared the environmental performance of Baypure® DS100 and CX100 to their conventional counterparts over the entire lifecycle of

the chemicals, from raw material procurement to production processes to their fate in the natural environment after use. The central challenge was to ensure not only that the chemical performs well for its desired function but that it sustains the natural environment throughout the product life cycle. EPEA found that, compared to their common counterparts, the Baypure® line of chemicals excels in their behavior in the environment, particularly after use and discharge into water. For example, the conventional counterpart of Baypure® CX100 is EDTA. EDTA is a commonly used complexing agent that degrades very slowly in the natural environment, mobilizes metals, and is problematic in the aquatic environment. In contrast, CX100 has low toxicity and is biodegradable, making it compatible with the natural environment.

Source: Thank you to Michael Braungart and EPEA Internationale Umweltforschung GmbH for permission to adapt the EPEA Baypure® case study for this publication.

thinking as these systems age and need to be replaced and as new systems are required. It is worth thinking of the designs (and design failures) of the past as prototypes for better and better designs. As illustrated by the products described in the boxes in this chapter, the design of the product determines the end-of-life material options. Engineers can influence the effectiveness of infrastructure by understanding and influencing product design and by not considering one without the other.

Discussion Questions

1. What value comes from human biological waste such as sewage? If waste equals food for human biological waste, what prevents us from optimizing its value?
2. What are the challenges to water and wastewater treatment systems as they are currently engineered? How could they be designed sustainably?

END NOTES

1. USEPA. 2008. What Is Green Engineering? Accessed at http://www.epa.gov/oppt/greenengineering/pubs/whats_ge.html on 8 December, 2008.

2. Abraham, Martin A. 2006. *Sustainability Science and Engineering: Defining the Principles*. Amsterdam: Elsevier.

3. Anastas, P. T., and J. B. Zimmerman. 2003. Design through the Twelve Principles of Green Engineering. *Env. Sci. and Tech.*, 37, No. 5.

4. Morse, J. L. 1972. The Holy Cross College football team hepatitis outbreak. *Journal of the American Medical Association* 219: 706–7.

5. Phelps, E. B. 1948. *Public Health Engineering.* New York: John Wiley & Sons.

6. Shrenk, H. H., H. Heimann, G. D. Clayton, W. M. Gafafer, and H. Wexler. 1949. *Air pollution in Donora, PA.* Public Health Bulletin No. 306. Washington, DC: U. S. Public Health Service.

7. Bartlett, L., and P. A. Vesilind. 1998. Expediency and human health: The regulation of environmental chromium. *Science and Engineering Ethics* 4: 191–201.

8. Vesilind, P. A. 2001. Assisting nature: William Dibdin and biological wastewater treatment. *Water Resources Impact* 2, No. 3.

9. Perlman, S. 1998. Barging into a trashy saga. *Newsday*, June 21.

10. ENV/EPOC/WGWPR/RD(2005)5/FINAL Organisation for Economic Co-operation and Development 27-Sep-2007 Working Group on Waste Prevention and Recycling Outcome of the First OECD Workshop on Sustainable Materials Management, Seoul, Korea, 28–30 November 2005.

11. http://www.naturalstep.org/com/nyStart/

12. McDonough, William, and M. Braungart, 2002. *Cradle to Cradle: Remaking the Way We Make Things.* New York: North Point Press

13. Personal communication. 2004. Roger McFadden, vice president technical services, Coastwide Laboratories.

14. http://www.epea.com/documents/EPEAProductCase_MirraChair.pdf

15. Clean Production Action. 2006. Healthy Business Strategies for Transforming the Toxic Chemical Exonomy. Accessed at http:// www.cleanproduction.org/library/CPA-HealthyBusiness-1.pdf on 8 December, 2008.

16. http://www.epea.com/documents/EPEAProductCase_Baypure.pdf

FUNDAMENTALS

© Keith and Susan Morgan

1820 Poinsett Bridge, South Carolina

Environmental engineers use fundamental concepts from a variety of disciplines—biology, chemistry, geology, geomorphology, physics, etc.—to design sound, long-lasting, sustainable solutions.

Engineering Decisions

Dam failure

Fish ladder

Environmental engineers have the immense responsibility of protecting both humans and the environment. Our decisions directly and indirectly affect lives and ecosystems, so we must take into consideration a multitude of factors when making decisions.

The implementation of any engineering project, large or small, entails a series of decisions made by engineers. Sometimes these decisions turn out to be poorly made. A far greater number of decisions, however, made hundreds of times a day by hundreds of thousands of engineers, are correct and improve the lot of human civilization, protect the global environment, and enhance the integrity of the profession. Because so few engineering decisions turn out poorly, engineering decision making is a little known and rarely discussed process. Yet, when a decision turns out to be wrong, the results are often catastrophic. As Gray pointed out, doctors typically can harm only one person at a time while engineers have the potential to harm thousands at a time through incorrectly designed systems.[1] (See Chapter 1.)

This chapter is a review of how environmental engineers make decisions, beginning with a short description of technical decisions and followed by a discussion of cost-effectiveness, possibly the second most commonly employed tool in environmental engineering decision making and the second most quantifiable. Next, the use of benefit/cost analysis is described, followed by a discussion of decisions based on risk analysis. Moving even further toward the more subjective forms of decision making, environmental impact analysis as an engineering tool is reviewed. This chapter concludes with an introduction to ethics and ethical decision making as applied to environmental engineering.

2.1   DECISIONS BASED ON TECHNICAL ANALYSES

In engineering, there seldom is one best way to design anything. If there ever *was* a best way, then engineering would become stagnant, innovation would cease, and technical paralysis would set in. Just as we recognize that there is no single perfect work of art, such as a painting, there also is no perfect water treatment plant. If there *were* a perfect plant or painting, all future water treatment plants would look alike and all paintings would look alike.

The undergraduate engineering student is often taught during the early years of an engineering education that each homework assignment and test question has a single right answer and that all other answers are wrong. But in engineering practice many technical decisions may be right, in that a problem may have several equally correct technical solutions. For example, a sewer may be constructed of concrete, cast iron, steel, aluminum, vitrified clay, glass, or many other materials. With correct engineering design procedures, such a sewer would carry the design flow and, thus, would be technically correct.

One characteristic of technical decisions is that they can be checked by other engineers. Before a design drawing leaves an engineering office, it is checked and rechecked to make sure that the technical decisions are correct; that is, the structure/machine/process will function as desired if built according to the specifications. Technical decisions, thus, are clearly quantifiable and can be evaluated and checked by other competent professional engineers.

E X A M P L E
2.1

Problem   A town with 1950 residents wants to establish a municipally owned and operated solid waste (garbage) collection program. They can purchase one of three possible trucks that have the following characteristics:

Truck A, 18 m^3 capacity
Truck B, 15 m^3 capacity
Truck C, 12 m^3 capacity

If the truck is to collect the refuse for one-fifth of the town each day during a 5-day work week, then every residence will be collected during the week, and the truck will have to make only one trip per day to the landfill. Which truck or trucks will have sufficient capacity?

Solution   Because there is no solid waste generation rate given for this town, a rate has to be assumed. Assume a rate of 1.5 kg/capita/d. The total tonnage of solid waste to be collected is then

$$(1950 \text{ people})(1.5 \text{ kg/capita/d})(7 \text{ days/week}) = 20,475 \text{ kg/wk}$$

Of that, one-fifth, or 4095 kg, will have to be collected every day. For efficiency, garbage trucks compact waste. Assume that the truck is able to compact the refuse to 300 kg/m^3. The required truck capacity is

$$(4095 \text{ kg/d})(300 \text{ kg/m}^3) = 13.65 \text{ m}^3\text{/d}$$

Both the 15 and 18 m^3 trucks have sufficient capacity while the 12 m^3 truck does not. Note that a 13 m^3 truck would also be insufficient; this case is one in which the number is rounded up rather than down.

When performing technical analyses, we often do not have all the information we need to make decisions. Therefore, we must make assumptions. These assumptions, of course, must be made using the best available data with a (sometimes liberal) sprinkling of good judgment. For example, when estimating the solid waste generation rate for the community in the problem above, it would have been best to collect data on that community's genera-tion (e.g., through analyzing waste haulers' records) rather than rely on national averages because every community is unique. In addition, engineers typically do not design sys-tems to last one or two years, so projections have to be made of the community's future population and waste-generation patterns.

Of course, the pressures of modern practice dictate that not only must the engineering decision be effective (it will do the job) but it must also be economical (it will do the job at minimum cost). For example, in Example 2.1, while both truck A and truck B have sufficient capacity to haul the garbage, the operation and maintenance costs might be very different. While technical calculations can answer technical questions, questions of cost require a different form of engineering decision making: *cost-effectiveness analysis.*

2.2   DECISIONS BASED ON COST-EFFECTIVENESS ANALYSES

Engineers typically find themselves working for an employer or client who requires that various alternatives for solving an engineering problem be analyzed on the basis of cost. For example, if a municipal engineer is considering purchasing refuse collection vehicles

FOCUS ON

Elegant Design

Contemporary artist Michael Singer transforms public art, architecture, landscape, and planning projects into successful models for urban and ecological renewal. He and Linnea Glatt designed the 27th Avenue Solid Waste Transfer and Recycling Center in Phoenix, Arizona, a 25-acre, 100,000 ft^2 (9290 m^2), $18 million facility for the transfer and recycling of garbage (Figure 2.1). The facility was built and came in $4 million under budget. The Singer–Glatt design invites public involvement and understanding of recycling and waste issues in an infrastructure facility that is normally closed to the public. Traditional designs of solid waste facilities aim to conceal the facility. In contrast, the Singer–Glatt design reveals the process of recycling as well as other issues related to the disposal of garbage. The concepts of renewal and transformation are integral to all elements of the design: buildings, roads, landscape, water, and wildlife habitat. The project won several awards and is credited with promoting aesthetic design excellence for infrastructure in the United States. In 1993, *The New York Times* chose Singer's design of the massive facility as one of the top eight design events of the year.

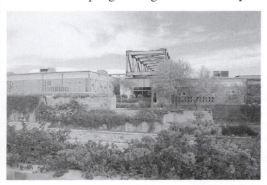

Figure 2.1 27th Avenue Solid Waste Transfer and Recycling Center in Phoenix, Arizona.
(© Michael Singer and Linnea Glatt, Photography by David Stansbury).

In 2004 the New York City Office of the Environmental Defense Fund asked Singer to explore ways that a solid waste transfer facility on the city's waterfront could become an amenity to surrounding communities rather than creating traffic congestion and air pollution. Mayor Michael Bloomberg had proposed establishing a network of marine transfer stations for solid waste. Singer designed a concept piece that included a number of strategies to transform an urban industrial waste transfer site into a place that is integral to its surrounding community in function and form (Figure 2.2). As designed, all aspects of the proposed Marine Transfer Station—walls, roofs, and interior spaces—support its function. Water collected on the road and roof surfaces is stored in wall cavities for cleaning the garbage trucks and facility. Plantings along the walls and on the roof treat wastewater, remove odors through air-to-soil filtration, and offer a flyway habitat stop for migrating birds. Solar photovoltaic panels generate power and reduce the energy load.

Figure 2.2 Concept for New York City Marine Transfer Station. (© Michael Singer and Environmental Defense Fund, Rendering by Trevor Lee).

To deal with the major community concern—the steady flow of smelly, noisy trucks in local streets—the proposal provided an automated ramp system that would accommodate more than 60 trucks. Upon entering the site, drivers would be required to turn off their engines to reduce noxious emissions. In addition, the land adjoining the marine transfer station could be transformed into a spacious waterfront park and gardens for public enjoyment. With ample interior space for the recycling and transfer station operations, some areas could be used for community activities, educational programs, and offices for local organizations.

While developed as a concept piece, such a visionary design challenges people to think beyond the status quo and to expand the scope of costs and considered benefits. In designing infrastructure, environmental engineers can think about how facilities can be integrated into their surroundings in a way that is beautiful and supports community and environmental health. The best designs—those that qualify as elegant—are usually those that solve multiple problems or that provide multiple benefits.

Source: *Clean and Green Infrastructure in New York City: Solid Waste Case Study* by Ramon J. Cruz, June 20, 2006, PRATT Center for Community Development, *Environmental Defense; Infrastructure and Community: How Can We Live With What Sustains Us?* by Michael Singer, Ramon J. Cruz, and Jason Bregman. 2007. Environmental Defense Fund and Michael Singer Studio.

and finds that he or she can buy either expensive trucks that achieve great compaction of the refuse, thereby making efficient trips to the landfill, or inexpensive trucks that require more trips to the landfill, how does the engineer know which is less expensive for the community? Obviously, the lowest total cost alternative (given *all* cost data) would be the most rational decision.

Besides the difficulties with estimating all the required costs, cost-effectiveness analysis is complicated by the fact that money changes in value with time. If a dollar today is invested in an interest-bearing account at 5%, this dollar will be a $1.05 a year from now. Thus, a dollar a year from now is not the same dollar that exists today, and the two dollars cannot be added directly. That is, instead of $2.00, there would be $2.05. Similarly, it does not make any sense to add the annual operating costs of a facility or a piece of equipment over the life of the equipment because, once again, the dollars are different, and adding such dollars is like adding apples and oranges since money changes in value with time. For example, if a community will spend $4000 to operate and maintain a trash collection truck during one year and $5000 the next year, the community needs less than $9000 invested to cover the expenses. (The same concept applies to saving for any future expenses, such as college or retirement.)

This issue can represent a problem for communities that are trying to understand just what it costs to build public facilities or operate public services. The technique used to get around this difficulty is to compare the costs of different alternatives based on either the *annual cost* or the *present worth* of the project. In the annual cost calculation, all costs represent the money the community needs annually to operate the facility and retire the debt. The operating costs are estimated from year to year, and the capital costs are calculated as the annual funds needed to retire the debt within the expected life of the project.

In the case of present worth calculations the capital costs are the funds needed to construct the facility, and the operating costs are calculated as if the money to pay for them is available today and is put into the bank to be used for the operation, again over the expected life of the facility. A project with a higher operating cost would require a larger initial investment to have sufficient funds to pay for the cost of operation.

FOCUS ON

The Pittsburgh Case

The powers of dilution were for a long time thought to be the answer to the wastewater disposal problems. The old saw was "the solution to pollution is dilution." George Fuller, an eminent sanitary engineer, observed during the first decade of the 1900s that "the disposal of sewage by dilution is a proper method when by dispersion in water the impurities are consumed by bacteria and larger forms of plant and animal life or otherwise disposed of so that no nuisance results."

But using the waterways as treatment systems was unacceptable to downstream communities, and many of them sued the upstream dischargers on the basis of the riparian water law, which states that the water flowing through one's land must not be diminished in quantity or impaired in quality. The suit upset the larger cities who were the major cause of pollution, and the magazine *City Hall* intoned in an editorial:

> The average city or individual considers a sewage disposal plant a novelty or unnecessary expense and they only build them when a threat of a damage suit arises and then to appease the wrath of the plaintiff or, in the case of some states, meet with the seemingly useless requirements of the State Board of Health. Then any sort of a plan or device prepared or proposed by an engineer or concern will do, as long as it is not too expensive.

But the state health departments thought otherwise and pushed to have cities construct effective wastewater treatment plants. The conflict between cities that polluted the rivers and the communities that used the rivers for a water supply was never more clearly etched than in the Pittsburgh case.

In 1905 Pennsylvania passed a law forbidding the discharge of untreated sewage into streams or rivers. In 1910 Dr. Samuel G. Dixon, the Pennsylvania commissioner of health, required Pittsburgh to submit a comprehensive plan for replacing the city's combined sewerage system with a separate system and a treatment plant (Figure 2.3). In

Figure 2.3 Pittsburgh, Pennsylvania. (© Margaret Bourke-White/Stringer/Time & Life Pictures/Getty Images).

response, the city hired two well-known sanitary engineers, Allen Hazen and George Whipple, to make recommendations. After a year of investigation, Hazen and Whipple submitted what the journal *Engineering Record* called "the most important sewerage and sewage disposal report made in the United States."

Hazen and Whipple estimated that replacing Pittsburgh's combined sewers with a separate system and building a treatment plant would cost Pittsburgh taxpayers a minimum of $46 million (in 1912 dollars) but that the 26 towns downstream from Pittsburgh on the Ohio River could provide filtered water for their residents for far less money. It was a no-brainer to them that the most efficient option was to have the 26 towns treat the water and allow the Ohio River to be an open conduit for the removal of human waste. No precedent existed, they argued, "for a city's replacing the combined system by a separate system for the purpose of protecting supplies of other cities." Hazen and Whipple concluded that "no radical change in the method of sewerage or of sewage disposal as now practiced by the City of Pittsburgh is now necessary or desirable."

Engineering opinion overwhelmingly supported the Hazen and Whipple report and viewed the controversy as an issue on "how far engineers are at liberty to exercise their own judgment as to what is best for their clients and how far they must give way to their medical colleagues." Uncertain of his ability to compel Pittsburgh to build a separate system and treat its own sewage, Dixon retreated and issued the city a temporary discharge permit. The state commissioner of health continued to issue such permits to cities until 1939.

Source: Based on Melrose, Martin. 2000. The Sanitary City. Baltimore: Johns Hopkins Press; and Tarr, J. A., and F. C. McMichael. 1977. Historic turning points in municipal water supply and waste-water disposal, 1850–1932. Civil Engineering, 47, No. 10. Reprinted from *Controlling Environmental Pollution*, Vesilind, P. A. and DiStefano, T. D., 2006. Lancaster, PA: DEStech Publications, Inc. Reproduced with permission of DEStech Publications, Inc.

Either annual cost or present worth is in most cases acceptable as a method of comparison among alternate courses of action. The conversion of capital cost to annual cost and the calculation of the present worth of operating cost is most readily performed by using tables (or preprogrammed hand calculators). Table 2.1 is a sample of the type of interest tables used in such calculations.

Table 2.1 Compound Interest Factors

Number of Years (n)	$i = 6\%$ Interest Capital Recovery Factor (C_R)	Present Worth Factor (C_P)
1	1.0600	0.9434
2	0.54544	1.8333
3	0.37411	2.6729
4	0.28860	3.4650
5	0.23740	4.2123
6	0.20337	4.9172
7	0.17914	5.5823
8	0.16104	6.2097
9	0.14702	6.8016
10	0.13587	7.3600
11	0.12679	7.8867
12	0.11928	8.3837
13	0.11296	8.8525
14	0.10759	9.2948
15	0.10296	9.711
16	0.09895	10.105
17	0.09545	10.477
18	0.09236	10.827
19	0.08962	11.158
20	0.08719	11.469

(continued)

Table 2.1 Continued

	$i = 8\%$ Interest	
Number of Years (n)	Capital Recovery Factor (C_R)	Present Worth Factor (C_P)
1	1.0800	0.9259
2	0.56077	1.7832
3	0.38803	2.5770
4	0.30192	3.3121
5	0.25046	3.9926
6	0.21632	4.6228
7	0.19207	5.2063
8	0.17402	5.7466
9	0.16008	6.2468
10	0.14903	6.7100
11	0.14008	7.1389
12	0.13270	7.5360
13	0.12642	7.9037
14	0.12130	8.2442
15	0.11683	8.5594
16	0.11298	8.8513
17	0.10963	9.1216
18	0.10670	9.3718
19	0.10413	9.6035
20	0.10185	9.8181

	$i = 10\%$ Interest	
Number of Years (n)	Capital Recovery Factor (C_R)	Present Worth Factor (C_P)
1	1.1000	0.9094
2	0.57619	1.7355
3	0.40212	2.4868
4	0.31547	3.1698
5	0.26380	3.7907
6	0.22961	4.3552
7	0.20541	4.8683
8	0.18745	5.3349
9	0.17364	5.7589
10	0.16275	6.1445
11	0.15396	6.4950
12	0.14676	6.8136
13	0.14078	7.1033
14	0.13575	7.3666
15	0.13147	7.6060
16	0.12782	7.8236
17	0.12466	8.0215
18	0.12193	8.2013
19	0.11955	8.3649
20	0.11746	8.5135

Annual costs are converted to present worth by calculating what the money spent annually would be worth at the present time. At 10% interest only $0.9094 needs to be invested today to have $1.00 to spend in one year (as shown in the "Present Worth Factor" column of Table 2.1). If another dollar is needed at the beginning of the second year, the total investment *today* would be $1.7355, rather than $2.00. In general terms, if a constant amount A is needed for n time periods at a rate of compound interest i, the present worth of this money, or the amount that needs to be invested, is $C_P \times A$, where C_P = present worth factor.

Capital costs are converted to annual costs by recognizing that the total outlay of capital would have earned interest at a given rate if the money had been invested at the prevailing interest rate. A $1.00 investment now at 10% interest would have earned $0.10 interest in one year and be worth $1.10. If a $1.00 capital cost is to be paid back over 2 years, the payment each year, as shown in the "Capital Recovery Factor" column of Table 2.1, is $0.57619. The amount of money necessary each year to pay back a loan of C dollars over n time periods at a compound interest rate of i is $C_R \times C$, where C_R = capital recovery factor.

Both types of conversion are illustrated in Example 2.2.

E X A M P L E

2.2   **Problem**   A municipality is trying to decide on the purchase of a refuse collection vehicle. Two types are being considered, A and B, and the capital and operating costs are as shown. Each truck is expected to have a useful life of 10 yr.

	Truck A	Truck B
Initial (capital) cost	$80,000	$120,000
Maintenance cost, per year	6,000	2,000
Fuel and oil cost, per year	8,000	4,000

Which truck should the municipality purchase based on these costs alone? Calculate the costs both on an annual and a present worth basis, assuming an interest rate of 8%.

Solution

Calculation of annual cost for Truck A:

From the 8% interest table (Table 2.1), the C_R for $n = 10$ years is 0.14903. Therefore, the annual cost for the capital is $80,000 \times 0.14903 = \$11,922$. The total annual cost for Truck A is then

Capital recovery	$11,922
Maintenance	6,000
Fuel	8,000
Total	$25,922

Calculation of annual cost for Truck B:

The C_R is the same because i and n are the same. The annual cost for the capital is then $120,000 \times 0.14903 = \$17,884$, and the total annual cost for Truck B is

Capital recovery	$17,884
Maintenance	2,000
Fuel	4,000
Total	$23,884

Based on an annual cost basis, Truck B, which has a higher capital cost, is the rational choice because its total annual cost to the community is lower than the total annual cost of Truck A.

Calculation of present worth for Truck A:

C_P for $n = 10$ and $i = 8\%$ is 6.7100 (Table 2.1)

The present worth of the annual maintenance and fuel costs is then ($6,000 + $8,000) × 6.7100 = $93,940, and the total present worth is

Capital cost	$80,000
Present worth of maintenance and fuel	93,940
Total	$173,940

Calculation of present worth for Truck B:

Again, the C_P is the same (6.7100) because i and n are the same. The present worth of the annual costs is ($2,000 + $4,000) × 6.7100 = $40,260, and the total present worth is

Capital cost	$120,000
Present worth of maintenance and fuel	40,260
Total	$160,260

Based on a present worth, Truck B is still the rational choice because, if the community were to borrow the money to operate the truck for 10 years, it would have to borrow less money than if it wanted to purchase Truck A.

◼

But suppose several alternate courses of action also have different benefits to the client or to the employer? Suppose in Example 2.2 one alternative open to the community would be to go from twice per week collection of refuse to one time per week collection. Now the level of service is also variable, and the analysis is no longer applicable. It is necessary to incorporate benefits into the cost-effectiveness analysis to be sure the most effective use is made of scarce resources.

2.3 DECISIONS BASED ON BENEFIT/COST ANALYSES

In the 1940s the Bureau of Reclamation and the U.S. Army Corps of Engineers battled for public dollars in their drive to dam up all the free-flowing rivers in America. To convince Congress of the need for major water storage projects, a technique called *benefit/cost analysis* was developed. At face value this is both useful and uncomplicated. If a project is contemplated, an estimate of the benefits derived is compared in ratio form to the cost incurred. Should this ratio be more than 1.0, the project is clearly worthwhile, and the projects with the highest benefit/cost ratios should be constructed first because these will provide the greatest returns on the investment. By submitting their projects to such an analysis, the Bureau and the Corps could argue for increased expenditure of public funds and could rank the proposed projects in order of priority.

As is the case with cost-effectiveness analyses, the calculations in benefit/cost analyses are in dollars, with each benefit and each cost expressed in monetary terms. For example, the benefits of a canal could be calculated as monetary savings in transportation costs. But some benefits and costs (such as clean air, flowers, whitewater rafting, foul odors, polluted groundwater, and littered streets) cannot be easily expressed in monetary terms. Yet these benefits and costs are very real and should somehow be included in benefit/cost analyses.

One solution is simply to force monetary values on these benefits. In estimating the benefits for artificial lakes, for example, recreational benefits are calculated by predicting what people would be willing to pay to use such a facility. There are, of course, many difficulties with this technique. The value of a dollar varies substantially from person to person, and some people benefit a great deal more from a public project than others, and yet all may share in the cost. Because of the problems involved in estimating such benefits, they can be bloated to increase the benefit/cost ratio. It is thus possible to justify almost any project because the benefits can be adjusted as needed.

In Example 2.3, monetary values are placed on subjective benefits and costs to illustrate how such an analysis is conducted. The reader should recognize that the benefit/cost analysis is a simple arithmetic calculation and that, even though the final value can be calculated to many decimal places, it is only as valuable as the weakest estimate used in the calculation.

EXAMPLE
2.3

Problem   A small community has for a number of years owned and operated a refuse collection service, consisting of one truck that is used to collect refuse once a week. That one truck is wearing out and must be replaced. It also appears that only one truck is no longer adequate and a second truck may have to be purchased. If the second truck is not purchased, the citizens will be asked to burn waste paper in their backyards so as to reduce the amount of refuse collected. There are two options:

1. The old truck is sold and two newer models purchased. This will allow for collection twice per week instead of only once.
2. The old truck will be sold and only one new truck bought, but the citizens of the community will be encouraged to burn the combustible fraction of the refuse in their backyards, thus reducing the quantity of refuse collected.

Decide on the alternative of choice using a benefit/cost analysis.

Solution   The benefits and costs of both alternatives are listed, using annual cost numbers from Example 2.2. For those costs that cannot be easily expressed in monetary terms, reasonable estimates are suggested. Next, the dollars are added.

Alternative 1:

Benefits

Collection of all refuse*	$250,000
Total benefits	$250,000

Costs

Two new vehicles (including operating costs)	$47,768
Increased noise and litter	10,000
Labor cost	200,000
Total annual costs	$257,768

$$\text{Benefit/cost} = \frac{\$250,000}{\$257,768} = 0.97$$

Alternative 2:

Benefits

Collection of 60% of refuse	$150,000
Total benefits	$150,000

Costs

New vehicle (including operating cost)	$23,884
Dirtier air	0
Labor costs	120,000
Total annual costs	$143,884

$$\text{Benefit/Cost} = \frac{\$150,000}{\$143,884} = 1.05$$

The benefit/cost of the first alternative is 0.97 whereas the benefit/cost ratio for the second alternative is 1.05. It would seem, therefore, that the second alternative should be selected.

■

Note that in Example 2.3 only some of the items could be quantified, such as the costs of the vehicles. Most of the other items are highly subjective estimates. Also note that there is no cost attached to dirtier air. If such a cost were included, the calculation might have resulted in a different conclusion. It is not unusual, however, for these types of calculations to include the benefits but not the shared costs of items such as clean air.

The concept of unequal sharing of costs and benefits is perhaps best illustrated by the example of the common as it was used in a medieval village.[2] The houses in the village surrounded the common, and everyone grazed their cows on the common. The common area was shared land, but the cows belonged to the individual citizens. It would not have taken too long for one of the citizens of the village to figure out that the benefits of having a cow were personal but the costs of keeping it on the common ground were shared by all. In true selfish fashion, it made sense for the farmer to graze several cows and thereby increase

*The money saved by each household not having to make the weekly trip to the landfill, multiplied by the total number of trips, could be considered a benefit.

FOCUS ON

The Walkerton, Ontario E. Coli *Outbreak*

On May 19, 2000, the public health office at Walkerton, Ontario received a call from the local hospital reporting two cases of severe diarrhea. The public health unit called the local water treatment plant and asked whether the water supply was the cause and was assured by the operators that all was well with the water. On the next day, the number of illnesses continued to rise, and the water treatment plant operators were again asked whether the water quality was safe, and they again assured the public health people that it was. On the third day, the first confirmed case of *E. coli* 157 infection was confirmed, and the public health personnel determined that the only reasonable source had to be the drinking water.

It turned out that the chlorinator at the water treatment plant had been down for some time and that the plant had stopped testing the water in its own lab. The independent laboratory that had responsibility for doing the bacteriological testing had informed the plant operators that excessive coliforms were in the water, a fact they would have suspected because the chlorinator was malfunctioning. Neither the laboratory nor the water treatment plant operators, however, contacted the public health agency to report the problem. As a result, there were 195 cases of *E. coli* infection and 118 cases of infection by other intestinal pathogens. Of those who were infected, 26 individuals developed hemolytic uremic syndrome, a serious complication of the *E. coli* 157 infection that results in swelling of the face and other extremities and abnormal kidney function, with some people requiring long-term dialysis.

The direct cause for the outbreak was the malfunctioning chlorinator, but the indirect cause was shown to be the 50% reduction in funding from the provincial government for water testing. As a result of the outbreak, the water lines in Walkerton were flushed with high chlorine solutions, and some lines that could not be adequately cleaned had to be replaced, resulting in a cost far exceeding the estimated savings from cutting the funding.

Source: Reprinted from *Controlling Environmental Pollution*, Vesilind, P. A. and DiStefano, T. D., 2006. Lancaster, PA: DEStech Publications, Inc. Reproduced with permission of DEStech Publications, Inc.

his wealth. But this action would have triggered a similar response by his neighbors. Why should *they* stay with one cow when they saw their friend getting wealthy? So they all would buy more cows, and all of these cows would graze on the common. Eventually, however, the number of cows would overwhelm the capacity of the common and all of the cows would have to be killed.

The sharing of the cost of dirty air is similar to the tragedy of the commons. Each one of us, in using clean air, uses it for personal benefit, but the cost of polluting it is shared by everyone. Is it possible for humans to rethink the way they live and to agree voluntarily to limit their pollutional activities? This is unlikely, and as a result, government has to step in and limit each one of us to only one cow.

Another significant problem with benefit/cost analyses is that they can and often are subverted by a technique known as "sunk cost." Suppose a government agency decides to construct a public facility and estimates that the construction costs will be $100 million. It argues that, because the benefits (however they might be calculated) are $120 million, the project is worth constructing because the benefit/cost ratio is greater than 1.0 ($120/$100). It then receives appropriations from Congress to complete the project.

Somewhere in the middle of the project, having already spent the original $100 million, the agency discovers that it underestimated the construction cost. It turns out that the project will actually cost $180 million. Now, of course, the benefit/cost ratio is 0.67 ($120/$180), which is less than 1.0, so the project is not economically justifiable. But, the agency has already spent $100 million on construction. This cost is considered a sunk cost, or money that will be lost forever if the project is not completed. Hence, the agency argues that the *true* cost of the project is the increment between what the estimated cost is and what has already been spent, or $80 million ($180 − $100). The benefit/cost ratio is then calculated as $120/$80 = 1.5, which is substantially greater than 1.0 and indicates that the project should be completed. It is absolutely astounding how many times this scam works, and nobody seems to ask the agencies why their *next* estimates should be believed if all the previous ones were wildly undervalued.

2.4 DECISIONS BASED ON RISK ANALYSES

Often the benefits of a proposed project are not such simple items as recreational values but the more serious concern of human health. When life and health enter benefit/cost calculations, the analyses are generally referred to as *risk*/benefit/cost analyses to indicate that people are at risk. They have become more widely known as simply *risk analyses*.

Risk analysis is further divided into risk assessment and risk management. The former involves a study and analysis of the potential effect of certain hazards on human health. Using statistical information, *risk assessment* is intended to be a tool for making informed decisions. *Risk management*, on the other hand, is the process of reducing risks that are deemed unacceptable.

In our private lives we are continually doing both. Smoking cigarettes is a risk to our health, and it is possible to calculate the potential effect of smoking. Quitting smoking is a method of risk management because the effect is to reduce the risk of dying of certain diseases.

In effect, the risk of dying of something is 100%. The medical profession has yet to save anyone from death. The question, then, becomes *when* death will occur and *what* the cause of death will be.

There are three ways of calculating risk of death due to some cause. First, risk can be defined as the ratio of the number of deaths in a given population exposed to a pollutant divided by the number of deaths in a population not exposed to the pollutant. That is,

$$\text{Risk} = \frac{D_1}{D_0}$$

where D_1 = number of deaths in a given population exposed to a specific pollutant, per unit time
D_0 = number of deaths in a similar sized population not exposed to the pollutant, per unit time.

EXAMPLE

2.4   **Problem**   Kentville, a community of 10,000 people, resides next to a krypton mine, and there is concern that the emissions from the krypton smelter have resulted in adverse effects. Specifically, kryptonosis seems to have killed 10 of Kentville's inhabitants last year. A neighboring community, Lanesburg, has 20,000 inhabitants and is far enough from the smelter to not be affected by the emissions. In Lanesburg only two people last year died of kryptonosis. What is the risk of dying of kryptonosis in Kentville?

Solution   If risk is so defined, then

$$\text{Risk of dying of kryptonosis} = \frac{\dfrac{10}{10,000}}{\dfrac{2}{20,000}} = 10$$

That is, a person is 10 times more likely to die of kryptonosis in Kentville than in a noncontaminated locality.

∎

Note, however, that, even though statistically there is a far greater chance of dying of kryptonosis in Kentville than in Lanesburg and even though Kentville just happens to have a krypton smelter, *we have not proven that the smelter is responsible*. All we have is statistical evidence of a relationship.

A second method of calculating risk is to determine the number of deaths due to various causes per population and compare these ratios. That is

$$\text{Relative risk of dying of cause A} = \frac{D_A}{P}$$

where D_A = number of deaths due to a cause A in a unit time
       P = population

EXAMPLE

2.5   **Problem**   The number of deaths in Kentville and their causes last year were

Heart attack	5
Accidents	4
Kryptonosis	10
Other	6

What is the risk of dying of kryptonosis relative to other causes?

Solution   The risk of dying of a heart attack in Kentville is 5/10,000, whereas the risk of dying of kryptonosis is 10/10,000. That is, the risk of dying of kryptonosis is twice as large as the chance of dying of a heart attack, 2.5 times the chance of dying of an accident, and 1.7 times the chance of dying of other causes. The risks may be different in Lanesburg, of course, and can be compared.

Finally, risk can be calculated as the number of deaths due to a certain cause divided by the total number of deaths, or

$$\text{Risk of dying of cause A} = \frac{D_A}{D_{total}}$$

where D_{total} = total number of deaths in the population in a unit time.

EXAMPLE
2.6

Problem What is the risk of dying of kryptonosis in Kentville relative to deaths due to other causes, using the data in Example 2.5?

Solution The total number of deaths from Example 2.5 is 25. Hence

$$\text{Risk of dying of kryptonosis} = \frac{10}{25} = 0.4$$

That is, of all the ways to go, the inhabitants of Kentville have a 40% chance of dying of kryptonosis.

Some risks we choose to accept while other risks are imposed upon us from outside. We choose, for example, to drink alcohol, drive cars, or fly in airplanes. Each of these activities has a calculated risk because people die every year as a result of alcohol abuse, traffic accidents, and airplane crashes. Most of us subconsciously weigh these risks and decide to take our chances. Typically, people seem to be able to accept such risks if the chances of death are on the order of 0.01, or 1% of deaths are attributed to these causes.

Some risks are imposed from without, however, and these we can do little about. For example, it has been shown that the life expectancy of people living in a dirty urban atmosphere is considerably shorter than that of people living identical lives but breathing clean air. We can do little about this risk (except to move), and it is this type of risk that people resent the most. In fact, studies have shown that the acceptability of an *involuntary* risk is on the order of 1000 times less than our acceptability of a *voluntary* risk. Such human behavior can explain why people who smoke cigarettes still get upset about air quality or why people will drive while intoxicated to a public hearing protesting the siting of an airport because they fear the crash of an airplane.

Some federal and state agencies use a modified risk analysis, wherein the benefit is a life saved. For example, if a certain new type of highway guardrail is to be installed, it might be possible that its use would reduce expected highway fatalities by some number. If a value is placed on each life, the total benefit can be calculated as the number of lives saved times the value of a life. Setting such a number is both an engineering as well as a public policy decision, answerable ideally to public opinion.

EXAMPLE

2.7   **Problem**   The 95% reduction of kryptonite emissions from a smelter will cost $10 million. Toxicologists estimate that such a pollution control scheme will reduce the deaths due to krypton poisoning from 10/yr to 4/yr. Should the money be spent?

Solution   Assume that each life is worth $1.2 million based on lifetime earnings. Six lives saved would be worth $7.2 million. This benefit is less than the cost of the control. Therefore, based on risk analysis, it is not cost-effective to install the pollution equipment.

But what about the assumption of a human life being worth $1.2 million? Is this really true? If in Example 2.7 a human life is assumed to be worth $5 million, then the pollution control is warranted. But this then means that the $10 million spent on pollution control could not be spent on expanding the plant or otherwise creating jobs that may increase the tax base that could provide money for other worthy projects, such as improving education, health, or transportation. More on this later.

Risk calculations are fraught with these types of great uncertainty. For example, the National Academy of Sciences' report on saccharin concluded that over 70 years the expected number of cases of human bladder cancer in the United States resulting from daily exposure to 120 mg of saccharin might range from 0.22 to 1.144 million cases. That is quite an impressive range, even for toxicologists. The problem, of course, is that we have to extrapolate data over many orders of magnitude, and often the data are not for humans but for other species, thus requiring a species conversion. Yet governmental agencies are increasingly placed in positions of having to make decisions based on such spurious data.

2.4.1   Environmental Risk Analysis Procedure

Environmental risk analysis takes place in discrete steps.

    I. Define the source and type of pollutant of concern. From where is it coming, and what is it?

    II. Identify the pathways and rates of exposure. How can it get to humans so it can cause health problems?

    III. Identify the receptors of concern. Who are the people at risk?

    IV. Determine the potential health impact of the pollutant on the receptor. That is, define the dose-response relationship, or the adverse effects observed at specific doses.

    V. Decide what impact is acceptable. What effect is considered so low as to be acceptable to the public?

    VI. Based on the allowable effect, calculate the acceptable level at the receptor, and then calculate the maximum allowable emission.

    VII. If the emission or discharge is presently (or planned to be) higher than the maximum allowable, determine what technology is necessary to attain the maximum allowable emission or discharge.

I. *Defining the source and type of pollutant* is often more difficult than it might seem. Suppose a hazardous waste treatment facility is to be constructed near a populated area. What types of pollutants should be considered? If the facility is to mix and blend various hazardous wastes in the course of reducing their toxicity, which products of such processes should be evaluated? In other cases the identification of both the pollutant of concern and the source are a simple matter, such as the production of chloroform during the addition of chlorine to drinking water or gasoline from a leaking underground storage tank.

II. *Identifying the pathway* may be fairly straightforward as in the case of water chlorination. In other situations, such as the effect of atmospheric lead, the pollutant can enter the body in a number of ways, including through food, skin, and water.

III. *Identifying the receptor* can cause difficulty because not all humans are of standard size and health. The USEPA has attempted to simplify such analyses by suggesting that all adult human beings weigh 70 kg, live for 70 years, drink 2 L of water daily, and breathe 20 m^3 air each day. These values are used for comparing risks.

IV. *Defining the effect* is one of the most difficult steps in risk analysis because this presumes a certain response of a human body to various pollutants. It is commonplace to consider two types of effects: cancerous and noncancerous.

The dose-response curve for toxic noncancerous substances is assumed to be linear with a threshold. As shown in curve A of Figure 2.4, a low dose of a toxin would not cause measurable harm, but any increase higher than the threshold would have a detrimental effect. It is considered acceptable, for example, to ingest a certain amount of mercury because it is impossible to show that this has any detrimental effect on human health. However, high doses have documented negative impacts.

Some toxins, such as zinc, are in fact necessary nutrients in our metabolic system and are required for good health. The absence of such chemicals from our diet can be detrimental, but high doses can be toxic. Such a curve is shown as B in Figure 2.4.

Some authorities suggest that the dose-response curve for carcinogens is linear, starting at zero effect at zero concentration, and the harmful effect increases linearly as shown by curve C in Figure 2.4. Every finite dose of a carcinogen can then cause a finite increase in the incidence of cancer. An alternative view is that the body is resistant to small doses

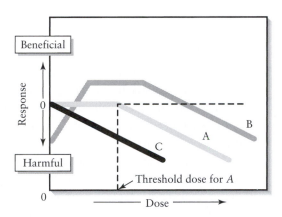

Figure 2.4   Three dose-response curves.

FOCUS ON

The Drake Chemical Supply Company Superfund Site

The Drake Chemical Company of Lock Haven, Pennsylvania, was a major producer of chemicals during World War II and continued to provide employment opportunities to the economically depressed town after the war. One of the waste chemicals that the company disposed of into an open pit was beta-napathylamine, a compound used as dye. Unfortunately, beta-napathylamine is also a potent carcinogen, having been found to cause bladder cancer. In 1962 the State of Pennsylvania banned the production of this chemical, but the damage to the groundwater had already been done with the disposal of beta-napathylamine into the uncontrolled pit. The order from the state caused Drake to stop manufacturing beta-napathylamine, but it continued to produce other equally toxic chemicals, without much concern for the environment or the health of the people in Lock Haven. Finally in 1981 the USEPA closed them down and took control of the property. They discovered several unlined lagoons and hundreds of often unmarked barrels of chemicals stored in makeshift buildings. After removing the drums and draining the lagoons, they found that the beta-napathylamine had seeped onto nearby property and into creeks, creating a serious health hazard.

The USEPA's attempts to clean the soil and the water were, however, met with public opposition. Much of the public blamed the USEPA for forcing Drake Chemical, a major local employer, to close the plant. In addition, the best way to treat the contaminated soil was to burn it in an incinerator, and the USEPA made plans to bring in a portable unit. Now the public, not at all happy

Figure 2.5 Drake Chemical Co. Superfund Site, Lock Haven, Pennsylvania. (© U.S. Environmental Protection Agency).

with USEPA being there in the first place, became concerned with the emissions from the incinerator. After many studies and the involvement of the U.S. Army Corps of Engineers, the incinerator was finally allowed to burn the soil, which was then spread out and covered with 3.5 ft (1.01m) of topsoil (Figure 2.5). The groundwater was pumped and treated until the levels of beta-napathylamine reached background concentrations. The project was finally completed in 1999 with the USEPA paying for the legal fees of the lawyers who argued against the cleanup.

Source: Reprinted from *Controlling Environmental Pollution*, Vesilind, P. A. and DiStefano, T. D., 2006. Lancaster, PA: DEStech Publications, Inc. Reproduced with permission of DEStech Publications, Inc.

of carcinogens and that there is a threshold below which there is no adverse effect (similar to curve A).

The USEPA chose the more conservative route and developed what it calls the *potency factor* for carcinogens. The potency factor is defined as the risk of getting cancer (not necessarily dying from it) produced by a lifetime average daily dose of 1 mg of the pollutant/kg body weight/day. The dose-response relationship is therefore

$$\text{Lifetime risk} = \text{Average daily dose} \times \text{Potency factor}$$

The units of average daily dose are mg pollutant/kg body weight/day. The units for the potency factor are therefore (mg pollutant/kg body weight/day)$^{-1}$. The lifetime risk is unitless. The dose is assumed to be a chronic dose over a 70-year lifespan. EPA has calculated the potency factors for many common chemicals and published these in the *Integrated Risk Information System (IRIS)* database (*www.epa.gov/iris*). Potency factors are listed for both ingestion and inhalation.

V. *Deciding what is acceptable risk* is probably the most contentious parameter in these calculations. Is a risk of one in a million acceptable? Who decides whether this level is acceptable? Certainly, if we asked that one person who would be harmed if the risk were acceptable, he or she would most definitely say no.

The entire concept of acceptable risk presupposes a value system used by environmental engineers and scientists in their professional work. Unfortunately, this value system often does not match the value system held by the public at large. Engineers tend to base their decisions so as to attain the greatest total good in all actions. For example, it is acceptable to place a value on a random human life to achieve some net good, and this allows the USEPA to calculate the risk associated with pollutants and to place an acceptable risk at a probability of one in a million. The cost is very small compared to the good attained, and thus (the argument goes) it is in the public's interest to accept the decision.

But there are strong arguments to be made that most people do not view their own welfare or even the welfare of others in the same light. Most people have the greatest interest in themselves while recognizing that it is also in their self-interest to behave morally toward other people. Many people consider unfair any analysis wherein costs are unequally assigned (like that one person in a million who gets hurt). Many people also believe that it is unethical to place a value on a human life and refuse to even discuss the one-in-a-million death for a given environmental contaminant. People are generally loath to take advantage of that single individual for the benefit of many.

In a public hearing the engineer may announce that the net detrimental effect of the emission from a proposed sludge incinerator is to increase cancer deaths by only one in a million and that this risk is quite acceptable given the benefit of the public funds saved. But members of the public, not appreciating this reasoning, will see this one death as grossly unethical. Such a disagreement has often been attributed by engineers to technical illiteracy on the part of those who disagree with them. But this reasoning seems to be a mistake. It is not technical illiteracy but a different ethical viewpoint, or value system, that is in question.

It is not at all clear how the public would make societal decisions, such as siting landfills, clear-cutting forests, using hazardous materials in consumer products, or selecting sludge disposal practices, if these decisions had to be made by the general public. If pressed, they may actually resort to a similar form of analysis but only after exhausting all other alternatives. The point is that they are not in the position to make such decisions and, thus, are free to criticize the engineer, who often does not understand the public's view and may get blindsided by what may appear to be irrational behavior.

   VI. *Calculating the acceptable levels of pollution* is the next step in the risk analysis process. This step is a simple arithmetic calculation because the value decisions have already been made.

   VII. Finally, it is necessary to *design treatment strategies* to meet this acceptable level of pollution.

   Example 2.8 illustrates how this process works.

E X A M P L E

2.8   **Problem**   The USEPA lists chromium VI as a carcinogen with an inhalation route potency factor of 41 $(\text{mg/kg-d})^{-1}$. A sludge incinerator with no air pollution control equipment is expected to emit chromium VI at a rate such that the airborne concentration at the plant boundary immediately downwind of the incinerator is 0.001 $\mu\text{g/m}^3$. Will it be necessary to treat the emissions so as to reduce the chromium VI to stay within the risk level of 1 additional cancer per 10^6 people?

Solution   The source is defined, and the pathway is the inhalation of chromium VI. The receptor is USEPA's "standard person" who weighs 70 kg, breathes 20 m^3 of air per day, and lives for 70 yr immediately downwind of the incinerator—never leaving to grab a breath of fresh air. (As an aside, consider the irrationality of this assumption, but how else could this be done?) The dose-response relationship is assumed to be linear, and the allowable chronic daily intake is calculated by relating the risk to the potency factor.

$$\text{Risk} = \text{Chronic daily intake} \times \text{Potency factor}$$

$$1 \times 10^{-6} = \text{CDI} \times 41(\text{mg/kg-d})^{-1}$$

$$\text{CDI} = 0.024 \times 10^{-6} \text{ mg/kg-d}$$

The allowable emission concentration is then calculated.

$$\text{CDI} = \text{Volume of air per day inhaled}$$

$$\times \text{ Concentration of chromium VI / kg body weight}$$

$$0.024 \times 10^{-6} \text{ mg/kg-d} = \frac{(20 \text{ m}^3/\text{day})(C\ \mu\text{g/m}^3)(10^{-3} \text{ mg/}\mu\text{g})}{70 \text{ kg}}$$

$$C = 0.085 \times 10^{-6} \text{ mg/m}^3 \quad \text{or} \quad 0.085 \times 10^{-3}\ \mu\text{g/m}^3$$

The system will emit 0.001 $\mu\text{g/m}^3$, which is more than 0.000085 $\mu\text{g/m}^3$; therefore, the system does not meet the requirements, so emission controls are necessary.

2.4.2 Environmental Risk Management

If it is the responsibility of government to protect the lives of its citizens against foreign invasion or criminal assault, then it is equally responsible for protecting the health and lives of its citizens from other potential dangers, such as falling bridges and toxic air pollutants. Government has a limited budget, however, and we expect that this money is distributed so as to achieve the greatest benefits to health and safety. If two chemicals are placing people at risk, then it is rational that funds and effort be expended to eliminate the chemical that results in the greatest risk.

But is this what we really want? Suppose, for example, it is cost-effective to spend more money and resources to make coal mines safer than it is to conduct heroic rescue missions if accidents occur. It might be more risk-effective to put the money we have into safety, eliminate all rescue squads, and simply accept the few accidents that will still inevitably occur. But since there would no longer be rescue teams, the trapped miners would then be left on their own. The net overall effect would be, however, that fewer coal miners' lives would be lost.

Even though this conclusion would be risk-effective, we would find it unacceptable. Human life is considered sacred. This value does not mean that infinite resources have to be directed at saving lives but rather that one of the sacred rituals of our society is the attempt to save people in acute or critical need, such as crash victims, trapped coal miners, and the like. Thus, purely rational calculations, such as the coal miners example above, might not lead us to conclusions that we find acceptable.[3]

In all such risk analyses the benefits are usually to humans only, and they are *short-term benefits*. Likewise, the costs determined in the cost-effectiveness analysis are real budgetary costs, money which comes directly out of the pocket of the agency. Costs related to environmental degradation and *long-term costs* that are very difficult to quantify are not included in these calculations. The fact that long-term and environmental costs cannot be readily considered in these analyses, coupled with the blatant abuse of benefit/cost analysis by governmental agencies, makes it necessary to bring into action other decision-making tools: alternatives assessment and environmental impact analysis.

2.5 DECISIONS BASED ON ALTERNATIVES ASSESSMENT

Ever since the 1983 publication of the National Research Council's so-called Red Book, also known as *Risk Assessment in the Federal Government: Understanding the Process*, chemical risk assessment has been institutionalized in the regulatory application of setting standards as a largely science-driven process that quantitatively evaluates the probability of risk.[4] But over the years people have come to question the effectiveness of risk assessment in achieving what it is intended to do: protect the health and safety of people and the environment. It is becoming clearer and clearer that risk assessment has limitations when used to support sustainability.

A common maxim in the field of toxicology, ascribed to Paracelsus in the 16th century, is that "the dose makes the poison." Actually, what he purportedly said was that "All things are poison and nothing is without poison; only the dose permits something not to be poisonous." In other words, all chemicals are poisonous at some dose; even water is hazardous if you drink—or worse—inhale enough of it. But, of course, some chemicals elicit their toxic effects at much lower doses than others. There is some truth in this maxim, but new studies are revealing that it is more complicated than that; it is not only the dose that matters but also the timing of exposure and more.

As described in the previous section, risk assessment typically assesses risks associated with one chemical at a time. It assesses the risk associated with chemical A separate from that of chemical B, unless A and B are known to be related chemicals with interactive effects. For example, toxic equivalency factors are calculated for congeners of dioxin when they are mixed. The values are then added to assess risk. (A congener refers to a closely related chemical.) But it does not consider the impact of exposure to the combination of chemicals A and B—never mind C, D, E, and F if they are not very similar molecules. Yet scientific research is demonstrating that impacts can be dependent not only dose but on dose in combination with other stressors on an organism, including the presence of very low levels of similar or dissimilar chemicals in the environment. In addition, the timing of exposure can be so critical that exposure within a brief developmental window can result in disruptions to normal development whereas exposure outside that window will not.

Scientists are also finding that effects can show up in unpredictable ways. Exposure to some chemicals at an early age may induce effects that do not occur until much later in life. Sometimes the impacts will not be evident in the exposed organism at all. Rather, the impacts may show up in the progeny one or even two generations later. Risk assessment as a standardized methodology does not deal well with constantly evolving scientific information, particularly when the implications of the identified hazards are not well understood and standardized test methods are unavailable. This is especially true for emerging areas of research, such as endocrine disruption, whereby very low levels of exposure to certain chemicals may interfere with the normal function of the endocrine system, resulting in a variety of possible impacts.

Although risk assessment is used to support decision making, it is fair to say that there are just too many variables to claim with confidence that risk assessment is comprehensive and certain. Yet conclusions from risk assessments are used to make important policy and regulatory decisions that affect human health and the environment. Often risk assessments have legal weight and they are treated as scientifically definitive, despite the fact that they are based on many assumptions and significant scientific uncertainty. In addition, risk assessments are very costly and time-consuming, and the more complexity one seeks to address in a risk assessment, the greater the cost gets.

Partly in response to the limitations of risk assessment, *alternatives assessment* is emerging as an approach to guide decision making toward improved human and environmental health, safety, and sustainability. Alternatives assessment for chemicals complements risk assessment by identifying options that have lower hazard. Risk assessment attempts to answer the question, "Is the risk acceptable?" Such a question results in a yes or no answer that does not encourage continual improvement. Alternatives assessment helps to answer the question, "Which option is safer?" This question supports a strategy of continual improvement.

The following story from Mary O'Brien's book, *Making Better Environmental Decisions: An Alternative to Risk Assessment*, illustrates a distinction between risk assessment and alternatives assessment.[5]

Imagine a woman standing by an icy mountain river, intending to cross to the other side. A team of four risk assessors stands behind her, reviewing her situation. The toxicologist says she ought to wade across the river because it is not toxic, only cold. The cardiologist says she ought to wade across the river because she looks to be young and not already chilled. Her risks of cardiac arrest, therefore, are low. The hydrologist says she ought to wade across the river because he has seen other rivers like this and estimates that this one is not more than 4 feet deep and probably has no whirlpools at this location. Finally, the EPA policy specialist says that the woman ought to wade across the river because, compared to global warming, ozone depletion, and loss of species diversity, the risks of her crossing are trivial.

The woman refuses to wade across. "Why?" the risk assessors ask. They show her their calculations, condescendingly explaining to her that her risk of dying while wading across is one in 40 million.

Still, the woman refused to wade across. "Why?" the risk assessors ask again, frustrated by this woman who clearly doesn't understand the nature of risks.

The woman points upstream and says "because there is a bridge."

The risk assessors in this story are basing their recommendations on the results of their evaluation of the risks of only one option: wading across an icy river. The woman is evaluating her alternatives, one of which is to use a bridge to cross the river.

The goal of alternatives assessment is to identify possible alternatives to existing chemicals, products, or processes and to choose among them with the intent to choose more health and environmentally protective options. Alternatives assessment is thus a comparative process intended to generate safer or less toxic products and practices. Achieving relatively greater safety depends on judgments about a variety of factors, which may include relative exposures, relative human health and environmental hazards, as well as life-cycle, social, and economic considerations.

To focus on green chemistry and green engineering, an alternatives assessment typically includes one or more of the components of a traditional risk assessment, but the focus is on comparing alternatives/options that (1) are available or (2) could be developed. That is, what is assessed are the *relative* utility/functionality (benefits), costs, hazards, exposures, and/or risks of the alternatives to discover or identify situations that provide win-win or win-win-win opportunities. Examples are (1) flame-retarding fabrics in which not only alternative flame-retarding chemicals are considered but also noncoated fabric using new weaving methods, which are more fire resistant or slow burning, and (2) nonfibrous alternatives to asbestos.

The emphasis is to reframe the public policy question: It is not, "What are the definitive risks, benefits, and costs of an existing hazardous product, process or approach?" The question is rather, "What are alternative ways that exist or that could be developed to achieve equal or better functionality at lower risk and possibly cost, reflecting a more socially optimal balance of risks, benefits, and costs?" This reframing would be informed by advances in green chemistry and engineering and would more easily allow the assessment of the *relative* risks, *relative* benefits, and *relative* costs of alternative technological options with the intent of continually raising the bar to achieve the highest level of human health and environmental benefit along with performance. A variety of approaches to alternatives assessment differ with regard to their purpose, scope, and other features.[6] One of the leading approaches is that of the USEPA's Design for the Environment Program.

The Design for the Environment (DfE) promotes pollution prevention and other risk reduction activities in industrial sectors by comparing human health and environmental risks and considering the performance and cost of existing and alternative products, processes, and practices. DfE does this in an innovative way: by forming partnerships with industry and other interested parties to develop information on environmental and human health impacts, performance, and cost of cleaner technologies and approaches. DfE then disseminates the information to help businesses design and redesign cost-effective products and processes that are cleaner and safer for workers and the public.

Through these partnerships, DfE supports organizations in gathering and assessing information that they may be unable to gather independently, whether due to cost or because the information is considered proprietary. Chemical manufacturers share proprietary formulations with DfE through these voluntary partnerships because of the way the hazard information is handled and presented. DfE receives full ingredient disclosure from the manufactures and holds the information in confidence. However, they make the results of the hazard assessments publicly accessible without revealing exact percentages of constituents or identifying proprietary chemicals. This process protects important confidential business information to preserve a company's competitiveness but gives consumers and other stakeholders the relevant human and environmental health and safety information that they need to make informed decisions.

An example of a successful DfE partnership is the Furniture Flame Retardancy Partnership.[7] Pentabromodiphenyl ether (pentaBDE) was the primary flame retardant used in low-density, flexible polyurethane furniture foam. Due to concerns over its use and the fact that the chemical was found widespread in the environment and in human tissue and breast milk, pentaBDE was voluntarily phased out of production by U.S. manufacturers in January 2004. The industry needed alternatives to meet furniture flame retardancy requirements but did not have the human and environmental health and safety information needed to compare the alternatives. DfE worked with the furniture manufacturers, foam manufacturers, and flame-retardant chemical suppliers, along with governmental and environmental groups, to evaluate possible alternatives.

Fourteen formulations of chemical alternatives were submitted to USEPA under confidentiality, and they were assessed based on numerous human health and ecotoxicity endpoints in addition to bioaccumulation potential and environmental persistence. They were also screened for potential exposure to workers, users, and the aquatic environment, and a distinction was made between flame retardants that are reacted into the foam and those that are added to the foam and can more easily leach out. Where data gaps existed, USEPA experts used models and chemical analogs to estimate the hazard for a particular endpoint. The literature and test data reviews were published in the final report, "Environmentally Preferable Options for Furniture Fire Safety: Low Density Furniture Foam," which is available on the DfE website. In addition, each hazard endpoint was ranked with a concern level (high, moderate, or low) based on the criteria used by the USEPA's New Chemicals Program to rate the concern level of new chemicals submitted under the Toxic Substance Control Act (TSCA). DfE noted where values were based on experimental data (using bold font) and where values were estimated based on models or chemical analogs (using italic font). A section of the table summarizing the screening results is presented in Figure 2.6.

L = Low hazard concern
M¹ = Moderate hazard concern
H = High hazard concern
L, M¹, or H = Endpoint assigned using estimated values and professional judgment (Structure Activity Relationships)

N = No
Y = Yes
P = Yes for pure chemical

*Ongoing studies may result in a change in this endpoint
▲Persistent degradation products expected²

Company	Chemical	% in Formulation³	Human Health Effects							Ecotoxicity		Environmental		Potential Routes of Exposure — Worker			Potential Routes of Exposure — General Population			Aquatic	Reactive or Additive?
			Cancer Hazard	Skin Sensitizer	Reproductive	Developmental	Neurological	Systemic	Genotoxicity	Acute	Chronic	Persistence	Bioaccumulation	Inhalation	Dermal	Ingestion	Inhalation	Dermal	Ingestion		
Supresta	**AB053**																				
	Tris(1,3-dichloro-2-propyl)Phosphate CAS # 13674-87-8		M	L	M	M	L	M	M	M	M	M	*L*	N	Y	Y	N	Y	Y	Y	Additive
Supresta	**AC003**																				
	Proprietary I Organic phosphate ester	92-99%	*L*	L	L	L	*L*	*M*	L	H	H	H	*L*	P	Y	Y	N	Y	Y	Y	Additive
	Triphenyl Phosphate CAS # 115-86-6	1-8%	*L*	L	L	L	L	M	L	H	*H*	L	*L*	Y	Y	Y	Y	Y	Y	Y	Additive
Supresta	**AC073**																				
	Triphenyl Phosphate CAS # 115-86-6	38-48%	*L*	L	L	L	L	M	L	H	*H*	L	*L*	Y	Y	Y	Y	Y	Y	Y	Additive
	Proprietary J Aryl phosphate	40-46%	*L*	*L*	L	*L*	L	M	M*	*L*	*H*	L	*L*	Y	Y	Y	Y	Y	Y	Y	Additive
	Proprietary K Aryl phosphate	12-18%	*L*	*L*	*L*	*L*	*L*	*M*	*L*	*L*	*L*	*L*	*L*	P	Y	Y	N	Y	N	N	Additive
	Proprietary L Aryl phosphate	1-3%	*L*	*L*	*L*	*L*	*L*	*M*	*L*	*L*	*L*	*L*	*L*	P	Y	Y	N	Y	N	N	Additive
Supresta	**Fyrol FR-2**																				
	Tris(1,3-dichloro-2-propyl)phosphate CAS # 13674-87-8	99%	M	L	M	M	L	M	M	M	M	M	*L*	N	Y	Y	N	Y	Y	Y	Additive

Figure 2.6 USEPA furniture flame retardants hazard assessment matrix. (U. S. Environmental Protection Agency.)

FOCUS ON

The Green Screen for Safer Chemicals

The Green Screen for Safer Chemicals (Green Screen) developed by Clean Production Action, a nongovernmental environmental and human health advocacy organization, builds on the work of the DfE Flame Retardancy Partnerships.[8] Because the DfE alternatives assessment approach was not intended to rank the flame retardants, questions were raised about how best to identify which alternative was indeed more benign. None of the options were considered to have low hazards across all endpoints. Each of them had one or more hazards associated with them. But not all hazards are considered equal. For example, would it be better to choose a flame retardant that causes cancer or skin irritation? Questions were raised about how to compare and prioritize hazards to identify safer alternatives—how to consider trade-offs between hazards. The Green Screen was designed to support this type of decision making by businesses, governments, and individuals concerned with the risks posed by chemicals as well as to advance the development of green chemistry and to be a building block on the path to sustainable product design and sustainable material flows.

The Green Screen defines four benchmarks on a path to safer chemicals, helping to provide perspective and to prioritize chemical alternatives based on hazards and combinations of hazards (Figure 2.7).

- Benchmark 1 characterizes chemical products as "Avoid—Chemical of High Concern."
- Benchmark 2 characterizes chemical products as "Use but Search for Safer Substitutes."
- Benchmark 3 characterizes chemicals as "Use but Still Opportunity for Improvement."
- Benchmark 4 characterizes chemicals as "Prefer—Safer Chemical."

Each benchmark consists of a set of hazard criteria that encompass a combination of

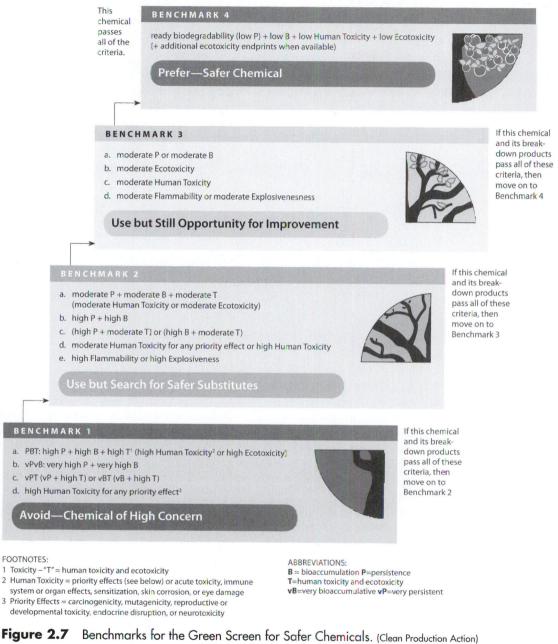

BENCHMARK 4

ready biodegradability (low P) + low B + low Human Toxicity + low Ecotoxicity (+ additional ecotoxicity endprints when available)

Prefer—Safer Chemical

This chemical passes all of the criteria.

BENCHMARK 3

a. moderate P or moderate B
b. moderate Ecotoxicity
c. moderate Human Toxicity
d. moderate Flammability or moderate Explosivenesness

Use but Still Opportunity for Improvement

If this chemical and its break-down products pass all of these criteria, then move on to Benchmark 4

BENCHMARK 2

a. moderate P + moderate B + moderate T (moderate Human Toxicity or moderate Ecotoxicity)
b. high P + high B
c. (high P + moderate T) or (high B + moderate T)
d. moderate Human Toxicity for any priority effect or high Human Toxicity
e. high Flammability or high Explosiveness

Use but Search for Safer Substitutes

If this chemical and its break-down products pass all of these criteria, then move on to Benchmark 3

BENCHMARK 1

a. PBT: high P + high B + high T[1] (high Human Toxicity[2] or high Ecotoxicity)
b. vPvB: very high P + very high B
c. vPT (vP + high T) or vBT (vB + high T)
d. high Human Toxicity for any priority effect[3]

Avoid—Chemical of High Concern

If this chemical and its break-down products pass all of these criteria, then move on to Benchmark 2

FOOTNOTES:
1 Toxicity – "T" = human toxicity and ecotoxicity
2 Human Toxicity = priority effects (see below) or acute toxicity, immune system or organ effects, sensitization, skin corrosion, or eye damage
3 Priority Effects = carcinogenicity, mutagenicity, reproductive or developmental toxicity, endocrine disruption, or neurotoxicity

ABBREVIATIONS:
B = bioaccumulation **P**=persistence
T=human toxicity and ecotoxicity
vB=very bioaccumulative **vP**=very persistent

Figure 2.7   Benchmarks for the Green Screen for Safer Chemicals. (Clean Production Action)

hazards and threshold values that a chemical, along with its known and predicted breakdown products (environmental degradation products and metabolites), must pass to progress to the next benchmark. The criteria for each benchmark become increasingly more demanding for environmental and human health and safety, with benchmark 4 representing the chemical with the least hazard.

2.6 DECISIONS BASED ON ENVIRONMENTAL IMPACT ANALYSES

On 1 January 1970, President Richard M. Nixon signed into law the National Environmental Policy Act (NEPA), which was intended to "encourage productive and enjoyable harmony between man and his environment." As with other imaginative and groundbreaking legislation, the law contained many provisions that were difficult to implement in practice. Nevertheless, it provided the model for environmental legislation soon adopted by most of the western world.

NEPA set up the Council on Environmental Quality (CEQ), which was to be a watchdog on federal activities that influence the environment; the CEQ reported directly to the President. The vehicle by which the CEQ would monitor significant federal activity impacting the environment was a report called the *environmental impact statement* (EIS). This lightly regarded provision in NEPA, tucked away in Section 102, stipulates that the EIS is to be an inventory, analysis, and evaluation of the effect of a planned project on environmental quality. The EIS is to be written first in draft form by the federal agency in question for each significant project, and then this draft is to be submitted for public comment. Finally, the report is rewritten, taking into account public sentiment and comments from other governmental agencies. When complete, the EIS is submitted to the USEPA.

The impact of Section 102 of NEPA on federal agencies was traumatic because they were not geared up in manpower or in training nor were they psychologically able to accept this new (what they viewed as a) restriction on their activities. Thus, the first few years of the EIS were tumultuous, with many environmental impact statements being judged for adequacy in courts of law.

Conflict, of course, arises when the cost-effective alternative, or the one with the highest benefit/cost ratio, also results in the greatest adverse environmental impact. Decisions have to be made, and quite often the B/C wins out over the EI. It is, nevertheless, significant that, since 1970, the effect of the project on the environment must be considered, whereas before 1970 these concerns were never even acknowledged, much less included in the decision-making process.

Governmental agencies tend to conduct internal EI studies and propose only those projects that have both a high B/C ratio and a low adverse environmental impact. Most EI statements are thus written as a justification for an alternative that has already been selected by the agency.

Although the CEQ developed guidelines for the EIS, the form of the EIS is still variable, and considerable judgment and qualitative information (some say prejudice) goes into every EIS. Each agency seems to have developed its own methodology within the constraints of the CEQ guidelines, and it is difficult to argue that any one format is superior to another. Because there is no standard EIS, the following discussion is a description of several alternatives within a general framework. It is suggested that the EIS should be in three parts: inventory, assessment, and evaluation.

2.6.1 Inventory

The first duty in the writing of any EIS is the gathering of data, such as hydrological, meteorological, or biological information. A listing of the species of plants and animals in the area of concern, for example, is included in the inventory. No decisions are made at this stage because *everything* properly belongs in the inventory.

2.6.2   Assessment

The second stage is the analysis part, commonly called the assessment. This is the mechanical part of the EIS in that the data gathered in the inventory are fed to the assessment mechanism, and the numbers are crunched accordingly. Numerous assessment methodologies have been suggested, only a few of which are described below.

Quantified checklist is possibly the simplest quantitative method of comparing alternatives. It involves first the listing of those areas of the environment that might be affected by the proposed project and then an estimation of

a. the *importance* of the impact
b. the *magnitude* of the impact
c. the *nature* of the impact (whether negative or positive).

Commonly, the importance is given numbers such as 0 to 5, where 0 means no importance whatever while 5 implies extreme importance. A similar scale is used for magnitude while the nature is expressed as simply -1 as negative (adverse) and $+1$ as positive (beneficial) impact. The environmental impact (EI) is then calculated as

$$\text{EI} = \sum_{i=1}^{n} (I_i \times M_i \times N_i)$$

where I_i = importance of ith impact
M_i = magnitude of ith impact
N_i = nature of ith impact, so that $N = +1$ if beneficial and $N = -1$
if detrimental
n = total number of areas of concern

Example 2.9 illustrates the use of the quantified checklist.

EXAMPLE
2.9

Problem   Continuing Example 2.3, the community has two alternatives: Increase the refuse collection frequency from 1 to 2 times per week or allow the burning of rubbish on premises. Analyze these two alternatives by using a quantified checklist.

Solution   First, the areas of environmental impact are listed. In the interest of brevity, only six areas are shown below; recognize, however, that a thorough assessment would include many other concerns. Following this, values for importance and magnitude are assigned (0 to 5), and the nature of the impact (+/−) is indicated. The three columns are then multiplied.

Alternative 1: Increasing collection frequency

Area of Concern	Importance (I)	Magnitude (M)	Nature (N)	Impact (I × M × N)
Air pollution (trucks)	4	2	−1	−8
Noise	3	3	−1	−9
Litter in streets	2	2	−1	−4
Odor	2	3	−1	−6
Traffic congestion	3	3	−1	−9
Groundwater pollution	4	0	−1	0

(*Note*: No new refuse will be landfilled.)

EI = −36

Alternative 2: Burning on premises

Area of Concern	Importance (I)	Magnitude (M)	Nature (N)	Impact (I × M × N)
Air pollution (burning)	4	4	−1	−16
Noise	0	0	−1	0
Litter	2	1	+1	+2

(*Note*: Present collection system causes litter.)

Odor	2	4	−1	−8
Traffic congestion	0	0	−1	0
Groundwater pollution	4	1	+1	+4

(*Note*: Less refuse will be landfilled.

EI = −18

On the basis of this analysis, burning the refuse would result in a lower adverse reaction.

Interaction Matrix

For simple projects, the quantified checklist is an adequate assessment technique, but it gets progressively unwieldy for larger projects that have many smaller actions all combining to produce the overall final product, such as the construction of a dam. The effect of each of these smaller actions should be judged separately with respect to its impact. Such an interaction between the individual actions and areas of concern gives rise to the *interaction matrix*, wherein once again the importance and the magnitude of the interaction are judged (such as the 0 to 5 scale previously used). There seems to be no agreement on what calculation should be made to produce the final numerical quantity. In some cases the importance is multiplied by the magnitude and the products summed as before while another procedure is simply to add all the numbers in the table. In Example 2.10, the products are summed into a grand sum shown on the lower-right corner of the matrix.

EXAMPLE
2.10   **Problem**   Continuing Example 2.3, use the interaction matrix assessment technique to decide on the alternatives presented.

Solution   Note again that these are incomplete lists used only for illustrative purposes. The results indicate once again that it makes more sense to burn the paper.

Alternative 1:
Increasing collection frequency

Actions

Area of concern	Collection by truck	Transport to landfill	Disposal in landfill	
Air pollution	−4 / 2	−2 / 2	−1 / 1	−13
Noise	−3 / 3	−2 / 1	−2 / 1	−13
Odor	−2 / 3	0 / 0	−2 / 1	−8
Groundwater pollution	0 / 0	0 / 0	−4 / 0	0
	−23	−6	−5	−34

Alternative 2:
Burning rubbish

Actions

Area of concern	Burning of rubbish	Less refuse to collect	Less refuse in landfill	
Air pollution	−4 / 4	0 / 0	0 / 0	−16
Noise	0 / 0	+1 / 1	+1 / 1	+2
Odor	−2 / 4	+1 / 1	+1 / 1	−6
Groundwater pollution	0 / 0	0 / 0	+4 / 1	+4
	−24	+2	+6	−16

Before moving on to the next technique, it should again be emphasized that the method illustrated in Example 2.10 can have many variations and modifications, none of which is right or wrong but depend on the *type of analysis* conducted, *for whom* the report is prepared, and *what* is being analyzed. Individual initiative is often the most valuable component in the development of a useful EIS.

Common parameter weighed checklist is another technique for environmental impact assessment. It differs from the quantified checklist technique only in that, instead of using arbitrary numbers for importance and magnitude, the importance term is called the effect (E) and is calculated from actual environmental data (or predicted quantitative values), and the magnitude is expressed as weighing factors (W).

The basic objective of this technique is to reduce all data to a common parameter, thus allowing the values to be added. The data (actual or predicted) are translated to the effect term by means of a function that describes the relationship between the variation of the measurable value and the effect of that variation. This function is commonly drawn for each interaction on a case-by-case basis. Three typical functions are illustrated in Figure 2.9. All these curves show that, as the value of the measured quantity increases, the adverse effect on the environment (E) also increases but that this relationship can take several forms. The value of E ranges from 0 to plus or minus 1.0, with the positive sign implying beneficial impact and the negative detrimental.

FOCUS ON

Tampa Bay Water

The Tampa Bay, Florida, area includes the city of Tampa on the land side and the venerable St. Petersburg on the bay side, connected by a long causeway (Figure 2.8). As the population of the region exploded during the past decades, the local governments needed to find water for the citizens. Desalinization is very expensive, and there are no large rivers to develop for surface supply, so the local water authority, Tampa Bay Water (TBW) decided that groundwater was the only option and constructed well fields throughout the Tampa area. No further freshwater was available from the St. Petersburg side because saltwater intrusion would have destroyed the remaining groundwater resource. (The seawater, being heavier than freshwater, seeps under the freshwater lens, and eventually all the wells start pumping seawater.)

So TBW started to pump an increasing amount of freshwater from the land side. The effects of their pumping can best be demonstrated by what happened to one family, Steve and Cathy Monsees. In 1988 they moved to Pasco County and bought a property with a 5-acre (20235 m^2) pond in the back and the 100-acre (404700 m^2) Prairie Lake in front. To their horror, they watched Prairie Lake dry up; by the early 1990s it was completely gone. They questioned the TBW about this and were told that the reason was a prolonged drought. Looking into the situation further, they found out that the past few years had been wetter than normal, so the dry lakebed could not have been caused by a drought. TBW continued to prevaricate, arguing variously that the cause was irrigation and then development. Finally, the Monsees found out that TBW had drilled a large well field about a mile (1.6 km) from the lake. The field produced 45 mgd (1971 L/s), a huge amount that could not be sustained by rainwater percolation. The groundwater table dropped to below the lake level, and the lake water all drained into the ground and finally ended up in the taps of the increasingly thirsty citizens of Tampa and St. Petersburg. The damage to the groundwater resources were found to be massive, with over half of the lakes in the region dried up. Damage to streams and creeks in the area precipitated widespread loss of wildlife. Giant trees that had taken water from the aquifer died and fell, and huge sinkholes were formed as the cavities in the ground collapsed.

Figure 2.8 Tampa Bay, Florida

fter many lawsuits, TBW offered an apology but did not stop the damage to the groundwater resources and the natural environment. They eventually realized that they could not indefinitely lower the groundwater table and constructed a $120-million desalinization plant, but it has not been enough to meet the growing demand. At what point does the availability of water dictate limits on population growth?

Source: Robert Glennon, *Water Follies*, Island Press, 2002.

Consider, for example, the presence of a toxic waste on the health and survival of a certain aquatic organism. The concentration of the toxin in the stream is the measured quantity while the health of the aquatic organism is the effect. The effect (detrimental) increases as the concentration increases. A very low concentration has no detrimental effect, whereas a very high concentration can be disastrous. But what type of function (curve) makes the most sense for this interaction? The straight line function (Figure 2.9A on the next page) implies that as the concentration of the toxin increases from zero, the detrimental effects are immediately felt. This is seldom true. At very low concentrations most toxins do not show a linear relationship with effect, so this function does not appear to be useful. Figure 2.9B, the next curve, is also incorrect. But Figure 2.9C seems much more reasonable because it implies that the effect of the toxin is very small at lower concentrations but, when it reaches a threshold level, it becomes very toxic quickly. As the level increases above the toxic threshold, there can be no further damage because the organisms are all dead, so the effect levels off at 1.0.

Once the effect (E) terms are estimated for each characteristic, they are multiplied by weighing factors (W), which are distributed among the various effects according to the importance of the effect. Typically, the weighing terms add to 100, but this is not important as long as an equal number of weighing terms are distributed to each alternative analyzed. The final impact is then calculated by adding the products of the effect terms (E) and weighing factors (W). Thus, for each alternative considered:

$$\text{Environmental impact} = \sum_{i=1}^{n}(E_i \times W_i)$$

where n = total number of environmental areas of concern considered. Remember that E can be negative (detrimental) or positive (beneficial).

EXAMPLE

2.11   **Problem**   Continuing Example 2.3, using only litter, odor, and airborne particulates (as an example of air quality), calculate the environmental impact of rubbish burning using the common parameter weighed checklist. (Again, this is only a very small part of the total impact assessment that would be necessary when using this technique.)

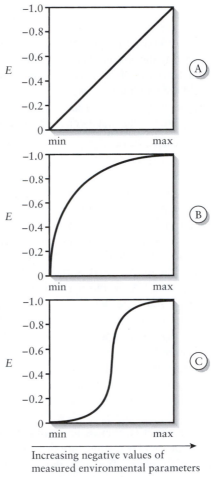

Figure 2.9   Three types of functions relating environmental characteristics to effects.

Solution   Assume that the three curves shown in Figure 2.10 are the proper functions relating the environmental characteristics of litter, odor, and airborne particulates, respectively. Assume that it has been estimated that the burning of rubbish will result in a litter level of 2 on a scale of 0 to 4, an odor number of 3, and an airborne particulate increase to 180 $\mu g/m^3$. Entering Figure 2.10A, B, and C at 2, 3, and 180, respectively, the effects (E) are read off as -0.5, -0.8 and -0.9. It is now necessary to assign weighing factors, and out of a total of 10, it is decided to assign 2, 3, and 5, respectively, implying that the most important effect is the air quality and the least important is the litter. Then the environmental impact is

$$EI = (-0.5 \times 2) + (-0.8 \times 3) + (-0.9 \times 5) = -7.9$$

A similar calculation would be performed for other alternatives and the EIs compared.

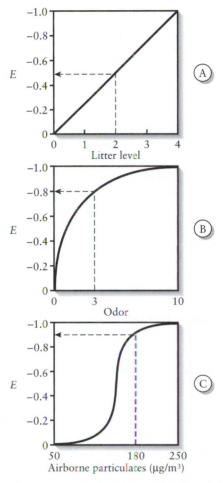

Figure 2.10   Three specific functions of environmental effects. See Example 2.11.

It should be clear that this technique is wide open to individual modifications and interpretations, and the example above should not be considered in any way a standard method. It does, however, provide a numerical answer to the question of environmental impact, and when this is done for several alternatives, the numbers can be compared. This process of comparison and evaluation represents the third part of an EIS.

2.6.3   Evaluation

The comparison of the results of the assessment procedure and the development of the final conclusions are all covered under evaluation. It is important to recognize that the previous two steps, inventory and assessment, are simple and straightforward procedures compared to the final step, which requires judgment and common sense. During this step in the writing of the EIS, the conclusions are drawn up and presented. Often, the reader of the EIS sees only the conclusions and never bothers to review all the assumptions that went into the assessment calculations, so it is important to include in the evaluation the flavor of these calculations and to emphasize the level of uncertainty in the assessment step.

But even when the EIS is as complete as possible and the data have been gathered and evaluated as carefully as possible, conclusions concerning the use of the analysis are open to severe differences. For example, the EIS written for the Alaska oil pipeline, when all the volumes are placed into a single pile, represents (4.3 meter) of work. And at the end of all that effort, good people on both sides drew diametrically opposite conclusions on the effect of the pipeline. The trouble was that they were arguing over the *wrong thing*.[9] They may have been arguing about how many caribou would be affected by the pipeline, but their disagreement was actually how deeply they *cared* that the caribou were affected by the pipeline. For a person who does not care one twit about the caribou, the impact is zero while those who are concerned about the herds and the long-range effects on the sensitive tundra ecology care very much. What then is the solution? How can engineering decisions be made in the face of conflicting *values*? Such decisions require another type of engineering decision making: an *ethical analysis*.

2.7 DECISIONS BASED ON ETHICAL ANALYSES

Before embarking on a discussion of ethical analysis, it is necessary to define quite clearly what is meant by ethics.[10] The popular opinion is that an ethical person, for example, is a "good" person, a person with high standards. Likewise, a moral person is thought to have certain conventional views on sex. These are both common misconceptions.

Morals are the values people choose to guide the way they ought to treat each other. One such moral value may be telling the truth, and some people will choose to be truthful. Such people are thought of as moral people with regard to truth because they would be acting according to their moral convictions. If a person does not value truthfulness, however, then telling the truth is irrelevant, and such a person does not have a moral value with regard to truthfulness. In fact, it is possible to hold a moral view that one always ought to lie, and in this case the person would be thought of as a moral person if he or she lied because he or she would be acting according to a moral conviction.

Most rational people will agree that it is much better to live in a society where people do not lie, cheat, or steal. Societies where people lie, cheat, and steal certainly exist, but given the choice, most people would not want to behave in such a manner and would choose to live in societies where everyone shares moral values that provide mutual benefit.

While it is fairly straightforward to agree that it is not acceptable to lie, cheat, or steal, and most people will not do so, it is a much more difficult matter to decide what to do when conflicts arise between values. For example, suppose it is necessary to lie to keep a promise. How are we to decide what to do when values conflict? In economic questions a similar situation occurs. For example, how are we to decide which project to undertake with limited resources? As already discussed, we use a benefit/cost analysis. How then can we make a decision when moral values conflict? We use an ethical analysis.

Ethics provide a systematized framework for making decisions where values conflict. The selection of the nature and function of that decision-making machinery depends on one's own moral values. Both the cost-effectiveness analysis and the benefit/cost analysis are methods for making decisions based (mostly) on money. Risk analysis calculates the potential damage to health, and environmental impact analysis provides a means for decision making based on long-term effects on resources. Ethics is similarly a framework for decision making, but the parameters of interest are not dollars or environmental data but values. It then follows that, because ethics is a system for decision making, an ethical person is one who makes decisions based on some ethical system. *Any system!* For example, if one chooses to observe a system of ethics that maximizes personal pleasure (hedonism), it would be correct (ethical) to make all decisions so that personal pleasure is maximized. One would, in that case, push old ladies off benches so one could sit down or cheat on tests because this decreases one's required study time and maximizes party time. Provided one adopts hedonism as the accepted mode of behavior (ethic), one would in these cases be acting ethically.

There are, of course, many other systems of ethics that result in actions most civilized people consider more acceptable norms of social behavior. The most important aspect of any ethical code or system one adopts is that one should then be prepared to defend it as a system that *everyone* should employ. If such a defense is weak or faulty, then the ethical system is considered inadequate, with the implication that a rational person would then abandon the system and seek one that can be defended as a system that ought to be adopted by everyone.

2.7.1   Utilitarianism and Deontological Theories

Most ethical thinking over the past 2500 years has been a search for the appropriate ethical theory to guide our behavior in human–human relationships. Some of the most influential theories in Western ethical thinking, theories that are most defensible, are based on consequences or on acts. In the former, moral dilemmas are resolved on the basis of what the consequences are. If it is desired to maximize good, then that alternative that creates the greatest good is correct (moral). In the latter, moral dilemmas are resolved on the basis of whether the alternative (act) is considered good or bad; consequences are not considered.

The most influential consequentialist ethical theory is *utilitarianism* by Jeremy Bentham (1784–1832) and John Stuart Mill (1806–1873). In utilitarianism, the pain and pleasure of all actions are calculated, and the worth of all actions is judged on the basis of the total happiness achieved, where happiness is defined as the highest pleasure/pain ratio. (Sound familiar?) The so-called utilitarian calculus allows for the calculation of happiness for all alternatives being considered. To act ethically, then, would be to choose the alternative that produces the highest level of pleasure and the lowest level of pain. Since the happiness of all human beings involved is summed, such calculations often dictate a decision wherein the moderate happiness of many results in the extreme unhappiness of a few. Benefit/cost analysis can be considered to be utilitarian in its origins because money is presumed to equate with happiness.

The supporters of consequentialist theories argue that these are the proper principles for human conduct because they promote human welfare and that to act simply on the basis of some set of rules without reference to the consequences of these actions is irrational. Agreeing with Aristotle, utilitarians argue that happiness is always chosen for its own sake and, thus, must be the basic good that we all seek. Consequently, because utilitarian calculus provides for that calculation, it is the proper tool for decision making wherein values are involved.

The second group of ethical theories is based on the notion that human conduct should be governed by the morality of acts and that certain rules (such as "do not lie") should always be followed. These theories, often called *deontological* theories, emphasize the goodness of the act and not its consequence. Supporters of these theories hold that acts must be judged as good or bad, right or wrong, *in themselves*, irrespective of the consequences of these acts. An early system of deontological rules is the Ten Commandments as these rules were meant to be followed *regardless of consequences*.

Possibly the best known deontological system is that of Immanuel Kant (1724–1804), who suggested the idea of the categorical imperative—the concept that one develops a set of rules for making value-laden decisions such that one would wish that all people obeyed the rules. Once these rules are established, one must always follow them, and only then can that person be acting ethically because it is the act that matters. A cornerstone of Kantian ethics is the principle of *universalizability*, a simple test for the rationality of a moral principle. In short, this principle holds that, if an act is acceptable for one person, it *must be equally acceptable for others*. For example, if one considers lying acceptable behavior for himself or herself, everyone should be allowed to lie, and everyone is in fact expected to lie. Similarly, if one person decides that cheating on an exam is acceptable, then he or she agrees, by the principle of universalizability, that it is perfectly acceptable for everyone else to cheat also. To live in a world where everyone is expected to lie or cheat would be a sorry situation, and our lives would be a great deal poorer as a result. It thus makes no sense to hold that lying or cheating is acceptable because these behaviors cannot be universalized.

Supporters of rule-based theories argue that consequentialist theories permit and often encourage the suffering of a few for the benefit of many and that this is clearly an injustice. For example, they assert that consequentialist theories would welcome the sacrifice of an innocent person if this death would prevent the death of others. They argue that, if killing is wrong, then the mere act of allowing one innocent person to die is wrong and immoral. The utilitarians counter by arguing that often a good act results in net harm. A trivial example would be the question from your roommate/spouse/friend, "How do you like my new hairdo/shirt/tie/etc.?" Even if you honestly think it is atrocious, the good act would be to tell the truth because one is never supposed to lie. Would a white lie not result in the greater good, ask the utilitarians? The deontologists respond that it is wrong to lie, even though it might hurt short-term feelings, because telling the truth may create a trust that would hold fast in times of true need.

There are, of course, many more systems of ethics that could be discussed and that have relevance to the environmental engineering profession, but it should be clear that traditional ethical thinking represents a valuable source of insight in one's search for a personal and professional lifestyle.

2.7.2 Environmental Ethics and Instrumental Value

All classical ethical systems are intended to provide guidance as to how human beings ought to treat each other. In short, the *moral community*, or those individuals with whom we would need to interact ethically, includes only humans, and the only *moral agents* within the moral community are human beings. Moral agency requires *reciprocity* in that each person agrees to treat one another in a mutually acceptable manner. But we obviously are not the only inhabitants on Earth, and is it not also important how we treat nonhuman animals? Or plants? Or places? Should the moral community be extended to include other animals? plants? inanimate objects, such as rocks, mountains, and even places? If so, should we also extend the moral community to our progeny, who are destined to live in the environment we will to them?

Such questions are being debated and argued in a continuing search for what has become known as *environmental ethics*, a framework that is intended to allow us to make decisions *within* our environment, decisions that will concern not only ourselves but the rest of the world as well. One approach to the formulation of an ethic that incorporates the environment is to consider environmental values as *instrumental values*, values that can be measured in dollars and/or the support nature provides for our survival (e.g., production of oxygen by green plants). The instrumental-value-of-nature view holds that the environment is useful and valuable to people, just like other desirable commodities—such as freedom, health, and opportunity.

This *anthropocentric* view of environmental ethics, the idea that nature is here only for the benefit of people, is, of course, an old one. Aristotle stated, "Plants exist to give food to animals, and animals to give food to men. . . . Since nature makes nothing purposeless or in vain, all animals must have been made by nature for the sake of men." Kant similarly incorporated nature into his ethical theories by suggesting that our duties to animals are indirect duties, which are really duties to our own humanity. His view is quite clear: "So far as animals are concerned, we have no direct duties. Animals are not self-conscious and are there merely as a means to an end. That end is man."[11] Thus, by this reasoning, the value of nonhuman animals can be calculated as their value to people. We would not want to kill off all the plains buffalo, for example, because they are beautiful and interesting creatures, and we enjoy looking at them. To exterminate the buffalo would mean that we are causing harm to other humans. However, William F. Baxter says that our concern for "damage to penguins, or to sugar pines, or geological marvels is . . . simply irrelevant . . . Penguins are important [only] because people enjoy seeing them walk about the rocks."[12]

We can agree that it is necessary to live in a healthy environment to be able to enjoy the pleasures of life, and, therefore, other aspects of the environment have instrumental value. One could argue that to contaminate the water or pollute the air or destroy natural beauty is taking something that does not belong to only a single person. Such pollution is stealing from others, plain and simple. It would also be unethical to destroy the natural environment because so many people enjoy hiking in the woods or canoeing down rivers, and we should preserve these for our benefit. In addition, we should not exterminate species since there is the possibility that they will somehow be useful in the future. An obscure plant or microbe might be essential in the future for medical research, and we should not deprive others of that benefit.

While the instrumental-value-of-nature approach to environmental ethics has merit, it also has a number of problems. First, this argument would not prevent us from killing or torturing individual animals as long as this does not cause harm to other people. Such a mandate is not compatible with our feelings about animals. We would condemn a person who causes unnecessary harm to any animal that could feel pain, and many of us do what we can to prevent these types of acts.

Second, this notion creates a deep chasm between humans and the rest of nature, a chasm with which most people are very uneasy. The anthropocentric approach suggests that people are the masters of the world and can use its resources solely for human benefit without any consideration for the rights of other species or individual animals. Such thinking led to the 19th century "rape of nature" in the United States, the effects of which are still with us. Henry David Thoreau (1817–1862), John Muir (1838–1914), and many others tried to put into words what many people felt about such destruction. They recognized that it is our alienation from nature, the notion that nature has only instrumental value, that will eventually lead to the destruction of nature. Clearly, the valuation of nature on only an instrumental basis is an inadequate approach to explain our attitudes toward nature.

2.7.3 Environmental Ethics and Intrinsic Value

Given these problems with the concept of the instrumental value of nature as a basis for our attitudes toward the environment, there has ensued a search for some other basis for including nonhuman animals, plants, and even things within our sphere of ethical concern. The basic thrust in this development is the attempt to attribute *intrinsic* value to nature and to incorporate nonhuman nature within our moral community. This theory is known as *extensionist* ethical thinking since it attempts to extend the moral community to include other creatures. Such a concept is perhaps as revolutionary as the recognition only a century ago that slaves are also humans and must be included in the moral community. Aristotle, for example, did not apply ethics to slaves because they were not, in his opinion, intellectual equals. We now recognize that this was a hollow argument, and today slavery is considered morally repugnant. It is possible that in the not too distant future the moral community will include the remainder of nature as well, and we will include nature in our ethical decision-making machinery.

The extensionist environmental ethic was initially publicized not by a philosopher but by a forester. Aldo Leopold (1887–1948) defined the environmental ethic (or as he called it, a land ethic) as an ethic that "simply enlarges the boundaries of the community to include soils, waters, plants and animals, or collectively the land."[13] He recognized that both our religious as well as secular training had created a conflict between humans and the rest of nature. Nature had to be subdued and conquered; it was something powerful and dangerous against which we had to continually fight. He believed that a rational view of nature would lead us to an environmental ethic that "changes the role of *Homo sapiens* from conqueror of the land community to plain member and citizen of it."

Leopold was, in fact, questioning the age-old belief that humans are special, that somehow we are not a part of nature but pitted against nature in a constant combat for survival, and that we have a God-given role of dominating nature, as specified in Genesis. Much as later philosophers (and people in general) began to see slavery as an untenable institution and recognized that slaves belonged within our moral community, succeeding generations may recognize that the rest of nature is equally important in the sense of having rights.

The question of admitting nonhumans to the moral community is a contentious one, and centers on whether or not nonhuman creatures have rights. If it can be argued that they have rights, then there is reason to include them in the moral community. The question of

FOCUS ON

The Asbestos Conspiracy

Asbestos is a mineral and a fiber and has the dual characteristics of being very strong and noncombustible. It found wide use in the insulation of boilers and other heating components as well as fireproofing and insulating buildings. It was widely used in automobile brakes and in asphalt tiles and shingles. Asbestos is not a health problem unless it gets into your lungs, where the fibers get caught in the alveoli (the small sacs where gas transfer takes place) and once there irritate the membranes, eventually producing a form of lung cancer called asbestosis.

Although the detrimental effects of asbestos became widely known only in the past few decades, insurance companies stopped insuring asbestos workers as early as 1918. In 1930, Johns Manville, the largest manufacturer of asbestos products, distributed an internal report outlining the negative health effects of asbestos, and in 1933 settled a confidential lawsuit brought by 11 workers with the understanding that the proceedings would be forever sealed, suggesting that the testimony would have been damaging to the company if it had been made public. In 1935 Johns Manville executives told the editor of the trade journal *Asbestos* not to publish anything about asbestosis. The conspiracy was most evident when the president of Johns Manville was quoted in 1942 to have described another asbestos manufacturer that shared health information with its workers as "a bunch of fools." In 1951 the Johns Manville management insisted that any reference to cancer be removed from any scientific publication resulting from research sponsored by the company.

When a class action suit was brought in 1977 against the company for workers who were exposed to asbestos in the 1940s, the documents showing the conspiracy became public. Because it takes decades for the effects of asbestos to be manifested, these former workers were only then experiencing health problems. A former Johns Manville manager testified that there was a "hush-hush policy" of not revealing what the company knew of the health dangers of asbestos. The extended lawsuit, which eventually ended in favor of the plaintiffs, caused Johns Manville to go into bankruptcy. The trial judge concluded that there had been a "conscious effort by the [asbestos] industry from the 1930s to downplay or arguably suppress the dissemination of information to employees and the public for fear of the promotion of lawsuits."

Source: Reprinted from *Controlling Environmental Pollution*, Vesilind, P. A. and DiStefano, T. D., 2006. Lancaster, PA: DEStech Publications, Inc. Reproduced with permission of DEStech Publications, Inc.

rights for nonhumans has drawn resounding no's from many philosophers, and these arguments are often based on reciprocity. For example, Richard Watson points out that "to say an entity has rights makes sense only if that entity can fulfill reciprocal duties, i.e. can act as a moral agent."[14] He goes on to argue that moral agency requires certain characteristics (such as self-consciousness, capability of acting, free will, and an understanding of moral principles) and that most animals do not fulfill any of these requirements so they cannot be moral agents and, therefore, cannot be members of the moral community. H. J. McClosky insists that "where there is no possibility of [morally autonomous] action, potentially or actually . . . and where the being is not a member of a kind which is normally capable of [such] action, we withhold talk of rights."[15] It, therefore, is not reasonable to extend our moral community to include anyone other than humans because of the requirement of reciprocity.

But is reciprocity a proper criterion for admission to the moral community? Do we not already include within our moral community human beings that cannot reciprocate—infants, the senile, the comatose, our ancestors, and even future people? Maybe we are making Aristotle's mistake again by our exclusionary practices. Perhaps being human is not a necessary condition for inclusion in the moral community, and other beings have rights similar to the rights that humans have. These rights may not be something that we necessarily give them but the rights they possess by virtue of being.

The concept of natural rights, those rights inherent in all humans, was proposed in the seventeenth century by John Locke[16] (1632–1704) and Thomas Hobbes[17] (1588–1679), who held that the right of life, liberty, and property should be the right of all, regardless of social status.* These rights are natural rights in that we humans cannot give them to other humans or we would be giving rights to ourselves, which makes no sense. Thus, all humans are "endowed with inalienable rights" that do not emanate from any human giver.

If this is true, then there is nothing to prevent nonhuman animals from having "inalienable rights" simply by virtue of their being, just as humans do. They have rights to exist and to live and to prosper in their own environment and not to have humans deny them these rights unnecessarily or wantonly. If we agree that humans have rights to life, liberty, and absence of pain, then it seems only reasonable that animals, who can feel similar sensations, should have similar rights. With these rights come moral agency, independent of the requirement of reciprocity. The entire construct of reciprocity is, of course, an anthropocentric concept that serves well in keeping others out of our private club. If we abandon this requirement, it is possible to admit more than humans into the moral community.

But if we crack open the door, what are we going to let in? What can legitimately be included in our moral community, or, to put it more crassly, where should we draw the line? If the moral community is to be enlarged, many people agree that it should be on the basis of sentience, or the ability to feel pain. This argument suggests that all sentient animals have rights that demand our concern.

Some of the classical ethical theories have recognized that animal suffering is an evil. Jeremy Bentham, for example, argues that animal welfare should somehow be taken into the utilitarian benefit/cost calculation because "the question is not, Can they *reason*? nor Can they *talk*?, but Can they *suffer*?"[18] Peter Singer believes that an animal is of value simply because it values its own life, and sentience is what is important, not the ability to reason. Equality is at the core of Singer's philosophy, and he believes that all sentient creatures have an equal right to life. Sentience, the capacity to have conscious experiences, such as pain and pleasure, is "the only defensible boundary of concern for the interests of others."[19]

To include animal suffering within our circle of concern, however, opens up a Pandora's box of problems. While we might be able to argue with some vigor that suffering is an evil and that we do not wish to inflict evil on any living being that suffers, we do not know for sure which animals (or plants) feel pain; therefore, we are unsure about who should be included. We can presume with fair certainty that higher animals can feel pain

---

*In the new United States of America, the revolution proclaimed life, liberty, and *the pursuit of happiness*, a modification to get around the sticky problem of slaves as both men and property.

because their reactions to pain resemble ours. A dog howls and a cat screams and tries to stop the source of the pain. Anyone who has put a worm on a hook can attest to the fact that the worm probably feels pain. But how about creatures that cannot show us in unambiguous ways that they are feeling pain? Does a butterfly feel pain when a pin is put through its body? An even more difficult problem is the plant kingdom. Some people insist that plants feel pain when they are hurt and that we are just too insensitive to recognize it.

If we use the utilitarian approach, we have to calculate the *amount* of pain suffered by animals and humans. If, for example, a human needs an animal's fur to keep warm, is it acceptable to cause suffering in the animal to prevent suffering in the human? It is clearly impossible to include such variables in the utilitarian calculus.

If we do *not* focus only on the pain and pleasure suffered by animals, then it is necessary either to recognize that the rights of the animals to avoid pain are equal to those of humans or to somehow list and rank the animals in order to specify which rights animals have under which circumstances. In the first instance trapping animals and torturing prisoners would have equal moral significance. In the second it would be necessary to decide that the life of a chickenhawk is less important than the life of a chicken and so on, making an infinite number of other comparisons.

Finally, if this is the extent of our environmental ethics, we are not able to argue for the preservation of places and natural environments, except as to how they might affect the welfare of sentient creatures. Damming up the Grand Canyon would be quite acceptable if we adopted the sentient animal criterion as our sole environmental ethic.

It seems, therefore, that it is not possible to draw the line at sentience, and the next logical step is simply to incorporate all life within the folds of the moral community. This step is not as outrageous as it seems, the idea having been developed by Albert Schweitzer, who called his ethic a "reverence for life." He concluded that limiting ethics to only human interactions is a mistake and that a person is ethical "only when life, as such, is sacred to him, that of plants and animals as that of his fellow men."[20] Schweitzer believed that an ethical person would not maliciously harm anything that grows but would exist in harmony with nature. He recognized, of course, that, in order to eat, humans must kill other organisms, but he held that this should be done with compassion and a sense of sacredness toward all life. To Schweitzer, human beings are simply a part of the natural system.

Charles Darwin is probably most responsible for the acceptance of the notion of humans not being different in kind from the rest of nature. If indeed humans evolved from less complex creatures, we are different only in degree and not kind from the rest of life and are simply another part of a long chain of evolution. As Janna Thompson points out, "Evolutionary theory, properly understood, does not place us at the pinnacle of the development of life on earth. Our species is one product of evolution among many others."[21]

Similarly, Paul Taylor[22] holds that all living things have an intrinsic good in themselves and, therefore, are candidates to be included in the moral community. He suggests that all living organisms have inherent worth, and as soon as we can admit that we humans are not superior, we will recognize that all life has a right to moral protection. What he labels the *biocentric* outlook depends on the recognition of common membership of all living things in Earth's community, that each organism is a center of life, and that all organisms are interconnected. For Taylor, humans are no more or less important than other organisms.

This approach to environmental ethics has a lot of appeal and quite a few proponents. Unfortunately, it fails to convince on several accounts. First, there is no way to determine where the line between living and nonliving really should be drawn. Viruses present the greatest problem here, and if Taylor's ideas are to be accepted, the polio virus might also be included in the moral community. Janna Thompson[23] points out that, based on most arguments for this position, there is also nothing to keep us from excluding organs (such as the liver or the kidney) from membership in the moral community.

Second, the problem of how to weigh the value of nonhuman animal life relative to the life of humans is unresolved. Should the life of all creatures be equal, and thus, a human life is equal to that of any other creature? If so, the squashing of a cockroach would be of equal moral significance to the murder of a human being. If this is implausible, then there must again be some scale of values, and each living creature must have a slot in the hierarchy of values as placed on them by humans.

If such a hierarchy is to be constructed, how would the value of the life of various organisms be determined? Are microorganisms of equal value to the value of polar bears? Is lettuce of the same order of importance as a gazelle?

Such ranking will also introduce impossible difficulties in determining what is and is not deserving of moral protection. "You, the amoeba, you're in. You, the paramecium, you're out. Sorry." That just doesn't compute. Drawing the line for inclusion in the moral community at all life, therefore, seems to be indefensible.

One means of removing the objection of knowing where to draw the line is simply to extend the line to include everything within the circle of moral concern that is important to the system within which individuals exist. Aldo Leopold is often credited with the initial idea for such an *ecocentric* environmental ethic. He suggested that ecosystems should be preserved because, without the ecosystem, nothing can survive. He stated that "a thing is right when it tends to preserve the integrity, stability, and beauty of the biotic community. It is wrong when it tends otherwise."

Val and Richard Routley[24] and Holmes Rolston III[25] recognize that, within the ecocentric environmental ethic, some creatures (such as family) take precedence over others (such as strangers) and that human beings take precedence over nonhuman animals. They view the ethic as a system of concentric rings, with the most important moral entities in the middle and the rings extending outward, incorporating others within the moral community but at decreasing levels of moral protection. The question of how the various creatures and places on Earth are to be graded in terms of their moral worth is not resolved and indeed is up to the people doing the valuing. This process is, of course, once again human centered, and the ecocentric environmental ethic is a form of the anthropocentric environmental ethic with fuzzy boundaries.

Tom Regan presents a similar concept as a preservation principle, a principle of "non-destruction, non-interference, and generally, non-meddling."[26] A school of thought embracing this idea is the *deep ecology* movement. Its most notable proponent is Arne Naess, who suggests that in nature humans are no more important than other creatures or the rest of the world.[27] Deep ecology centers on the idea that humans are part of the total cosmos and are made of the same raw materials as everything else; therefore, humans should live so that they respect all of nature and should recognize the damage that *Homo sapiens* have done to the planet. Deep ecologists call for a gradual reduction of the human population as well as changes in lifestyle to use fewer resources. Deep ecology, named to distinguish the philosophy from "shallow ecologists" who value nature instrumentally, eliminates the problem of where to draw the line between those that are in and out of our moral community because everyone and everything is included, but it again presents us with the necessity of valuing all of nature equally, so we are back to the original problem of judging everything by human standards.*

2.7.4 Environmental Ethics and Spirituality

There is a third approach to environmental ethics—recognizing that we are, at least at the present time, unable to explain rationally our attitudes toward the environment and that these attitudes are deeply felt, not unlike a feeling of spirituality. Why don't we then simply admit that these attitudes are grounded in spirituality? This suggestion may not be as outrageous as it might sound at first cut (but certainly is outrageous in an engineering textbook!). Although we are deeply imbedded in the Western culture, other cultures exist whose approaches to nature may be instructive.

Many older religions, including the Native American religions, are animistic, recognizing the existence of spirits within nature. These spirits do not take human form, as in the Greek, Roman, and Judaic religions. They simply are within the tree or the brook or the sky. It is possible to commune with these spirits—to talk to them, to feel close to them.

Is it too farfetched to hope that future people will live in harmony with the world because we will experience, in Wendell Berry's words, a "secular pilgrimage"?[28] John Stewart Collis had an optimistic view of our future. He wrote:[29]

> Both polytheism and monotheism have done their work. The images are broken; the idols are all overthrown. This is now regarded as a very irreligious age. But perhaps it only means that the mind is moving from one state to another. The next stage is not a belief in many gods. It is not a belief in one god. It is not a belief at all—not a conception in the intellect. It is an extension of consciousness so that we may *feel* God.

A version of the Christian Bible, the Green Bible, has been published. It highlights verses in green ink that reinforce the message of the importance of caring for the earth.

In all likelihood the spiritual alternative is the least likely to withstand rational scrutiny. Yet does this not best explain how we feel toward nature? How do we explain why some people "may avoid making unnecessary noise in the forest, out of respect for the forest and its nonhuman inhabitants"[30] if it cannot be explained on the basis of spiritual feelings?

It is not at all obvious why we should have protective, caring attitudes for an organism or thing only if these attitudes can be reciprocated. Perhaps we are hung up on the idea of acceptance into the moral community, and the logjam can be broken by thinking of it as the inclusion of all things in a community of concern. In this community, reciprocity is not necessary; what matters is loving and caring for others simply because of their presence in the community of concern. The amount of love and care is proportional to the ability to give it and demands nothing in return.

2.7.5 Concluding Remarks

One aspect of the profession of environmental engineering (oft unstated, as if we are embarrassed about it) is that the environmental engineer is engaged in a truly worthwhile

---

*It seems most unfortunate that we refer to organisms as "higher" or "lower," with the implication of superior or inferior. Quite clearly, an earthworm does very well what it is supposed to do, as does a cheetah. It would be very difficult for humans either to wiggle their way through the earth or to catch an antelope on foot. Yes, we can construct machines to do these things, but this is due to our skill in thinking. We lack many other skills, and therefore we cannot claim to be higher than other creatures. Similarly, it is nonsense to talk of a cheetah as higher and an earthworm as lower.

mission. The environmental engineer is the epitome of the *solution* as opposed to the *problem*, and we should feel good about that. Our client, in the broadest sense, is the environment itself, and our objective is to preserve and protect our global home, for the sake of our progeny as well as Mother Earth herself.

2.8   CONTINUITY IN ENGINEERING DECISIONS

The methods of decision making available to engineers stretch from the most objective (technical) to the most subjective (ethical). The inherent method of decision making is the same in all cases. The problem is first analyzed—taken apart and viewed from many perspectives. When all the numbers are in and the variables are evaluated, the information is synthesized into a solution. Then this solution is looked at in its entirety to see if it makes sense or, perhaps most importantly, feels right. This process is especially true in ethical decisions, where there seldom are numbers for comparison.

As engineering decision making stretches from the technical to the ethical, decisions become increasingly less quantitative, and subject to the personal tastes, prejudices, and concerns of the decision maker. Is it reasonable to suggest that at some point these decisions cease being true engineering decisions? Not a few prominent engineers have argued eloquently that the only true engineering decisions are technical decisions, and other concerns should be left to some undefined decision maker for whom the engineer works and who presumably has the training and background for making these decisions of which the engineer is not perhaps capable and certainly not responsible. Such a view would, of course, free you, the engineer, of all judgment (other than technical) and make you a virtual smart robot, working at the behest of your client or employer. By this argument the social consequences of your actions (how they affect society at large) are of little interest as long as your client or employer is well served.

Fortunately, most engineers do not accept such a cop-out. We recognize that engineering, perhaps more than other professions, *can make a difference*. Projects involving environmental change or manipulation will invariably need the services of the professional engineer. We are, thus, morally obligated, as perhaps the one indispensable cog in the wheel of progress, to seek the best solutions not only technically but also economically and ethically.

PROBLEMS

2.1   Chapel Hill, North Carolina, has decided to build a 25 km raw water pipeline to allow it to purchase water from Hillsborough, North Carolina, during times of drought. The engineer has recommended that a 40 cm pipeline be built. The basis for the decision is as follows:

Diameter (cm)	Cost of Pipe in the Ground ($/m)	Capital Cost of Pumping Station	Expected Annual Power Cost
20	1.5	$150,000	$10,000
25	2.5	145,000	8,000
30	3.5	140,000	7,000
40	4.0	120,000	6,000

Interest rate is 8%; expected life is 20 yr.

a. Compare the four alternatives based on annual cost.
b. If the engineer is paid on the basis of 6% of the capital cost of the total job, which alternative might he recommend if he based all decisions on hedonistic ethics?
c. Who might the engineer consider to be the "clients" in this situation? Defend each choice with a few sentences.

2.2 The local nuclear power plant has decided that the best place to store its high-level nuclear waste (small in volume but highly radioactive) would be in an unoccupied field next to your college campus. They propose to build storage sheds, properly shielded so that the level of radioactivity at the site boundary is equal to background radiation, and use these sheds for the next 20 yr. The power plant serves a population of 2 million and it is presently costing the power company $1.2 million/yr to dispose of the spent fuel in Washington state. The new facility will cost $800,000 to construct and $150,000/yr to operate. The power industry pays 8% interest on its borrowed money. The power company is willing to pay the college $200,000/yr rent.

a. Will the power company save money?
b. If you are the college president, decide if this is an acceptable scheme by using

  a. a cost-effectiveness analysis
  b. a benefit/cost analysis
  c. a risk analysis
  d. an environmental impact analysis
  e. an ethical analysis.

If additional information is needed, assume reasonable values and conditions.

2.3 The following cost information was calculated for a proposed birdhouse:

Costs

	Capital	Operating
Lumber	$4.00	
Nails	0.50	
Additional birdseed required		1.50/wk

Benefits

Joy of watching birds	5.00/wk

Expected life of the birdhouse is 2 yr. Assume 6% interest. Calculate the benefit/cost ratio. Should you build the birdhouse?

2.4 One definition of pollution is unreasonable interference with another beneficial use. On the basis of this definition, defend the use of a stream as a conduit for wastewater disposal. Use any of the decision-making tools presented in this chapter. Then argue against this definition.

2.5 You have been given the responsibility of designing a large trunk sewer. The sewer alignment is to follow a creek that is widely used as a recreational facility. It is a popular place for picnics, nature walks have been constructed by volunteers along its banks, and the local community wants to eventually make it a part of the state park system. The trunk sewer will badly disrupt the creek, destroy its ecosystem, and make it unattractive for recreation. What thoughts would you have on this assignment? Write a journal entry as if you were keeping a personal journal (diary).

2.6 Would you intentionally run over a box turtle trying to cross a road? Why or why not? Present an argument for convincing others that your view is correct.

2.7 You have decided to start a ranch for growing Tasmanian devils. (You may wish to look up something about these unusual creatures.) Your ranch will be located in a residential neighborhood, and you have discovered that the ordinances have nothing to say about growing the devils, so you believe your ranch will be legal.

You plan to sell the devils to the local land-fill operator, who will use them as scavengers. Discuss your decision from the standpoint of

a. benefit/cost to yourself
b. risk to yourself and others
c. environmental impact.

2.8 The water authority for a small community argues that a new dam and lake are needed, and presents an initial construction cost estimate of $1.5 million. The authority calculates a benefit (recreation, tourism, water supply) of $2 million. The community agrees to the project and floats bonds to pay for the construction. As construction proceeds, however, and $1.0 million has been spent, it becomes evident that the dam will actually cost $3.0 million, and the extra cost is due to change orders and unforeseen problems, not a responsibility of the contractor. The half-finished dam is of no use to the community, and $1.5 million additional funds are needed to finish the project. Use benefit/cost analysis to argue whether the community should continue with this project.

2.9 Name one Department of Defense project that has not used the technique described in Problem 2.8 to argue for its completion.

2.10 Often engineering decisions affect people of the next generation, and even several generations to come. A major bridge just celebrated its 100th anniversary, and it may be able to stand for another century. This bridge is a mere piker compared to some of the Roman engineering feats, of course.

Engineering decisions such as nuclear waste disposal also affect future generations, but in a negative way. We are willing this problem to our kids, their kids, and untold generations down the line. But should this be our concern? What, after all, have future generations ever done for us?

Philosophers have been wrestling with the problem of concern for future generations, and without much success. Listed below are some arguments often found in literature about why we should *not* worry about future people. Think about these arguments, and construct opposing arguments if you disagree. Be prepared to discuss these in class.

We do not have to bother worrying about future generations:

a. These people do not even exist, so they have no claim to any rights such as moral consideration. We cannot allow moral claims by people who do not exist, and may not ever exist.
b. We have no idea what the future generations may be like, and what their problems will be, so it makes no sense to plan for them. To conserve (not use) nonreplenishable resources just in case they may need them makes no sense whatever.
c. If there will be future people, we have no idea how many, so how can we plan for them?
d. Future generations will be better off than each previous generation, given the advance of technology, so we have no moral obligation to them.
e. If we discount the future, we find that reserves of resources are of little use to future generations. For example, if a barrel of oil will cost, say $100, 50 yr from now, how much would that barrel cost today? Assume an interest rate of 5%. Clearly, from your answer it makes no sense to conserve a barrel of oil today that is worth only the pennies you calculated.

2.11 A particularly bombastic book painting the environmental movement as a bunch of nincompoops is *Disaster Lobby*.[31] Chapter 6 begins with this sentence: "Since 1802, the Corps of Engineers of the United States Army has been busy helping to carve a dry, safe, habitable environment out of the savagely hostile wilderness that once was America." Consider the language used in that one sentence, and write a one-page analysis of the values held by the author.

2.12 Chlorides enter the Great Lakes from human activity, but the most important source is from the salting of roads in the winter to make them safe for driving. High chloride concentrations (salty water) are damaging to the aquatic ecosystem. If, the Great Lakes would have chloride levels of >20 mg/L, why would this be a problem? What uses of the lakes might be threatened by such high chloride concentrations? How do you think the lakes will be changed as a result of such high chloride concentrations?

Suppose you are the chief environmental engineer in charge of the joint Canadian–U.S. Great Lakes Water Quality Commission. You have to set water quality limits for chlorides, recognizing that the limits you set would be considerably less than 20 mg/L and that changes in human lifestyle would be needed if the chloride concentration is to remain below these standards. What type of decisions would you be making? (There may be more than one, and there is no "correct" answer to this question.)

2.13 One of the alternatives to fossil-fueled power plants is to dam up rivers and use the water power to produce electricity. Suppose you are working for a power company and the decision has to be made to construct one of three new facilities: a large dam on a scenic river, a coal-fired power plant, or a nuclear power plant. Suppose the three plants would all produce about the same amount of electricity, but the nuclear plant is the most expensive, followed by the dam, and finally the coal-fired plant.

What factors would go into your decision making, and what types of decisions would the management of the *power company* probably make to decide which plant to construct? Why?

2.14 A wastewater treatment plant for a city discharges its treated plant effluent into a stream, but the treatment is inadequate. You are in charge of the city's public works, and you hire a consulting engineer to assess the problem and to offer solutions. She estimates that expanding the capacity of the treatment plant to achieve the required effluent quality will be an expensive proposition. She figures out that the city can meet the downstream water quality standard by building a large holding basin for the plant effluent (discharge) and holding back the treated wastewater during dry weather (low river flow) and discharging only during high streamflows (rainy weather). The amount of organic pollution being discharged would remain the same, of course, but now the stream standards would be met, the river water quality would be acceptable for aquatic life, and the city would be off the hook.

Your consulting engineer did some calculations before making her recommendation, of course. The cost of increasing the plant capacity is $1.5 million and the cost of the holding basin is $1.8 million. The annual operating cost of the treatment plant expansion is $400,000, and the operating cost of the holding pond is $100,000/yr. The city can borrow money at 6% interest; the expected life of the treatment plant expansion is 10 years, and the expected life of the holding pond is 20 yr. On the basis of economics alone, is the engineer right? Would you recommend this solution to the city council? What type of decisions would you make to determine your final recommendation?

2.15 A farmer who has a well for his irrigation water hires you as a consultant, and asks you if he can withdraw more than he is presently withdrawing. You respond honestly that yes, he can, but that the groundwater reserves will be depleted and that his neighbors' wells may go dry. He asks you if it would be possible for anyone to know that he is withdrawing more water than the maximum rate, and you again tell him honestly that it is unlikely that anyone will know. He then tells you that he plans to double the withdrawal rate, run as long as he can, and when the water gives out, abandon the farm and move to Florida. You appear to be the only person who knows of the farmer's plans.

Driving home, you reach a decision on what you will do. What kind of decision did you make, and how did you make it?

2.16 Consider the lifetime health risk of eating 100-g apples contaminated with 1 ppb (part per billion) heptachlor (a pesticide and probable human carcinogen). The potency factor for heptachlor has been estimated by EPA as $3.4 \ (mg/kg/d)^{-1}$.

   a. On the basis of 70-kg adults who eat one apple per day for 70 yr, roughly estimate the annual number of additional heptachlor-related cancers one might expect in a population of 100,000.
   b. Aside from the obvious consumption, lifetime, and body-weight assumptions, briefly discuss at least two other assumptions that underlie the estimate calculated in part a.

2.17 Engineer Diane works for a large international consulting firm that has been retained by a federal agency to assist in the construction of a natural gas pipeline in Arizona. Her job is to lay out the centerline of the pipeline according to the plans developed in Washington.

   After a few weeks on the job, she is approached by the leaders of a local Navajo village and told that it seems that the gas line will traverse an ancient sacred Navajo burial ground. She looks on the map and explains to the Navajo leaders that the initial land survey did not identify any such burial grounds.

   "Yes, although the burial ground has not been used recently, our people believe that in ancient times this was a burial ground, even though we cannot prove it. What matters is not that we can show by archeological digs that this was indeed a burial ground, but rather that the people believe that it was. We therefore would like to change the alignment of the pipeline to avoid the mountain."

   "I can't do that by myself. I have to get approval from Washington. And whatever is done would cost a great deal of money.

I suggest that you not pursue this further," replies Diane.

   "We already have talked to the people in Washington, and as you say, they insist that in the absence of archeological proof, they cannot accept the presence of the burial ground. Yet, to our people the land is sacred. We would like you to try one more time to divert the pipeline."

   "But the pipeline will be buried. Once construction is complete, the vegetation will be restored, and you'll never know the pipeline is there," suggests Diane.

   "Oh, yes. We will know it is there. And our ancestors will know it is there."

   The next day Diane is on the telephone with Tom, her boss in Washington, and she tells him about the Navajo visit.

   "Ignore them," advises Tom.

   "I can't ignore them. They truly *feel* violated by the pipeline on their sacred land," replies Diane.

   "If you are going to be so sensitive to every whim and wish of every pressure group, maybe you shouldn't be on this job," suggests Tom.

   What responsibility do engineers have for public attitude? Should they take into account the feelings of the public, or should they depend only on hard data and quantitative information? Should Diane simply tell the Navajo leaders that she has called Washington and that they are sorry, but the alignment cannot be changed? If she feels that the Navajo people have been wronged, what courses of action does she have? How far should she stick her neck out?

   Respond by writing a two-page paper analyzing the ethics of this case.

2.18 The selection of what scientists choose to study is the subject of a masterful essay by Leo Tolstoy, "The Superstitions of Science," first published in 1898.[32] He recounts how

   . . .a simple and sensible working man holds in the old-fashioned and sensible way that if people who study during their whole lives, and, in return for the food and support he gives them, think for him, then these thinkers are probably occupied with what is necessary for people, and he expects from science a solution to those questions on which his well-being and the well-being of all people depends. He expects that science will teach him how to live, how to act toward members of his family, towards his neighbors, towards foreigners; how to battle with his passions, in what he should or should not believe, and much more. And what does our science tell him concerning all these questions?

It majestically informs him how many million miles the sun is from the earth, how many millions of ethereal vibrations in a second constitute light, how many vibrations in the air make sound; it tells him of the chemical constitution of the Milky Way, of the new element helium, of microorganisms and their waste tissue, of the points in the hand in which electricity is concentrated, of X-rays, and the like. But, protests the working man, I need to know today, in this generation, the answers to how to live.

Stupid and uneducated fellow, science replies, he does not understand that science serves not utility, but science. Science studies what presents itself for study, and cannot select subjects for study. Science studies everything. This is the character of science.

And men of science are really convinced that this quality of occupying itself with trifles, and neglecting what is more real and important, is a quality not of themselves, but of science; but the simple, sensible person

© Jim Rearden

Figure 2.11 Drenched and shivering after nearly drowning, a [caribou] calf stranded on an island gets a tender toweling from Don Frickie of the Arctic National Wildlife Range. Deposited on the stream's bank, the 2-day-old was quickly reunited with its frantic mother.

begins to suspect that this quality belongs not to science, but to people who are inclined to occupy themselves with trifles, and to attribute to these trifles a high importance.

In the formulation and planning of your career as an *engineer*, not a scientist, how would you answer Tolstoy?

2.19 Read the second chapter of Genesis, then look at Figure 2.11. In a one-page paper, express your thoughts relative to the "stewardship" concept of environmental ethics.

2.20 How would you estimate the data required for Example 2.1?

2.21 The following statement was made in Section 2.2: "Obviously, the lowest total cost alternative (given *all* cost data) would be the most rational decision." What is meant by "all cost data"?

2.22 You, as the city engineer, are checking the work of Chris, the newest Public Works Department employee. Chris performed the calculations in Example 2.2 for you to use in your presentation to the city council. Do you agree or disagree with Chris' calculations?

2.23 Will it make a difference in your risk analysis if that one person in a million who may be harmed is you or someone you love?

2.24 List other impacts that could be included in Example 2.9.

END NOTES

1. Gray, R. G. 2000. Letter to the editor: Education system isn't working. *ASCE News* 25, No. 9:8.

2. The story is credited to Garrett Hardin, Tragedy of the commons. *Science*, 162 (December 1968).

3. MacLean, D., ed. 1986. *Values at Risk*. Totowa, NJ: Rowman and Littlefield.

4. National Research Council (NRC). 1983. *Risk Assessment in the Federal Government: Understanding the Process*. Washington DC: National Academy Press.

5. O'Brien, Mary. 2000. *Making Better Environmental Decisions: An Alternative to Risk Assessment*. Cambridge, MA: MIT Press.

6. California Green Chemistry Initiative Final Report. December 2008. California Environmental Protection Agency. Accessed at http://www.dtsc.ca.gov/PollutionPrevention/GreenChemistryInitiative/index.cfm on January 8, 2009.

7. Accessed at http://www.epa.gov/dfe/pubs/flameret/ffr-alt.htm in September 2008.

8. Accessed at http://www.cleanproduction.org/library/Green%20Screen%20Report.pdf in September 2008.

9. Petulla, J. M. 1980. *American Environmentalism*. College Station, TX: Texas A&M University Press.

10. Much of this discussion is based on Gunn, A. S., and P. A. Vesilind. 1986. *Environmental Ethics for Engineers*. Chelsea, MI: Lewis Publishers.

11. Immanuel Kant. Duties toward animals and spirits. *Lecture on ethics*, p. 240. Quoted by Mary Midgley, in "Duties concerning islands," in Elliot R., and A. Gare (eds) 1983. *Environmental Philosophy*. State College, PA: Pennsylvania State University Press.

12. Baxter, William F. 1974. *People or Penguins: The Case for Optimal Pollution*. New York: Columbia University Press, p. 5.

13. Leopold, A. 1966. *A Sand County Almanac*. New York: Oxford University Press.

14. Watson, Richard A. 1979. Self-consciousness and the rights of nonhuman animals and

nature. *Environmental Ethics*, 1, No. 2: 99–129.

15. McClosky, H. J. 1983. *Ecological Ethics and Politics*. Totowa, NJ: Rowman and Littlefield, p. 29.

16. Locke, John. 1967. *Two Treatises on Government*, 2nd ed. Edited by Peter Laslett. Cambridge, UK: Cambridge University Press.

17. Hobbes, Thomas. 1885. *Leviathan*. Edited by Henry Morley. London: Routledge

18. Bentham, Jeremy. 1948. *An Introduction to the Principles of Morals and Legislation*. Edited by Laurence J. LaFleur. New York: Hafner Press.

19. Singer, Peter. 1979. *Practical Ethics*. Cambridge, UK: Cambridge University Press, p. 50.

20. Schweitzer, Albert. 1933. *Out of My Life and Thought: An Autobiography*. New York.

21. Thompson, Janna. 1983. Preservation of wilderness and the good life. In *Environmental Philosophy*. Edited by R. Elliot and A. Gare. State College, PA: Pennsylvania State University Press, p. 97.

22. Taylor, Paul W. 1986. *Respect for Nature: A Theory of Environmental Ethics*. Princeton, NJ: Princeton University Press.

23. Thompson, Janna. 1990. "A refutation of environmental ethics." *Environmental Ethics*, 12, No. 4: 147–60.

24. Routley, Val and Richard. 1987. Against the inevitability of human chauvinism. In *Ethics and the Problems of the 21st Century*. Edited

by Kenneth Goodpaster and Kenneth Sayre. Notre Dame, IN: University of Notre Dame Press.

25. Rolston, Holmes III. 1988. *Environmental Ethics*. Philadelphia: Temple University Press.

26. Regan, Tom. 1981. The nature and possibility of an environmental ethic. *Environmental Ethics*, 3, No. 1: 31–2.

27. Naess, Arne. 1985. Basic principles of deep ecology. In *Deep Ecology*. Edited by Bill Devall and George Sessions. Salt Lake City: Gibbs Smith

28. Berry, Wendell. 1973. A secular pilgrimage. In *Western Man and Environmental Ethics*. Edited by Ian Barbour. Reading, MA: Addison-Wesley.

29. Collis, John Stewart. 1954. *The Triumph of the Tree*. New York: William Sloane Associates.

30. Routley, Richard and Val. 1980. Human chauvinism and environmental ethics. In *Environmental Philosophy*. Edited by D. Mannison, M. McRobbie, and R. Routley. Canberra: Research School of Social Sciences, Australian National University, p. 130.

31. Grayson, M. J., and T. R. Shepard Jr. 1973. *Disaster Lobby*. Chicago: Follett Publishing Co.

32. Tolstoy, Leo. 1898. The superstitions of science. In *The Arena*, 20. Reprinted in *The New Technology and Human Values*. Edited by J. G. Burke. 1966. Belmont, CA: Wadsworth Publishing Co.

Engineering Calculations

Gateway Arch, Missouri

Snake, South Carolina

Environmental engineers perform numerous calculations while analyzing and designing. We double-
and even triple-check key steps. But it's also important to pay attention to those seemingly minor details,
such as units, to avoid being bitten by a simple error.

The ability to solve problems by using engineering calculations represents the very essence of engineering. While certainly not all engineering problems can be solved by using numerical calculations, such calculations are absolutely necessary for the development of technical solutions. Engineering calculations make it possible to describe the physical world in terms of units and dimensions that are understood by all those with whom communication takes place.

The first section of this chapter is devoted to a review of units and dimensions used in engineering. The second part describes some basic principles of performing some back-of-the-envelope calculations in the face of incomplete or unobtainable information, and the last section is about how to manage information.

3.1   ENGINEERING DIMENSIONS AND UNITS

A *fundamental dimension* is a unique quantity that describes a basic characteristic, such as force (F), mass (M), length (L), and time (T). *Derived dimensions* are calculated by an arithmetic manipulation of one or more fundamental dimensions. For example, velocity has the dimensions of length per time (L/T), and volume is L^3.

Dimensions are descriptive but not numerical. They cannot describe *how much*; they simply describe *what*. *Units* and the *values* of those units are necessary to describe something quantitatively. For example, the length (L) dimension may be described in units as meters, yards, or fathoms. Adding the value, we have a complete description, such as 3 meters (m), 12.6 yards (yd), or 600 fathoms (fth).

Three systems of units are in common use: the SI system, the American engineering system, and the cgs system. Developed in 1960 in an international agreement, the SI system (for System International d'Unites) is based on meter for length, second for time, kilogram for mass, and Kelvin for temperature. Force is expressed in Newtons. The tremendous advantage of the SI system over the older English (and now American) system is that it works on a decimal basis, with prefixes decreasing or increasing the units by powers of ten. Although the SI units are now used throughout the world, most contemporary American engineers still use the old system of feet, pounds (mass), and seconds, with force expressed as pounds (force). This book primarily uses the SI system.

3.1.1   Density

The *density* of a substance is defined as its mass divided by a unit volume, or

$$\rho = \frac{M}{V}$$

where ρ = density
M = mass
V = volume

In the SI system the base unit for density is kg/m^3.

FOCUS ON

Units

The importance of keeping track of units in your calculations is epitomized in the failure of the U.S. National Aeronautics and Space Administration's (NASA's) $327.6 million Mars Climate Orbiter mission (Figure 3.1).[1,2] One team of engineers used the SI system (N-s) in the calculation of the power of the thrusters, while another team used the American system (lb-s). The difference in units was enough to miscalculate the orbiter's trajectory by 100 km (60 mi), which resulted in its destruction by friction with Mars' atmosphere. Although in this case the mistake resulted in the loss of an expensive piece of equipment, such mistakes in engineering practice can result in much higher prices, including lost lives. To avoid making such easily avoidable mistakes, it is good practice to be in the habit of explicitly stating and keeping track of all units in all your calculations. In fact, you'll find that many

Figure 3.1   The Mars Climate Orbiter in place for its spin test. (© NASA Kennedy Space Center (NASA-KSC). KSC-98pc-1723)

environmental engineering calculations are essentially unit conversions and are simple if you keep track of your units.

Water in the SI system has a density of 1×10^3 kg/m^3, which is equal to 1 g/cm^3.

3.1.2 Concentration

The derived dimension *concentration* is usually expressed gravimetrically as the mass of a material A in a unit volume consisting of material A and some other material B. The concentration of A in a mixture of A and B is

$$C_A = \frac{M_A}{V_A + V_B}$$

(3.1)

where C_A = concentration of A
M_A = mass of material A
V_A = volume of material A
V_B = volume of material B

In the SI system the basic unit for concentration is kg/m^3. However, the most widely used concentration term in environmental engineering is milligrams per liter (mg/L), which is also the unit we will use in this book.

EXAMPLE

3.1

Problem   Plastic beads with a volume of 0.04 m³ and a mass of 0.48 kg are placed into a container, and 100 liters of water are poured into the container. What is the concentration of plastic beads, in mg/L?

Solution   Using Equation 3.1, where A represents the beads and B the water:

$$C_A = \frac{M_A}{V_A + V_B}$$

$$C_A = \frac{0.48 \text{ kg}}{0.04 \text{ m}^3 + (100 \text{ L} \times 10^{-3} \text{ m}^3/\text{L})}$$

$$C_A = 3.43 \text{ kg/m}^3 = 3.43 \frac{10^6 \text{ mg/kg}}{10^3 \text{ L/m}^3} = 3,430 \text{ mg/L}$$

Note that in the previous example the volume of water is added to the volume of the beads. If the plastic beads with a volume of 0.04 m³ are placed into a 100-liter container and the container filled to the brim with water, the total volume is

$$V_A + V_B = 100 \text{ L}$$

and the concentration of beads, C_A, is 4,800 mg/L. The concentration of beads is higher because the total volume is lower.

Another measure of concentration is *parts per million* (ppm). This is numerically equivalent to mg/L if the fluid in question is water because one milliliter (mL) of water weighs 1 g (i.e., the density is 1.0 g/cm³). This fact is demonstrated by the following conversion:

$$\frac{1 \text{ mg}}{\text{L}} = \frac{0.001 \text{ g}}{1000 \text{ mL}} = \frac{0.001 \text{ g}}{1000 \text{ cm}^3} = \frac{0.001 \text{ g}}{1000 \text{ g}} = \frac{1 \text{ g}}{1,000,000 \text{ g}}$$

or 1 gram in a million grams, or 1 ppm.

Some material concentrations are most conveniently expressed as percentages, usually in terms of mass:

$$\Phi_A = \frac{M_A}{M_A + M_B} \times 100 \tag{3.2}$$

where Φ_A = percentage of material A
M_A = mass of material A
M_B = mass of material B

Φ_A can, of course, also be expressed as a ratio of volumes.

EXAMPLE

3.2

Problem   A wastewater sludge has a solids concentration of 10,000 ppm. Express this in percent solids (mass basis), assuming that the density of the solids is 1 g/cm³.

Solution

$$10,000 \text{ ppm} = \frac{1 \times 10^4 \text{ parts}}{1 \times 10^6 \text{ parts}} = \frac{1}{100} = 0.01 \text{ or } 1\%$$

This example illustrates a useful relationship:

$$10,000 \text{ mg/L} = 10,000 \text{ ppm (if density} = 1) = 1\% \text{ (by weight)}$$

Many wastewaters are assumed to be dilute, so their density can be assumed to be approximately 1.

In air pollution control, concentrations are generally expressed gravimetrically as mass of pollutant per volume of air at standard temperature and pressure. For example, the national air quality standard for lead is 0.15 μg/m^3 (one microgram $= 10^{-6}$ gram). Occasionally, air quality is expressed in ppm, and in this case the calculations are in terms of volume/volume, or 1 ppm $=$ 1 volume of a pollutant per 1×10^6 volumes of air. Conversion from mass/volume (μg/m^3) to volume/volume (ppm) requires knowledge of the molecular weight of the gas. At standard conditions—0°C and 1 atmosphere of pressure—1 mol of a gas occupies a volume of 22.4 L (from the ideal gas law). One mole is the amount of gas in grams numerically equal to its molecular weight. The conversion is therefore

$$\mu\text{g/m}^3 = \frac{1 \text{ m}^3 \text{ pollutant}}{10^6 \text{ m}^3 \text{ air}} \times \frac{\text{Molecular weight (g/mole)}}{22.4 \times 10^{-3} \text{ m}^3/\text{mole}} \times 10^6 \text{ } \mu\text{g/g}$$

or simplifying:

$$\mu\text{g/m}^3 = (\text{ppm} \times \text{Molecular weight} \times 10^3)/22.4 \quad \text{at 0°C and 1 atm}$$

If the gas is at 25°C and 1 atm, as is common in air quality standards, the conversion is

$$\mu\text{g/m}^3 = (\text{ppm} \times \text{Molecular weight} \times 10^3)/24.45 \quad \text{at 25°C and 1 atm} \qquad (3.3)$$

3.1.3 Flow Rate

In engineering processes the *flow rate* can be either *gravimetric (mass) flow rate* or *volumetric (volume) flow rate*. The former is in kg/s while the latter is expressed as m^3/s. The mass and volumetric flow rates are not independent quantities because the mass (M) of material passing a point in a flow line during a unit time is related to the volume (V) of that material:

$$\text{Mass} = \text{Density} \times \text{Volume}$$

Thus, a volumetric flow rate (Q_V) can be converted to a mass flow rate (Q_M) by multiplying by the density of the material:

$$Q_M = Q_V \rho \qquad (3.4)$$

where Q_M = mass flow rate
$\qquad Q_V$ = volume flow rate
$\qquad \rho$ = density

The symbol Q is almost universally used to denote flow rate.

The relationship between mass flow of some component A, concentration of A, and the total volume flow (A plus B) is

$$Q_{M_A} = C_A \times Q_{V_{(A+B)}} \tag{3.5}$$

Note that Equation 3.5 is not the same as Equation 3.4, which is applicable to only one material or one component in a flow stream. Equation 3.5 relates to two *different* materials or components in a flow. For example, a mass flow rate of plastic balls moving along and suspended in a stream is expressed as kilograms of these balls per second passing some point, which is equal to the concentration (kg balls/m^3 total volume, balls plus water) times the stream flow (m^3/s of balls plus water).

**EXAMPLE
3.3**

Problem   A wastewater treatment plant discharges a flow of 1.5 m^3/s (water plus solids) at a solids concentration of 20 mg/L (20 mg solids per liter of flow, solids plus water). How much solids is the plant discharging each day?

Solution   Use Equation 3.5.

Mass flow = Concentration × Volume flow

$$Q_{M_A} = C_A \times Q_{V_{(A+B)}}$$
$$= \left(20 \text{ mg/L} \times 1 \times 10^{-6} \text{ kg/mg}\right) \times \left(1.5 \text{ m}^3/\text{s} \times 10^3 \text{ L/m}^3 \times 86{,}400 \text{ s/d}\right)$$
$$= 2592 \text{ kg/day} \approx 2600 \text{ kg/d}$$

**EXAMPLE
3.4**

Problem   A wastewater treatment plant discharges a flow of 1498 L/s (Liters per second) at a solids concentration of 0.002% solids (by weight). How many kilograms per day of solids does it discharge?

Solution   Use Equation 3.5.

Mass flow = Concentration × Volume flow

$$Q_{M_A} = C_A \times Q_{V_{(A+B)}}$$

Assume that $\rho = 1$ g/cm^3, so 0.002% = 20 mg/L, and assume the stated volume flow rate includes solids plus water. Then

$$Q_{MA} = (20 \text{ mg/L}) \ (1498 \text{ L/s}) = 29960 \text{ mg/s}$$

$$= (29960 \text{ mg/s}) \ (86400 \text{ s/day}) \left(\frac{1 \text{kg}}{10^6 \text{mg}} \right)$$

$$= 2589 \text{ kg/d}$$

E X A M P L E

3.5   **Problem**   A drinking water treatment plant adds fluorine at a concentration of 1 mg/L. The average daily water demand is 68 million L. How much fluorine must the community purchase?

Solution   Use Equation 3.6.

$$68 \text{ million L/d} \times 1 \text{ mg/L} \times 1 \text{kg/10}^6 \text{mg} = 68 \text{ kg/d}$$

3.1.4   Retention Time

One of the most important concepts in treatment processes is retention time, also called detention time or even residence time. *Retention time* is the time an average particle of the fluid spends in the container through which the fluid flows (which is the time it is exposed to treatment or a reaction). An alternate definition is the time it takes to fill the container.

Mathematically, if the volume of a container, such as a large holding tank, is V (L^3), and the flow rate into the tank is Q (L$^3/t$), then the residence time is

$$\bar{t} = \frac{V}{Q} \tag{3.6}$$

The average retention time can be increased by reducing the flow rate Q or increasing the volume V, and decreased by doing the opposite.

E X A M P L E

3.6   **Problem**   A lagoon has a volume of 1500 m^3, and the flow into the lagoon is 3 m^3/h. What is the retention time in this lagoon?

Solution   Use Equation 3.7.

$$\bar{t} = \frac{1500 \text{ m}^3}{3 \text{ m}^3/\text{h}} = 500 \text{ h}$$

3.2 APPROXIMATIONS IN ENGINEERING CALCULATIONS

Engineers are often called on to provide information not in its exact form but as approximations. For example, an engineer may be asked by a client, such as a city manager, what it might cost to build a new wastewater treatment plant for the community. The manager is not asking for an exact figure but a ball park estimate. Obviously, the engineer cannot in a few minutes conduct a thorough cost estimate. She would recognize the highly variable nature of land costs, construction costs, required treatment efficiency, etc. Yet the manager wants a preliminary estimate—a *number*—and quickly!

In the face of such problems the engineer has to draw on whatever information might be available. For example, she might know that the population of the community to be served is approximately 100,000. Next, she estimates, based on experience, that the domestic wastewater flow might be about 400 L per person per day, thus requiring a plant of about 440 L/s capacity. With room for expansion, industrial effluents, storm inflow, and infiltration of groundwater into the sewers, she may estimate that a 660 L/s capacity may be adequate.

Next, she evaluates the potential treatment necessary. Knowing that the available watercourses for discharging the effluent are all small streams that may dry up during droughts, a high degree of treatment is required. She figures that nutrient removal will be needed. Such treatment plants, she is aware, cost about $800,000 per million liters of influent to construct. She calculates that the plant would cost about $45 million. Giving herself a cushion, she could respond by saying, "About $50 million."

This is exactly the type of information the manager seeks. He has no use for anything more accurate because he might be trying to decide whether to ask for a bond issue of around $100 million or $200 million. There is time enough for more exact calculations later.

3.2.1 Procedure for Calculations with Approximations

Problems not requiring exact solutions can be solved by

1. carefully defining the problem
2. introducing simplifying assumptions
3. calculating an answer
4. checking the answer, both systematically and realistically.

Defining the Problem

The engineer in the previous case is asked for an estimate, not an exact figure. She recognizes that the use of this figure would be for preliminary planning purposes, and thus, valid approximations are adequate. She also recognizes that the manager wants a *dollar* figure answer, thus establishing the units.

Simplifying

This step is perhaps the most exciting and challenging of the entire process because intuition and judgment play an important role. For example, the engineer has to first estimate the population served and then consider the average flow. What does she ignore? Obviously, a great deal, such as daily transient flows, variability in living standards, and seasonal variations. A thorough estimate of potential wastewater flows requires a major study. She has to simplify her problem and choose to consider only an estimate of the population and an average per capita discharge.

Calculating
In the case of this problem the calculations are straightforward.

Checking
More important is the process of checking. There are two kinds of checks: systematic and realistic. In systematic checking the units are first checked to see whether they make sense. For example,

$$\text{persons} \times \frac{L}{\text{persons}} = L$$

makes sense, whereas

$$\frac{\text{persons}}{L/\text{person}} = \frac{\text{persons}^2}{L}$$

is nonsense. If the units check out, the numbers can be recalculated to check for mistakes. It is wise always to write your units as you do your calculations, making this check as you go.

Finally, a reality check is necessary. Possibly no practicing engineer will explicitly recognize that they perform reality checks day in and day out, but such checks are central to good engineering. Consider, for example, if the engineer had made a mistake and thought (erroneously) that a wastewater treatment plant of the type needed by the community costs $3,000 per million liters of influent. Her calculations would have checked, but her answer would have been $50,000 instead of $50 million. Such an answer should have immediately been considered ludicrous, and a search for the error initiated. Reality checks, when routinely performed, will save considerable pain and embarrassment.

3.2.2 Use of Significant Figures
Finally, note that significant figures in the answer reflect the accuracy of the data and the assumptions. Consider how silly it would sound to say that the treatment plant would cost *about* $5,028,467.19. Many problems require answers to only one significant figure or even to an order of magnitude. Nonsignificant figures tend to accumulate in the course of calculations like mud on a boot and must be wiped off at the end.[3]

Suppose you are asked to estimate the linear meter of fence posts needed for a pasture and are told that there will be 87 posts with an average height of 117.6 cm. You multiply and get 102.313 m. Now it is time to scrape the mud off because the most accurate of your numbers has only three significant figures but your answer has six significant figures. So you report 102 m.

Significant figures are those that transfer information based on the value of that digit. Zeros that merely hold place are not significant because they can be eliminated without loss of information. For example, in the number 0.0036 the two zeros are only holding a place and can be eliminated by writing $0.0036 = 3.6 \times 10^{-3}$.

Zeros at the end of a number are a problem, however. Suppose the newspaper reports that 46,200 fans attended a football game. The last two digits (zeros) may be significant if every person was counted, and indeed there were *exactly* 46,200 fans in the stadium. If, however, one were to estimate the number of people as 46,200 fans, then the last two zeros are simply holding places and are not significant. To avoid confusion when reporting numbers, it is useful to say "about" or "approximately" if that is what is meant. When using numbers of unknown significant figures, erring on the side of caution (fewer significant figures) is usually best.

EXAMPLE
3.7

Problem   A community of approximately 100,000 people has about 20,000 m² of landfill left that can be filled to about 10 m deep with refuse compacted to somewhere between 350 kg/m³ and 500 kg/m³. What is the remaining life of the landfill?

Solution   Using the procedure just outlined, the first step is to define the problem. Clearly, the answer does not require high precision because the data are not precise. In addition, the definition of the problem requires an answer in time units.

The second step is to simplify the problem. There is no need to consider commercial or industrial wastes. Estimate only refuse generated by individual households.

The third step is to calculate an answer. All the necessary data are available except the per capita production of refuse. Suppose a family of 4 fills up 3 garbage cans per week. If each can is about 0.25 m³ and if we assume the uncompacted garbage is at about one-fourth of the compacted density, say at 120 kg/m³, it is reasonable to calculate the per capita production as

$$\frac{0.25 \text{ m}^3}{\text{can}} \times \frac{3 \text{ cans}}{4 \text{ people}} \times 120 \text{ kg/m}^3 \times \frac{1 \text{ week}}{7 \text{ days}} = 3.2 \frac{\text{kg}}{\text{person/d}}$$

If there are about 100,000 people, the city produces

$$3.2 \frac{\text{kg}}{\text{person/d}} \times 100,000 \text{ people} \times \frac{1 \text{ m}^3}{400 \text{ kg}} = 800 \frac{\text{m}^3}{\text{d}}$$

The total available volume is

$$20,000 \text{ m}^2 \times 10 \text{ m} = 200,000 \text{ m}^3$$

Thus, the expected life is

$$\frac{200,000 \text{ m}^3}{800 \text{ m}^3/\text{d}} = 250 \text{ d}$$

Remember the fourth step. Is this reasonable? Pretty much so. The calculations may have overestimated refuse production as the average national per capita production is closer to 2.5 kg/capita/d, so in actuality the landfill may last a year. But considering the extreme difficulties of siting additional landfills, the town is clearly already in a crisis situation.

In professional engineering it is necessary to carry around in one's head a suitcase full of numbers and approximations. For example, most people would know that a meter is a few inches longer than a yard. We may not know *exactly* how much longer, but we could make a pretty fair guess. Similarly, we know what (90 m) looks like (from goal line to goal line). In a similar way, an environmental engineer in practice knows instinctively what a flow of 440 L/s looks like because he or she has been working in plants that received that magnitude of flow. Such knowledge becomes second nature and is often the reason why engineers can avoid stupid and embarrassing mistakes. A feel for units is a part of engineering and is the reason why a change of units from mgd to m³/s for American engineers is so difficult. It would be an unusual American engineer who would know what a flow of 10 m³/s looks like (without doing some quick mental approximate conversions!).

3.3 INFORMATION ANALYSIS

Not only do engineers seldom know anything accurately, we seldom know anything for sure! Although there may never before have been a flood in August, as soon as a million-dollar construction project depends on dry weather, it'll rain for 40 days and nights. Murphy's law—"if anything can go wrong, it will"—and its corollary—"at the worst possible time"—are as central to engineering as are Newton's laws of motion. Engineers take chances; we place probabilities on events of interest, and decide on prudent actions. We know that a flood might occur in August, but if it hasn't in the last 100 years, chances are fairly good it won't next year. But what if the engineer has only 5 years of data? How much risk is he or she taking? How sure can the engineer be that a flood won't occur next year?

The concept illustrated in Example 3.7 is called *probability*, and probability calculations are central to many engineering decisions. Related to probability is the analysis of incomplete data using *statistics*. A piece of data (e.g., stream flow for one day) is valuable in itself, but when combined with hundreds of other daily stream flow data points, the information becomes even more useful, but only if it can be somehow manipulated and reduced. For example, if it is decided to impound water from a stream for a water supply, the individual daily flow rates would be averaged (a statistic), and this would be one useful number in deciding just how much water would be available.

This section introduces the central ideas of probability and statistics first and describes some useful tools engineers have available in the analysis of information. Neither probability nor statistics is developed from basic principles here, and a proper statistics and probability course is highly recommended. In this introduction enough material is presented to allow for the solution of simple environmental engineering problems involving frequency distributions.

Much of engineering statistical analysis is based on the bell-shaped curve shown in Figure 3.2. The assumption is that a data set (such as annual rainfall) can be described by a bell-shaped curve and that the location of the curve can be defined by the *mean*, or average, and some measure of how widely the data are distributed, or the spread of the

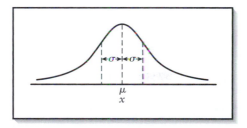

Figure 3.2   Bell-shaped curve suggesting normal distribution.

curve. An ideal form of the function describing the bell-shaped curve is called the *Gaussian distribution*, or the *normal distribution*, expressed as

$$P_{\sigma(x)} = \frac{1}{\sigma\sqrt{2\pi}} \exp\left[-\frac{1}{2}\left(\frac{x-\mu}{\sigma}\right)\right]^2$$

where μ = mean, estimated as $\bar{x}$

$\bar{x}$ = observed sample mean $= \dfrac{\Sigma x}{n}$

σ = standard deviation, estimated as s

s = observed standard deviation $= \left[\dfrac{\Sigma x_i^2}{n-1} - \dfrac{(\Sigma x_i)^2}{n(n-1)}\right]^{1/2}$

n = sample size

The standard deviation is a measure of the curve's spread, defined as the value of x that encompasses 68.3% of all values of x centered on the mean, μ. A small standard deviation indicates that all the data are closely bunched, so there is little variability in the data. In contrast, a large standard deviation indicates that the data are widely spread.

A further useful statistic in engineering work is called the *coefficient of variation*, defined as

$$C_v = \frac{\sigma}{\mu}$$

and estimated in practice as

$$C_v = \frac{s}{\bar{x}}$$

where s = observed standard deviation
$\bar{x}$ = estimate of the mean

Engineers often plot data described by a bell-shaped curve as a *cumulative* function, wherein the ordinate (vertical axis) is the cumulative fraction of observations. These curves are commonly used in hydrology, soils engineering, and resource recovery. The construction of a cumulative curve is shown in Example 3.8.

EXAMPLE

3.8

Problem   One hundred kilograms of glass are recovered from municipal refuse and processed in a resource recovery plant. The glass is crushed and run through a series of sieves,
as shown in the following table. Plot the cumulative distribution of particle sizes.

Sieve Size (mm)	Glass Weight Caught (kg)
4	10
3	25
2	35
1	20
Pan (no holes)	10

Solution   These data can be plotted as shown in Figure 3.3, and a rough approximation
of a normal distribution is noted. The data can then be tabulated as a cumulative function.

Sieve Size (mm)	Fraction of Glass Retained on Sieves	Fraction of Glass Smaller Than This Sieve Size
4	$\frac{10}{100} = 0.10$	$1.0 - 0.10 = 0.90$
3	$\frac{25}{100} = 0.25$	$0.90 - 0.25 = 0.65$
2	$\frac{35}{100} = 0.35$	$0.65 - 0.35 = 0.30$
1	$\frac{20}{100} = 0.20$	$0.30 - 0.20 = 0.10$
<1.0	$\frac{10}{100} = 0.10$	$0.10 - 0.10 = 0$

These data are plotted in Figure 3.4. Note that the last point cannot be plotted because
the sieve size is less than 1.0 mm, and that the total fraction has to add up to 1.00. The
result is a classical S-curve.

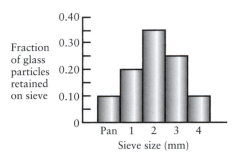

Figure 3.3   Histogram showing sizes of glass particles.

Figure 3.4 is plotted with the *independent variable* as the abscissa (*x* axis) and the
dependent variable as the ordinate (*y* axis). A variable is independent if its value is chosen,

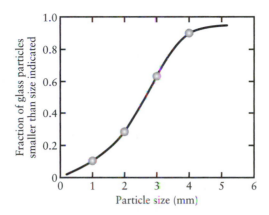

Figure 3.4   Cumulative glass particles size distribution. See Example 3.8.

such as the size of a sieve. The value of a dependent variable is obtained by experimentation. With very few exceptions, engineering graphs are drawn using the convention of the independent variable plotted as the x axis and the dependent variable as the y axis.

Often it is simpler to work with data if the classical S-curve shown in Figure 3.4 can be plotted as a straight line. This often occurs on arithmetic probability paper, originally invented by Alan Hazen, a prominent sanitary engineer. The same data plotted on probability paper are shown in Figure 3.5. Note that the ordinate in probability paper does not have a 0% or 100%, just as the normal distribution curve has no zero or 1.0.

A straight line on probability paper implies that the data are *normally distributed* and that statistics, such as the mean and standard deviation, can be read off the plot. The mean is at 0.5, or 50%, while the standard deviation is 0.335 on either side of the mean.

The standard deviation can also be approximated as

$$s \approx \frac{2}{5}(x_{90} - x_{10}) \tag{3.7}$$

where x_{90} and x_{10} are the abscissa values at P_{90} and P_{10}, respectively.

Such a plot also makes it convenient to estimate the value of any other interval. For example, if it is necessary to estimate the values of x such that 95% of the values fall within these bounds (the "95% confidence interval"), it is possible to read the values at 2.5% and 97.5%. Enclosed within these bounds should be 95% of the data.

E X A M P L E

3.9   **Problem**   Given the plot in Figure 3.5 on the next page, what is the (a) mean, (b) standard deviation, and (c) 95% interval?

Solution

a. The mean is read at 0.5 (which, remember, is the cumulative fraction of glass particles smaller than the size indicated). It is 2.5 mm.

b. The standard deviation is the spread so that 67% of the observations lie within the range centered on the mean (or 33.5% on either side of the mean). From Figure 3.5 the

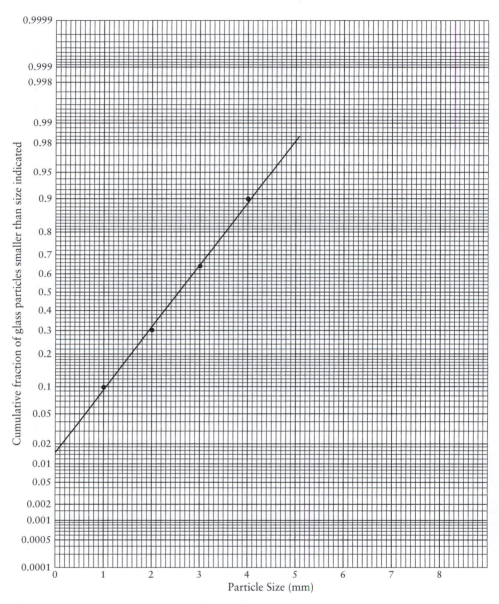

Figure 3.5 Glass particle size distribution on an arithmetic probability plot. See Example 3.9.

estimated standard deviation is calculated as the difference between the mean and the point $0.50 + 0.335 = 0.84$, which corresponds to 3.7 mm. Therefore, $s = 3.7 - 2.5 = 1.2$ mm. Alternatively, using Equation 3.8:

$$s \approx \frac{2}{5}(4.05 - 1.05) = 1.2$$

where 4.05 is the sieve size passing 90% of the glass and 1.05 is the size passing 10% of the glass.

When reporting a mean, it is good practice to include the standard deviation, for example, 2.5 ± 1.2 mm.

c. The 95% interval means that 95% of the data can be expected to be in this range. The points corresponding to 0.025 and 0.975 on the graph are 0.2 mm and 4.9 mm, respectively. Therefore, the 95% interval is 0.2 mm to 4.9 mm.

■

The data in Example 3.9 can be *normalized* by dividing each dependent data point by a constant—the total weight of the glass being sieved. Thus, each data point is converted from a number with dimensions (kg) to a dimensionless fraction (kg/kg). This same trick is sometimes performed on the independent variable when time frequencies are required.

It is often necessary to calculate the probabilities that certain events will recur. Hydrologists express this in terms of the *return period*, or how often the event is expected to recur. If the annual probability of an event occurring is 5%, then the event can be expected to recur once in 20 years, or have a return period of 20 years. Stated in equation form:

$$\text{Return period} = \frac{1}{\text{Fractional probability}}$$

When time variant data are analyzed by a frequency distribution, the data must first be ranked (smallest to largest, or largest to smallest). The probabilities are then calculated, the data plotted, and the plot used to determine the desired statistics.

EXAMPLE
3.10

Problem   A quality of a municipal wastewater treatment plant effluent (discharge) is measured daily by its biochemical oxygen demand (BOD). Because of variations in influent (flow coming into the plant) as well as the treatment plant dynamics, the effluent quality can be expected to vary. Does this variation fit the normal distribution, and if so, can this be used to calculate the mean and the standard deviation? What is the worst effluent quality to be expected once a month (30 days)? The data are given below.

Sample Day	Effluent BOD	Sample Day	Effluent BOD
1	56	10	35
2	63	11	65
3	57	12	35
4	33	13	31
5	21	14	21
6	17	15	21
7	25	16	28
8	49	17	41
9	21	18	36

Solution The first step is to rank the data. The total number of data points is $n = 18$. The rankings (m) are from the lowest BOD (17 mg/L) to the highest (65 mg/L).

Rank (m)	Effluent BOD (mg/L)	m/n
1	17	0.056
2	21	0.111
3	21	0.167
4	21	0.222
5	21	0.278
6	25	0.333
7	28	0.389
8	31	0.444
9	33	0.500
10	35	0.556
11	35	0.611
12	36	0.667
13	41	0.722
14	49	0.778
15	56	0.833
16	57	0.889
17	63	0.944
18	65	1.00

The last column can be thought of as the fraction of time the BOD can be expected to be less than this value. That is, 0.056, or 5.6%, of the time the BOD should be equal to or less than 17 mg/L.

A plot of the second and third columns gives a reasonably straight line on a probability plot as shown in Figure 3.6 on the next page. The estimated mean is read off the plot at $(m/n) = 0.5$ and is 35 mg/L BOD; the standard deviation is 17 mg/L. The worst effluent quality expected every month is found at the fraction of time of 29/30 = 0.967 (that is, for 29 days out of 30 the BOD is less). Enter the plot at fraction of time = 0.967, and read the BOD = 66 mg/L.[4]

Note that the last data point (rank no. 18) cannot be plotted because all the readings are less than this value. This problem can be avoided by dividing all ranks (m) by $n + 1$ instead of n. As the number of data points increases (n gets bigger), this becomes less and less of a problem.

Another method of handling data, especially if large numbers are involved, is to group them. In this case the mean of the group becomes the variable plotted, and the probability is calculated as

$$P = \frac{\Sigma r}{n}$$

Figure 3.6   BOD of a wastewater treatment plant effluent. See Example 3.10.

where r = sum of the number of data points equal to or greater
             than the value indicated
         n = total data points

The probability of an event occurring is again read from the graph.

E X A M P L E

3.11   **Problem**   Using the data from Example 3.10, estimate the highest expected BOD to occur once every 30 days. Use grouped data analysis.

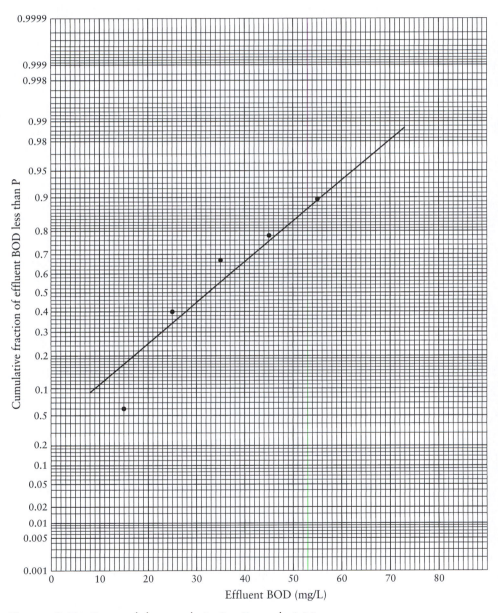

Figure 3.7   Grouped data analysis. See Example 3.11.

Solution   The first step is to define the groups of BOD values. In this case a 10 mg/L step is chosen simply because it is convenient. The *number* of data points falling into each group is noted (r).

Effluent BOD (mg/L)	Mean of Group	Number of Days BOD Falls into Group (r)	Σr	$P = \Sigma r/n$
0–9	5	0	0	0
10–19	15	1	1	0.06
20–29	25	6	7	0.39
30–39	35	5	12	0.67
40–49	45	2	14	0.78
50–59	55	2	16	0.89
60–69	65	2	18	1.00

The number of data points are progressively summed (Σr), and this value is then divided by the total data points, $n = 18$. These now become single points on a graph of mean BOD group vs P, as in Figure 3.7. The probabilities can be read off the curve as before. As in Example 3.10, if we want to know the worst effluent quality in 30 days, we use $P_{29/30} = 0.967$ and read 66 mg/L BOD. ∎

PROBLEMS

3.1   The following table shows the frequency distribution of influent biochemical oxygen demand (BOD) in a wastewater treatment plant:

BOD (mg/L)	No. of Samples
110–129	18
130–149	31
150–169	23
170–189	20
190–209	7
210–229	1
230–249	2

Compute the following:

a. P_{10} (probability that 10% of the time the BOD will be less than this value)

b. P_{50}

c. P_{95}

3.2   The following are characteristics of annual precipitation for various cities.

	Years of Record	Annual Rainfall $\bar{x}$ (cm/yr)	Annual Rainfall s (cm/yr)
Cheyenne, WY	70	37.11	9.17
Pueblo, CO	52	29.24	13.44
Kansas City, MO	63	91.69	16.87

a. Which city's rainfall is most variable?

b. If it is assumed that rainfall data are normally distributed, how many years in the next 50 would we expect rainfall to be above 50 cm in each city?

c. Assuming a normal distribution, how many years in the next 50 would we expect rainfall to be below 30 cm in each city?

d. What is the probability of the annual rainfall exceeding 92 cm in Kansas City?

3.3 Find the won/lost records for your college basketball team for the past 10 years. Calculate the winning percentage for each year. How often can the team be expected to have an equal number of wins and losses (play at least 500 ball)?

3.4 A normal distribution has a standard deviation of 30 and a mean of 20. Find the probability that

a. $x \geq 80$
b. $x \leq 80$
c. $x \leq -70$
d. $50 \leq x \leq 80$

Graphical means are recommended.

3.5 A lake has the following dissolved oxygen readings:

Month	DO (mg/L)
1	12
2	11
3	10
4	9
5	10
6	8
7	9
8	6
9	10
10	11
11	10
12	11

a. What is the average DO?
b. What is the standard deviation?
c. What fraction of time can the DO be expected to be above 10 mg/L?
d. What fraction of time will the DO be less than 8 mg/L?

3.6 Here are the return periods and flood levels for Morgan Creek:

Return Period (years)	Flood level (m³/s)
1	108
5	142
10	157
20	176
40	196
60	208
100	224

Estimate the flood level for a return period of 500 yr.

3.7 The number of students taking a course for the past five years is as follows:

Year	Number of Students
1	13
2	18
3	17
4	15
5	20

How many years in the next 20 years should the class be 10 students or smaller?

3.8 How much beer is consumed in the United States every day?

3.9 Coal has about 2% sulfur. How many kilograms of sulfur would be emitted from coal-fired power plants in the United States in 1 yr? (U.S. electricity production is about 0.28×10^{12} watts, and coal has an energy content of about 30×10^6 J/kg. Note that 1 J = 1 W-s.)

3.10 How many automobile tires are sold annually in the United States? Make reasonable assumptions.

3.11 The EPA expresses the emission rate of gaseous pollutants from cars in terms of grams per kilometer (g/km). What do you think of this constructed unit of measurement?

3.12 One gram of table salt is put into an 0.25 L glass, and the glass is filled with tap water.

What is the concentration of salt in mg/L? Why might you not be so sure about your answer?

3.13 Metal concentrations in wastewater sludges are often expressed in terms of grams of metal per kilogram of total dry solids. A wet sludge has a solids concentration of 200,000 mg/L, and 8000 mg/L of these solids are zinc.

   a. What is the concentration of zinc as g Zn/g dry solids?
   b. If this sludge is to be applied to farmland and spread on pastures used by cows, explain how you might monitor the project with regard to the potential environmental or health effects caused by the presence of the zinc.

3.14 One gram of pepper is placed into a 100-mL beaker, and the beaker is filled with water to the 100-mL mark. What is the concentration of pepper in mg/L?(What should you assume in this problem? What is the density of pepper? What is the volume of 1 g of pepper?)

3.15 A wastewater treatment plant receives 440 L/s of flow. This wastewater has a solids concentration of 192 mg/L. How many pounds of solids enter the plant every day?

3.16 Ten grams of plastic beads with a density of 1.2 g/cm^3 are added to 500 mL of an organic solvent with a density of 0.8 g/cm^3. What is the concentration of the plastic beads in mg/L? (This is a curve ball!)

3.17 A stream flowing at 240 L/min carries a sediment load of 2000 mg/L. What is the sediment load in kilograms per day?

3.18 A water treatment plant produces water that has an arsenic concentration of 0.05 mg/L.

   a. If the average consumption of water is 8 L of water each day, how much arsenic is each person drinking?
   b. If you are the water treatment plant operator and the chairman of the town council asks you during a public hearing whether there is arsenic in the town's drinking water, how do you respond?
   c. Analyze your answer to part b. from a moral standpoint. Did you do the right thing by answering as you did? Why do you think so?

3.19 A power plant emits 55 kg of flyash per hour up the stack. The flow rate of the hot gases in the stack is 1 m^3/s. What is the concentration of the flyash in micrograms per cubic meter?

3.20 If electricity costs $0.05/kW· h, how much does your university pay annually on its power bill due to the use of desk lamps? (You need to make several assumptions here specific to your university.)

3.21 If tuition were to be paid in the form of tickets to lectures, how much would each ticket to one lecture cost at your university?

3.22 A university has about 12,600 undergraduate students, and graduates 2,250 students each year. Suppose everyone enrolled at the university graduates eventually. What is the approximate retention time at the university?

3.23 A water treatment plant has 6 settling tanks that operate in parallel (the flow gets split into six equal flow streams), and each tank has a volume of 600 m^3. If the flow to the plant is 440 L/s, what is the retention time in each of the settling tanks? If, instead, the tanks operated in series (the entire flow goes first through one tank, then the second, and so on), what would be the retention time in each tank?

3.24 Some of the most confusing statistics are in sports (see, for example, Figure 3.8). Define the following:

   a. ERA
   b. RBI
   c. slugging average
   d. average hang time
   e. on base percentage

Figure 3.8   Example of sports statistics. See Problem 3.24. (SHOE-NEW BUSINESS MACNELLY. DISTRIBUTED BY KING FEATURES SYNDICATE.)

f. average points per game (ice hockey)

g. others?

3.25 You are working as a recruiter for your consulting engineering firm, and an applicant asks you the following question:

> If I work with your firm and I find that I object, on ethical grounds, to a project the firm has been hired to do, can I request not to work on that project, without penalizing my future career with the firm?

Give three responses you might make to such a question, and then respond to these statements according to what you believe the student interviewee would think. Would he/she believe you? Why or why not?

3.26 How should Chris report the estimate of the construction cost? Why?

a. $56,078.34

b. $56,078

c. $56,080

d. $56,000

3.27 Why is the number of significant figures in your final answer important? In other words, what does the number of significant figures in your final answer reflect?

3.28 Example 3.7 mentions precision. What is the difference between precision and accuracy?

3.29 Should you round throughout a calculation or at the end (i.e., the final answer) or both? Why?

END NOTES

1. "Mars Climate Orbiter," Microsoft® Encarta® Online Encyclopedia 2008. Accessed at http://encarta.msn.com on 8 December, 2008.

2. Articles on Mars Climate Orbiter. U.S. National Aeronautics and Space Administration. Accessed at http://mars.jpl.nasa.gov/msp98/orbiter/ on 8 December, 2008.

3. As observed by John Hart in a wonderful book on environmental calculations: *Consider a Spherical Cow*. 1985. Los Altos, CA: William Kaufmann, Inc.

4. This example is adapted from Middlebrooks, E. Joe. 1976. *Statistical Calculations*. Ann Arbor, MI: Ann Arbor Science Pub.

Material Balances and Separations

Dog Fence, Australia

Red Banks Conservation Park, Australia

Environmental engineers need to know how big to draw the "box", for example, where to draw the line and what to include and exclude. If the "box" is too small, unintended consequences may result. If the "box" is too big, time and resources may be wasted.

The alchemists, perhaps the original practitioners of science for profit, tried to develop processes for making gold out of less expensive metals. Viewed from the vantage point of modern chemistry, it is clear why they failed. Except for processes involving nuclear reactions, with which we are not concerned in this text, a pound of any material, such as lead, in the beginning of any process will yield a pound of that material in the end, although perhaps in a very different form. This simple concept of the conservation of mass leads to a powerful engineering tool, the *material balance*. In this chapter the material balance around a black box unit operation is introduced first. Then these black boxes are identified as actual unit operations that perform useful functions. Initially, these black boxes have nothing going on inside them that affects the materials flow. Then it is presumed that material quantities are produced or consumed within the box. In all cases the flow is assumed to be at steady state, that is, not changing with time. This constraint is lifted in subsequent chapters.

4.1 MATERIAL BALANCES WITH A SINGLE MATERIAL

Material flows can be most readily understood and analyzed by using the concept of a black box. These boxes are schematic representations of real processes or flow junctions, and it is not necessary to specify just what this process is to be able to develop general principles about the analysis of flows.

Figure 4.1 shows a black box into which some material is flowing. All flows into the box are called *influents* and represented by the letter X. If the flow is described as mass per unit time, X_0 is a mass per unit time of material X flowing into the box. Similarly, X_1 is the outflow, or *effluent*. If no processes are going on inside the box that will either make more of the material or destroy some of it and if the flow is assumed not to vary with time (that is, to be at *steady state*), then it is possible to write a material balance around the box as

$$\begin{bmatrix} \text{Mass per unit} \\ \text{time of } X \\ \text{IN} \end{bmatrix} = \begin{bmatrix} \text{Mass per unit} \\ \text{time of } X \\ \text{OUT} \end{bmatrix}$$

or

$$[X_0] = [X_1]$$

The black box can be used to establish a volume balance and a mass balance if the density does not change in the process. Because the definition of density is mass per unit volume, the conversion from a mass balance to a volume balance is achieved by dividing each term by the density (a constant). It is generally convenient to use the volume balance for liquids and the mass balance for solids.

Figure 4.1   A black box with one inflow and one outflow.

Figure 4.2   A separator with one inflow and two outflows.

4.1.1   Splitting Single-Material Flow Streams

A black box shown in Figure 4.2 receives flow from one feed source and separates this into two or more flow streams. The flow into the box is labeled X_0, and the two flows out of the box are X_1 and X_2. If again it is assumed that steady state conditions exist and that no material is being destroyed or produced, then the material balance is

$$\begin{bmatrix} \text{Mass per unit} \\ \text{time of } X \\ \text{IN} \end{bmatrix} = \begin{bmatrix} \text{Mass per unit} \\ \text{time of } X \\ \text{OUT} \end{bmatrix}$$

or

$$[X_0] = [X_1] + [X_2]$$

The material X can, of course, be separated into more than two fractions, so the material balance can be

$$[X_0] = \sum_{i=1}^{n} [X_i]$$

where there are n exit streams, or effluents.

E X A M P L E

4.1   **Problem**   A city generates 102 tonnes/d of refuse, all of which goes to a transfer station. At the transfer station the refuse is split into four flow streams headed for three incinerators and one landfill. If the capacity of the incinerators is 20, 50, and 22 tonnes/d, how much refuse must go to the landfill?

Solution   Consider the situation schematically as shown in Figure 4.3. The four output streams are the known capacities of the three incinerators plus one landfill. The input stream is the solid waste delivered to the transfer station. Using the diagram, set up the mass balance in terms of tons per day:

$$\left[\text{Mass per unit of refuse IN}\right] = \left[\text{Mass per unit of refuse OUT}\right]$$
$$102 = 20 + 50 + 22 + M$$

where M = mass of refuse to the landfill. Solving for the unknown, $M = 10$ tonnes/d.

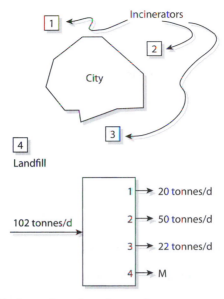

Figure 4.3   Material balance for refuse disposal.

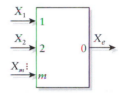

Figure 4.4   A blender with several inflows and one outflow.

4.1.2   Combining Single-Material Flow Streams

A black box can also receive numerous influents and discharge one effluent, as shown in Figure 4.4. If the influents are labeled $X_1, X_2, \ldots, X_m$, the material balance would yield

$$\sum_{i=1}^{m} [X_i] = [X_e]$$

EXAMPLE
4.2

Problem   A trunk sewer, shown in Figure 4.5 on the next page, has a flow capacity of 4.0 m³/s. If the flow to the sewer is exceeded, it will not be able to transmit all the sewage through the pipe, and backups will occur. Currently, three neighborhoods contribute to the sewer, and their maximum (peak) flows are 1.0, 0.5, and 2.7 m³/s. A builder wants to construct a development that will contribute a maximum flow of 0.7 m³/s to the trunk sewer. Would this cause the sewer to exceed its capacity?

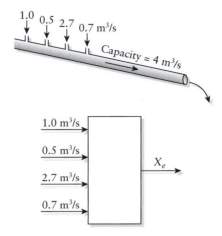

Figure 4.5 Capacity of a trunk sewer.

Solution Set up the material balance in terms of m^3/s.

[Volume/Unit time of sewage IN]   = [Volume/Unit time sewage OUT]

$[1.0 + 0.5 + 2.7 + 0.7]$ $= X_e$

where X_e is the flow in the trunk sewer. Solving yields $X_e = 4.9 \ m^3/s$, which is greater than the capacity of the trunk sewer, so the sewer would be overloaded if the new development is allowed to attach onto it. Even now, the system is overloaded (at $4.2 \ m^3/s$), and the only reason disaster has not struck is that not all the neighborhoods produce the maximum flow at the same time of day.

■

4.1.3 Complex Processes with a Single Material

The preceding simple examples illustrate the basic principle of material balances. The two assumptions used to approach the analysis above are that the flows are in steady state (they do not change with time) and that no material is being destroyed (consumed) or created (produced). If these possibilities are included in the full material balance, the equation reads

$$
\begin{bmatrix} \text{Material} \\ \text{per} \\ \text{unit time} \\ \text{ACCUMULATED} \end{bmatrix} = \begin{bmatrix} \text{Material} \\ \text{per} \\ \text{unit time} \\ \text{IN} \end{bmatrix} - \begin{bmatrix} \text{Material} \\ \text{per} \\ \text{unit time} \\ \text{OUT} \end{bmatrix}
$$

$$
+ \begin{bmatrix} \text{Material} \\ \text{per} \\ \text{unit time} \\ \text{PRODUCED} \end{bmatrix} - \begin{bmatrix} \text{Material} \\ \text{per} \\ \text{unit time} \\ \text{CONSUMED} \end{bmatrix}
$$

If the material in question is labeled A, the mass balance equation reads

$$
\begin{bmatrix} \text{Mass of} \\ A \text{ per} \\ \text{unit time} \\ \text{ACCUMULATED} \end{bmatrix} = \begin{bmatrix} \text{Mass of} \\ A \text{ per} \\ \text{unit time} \\ \text{IN} \end{bmatrix} - \begin{bmatrix} \text{Mass of} \\ A \text{ per} \\ \text{unit time} \\ \text{OUT} \end{bmatrix}
$$

$$
+ \begin{bmatrix} \text{Mass of} \\ A \text{ per} \\ \text{unit time} \\ \text{PRODUCED} \end{bmatrix} - \begin{bmatrix} \text{Mass of} \\ A \text{ per} \\ \text{unit time} \\ \text{CONSUMED} \end{bmatrix}
$$

or, provided the density does not change, in volume terms as

$$
\begin{bmatrix} \text{Volume} \\ \text{of } A \text{ per} \\ \text{unit time} \\ \text{ACCUMULATED} \end{bmatrix} = \begin{bmatrix} \text{Volume} \\ \text{of } A \text{ per} \\ \text{unit time} \\ \text{IN} \end{bmatrix} - \begin{bmatrix} \text{Volume} \\ \text{of } A \text{ per} \\ \text{unit time} \\ \text{OUT} \end{bmatrix}
$$

$$
+ \begin{bmatrix} \text{Volume} \\ \text{of } A \text{ per} \\ \text{unit time} \\ \text{PRODUCED} \end{bmatrix} - \begin{bmatrix} \text{Volume} \\ \text{of } A \text{ per} \\ \text{unit time} \\ \text{CONSUMED} \end{bmatrix}
$$

Mass or volume per unit time can be simplified to *rate*, where rate simply means the flow of mass or volume. Thus, the material balance equation for either mass or volume reads

$$
\begin{bmatrix} \text{Rate of } A \\ \text{ACCUMULATED} \end{bmatrix} = \begin{bmatrix} \text{Rate of } A \\ \text{IN} \end{bmatrix} - \begin{bmatrix} \text{Rate of } A \\ \text{OUT} \end{bmatrix}
$$

$$
+ \begin{bmatrix} \text{Rate of } A \\ \text{PRODUCED} \end{bmatrix} - \begin{bmatrix} \text{Rate of } A \\ \text{CONSUMED} \end{bmatrix} \tag{4.1}
$$

As noted, many systems do not change with time; the flows at one moment are exactly like the flows later. This means that nothing can be accumulating in the black box, either positively (material builds up in the box) or negatively (material is flushed out). The material balance equation, when the *steady state assumption* holds, is

$$
0 = \begin{bmatrix} \text{Rate of } A \\ \text{IN} \end{bmatrix} - \begin{bmatrix} \text{Rate of } A \\ \text{OUT} \end{bmatrix} + \begin{bmatrix} \text{Rate of } A \\ \text{PRODUCED} \end{bmatrix} - \begin{bmatrix} \text{Rate of } A \\ \text{CONSUMED} \end{bmatrix}
$$

In many simple systems the material in question is not being consumed or produced. If this *conservation assumption* is applied to the material balance equation, it simplifies to

$$
0 = \begin{bmatrix} \text{Rate of} \\ \text{mass or volume} \\ \text{IN} \end{bmatrix} - \begin{bmatrix} \text{Rate of} \\ \text{mass or volume} \\ \text{OUT} \end{bmatrix} + 0 - 0
$$

or

$$
\begin{bmatrix}
\text{Rate of} \\
\text{mass or volume} \\
\text{IN}
\end{bmatrix}
=
\begin{bmatrix}
\text{Rate of} \\
\text{mass or volume} \\
\text{OUT}
\end{bmatrix}
$$

the same equation used in the previous examples.

EXAMPLE

4.3

Problem   A sewer carrying stormwater to Manhole 1 (Figure 4.6) has a constant flow of 2000 L/min (Q_A). At Manhole 1 it receives a constant lateral flow of 100 L/min (Q_B). What is the flow to Manhole 2 (Q_C)?

Solution   Think of Manhole 1 as a black box, as shown in Figure 4.6. Write the material balance equation for water (Equation 4.1):

$$
\begin{bmatrix}
\text{Rate of} \\
\text{water} \\
\text{ACCUMULATED}
\end{bmatrix}
=
\begin{bmatrix}
\text{Rate of} \\
\text{water} \\
\text{IN}
\end{bmatrix}
-
\begin{bmatrix}
\text{Rate of} \\
\text{water} \\
\text{OUT}
\end{bmatrix}
$$

$$
+
\begin{bmatrix}
\text{Rate of} \\
\text{water} \\
\text{PRODUCED}
\end{bmatrix}
-
\begin{bmatrix}
\text{Rate of} \\
\text{water} \\
\text{CONSUMED}
\end{bmatrix}
$$

Figure 4.6   Sewer with two flows entering Manhole 1, combining, and moving toward Manhole 2.

Since no water accumulates in the black box, the system is defined as being in steady state. The first term of this balance equation then goes to zero. Similarly, no water is produced or consumed, so the last two terms are zero. The equation then reads

$$0 = (Q_A + Q_B) - Q_C + 0 - 0$$

Now substitute the given flows in liters per minute:

$$0 = (2000 + 100) - Q_C$$

and solve:

$$Q_C = 2100 \text{ L/min}$$

Where the flow is contained in pipes or channels, it is convenient to place a black box at any junction with 3 or more flows. For example, the manhole in Example 4.3 is a junction receiving two flows and discharging one, for a total of three.

Sometimes the flows may not be contained in conduits or rivers. In such cases it is useful to first visualize the system as a pipe flow network. Rainwater falling to earth, for example, can either percolate into the ground or run off in a watercourse. This system can be pictured as shown in Figure 4.7 and a material balance performed on the imaginary black box.

A system may contain any number of processes or flow junctions, all of which could be treated as black boxes. For example, Figure 4.8A on the next page shows a schematic of the hydrologic cycle. Precipitation falls to earth; some of it runs off, and some of it percolates into the ground, where it joins a groundwater reservoir. If the water is used for irrigation, it is taken out of the groundwater reservoir through wells. The irrigation water

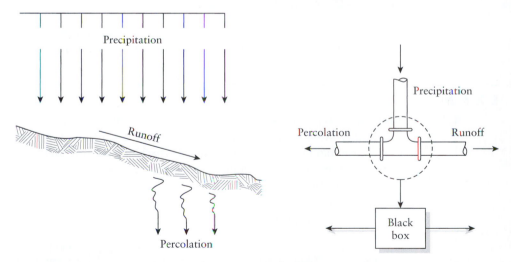

Figure 4.7   Precipitation, runoff, and percolation can be visualized as a black box.

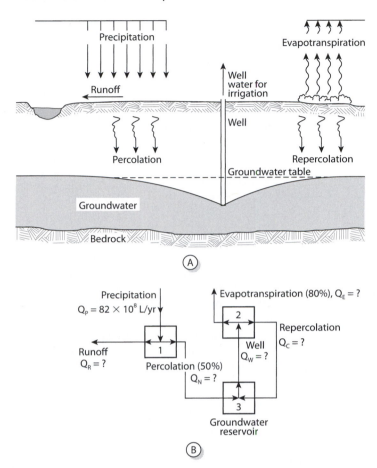

Figure 4.8 A simplified hydrologic cycle.

either percolates back into the ground or goes back to the atmosphere through evaporation or transpiration (water released to the atmosphere from plants), commonly combined into one process: evapotranspiration.

EXAMPLE

4.4

Problem   Suppose the rainfall is 102 cm/yr, of which 50% percolates into the ground. The farmer irrigates crops using well water. Of the extracted well water, 80% is lost by evapotranspiration; the remainder percolates back into the ground. How much groundwater could a farmer on a 8 km² farm extract from the ground per year without depleting the groundwater reservoir volume?

Solution   Recognizing this as a material balance problem, first convert the rainfall to a rate; 102 cm/yr over 8 km² is

$$102 \ \frac{cm}{yr} \times 8 \ km^2 \left(\frac{1 \ m}{100 \ cm} \right) \left(\frac{1000 \ m}{1 \ km} \right) \left(\frac{1000 \ m}{km} \right) \left(\frac{1000 \ L}{1 \ m^3} \right) = 82 \times 10^8 \ L/yr$$

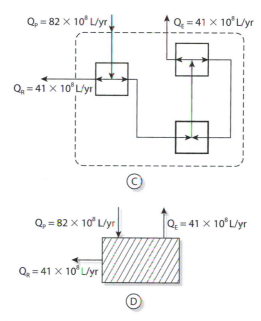

Figure 4.8 Continued.

The system in question is drawn as Figure 4.8B, and the given information is added to the sketch. Unknown quantities are noted by variables.

It is convenient to start the calculations by constructing a balance on the first black box (1), a simple junction with precipitation coming in and runoff and percolation going out. The volume rate balance is

$$\begin{bmatrix} \text{Volume} \\ \text{rate of} \\ \text{water} \\ \text{ACCUMULATED} \end{bmatrix} = \begin{bmatrix} \text{Volume} \\ \text{rate of} \\ \text{water} \\ \text{IN} \end{bmatrix} - \begin{bmatrix} \text{Volume} \\ \text{rate of} \\ \text{water} \\ \text{OUT} \end{bmatrix}$$
$$+ \begin{bmatrix} \text{Volume} \\ \text{rate of} \\ \text{water} \\ \text{PRODUCED} \end{bmatrix} - \begin{bmatrix} \text{Volume} \\ \text{rate of} \\ \text{water} \\ \text{CONSUMED} \end{bmatrix}$$

Since the system is assumed to be in steady state, the first term is zero. Likewise, the last two terms are zero as water, again, is not produced or consumed (e.g., in reactions):

$$0 = \left[82 \times 10^8 \frac{L}{yr} \right] - [Q_R + Q_N] + 0 - 0$$

As stated in the problem, half the water percolates into the ground; therefore, half the water also runs off:

$$Q_R = 0.5 Q_P = Q_N$$

so plugging this information into the material balance yields

$$0 = 82 \times 10^8 - 2Q_R$$

Solve for Q_R and Q_N:

$$Q_R = 41 \times 10^8 \frac{\text{L}}{\text{yr}} = Q_N$$

A balance on the second black box (2) yields

$$0 = [Q_W] - [Q_E + Q_C] + 0 - 0$$

As stated in the problem, 80% of the irrigation water is lost by evapotranspiration; therefore, 20% of the irrigation water percolates back into the ground:

$$0 = Q_W - 0.8Q_W - Q_C$$
$$Q_C = 0.2(Q_W)$$

Finally, a material balance on the groundwater reservoir (3) can be written if the quantity of groundwater in the reservoir is assumed not to change:

$$0 = [Q_N + Q_C] - [Q_W] + 0 - 0$$

From the first material balance, $Q_N = 41 \times 10^8$ L/yr and from the second, $Q_C = 0.2Q_W$. Plug this information into the third material balance and solve:

$$0 = 1.45 \times 10^8 + 0.2Q_W - Q_W$$
$$Q_W = 51 \times 10^8 \text{ L/yr}$$

This is the maximum safe yield of well water for the farmer.

As a check, consider the entire system as a black box. This is illustrated in Figure 4.8C. There is only one way water can get into this box (precipitation) and two ways out (runoff and evapotranspiration). Representing this black box as Figure 4.8D, it is possible to write the material balance in liters per year of water:

$$0 = (82 \times 10^8) - (41 \times 10^8) - (41 \times 10^8)$$

The balance of the overall system checks the calculations.

The last check illustrates a useful and powerful tool in material balance calculations. When the flows in a system of black boxes are balanced, it is possible to draw boundaries (dashed lines) around any one or any combination of black boxes, and this combination is itself a black box. Returning to the above example, if a dashed line is drawn around the entire system (Figure 4.8C), it is useful to picture it as a single black box (Figure 4.8D). If the calculations are correct, the flows for this box should also balance (and they do).

Incidentally, what would happen if the farmer decided to take more than that amount of water out of the ground? An *unsteady* situation would occur, and in time the ground-water reservoir would run dry. This is exactly what is occurring in the Great Plains today. The Ogallala ($\bar{O}g$ *law law* or $\bar{O}g$ *lowl law*) Aquifer, a vast reservoir that has provided water for farmers since 1930, is being drained dry. The annual overdraft is nearly equal to the flow of the Colorado River. A major engineering study warned that 20,000 million square meters of irrigated farms will dry up by the year 2020 if this trend continues.

Excessive use of fertilizers represents another serious environmental problem. We can feed the world population only if we continue to extract huge quantities of phosphorus and other nutrients out of the ground. We know that eventually these reserves will be depleted and that the yield from our farms will be greatly reduced. If these events come to pass, starvation will be common in much of the world. With not enough food to go around and too many people desperate for it, how will it be distributed? What if we simply ignore those starving people and wait for starvation to reduce the global population so that the food we produce will be in balance with the survivors? We could argue that the amount of food available would not feed everyone anyway and that taking away our food may cause starvation here as well so that eventually *all* people would die of starvation. It is best, so goes the argument, to simply allow nature to take its course. Any food we give now to needy countries just encourages them to produce more people, and this will eventually lead to tragedy. Such thinking is known as "lifeboat ethics," the name coming from the idea that a lifeboat can only hold so many people. If more than the capacity try to get in, the boat will be swamped, and *everyone* will drown.

Do we, therefore, have an obligation to save our own skin and not worry about our neighbors? This seems like a reasonable thing to do, until we consider the possibility that perhaps *we* are the ones without food, and our neighbors decide to allow *us* to starve. What should/could/would we do, if anything?

This seems like a science-fiction scenario, but unfortunately, it may be all too true. Resource deprivation is predictable and assured if our global system continues in a non-steady state. You are perhaps the last lucky generation. Your children may not be so fortunate.

But back to the black boxes. Thus far, the black boxes have had to process only one feed material; for example, water is the only material balanced in the previous two examples. Material balances become considerably more complicated (and useful) when several materials flow through the system.

However, before introducing examples of systems with multiple materials, it is appropriate to establish some general rules that are useful in attacking all material balance problems. The rules are

1. Draw the system as a diagram, including all flows (inputs and outputs) as arrows.
2. Add the available information, such as flow rates and concentrations. Assign symbols to unknown variables.
3. Draw a continuous dashed line around the component or components that are to be balanced. This could be a unit operation, a junction, or a combination of these. Everything inside this dashed line becomes the black box.

4. Decide what material is to be balanced. This could be a volumetric or mass flow rate.
5. Write the material balance equation by starting with the basic equation:

$$\begin{bmatrix} \text{Mass or} \\ \text{volume} \\ \text{rate} \\ \text{ACCUMULATED} \end{bmatrix} = \begin{bmatrix} \text{Mass or} \\ \text{volume} \\ \text{rate} \\ \text{IN} \end{bmatrix} - \begin{bmatrix} \text{Mass or} \\ \text{volume} \\ \text{rate} \\ \text{OUT} \end{bmatrix}$$

$$+ \begin{bmatrix} \text{Mass or} \\ \text{volume} \\ \text{rate} \\ \text{PRODUCED} \end{bmatrix} - \begin{bmatrix} \text{Mass or} \\ \text{volume} \\ \text{rate} \\ \text{CONSUMED} \end{bmatrix}$$

6. If only one variable is unknown, solve for that variable.
7. If more than one variable is unknown, repeat the procedure, using a different black box or a different material for the same black box.

Thus armed with a set of guidelines, we now turn to the problems wherein more than one material is involved.

4.2 MATERIAL BALANCES WITH MULTIPLE MATERIALS

Mass and volume balances can be developed with multiple materials flowing in a single system. In some cases the process is one of mixing, where several inflow streams are combined to produce a single outflow stream, while in other cases a single inflow stream is split into several outflow streams according to some material characteristics.

4.2.1 Mixing Multiple-Material Flow Streams

Because the mass balance and volume balance equations are actually the same equation, it is not possible to develop more than one material balance equation for a black box, unless there is more than one material involved in the flow. Consider Example 4.5, wherein silt flow in rivers (expressed as mass of solids per unit time) is analyzed.

EXAMPLE
4.5

Problem   The Allegheny and Monongahela Rivers meet at Pittsburgh to form the mighty Ohio. The Allegheny, flowing south through forests and small towns, runs at an average flow of $10\,\text{m}^3/\text{s}$ (cubic meter per second) and has a low silt load, 250 mg/L. The Monongahela, on the other hand, flows north at a flow of $13\,\text{m}^3/\text{s}$ cfs through old steel towns and poor farm country, carrying a silt load of 1500 mg/L.

a. What is the average flow in the Ohio River?
b. What is the silt concentration in the Ohio?

Solution   Follow the general rules.

Step 1. Draw the system. Figure 4.9 shows the confluence of the rivers with the flows identified.

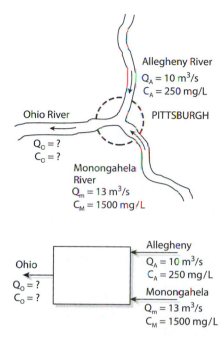

Figure 4.9   Confluence of the Allegheny and Monongahela to form the Ohio River.

Step 2. All the available information is added to the sketch, including the known and unknown variables.

Step 3. The confluence of the rivers is the black box, as shown by the dashed line.

Step 4. Water flow is to be balanced first.

Step 5. Write the balance equation:

$$\begin{bmatrix} \text{Rate of} \\ \text{water} \\ \text{ACCUMULATED} \end{bmatrix} = \begin{bmatrix} \text{Rate of} \\ \text{water} \\ \text{IN} \end{bmatrix} - \begin{bmatrix} \text{Rate of} \\ \text{water} \\ \text{OUT} \end{bmatrix}$$

$$+ \begin{bmatrix} \text{Rate of} \\ \text{water} \\ \text{PRODUCED} \end{bmatrix} - \begin{bmatrix} \text{Rate of} \\ \text{water} \\ \text{CONSUMED} \end{bmatrix}$$

Because this system is assumed to be in steady state, the first term is zero. Also, because water is neither produced nor consumed, the last two terms are zero. Thus, the material balance becomes

$$0 = [\text{Water IN}] - [\text{Water OUT}] + 0 - 0$$

Two rivers flow in and one out, so the equation in cfs reads

$$0 = [10 + 13] - [Q_O] + 0 - 0$$

where Q_O = flow in the Ohio.

Step 6. Solve for the unknown.

$$Q_O = 23 \text{ m}^3/\text{s}$$

The solution process must now be repeated for the silt. Recall that mass flow is calculated as concentration times volume, or

$$Q_{Mass} = C \times Q_{Volume} = \frac{mg}{L} \times \frac{L}{s} = \frac{mg}{s}$$

Starting with Step 5, the mass balance is

$$\begin{bmatrix} \text{Silt} \\ \text{ACCUMULATED} \end{bmatrix} = \begin{bmatrix} \text{Silt} \\ \text{IN} \end{bmatrix} - \begin{bmatrix} \text{Silt} \\ \text{OUT} \end{bmatrix} + \begin{bmatrix} \text{Silt} \\ \text{PRODUCED} \end{bmatrix} - \begin{bmatrix} \text{Silt} \\ \text{CONSUMED} \end{bmatrix}$$

Again the first and last two terms are assumed to be zero, so the equation becomes

$$0 = [\text{Silt IN}] - [\text{Silt OUT}] + 0 - 0$$
$$0 = [(C_A Q_A) + (C_M Q_M)] - [C_O Q_O] + 0 - 0$$

where C = concentration of silt

$\hat{A}, \hat{M}, \hat{O}$ = the three rivers

Substituting the known information yields

$$0 = [(250 \text{ mg/L} \times 10 \text{ m}^3/\text{s}) + (1500 \text{ mg/L} \times 13 \text{ m}^3/\text{s})] - [C_O \times 23 \text{ m}^3/\text{s}]$$

Note that the flow of the Ohio is 23 m³/s as calculated from the volume balance (which is why the water balance was done first).

Solve the equation:

$$C_O = 969 \text{ mg/L} \approx 970 \text{ mg/L}$$

In Example 4.5 all unknowns except one are defined, resulting in a simple calculation for the remaining unknown term. In most applications, however, it is necessary to work a little harder for the answer, as illustrated in Example 4.6.

EXAMPLE
4.6

Problem Suppose the sewers shown in Figure 4.10 have $Q_B = 0$ and Q_A and Q_C are unknowns. By sampling the flow at the first manhole, it is found that the concentration of dissolved solids in the flow coming into Manhole 1 is 50 mg/L. An additional flow, $Q_B = 100$ L/min, is added to Manhole 1, and this flow contains 20% dissolved solids.

Figure 4.10   A manhole as a black box with solid and liquid flows.

(Recall that 1% = 10,000 mg/L.) The flow through Manhole 2 is sampled and found to contain 1000 mg/L dissolved solids. What is the flow rate of wastewater in the sewer (Q_A)?

Solution   Again, follow the general rules.

Step 1.   Draw the diagram, which is given in Figure 4.10.
Step 2.   Add all information, including concentrations.
Step 3.   Manhole 1 is the black box.
Step 4.   What is to be balanced? If the flows are balanced, there are two unknowns. Can something else be balanced? Suppose a balance is run in terms of the solids?
Step 5.   Write the material balance for solids.

$$\begin{bmatrix} \text{Solids} \\ \text{ACCUMULATED} \end{bmatrix} = \begin{bmatrix} \text{Solids} \\ \text{IN} \end{bmatrix} - \begin{bmatrix} \text{Solids} \\ \text{OUT} \end{bmatrix}$$
$$+ \begin{bmatrix} \text{Solids} \\ \text{PRODUCED} \end{bmatrix} - \begin{bmatrix} \text{Solids} \\ \text{CONSUMED} \end{bmatrix}$$

A steady state assumption allows the first term to be zero, and because no solids are produced or consumed in the system, the last two terms are zero, so the equation reduces to

$$0 = [\text{Solids flow in}] - [\text{Solids flow out}] + 0 - 0$$
$$0 = [Q_A C_A + Q_B C_B] - [Q_C C_C] + 0 - 0$$
$$0 = \left[Q_A \left(50 \, \frac{\text{mg}}{\text{L}} \right) + \left(100 \, \frac{\text{L}}{\text{min}} \right) \left(200{,}000 \, \frac{\text{mg}}{\text{L}} \right) \right] - \left[Q_C \left(1000 \, \frac{\text{mg}}{\text{L}} \right) \right]$$

Note that (L/min) × (mg/L) = (mg solids/min). This results in an equation with two unknowns, so it is necessary to skip step 6 and proceed to step 7.

Step 7. If more than one unknown variable results from the calculation, establish another balance. A balance in terms of the volume flow rate is

$$\begin{bmatrix} \text{Volume} \\ \text{ACCUMULATED} \end{bmatrix} = \begin{bmatrix} \text{Volume} \\ \text{IN} \end{bmatrix} - \begin{bmatrix} \text{Volume} \\ \text{OUT} \end{bmatrix}$$
$$+ \begin{bmatrix} \text{Volume} \\ \text{PRODUCED} \end{bmatrix} - \begin{bmatrix} \text{Volume} \\ \text{CONSUMED} \end{bmatrix}$$

Again, the first term and the last two terms are assumed to be zero, so

$$0 = [Q_A + Q_B] - [Q_C]$$

and

$$0 = Q_A + 100 - Q_C$$

There now are two equations with two unknowns. Substitute $Q_A = (Q_C - 100)$ into the first equation:

$$50(Q_C - 100) + 200,000(100) = 1000Q_C$$

and solve:

$$Q_C = 21,047 \text{ L/min} \approx 21,000 \text{ L/min and } Q_A = 20,947 \text{ L/min} \approx 20,900 \text{ L/min}$$

Sometimes in material balance calculations the black box is literally a box. For example, a crude but useful means of estimating the relationship between the emission of air pollutants and the air quality above a city is the *box model*. In this analysis a box is assumed to sit above the city, and this box is as high as the *mixing depth* of pollutants and as wide and long as the city boundaries. Wind blows through this box and mixes with the pollutants emitted, resulting in a concentration term. The box model can be analyzed by using the same principles just discussed.

EXAMPLE
4.7

Problem   Estimate the concentration of SO_2 in the urban air above the city of St. Louis if the mixing height above the city is 1210 m, the width of the box perpendicular to the wind is 10^5 m, the average annual wind speed is 15,400 m/hr, and the amount of sulfur dioxide discharged is 6242×10^8 g/yr.

Solution   First, construct the box above St. Louis, as shown in Figure 4.11.

The volume of air moving into the box is calculated as the velocity times the area through which the flow occurs, or $Q = Av$, where $v =$ wind velocity and $A =$ area of the side of the box (mixing depth times width).

$$Q_{\text{air}} = (1210 \text{ m} \times 10^5 \text{ m})(15,400 \text{ m/hr}) = 1.86 \times 10^{12} \text{ m}^3/\text{hr}$$

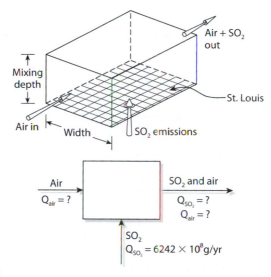

Figure 4.11   Air quality box model. See Example 4.7.

A simplified box is also shown in the figure. It is clear that a simple application of the volume balance equation would show [Air IN] = [Air OUT] so Q_{air} out is 1.86×10^{12} m³/hr. Remembering that *mass flow* can be expressed as [Concentration] × [Volume flow], a mass balance in terms of SO₂ can be written:

$$\begin{bmatrix} \text{Rate of SO}_2 \\ \text{ACCUMULATED} \end{bmatrix} = \begin{bmatrix} \text{Rate of SO}_2 \\ \text{IN} \end{bmatrix} - \begin{bmatrix} \text{Rate of SO}_2 \\ \text{OUT} \end{bmatrix}$$
$$+ \begin{bmatrix} \text{Rate of SO}_2 \\ \text{PRODUCED} \end{bmatrix} - \begin{bmatrix} \text{Rate of SO}_2 \\ \text{CONSUMED} \end{bmatrix}$$

Assuming steady state and no reactions, the balance becomes

$$0 = [7.126 \times 10^{13} \ \mu g/hr] - [(C) (1.86 \times 10^{12} \ m^3/hr)]$$

where

$$6242 \times 10^8 g/yr \ \times 10^6 \ \frac{\mu g}{g} \times \frac{1}{8760} \ \frac{yr}{hr} = 7.126 \times 10^{13} \ \frac{\mu g}{hr}$$

Solving gives $C = 38 \ \mu g/m^3$.[1]

Sulfur dioxide has a long history as a problem air pollutant. Recall the 1948 Donora, Pennsylvania, episode discussed in Chapter 1. Following the passage of national air pollution legislation, many United States companies moved offshore to islands where pollution control is less restrictive. As long as pollution is considered a local problem, the relocation of plants offshore makes sense, except for the jobs lost, of course. Now, however, when we recognize that pollution is of global concern, the mere moving of a plant a few thousand miles may make no difference whatever as to its effect on the global atmosphere. Earth has become our black box.

4.2.2 Separating Multiple-Material Flow Streams

So far, we have discussed black boxes in which the characteristics of the material flows are not altered by the black boxes. In this section the black boxes receive flows made up of mixtures of two or more materials, and the intent of the black box is to change the concentration of these influents.

The objective of a material *separator* is to split a mixed-feed material into the individual components by exploiting some difference in the material properties. For example, suppose it is necessary to design a device for a solid waste processing plant that will separate crushed glass into two types of glass: transparent, or clear glass (flint), and opaque (amber or brown) glass. First, decide what material property to exploit. This becomes the *code*, or signal, that can be used to tell the machine how to divide the individual particles in the feed stream. In the case of the glass the code could obviously be *transparency*, and this property is used in the design of the separating system, such as the device shown in Figure 4.12. In this device the glass pieces drop off a belt conveyor and pass through a light beam. The amount of light transmitted across the box is read by a photocell. If the light beam is interrupted by an opaque piece of glass, the photocell receives less light and an electromagnet is activated that pulls a gate to the left. A transparent piece of glass will not interrupt the light beam and the gate will not move. The gate, activated by the signal from the photocell, is the *switch*, separating the material according to the code.

This simple device illustrates the nature of coding and switching: reading a property difference and then using that signal to achieve separation. Material separation using

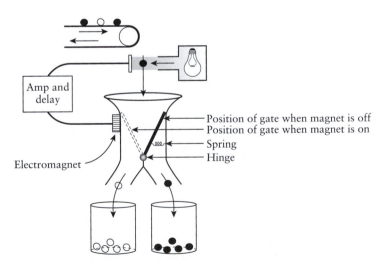

Figure 4.12 Device for separating colored glass from clear glass; a binary separator.

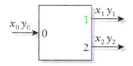

Figure 4.13   A binary separator.

the principle of coding and switching is an integral and central part of environmental engineering.

Material separation devices are not infallible; they make mistakes. It cannot be assumed that the light will always correctly identify the clear glass. Even for the most sophisticated and carefully designed devices, mistakes will occur, and an opaque piece of glass may end up in a box full of transparent pieces or the other way around. The measure of how well the separation works uses two parameters: *recovery* and *purity*.

Consider a black box separator shown in Figure 4.13. The two components x and y are to be separated so that x goes to product stream 1 and y goes to product stream 2. Unfortunately, some of the y mistakenly ends up in product stream 1 and some of the x comes out in stream 2. The *recovery* of component x in product stream 1 is defined as

$$R_{x_1} = \frac{x_1}{x_0} \times 100 \qquad (4.2)$$

where $x_1 =$ material flow in effluent stream 1
$\quad\quad\ x_0 =$ material flow in the influent stream

Similarly, the recovery of material y in product stream 2 can be expressed as

$$R_{y_2} = \frac{y_2}{y_0} \times 100$$

where $y_2 =$ material flow in effluent stream 2
$\quad\quad\ y_0 =$ material flow in the influent stream

If the recovery of both x and y is 100%, the separator is a perfect device.

Suppose it is now necessary to maximize the recovery of x in product stream 1; it would make sense to reduce x_2 to the smallest possible amount. This can be effectively accomplished by shutting off product stream 2, diverting the entire flow to product stream 1, and achieving $R_x = 100\%$! Obviously, we have a problem. We are getting 100% recovery, but we aren't achieving anything useful. Recovery cannot be the sole criterion for judging the performance of a separator. Therefore, a second parameter used to describe separation is *purity*. The purity of x in exit stream 1 is defined as

$$P_{x_1} = \frac{x_1}{x_1 + y_1} \times 100$$

A combination of recovery and purity will yield the performance criteria we need.

E X A M P L E

4.8   **Problem**   Assume that an aluminum can separator in a local recycling plant processes 400 cans/min. The two product streams consist of the following:

	Total in Feed	Product Stream 1	Product Stream 2
Aluminum cans/min	300	270	30
Steel cans/min	100	0	100

Calculate the recovery of aluminum cans and the purity of the product.

Solution   Note first that a material balance for aluminum cans per minute yields

$$0 = [IN] - [OUT] = 300 - (270 + 30)$$

$$0 = 0 \text{ check}$$

$$R_{\text{Al cans}_1} = \frac{270}{300} \times 100 = 90\%$$

$$P_{\text{Al cans}_1} = \frac{270}{270 + 0} \times 100 = 100\%$$

This separator has a high recovery and a very high purity. In fact, according to these data, there is no contamination (i.e., steel cans) in the final aluminum can stream (product stream 1).

■

The same principles can be applied to a more complex environmental process, such as gravitational thickening.

E X A M P L E

4.9   **Problem**   Figure 4.14A shows a gravitational thickener used in numerous water and wastewater treatment plants. This device separates suspended solids from the liquid (usually water) by taking advantage of the fact that the solids have a higher density than the water. The code for this separator is density, and the thickener is the switch, allowing the denser solids to settle to the bottom of a tank from which they are removed. The flow into the thickener is called the influent, or feed; the low solids exit stream is the overflow, and the heavy, or concentrated, solids exit stream is the underflow. Suppose a thickener in a metal plating plant receives a feed of 40 m³/hr of precipitated metal plating waste with a suspended solids concentration of 5000 mg/L. If the thickener is operated in a steady state mode so that 30 m³/hr of flow exits as the overflow, and this overflow has a solids concentration of 25 mg/L, what is the underflow solids concentration, and what is the recovery of the solids in the underflow?

Solution   Once again consider the thickener as a black box (Figure 4.14B) and proceed stepwise to balance first the volume flow and then the solids flow. Assuming steady state, the volume balance in cubic meters per hour is

$Q_i = 40$ m³/hr $Q_o = 30$ m³/hr
$C_i = 5000$ mg/L $C_o = 25$ mg/L

$Q_u = ?$
$C_u = ?$

(A)

$Q_o = 30$ m³/hr
$C_o = 25$ mg/L

$Q_i = 40$ m³/hr
$C_i = 5000$ mg/L

$Q_u = ?$
$C_u = ?$

(B)

Figure 4.14   A gravity thickener for sludge thickening. See Example 4.9.

$$\begin{bmatrix} \text{Volume} \\ \text{ACCUMULATED} \end{bmatrix} = \begin{bmatrix} \text{Volume} \\ \text{IN} \end{bmatrix} - \begin{bmatrix} \text{Volume} \\ \text{OUT} \end{bmatrix}$$
$$+ \begin{bmatrix} \text{Volume} \\ \text{PRODUCED} \end{bmatrix} - \begin{bmatrix} \text{Volume} \\ \text{CONSUMED} \end{bmatrix}$$

$$0 = 40 - (30 + Q_u) + 0 - 0$$
$$Q_u = 10 \text{ m}^3/\text{h}$$

For the solids the mass balance is

$$0 = (C_i Q_i) - [(C_u Q_u) + (C_o Q_o)] + 0 - 0$$
$$0 = (5000 \text{ mg/L})(40 \text{ m}^3/\text{h}) - [C_u(10 \text{ m}^3/\text{hr}) + (25 \text{ mg/L})(30 \text{ m}^3/\text{h})]$$
$$C_u = 19,900 \text{ mg/L}$$

The recovery of solids is

$$R_u = \frac{C_u Q_u}{C_i Q_i} \times 100$$

$$R_u = [(19,900 \text{ mg/L})(10 \text{ m}^3/\text{h}) \times 100]/[(5000 \text{ mg/L})(40 \text{ m}^3/\text{h})] = 99.5\%$$

In many engineering studies it is inconvenient to have two parameters, such as recovery and purity, that describe the performance of a unit operation. For example, the performance of two competing air classifiers used for producing refuse-derived fuel may be advertised as follows:

	R (%)	P (%)
Air Classifier 1	90	92
Air Classifier 2	93	87

It is difficult to state which is the better device. Accordingly, a number of single-value parameters have been suggested. For a binary separator, two suggestions are

$$E_{WS} = \left[\frac{x_1}{x_0} \times \frac{y_2}{y_0} \right]^{1/2} \times 100$$

where E_{WS} = Worrell–Stessel efficiency, and

$$E_R = \left| \frac{x_1}{x_0} - \frac{y_1}{y_0} \right| \times 100$$

where E_R = Rietema efficiency.

Both of these allow for comparisons using a single value, and both have the benefit of ranging from 0 to 100%. The use of these definitions for efficiency can be illustrated by using the above data for the two air classifiers.

EXAMPLE

4.10

Problem   Calculate the efficiencies for the following data, using the Worrell–Stessel and Rietema formulas.

	Feed (tonnes/day) Organics/Inorganics	Product Stream 1 (tonnes/day) Organics/Inorganics	Product Stream 2 (tonnes/day) Organics/Inorganics
Air Classifier 1	80/20	72/6	8/14
Air Classifier 2	80/20	76/8	4/12

Solution   For the first air classifier:

$$E_{WS} = \left[\frac{72}{80} \times \frac{14}{20} \right]^{1/2} \times 100 = 79\%$$

$$E_R = \left| \frac{72}{80} - \frac{6}{20} \right| \times 100 = 60\%$$

For the second air classifier, $E_{WS} = 75\%$ and $E_R = 55\%$. Clearly, the first classifier is superior.

These processes illustrate the separation of a single material into two product streams. These so-called *binary separators* are used to split a feed into two parts according to a

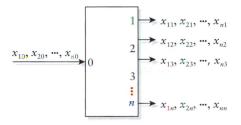

Figure 4.15   A polynary separator.

material property, or code. It is possible to envision also a *polynary separator*, which divides a mixed material into three or more components. The performance of a polynary separator can also be described by recovery, purity, and efficiency. With reference to Figure 4.15, the recovery of component x_1 in effluent stream 1 is

$$R_{x_{11}} = \frac{x_{11}}{x_{10}} \times 100$$

where $x_{11} =$ component x_1 in effluent stream 1
 $x_{10} =$ component x_1 in the influent

The purity of product stream 1 with respect to component x_1 is

$$P_{x_{11}} = \frac{x_{11}}{x_{11} + x_{21} + \cdots + x_{n1}} \times 100$$

The Worrell–Stessel and Rietema efficiencies are, respectively:

$$E_{\text{WS}} = \left[\frac{x_{11}}{x_{10}} \cdot \frac{x_{22}}{x_{20}} \cdot \frac{x_{33}}{x_{30}} \cdots \frac{x_{nn}}{x_{n0}} \right]^{1/n} \times 100$$

$$E_{\text{R}_1} = \left| \frac{x_{11}}{x_{10}} - \frac{x_{12}}{x_{20}} - \frac{x_{13}}{x_{30}} \cdots \frac{x_{1n}}{x_{n0}} \right| \times 100$$

In this polynary separator, it is assumed that the feed has n components and that the separation process has n product streams. A more general condition would be if there were m product streams for a feed with n components, as shown in Figure 4.16. The equations for recovery, purity, and efficiency can be constructed from the definitions as before.

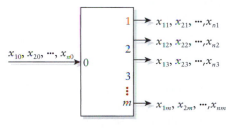

Figure 4.16   A polynary separator for *n* components and *m* product streams.

Large-Scale Applications of Material Balances

Material balances, also called mass balances, can be applied at any scale—part of the reason that the concept is so useful. In Britain, Biffaward, an organization created to use landfill taxes to fund environmental grants, supported an initiative called The Mass Balance Movement.[2] This initiative produced a number of research projects and publications that shed light on the value of material balance analyses.

The material balance projects examined resource flows in the United Kingdom in different ways. Resources were considered from the point at which they were extracted or imported, manufactured into products, used and disposed of, or recycled. Some studies focused on the production of specific materials. Others focused on specific economic sectors, and others focused on specific geographical regions.

The material balance concept is based on the fundamental principle that matter cannot be created or destroyed. Therefore, the mass of inputs, whether to a process, an industry, or a region, must balance the mass of outputs as products, emissions, and wastes, and it must account for any changes in the accumulated mass (referred to as stock, such as building infrastructure) (Figure 4.17).

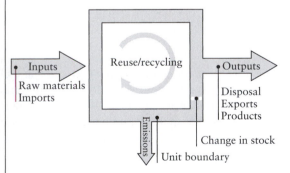

Figure 4.17   General material balance.[6]
(By permissin of Biffaward Mass Balance Project, RSWT.)

When applied in a systematic manner, this simple and straightforward concept can help industry, government, and others make more informed decisions and prioritize both problems and opportunities.

For policymakers:

- Material balance analyses help policymakers determine whether a landfill ban is feasible or whether alternative means of materials management are needed—and where.
- Data showing that schools can save £17.85 per pupil each year by using energy efficiency measures can help support government program priorities.
- Data showing that only 34% of glass in the United Kingdom is recycled (compared to 92% in Switzerland) can drive a study to compare policy instruments in the United Kingdom and Switzerland to see how the recycling rate can be greatly increased.
- Data showing that the paper and board industry uses 6.5 million tonnes of pulp each year, of which 4.5 million tonnes is derived from waste paper (a recycling rate of 69%, increased from 59% in 1996) and that it also reduced its water use by 70% over the last 20 yr can result in opportunities for positive public recognition and future target setting.

For businesses:

- Material balance analyses can help businesses see where valuable recyclable resources (such as copper) are leaking out of the economic system, such as whether resources are being exported as product or lost as waste. Competitiveness may drive businesses to consider estimates of global stocks and flows and to recapture all that they can to ensure that they will have adequate resources for the future.
- Data showing that around 573 million magazines are unsold each year might drive

innovative ways to publish, such as on-demand, to avoid having to manage unsold magazines.

For engineers:

- Knowing that animal digestive processes and manures release about 40% of the United Kingdom's methane might lead to the construction of manure digesters that allow for energy recovery from methane to help achieve CO_2 emission targets.
- Results from two studies identifying agricultural resource flows as 22% of total UK waste might trigger further evaluation of the wastes for value recovery options.

- Knowing that over 50 million tires were scrapped in the United Kingdom in 2001 might lead to consideration of value recovery options and even product design changes to optimize value recovery without creating collateral pollution.

The National Research Council in the United States made a case for materials flow analysis similar to the Mass Balance Movement. Both argued that better tools and data for analyzing material flows at the local, national, and global scales are sorely needed.[3] The European Union is leading the development of standardized methods for measuring material flows.[4,5]

4.2.3   Complex Processes with Multiple Materials

The preceding principles of mixing and separating using black boxes can be applied to a complex system with multiple materials by analyzing the system as a combination of several black boxes. The stepwise procedure can be applied to this system as well. In Example 4.11, two black boxes are employed in addition to the dewatering unit operation, which is a separation step.

EXAMPLE

4.11   **Problem**   Consider the system pictured in Figure 4.18 on the next page. A sludge of solids concentration $C_O = 4\%$ is to be thickened to a solids concentration $C_E = 10\%$ by using a centrifuge. Unfortunately, the centrifuge produces a sludge at 20% solids from the 4% feed sludge. In other words, it works too well. It has been decided to bypass some of the feed sludge flow and blend it later with the dewatered (20%) sludge so as to produce a sludge with exactly 10% solids concentration. The question is, then, how much sludge to bypass. The influent flow rate (Q_O) is 4 L/min at a solids concentration (C_O) of 4%. It is assumed here, and in the following problems, that the specific gravity of the sludge solids is 1.0 g/cm^3. That is, the solids have a density equal to that of water, which is usually a good assumption. The centrifuge produces a centrate (effluent stream of low solids concentration) with a solids concentration (C_C) of 0.1% and a cake (the high solids concentrated effluent stream) with a solids concentration (C_K) of 20%. Find the required flow rates.

Solution

1. Consider first the centrifuge as a black box separator (Figure 4.18B). A volume balance yields

Figure 4.18 Centrifugal dewatering of sludge. See Example 4.11.

$$\begin{bmatrix} \text{Volume} \\ \text{of sludge} \\ \text{ACCUMULATED} \end{bmatrix} = \begin{bmatrix} \text{Volume} \\ \text{of sludge} \\ \text{IN} \end{bmatrix} - \begin{bmatrix} \text{Volume} \\ \text{of sludge} \\ \text{OUT} \end{bmatrix}$$
$$+ \begin{bmatrix} \text{Volume} \\ \text{of sludge} \\ \text{PRODUCED} \end{bmatrix} - \begin{bmatrix} \text{Volume} \\ \text{of sludge} \\ \text{CONSUMED} \end{bmatrix}$$

Assuming steady state,

$$0 = Q_A - [Q_C + Q_K] + 0 - 0$$

Note that the volume includes the volume of the sludge solids and the volume of the surrounding liquid. Assuming steady state, a solids balance on the centrifuge gives

$$0 = [Q_A C_A] - [Q_K C_K + Q_C C_C] + 0 - 0$$
$$0 = Q_A\,(4\%) - Q_K\,(20\%) - Q_C\,(0.1\%)$$

Obviously, there are only two equations and three unknowns. For future use, solve both of these in terms of Q_A, as $Q_C = 0.804 Q_A$ and $Q_K = 0.196 Q_A$.

2. Next, consider the second junction, in which two streams are blended (Figure 4.18C). A volume balance assuming steady state yields

$$0 = [Q_B + Q_K] - [Q_E] + 0 - 0$$

and a solids mass balance is

$$0 = [Q_B C_B + Q_K C_K] - [Q_E C_E] + 0 - 0$$
$$0 = Q_B \, (4\%) + Q_K \, (20\%) - Q_E \, (10\%)$$

Substituting and solving for Q_K yields

$$Q_K = 0.6 Q_B$$

Above it was found that $Q_K = 0.196 Q_A$. Hence

$$0.6 Q_B = 0.196 Q_A \quad \text{or} \quad Q_B = 0.327 Q_A$$

3. Now consider the first separator box, (Figure 4.18D). A volume balance assuming steady state yields

$$0 = [Q_O] - [Q_B + Q_A] + 0 - 0$$

From above, $Q_B = 0.327 Q_A$. Substituting:

$$0 = 4 \text{ L/min} - 0.327 Q_A - Q_A$$

or

$$Q_A = 3 \text{ L/min}$$

and

$$Q_B = 0.327 \, (3 \text{ L/min}) = 1 \text{ L/min}$$

which is the answer to the problem. Further, from the centrifuge balance:

$$Q_C = 0.804 \, (3 \text{ L/min}) = 2.4 \text{ L/min}$$
$$Q_K = 0.196 \, (3 \text{ L/min}) = 0.6 \text{ L/min}$$

and from the blender box the balance is

$$0 = Q_K + Q_B - Q_E$$

or

$$Q_E = Q_B + Q_K = 1.6 \text{ L/min}$$

A check is also available. Using the entire system and assuming steady state, a volume balance in gpm gives

$$0 = Q_O - Q_C - Q_E$$
$$0 = 4.0 - 2.4 - 1.6 = 0 \qquad \text{check}$$

Most environmental engineering processes involve a series of unit operations, each intended to separate a specific material. Often the order in which these unit operations are employed can influence the cost and/or the efficiency of operation. This is shown by Example 4.12.

EXAMPLE

4.12

Problem Suppose seven tons of shredded waste plastics are classified into individual types of plastic. The four different plastics in the mix have the following quantities and densities:

Plastic	Symbol	Quantity (tonnes)	Density (g/cm³)
Polyvinylchloride	PVC	4	1.313
Polystyrene	PS	1	1.055
Polyethylene	PE	1	0.916
Polypropylene	PP	1	0.901

Determine the best method of separation.

Solution Although the differences are small, it still might be possible to use density as the code, and a float/sink apparatus can be used as a switch. By judiciously choosing a proper fluid density, a piece of plastic can be made to either sink or float.

In this case the easiest separation is the removal of the two heavy plastics (PVC and PS) from the two lighter ones (PE and PP). This can be done by using a fluid with a density of 1.0 g/mL, most likely plain water. Following this step, the two streams each must be separated further. The trickiest separation, the splitting of PE and PP, must be done with great care, and is made somewhat easier by already having removed the PVC and PS. The entire process train is shown in Figure 4.19.

But is this the best order of separation? Suppose instead of the process train in Figure 4.19, the PVC is removed first, resulting in the process shown in Figure 4.20. In the case of the first process train, the first separator receives all the flow, or 7 tonnes. The following two units receive 5 and 2 tonnes each, for a total of 14 tonnes handled. In the alternate processing system, the first separation device must handle 7 tonnes, the next one

Figure 4.19 A process train for the separation of four types of plastics by float/sink. See Example 4.12.

Figure 4.20   An alternative process train for the separation of plastics by float/sink.

3 tonnes (4 are removed) and the last one 2 tonnes, for a total of 12 tonnes. Thus, the second process train is to be preferred in terms of the least total quantity of material handled.[6]

In Example 4.12, various fluids must be used to make the separation of solids possible. If a fluid of a desired density is to be used, economics would dictate that it be separated from the plastic particles after the plastic particles have been separated from each other. This requires an extra step (perhaps screening) and invariably results in the loss of some of the fluid, which must be made up, thereby increasing the cost of the separation.

Following is a set of suggested general rules for the placement of unit operations in a process train.

1. Decide which material properties are to be exploited (e.g., magnetic vs. nonmagnetic or big vs. small). This becomes the code.
2. Decide how the code is to activate the switch.
3. If more than one material is to be separated, try to separate the easy one first.
4. If more than one material is to be separated, try to separate the one in greatest quantity first. (This rule may contradict rule 3, and engineering judgment will have to be exercised.)
5. If at all possible, do not add any materials to facilitate separation because this often involves the use of another separation step to recover the material.

4.3   MATERIAL BALANCES WITH REACTORS

To this point we have assumed not only that the system is in steady state but that there is no production or destruction of the material of interest. So we consider next a system wherein the material is being destroyed or produced in a reactor but in which the steady state assumption is maintained. That is, the system does not change with time so that, if the flows are sampled at any give moment, the results will always be the same.

EXAMPLE
4.13   **Problem**   In wastewater treatment, microorganisms are often used to convert dissolved organic compounds to more microorganisms, which are then removed from the flow stream by such processes as thickening. (See Example 4.11.) One such operation is known as the *activated sludge system* shown in Figure 4.21A on the next page.

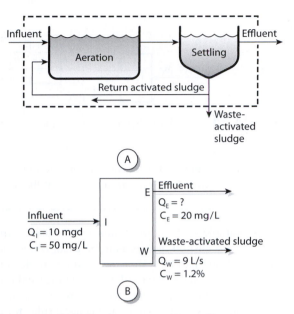

Figure 4.21 The activated sludge system. See Example 4.13.

Because the activated sludge system converts dissolved organics to suspended solids, it increases the solids concentration. These solids are the microorganisms. Unfortunately, the process is so efficient that some of these microorganisms (solids) must be wasted. The slurry of these solids removed from the activated sludge system is called *waste activated sludge*.

Suppose an activated sludge system has an influent (feed) of 440 L/s at a suspended solids concentration of 50 mg/L. The waste activated sludge flow rate is 9 L/s at a solids concentration of 1.2%. The effluent (discharge) has a solids concentration of 20 mg/L. What is the yield of waste activated sludge in pounds per day, or in other words, what is the rate of solids production in the system? Assume steady state.

Solution   With reference to Figure 4.21B, the entire system is treated as the black box, and the variables are added.

The first balance is in terms of the volume in mgd, producing

$$0 = 440 - [9 + Q_E] + 0 - 0$$
$$Q_E = 431 \text{ L/s}$$

The second balance is in terms of the suspended solids:

$$\begin{bmatrix} \text{Solids} \\ \text{ACCUMULATED} \end{bmatrix} = \begin{bmatrix} \text{Solids} \\ \text{IN} \end{bmatrix} - \begin{bmatrix} \text{Solids} \\ \text{OUT} \end{bmatrix}$$
$$+ \begin{bmatrix} \text{Solids} \\ \text{PRODUCED} \end{bmatrix} - \begin{bmatrix} \text{Solids} \\ \text{CONSUMED} \end{bmatrix}$$

$$0 = [Q_I C_I] - [Q_E C_E + Q_W C_W] + [X] - 0$$

where X is the rate at which solids are produced in the aeration basin. Because all the known terms are in units of L/s × mg/L, they must all be converted to kg/d, the unit of X.

$$0 = [Q_I C_I] - [Q_E C_E + Q_W C_W] + [X]$$
$$0 = [440 \times 50] - \big[[431 \times 20] + [9 \times 12{,}000]\big] + [X]$$
$$[X] = 94620 \text{ mg/s}$$
$$[X] \approx 8200 \text{ kg/d}$$

∎

While the activated sludge system produces suspended solids, other processes in wastewater treatment, such as anaerobic digestion, are designed to destroy suspended solids. The anaerobic digestion system can also be analyzed by using mass balances.

**EXAMPLE
4.14**

Problem   In wastewater treatment the digestion of sludge produces a useful gas (carbon dioxide and methane), and in the course of digestion the sludge becomes less odoriferous. Raw sludges can contain pathogenic microorganisms, and digestion to some extent disinfects the sludge and may make it easier to dewater. Typically, digestion is carried out in a two-stage system shown in Figure 4.22A on the next page. The first digester, called the *primary digester*, is the main reactor where the anaerobic (without oxygen) microorganisms convert the high-energy organic solids (called *volatile solids*) to methane gas, carbon dioxide, and lower-energy solids that make up the digested sludge. The second tank, the *secondary digester*, is used as both a gas storage tank and a solids separator, where the heavier solids settle to the bottom and a supernatant is drawn off the sludge and reintroduced to the treatment process.

Assuming steady state conditions, let the feed solids be 4%, of which 70% are volatile solids, and the feed flow rate be 0.1 m³/s. The gas, methane and carbon dioxide, contains no solids. The supernatant solids are 2%, with 50% being volatile, and the digested sludge is at a solids concentration of 6%, with 50% volatile. Find the flow rate of the supernatant and digested sludge.

Solution   First, redraw the system as shown in Figure 4.22B. There are several possible balances. The following three are convenient:

- volatile solids
- total solids
- sludge volume (solids plus liquid).

1. The mass balance in terms of volatile solids reads

$$\begin{bmatrix} \text{Volatile} \\ \text{solids} \\ \text{ACCUMULATED} \end{bmatrix} = \begin{bmatrix} \text{Volatile} \\ \text{solids} \\ \text{IN} \end{bmatrix} - \begin{bmatrix} \text{Volatile} \\ \text{solids} \\ \text{OUT} \end{bmatrix}$$
$$+ \begin{bmatrix} \text{Volatile} \\ \text{solids} \\ \text{PRODUCED} \end{bmatrix} - \begin{bmatrix} \text{Volatile} \\ \text{solids} \\ \text{CONSUMED} \end{bmatrix}$$

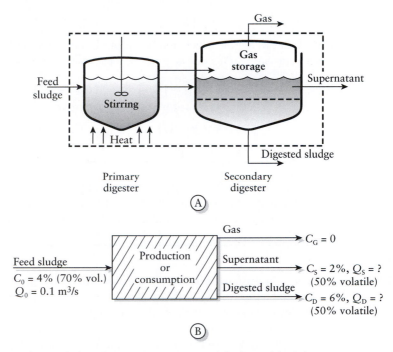

Figure 4.22   A sludge digestion system, simplified to a black box. See Example 4.14.

The volatile solids enter as the feed and may leave as the volatile solids in the gas, supernatant, and digested sludge. In addition, some volatile solids are consumed by the process.

Recall that in a mass balance the mass of the solids can be expressed as the product of the flow and concentrations. Because no volatile solids are produced and steady state is assumed:

$$0 = [Q_0 C_0] - [Q_G C_G + Q_S C_S + Q_D C_D] + 0 - [X]$$

where

X = consumption of volatile solids, g/s

$\hat{0}, \ \hat{G}, \ \hat{S}, \hat{D}$ = feed, gas, supernatant, and digested sludge, respectively.

The gas, of course, has no volatile solids, so $Q_G C_G = 0$. The equation becomes

$$0 = [(0.1 \ \text{m}^3/\text{s})(40{,}000 \ \text{mg/L})(0.7)] - [0 + Q_S(20{,}000 \ \text{mg/L})(0.5)$$
$$+ \ Q_D(60{,}000 \ \text{mg/L})(0.5)] + 0 - [X]$$

(Note that the conversion factors required to obtain grams per second cancel, so they are not included.)

Simplified:

$$0 = 2{,}800 - 10{,}000 Q_S - 30{,}000 Q_D - X$$

2. The balance in terms of total solids is

$$0 = [(0.1 \text{ m}^3/\text{s})(40{,}000 \text{ mg/L})] - [0 + Q_S(20{,}000 \text{ mg/L})$$
$$+ Q_D(60{,}000 \text{ mg/L})] + [0] - [X]$$

3. The liquid volume balance can be approximated as

$$0 = Q_0 - Q_S - Q_D$$

because the volume of liquid in the gas is very small compared to the other effluent streams. There is no production or consumption of the liquid, so the last terms are zero, and

$$0 = (0.1 \text{ m}^3/\text{s}) - Q_S - Q_D$$

We now have three equations and three unknowns.

1. $0 = 2{,}800 - 10{,}000Q_S - 30{,}000Q_D - X$
2. $0 = 4{,}000 - 20{,}000Q_S - 60{,}000Q_D - X$
3. $0 = 0.1 - Q_S - Q_D$

Solving these: $Q_D = 0.01 \text{ m}^3/\text{s}$, $Q_S = 0.09 \text{ m}^3/\text{s}$, and $X = 1600$ g of volatile solids removed per second. ∎

The next assumption to be considered is the steady state problem. This step requires some preparation in concepts, such as reactions, reactors, and other process dynamics, which are discussed in Chapter 5.

PROBLEMS

4.1 A pickle-packing plant produces and discharges a waste brine solution with a salinity of 13,000 mg/L NaCl, at a rate of 400 L/min, discharged into a stream with a flow rate above the discharge of 5 million L/d and a salinity of 20 mg/L. Below the discharge point is a prime sport fishing spot, and the fish are intolerant to salt concentrations above 200 mg/L.

a. What must the level of salt in the effluent be to reduce the level in the stream to 200 mg/L?
b. If the pickle-packing plant has to spend $8 million to reduce its salt concentration to this level, it cannot stay in business. Over 200 people will lose their jobs. All this because of a few people who like to fish in the stream? Write a letter to the editor expressing your views, pro or con, concerning the action of the state in requiring the pickle plant to clean up its effluent. Use your imagination.
c. Suppose now that the pickle plant is 3 km upstream from a brackish estuary, and asks permission to pipe its salty wastewater into the estuary where the salinity is such that the waste would actually *dilute* the estuary (the waste has a lower salt concentration than the estuary water). As plant manager, write a letter addressed to the director of the State Environmental Management Division

requesting a relaxation of the salinity efflu-ent standard for the pickle plant. Think about how you are going to frame your arguments.

4.2 Raw primary sludge at a solids concentration of 4% is mixed with waste activated sludge at a solids concentration of 0.5%. The flows are 80 L/min and 96 L/min, respectively. What is the resulting solids concentration? This mixture is next thickened to a solids concentration of 8% solids. What are the quantities (in liters per minute) of the thickened sludge and thickener overflow (water) produced? Assume perfect solids capture in the thickener.

4.3 A 3.78×10^7 L/d wastewater treatment plant's influent and effluent concentrations of several metals are as follows:

Concentration of the Metal

Metal	Influent (mg/L)	Effluent (mg/L)
Cd	0.012	0.003
Cr	0.32	0.27
Hg	0.070	0.065
Pb	2.42	1.26

The plant produces a dewatered sludge cake at a solids concentration of 22% (by weight). Plant records show that 45,000 kg of sludge (wet) are disposed of on a pasture per day. The state has restricted sludge disposal on land only to sludges that have metal concentrations less than the following:

Maximum Allowable Metal Concentration

Metal	(mg metal/kg dry sludge solids)
Cd	15
Cr	1000
Hg	10
Pb	1000

Is the sludge meeting the state standards?

a. Assume that the flow rate (mgd of sludge) is negligible compared to the flow rate of the influent. That is, assume $Q_{influent} = Q_{effluent}$, where $Q =$ flow rate in mgd. Check to see whether it was a valid assumption.

b. Do not make the preceding assumption, but assume that the density of the sludge solids is about that of water, so that 1 kg of sludge represents a volume of 1 L.

4.4 Two flasks contain 40% and 70% by volume formaldehyde, respectively. If 200 g out of the first flask and 150 g out of the second are mixed, what is the concentration (expressed as percent formaldehyde by volume) of the formaldehyde in the final mixture?

4.5 A textile plant discharges a waste that con-tains 20% dye from vats. The color intensity is too great, and the state has told the plant manager to reduce the color in the discharge. The plant chemist tells the manager that color would not be a problem if they could have no more than 8% dye in the wastewater. The plant manager decides that the least expensive way of doing this is to dilute the 20% waste stream with drinking water so as to produce an 8% dye waste. The 20% dye wastewater flow is 3600 L/min.

a. How much drinking water is necessary for the dilution?

b. What do you think of this method of pollution control?

c. Suppose the plant manager dilutes his waste and the state regulatory personnel assume that the plant is actually removing the dye before discharge. The plant manager, because he has run a profitable operation, is promoted to corporate headquarters. One day he is asked to prepare a presentation for the corporate board (the big cheeses) on how he was able to save so much money on wastewater treatment. Develop a short play for this meeting, starting with the pre-sentation by the former plant manger. He will, of course, try to convince the board that he did the right thing. What will be the board's reaction? Include in your script the company president, the treasurer, the legal counsel, and any other characters you want to invent.

4.6 A capillary sludge drying system operates at a feed rate of 200 kg/h and accepts a sludge with a solids content of 45% (55% water). The dried sludge is 95% solids (5% water), and the liquid stream contains 8% solids. What is the quantity of dried sludge, and what is the liquid and solids flow in the liquid stream?

4.7 A solid waste processing plant has two classifiers that produce a refuse-derived fuel (RFD) from a mixture of organic (A) and inorganic (B) refuse. A portion of the plant schematic and the known flow rates are shown in Figure 4.23.

a. What is the flow of A and B from Classifier I to Classifier II [Q_{A2} and Q_{B2}]
b. What is the composition of the Classifier II exit stream (Q_{A3} and Q_{B3})?
c. What is the purity of the RFD and the recovery of component A?

4.8 A separator accepts waste oil at 70% oil and 30% water by weight. The top product stream is pure oil while the bottom underflow contains 10% oil. If a flow of 80 L/min is fed to the tank, how much oil is recovered?

4.9 The Mother Goose Jam Factory makes jam by combining black currants and sugar at a weight ratio of 45:55. This mixture is heated to evaporate the water until the final jam contains one-third water. If the black currants originally contain 80% water, how many kilograms of berries are needed to make 1 kg of jam?

4.10 The flow diagram in Figure 4.24 on the next page shows the material flow in a heat-recovery incinerator.

a. How much fly ash will be emitted out of the stack (flue) per tonnes of refuse burned?
b. What is the concentration of the particulates (fly ash) in the stack, expressed as $\mu g/m^3$ (μg of fly ash per m^3 of flue gases emitted from the stack)? Assume a density of 1 tonnes/500 m^3.

4.11 An electrostatic precipitator for a coal-fired power plant has an overall particulate recovery of 95%. (It removes 95% of all the flue gas particulates coming to it.) The company engineer decides that this is too good, that it is not necessary to be quite this efficient, and proposed that part of the flue gas be bypassed around the electrostatic precipitator so that the recovery of fly ash (particulates) from the flue gas would be only 85%.

a. What fraction of the flue gas stream would need to be bypassed?
b. What do you think the engineer means by "too good?" Explain her thought processes. Can any pollution control device ever be

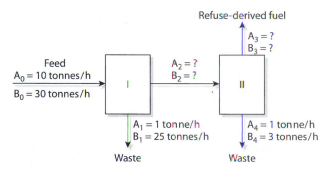

Figure 4.23 Air classifiers producing a refuse-derived fuel. See Problem 4.7.

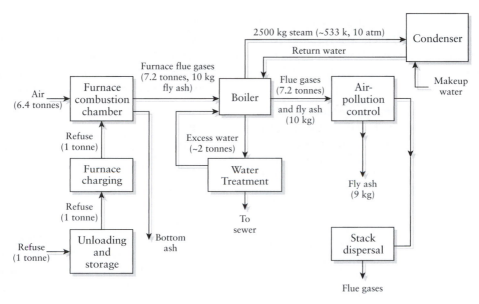

Figure 4.24 Material flow in a solid waste incinerator. See Problem 4.10.

"too good?" If so, under what circumstances? Write a one-page essay entitled "Can Pollution Control Ever Be Too Effective?" Use examples to argue your case.

4.12 Two air classifier manufacturers report the following performance for their units:

Manufacturer A: Recovery of organics = 80%
                 Recovery of inorganics = 80%
Manufacturer B: Recovery of organics = 60%
                 Purity of the extract = 95%

Assume the feed consists of 80% organics. Which unit is better? *Note*: The air classifier is supposed to separate organics from inorganics.

4.13 A package water treatment plant consists of filtration and settling, both designed to remove solids, but solids of different sizes. Suppose the solids removals of the two processes are as follows:

	Small Solids	**Large Solids**
Filter	90%	98%
Settling tank	15%	75%

If the two types of solids are each 50% by weight, would it be better to put the filter before the settling tank in the process train? Why?

4.14 A cyclone used for removing fly ash from a stack in a coal-fired power plant has the following recoveries of the various size particulates:

Size (μm)	Fraction of Total, by Weight	Recovery
0–5	0.60	5
5–10	0.18	45
10–50	0.12	82
50–100	0.10	97

What is the overall recovery of fly ash particulates?

4.15 In the control of acid rain, power plant sulfur emissions can be controlled by scrubbing and spraying the flue gases with a lime slurry, $Ca(OH)_2$. The calcium reacts with the sulfur dioxide to form calcium sulfate, $CaSO_4$, a white powder suspended in water. The volume of this slurry is quite large, so it has to be thickened. Suppose a power plant produces 27 metric tonnes of $CaSO_4$ per day, and that this is suspended in 108 metric tons of water. The intention is to thicken this to a solids concentration of 40% solids.

a. What is the solids concentration of the $CaSO_4$ slurry when it is produced in the scrubber?
b. How much water (in tonnes per day) will be produced as effluent in the thickening operation?
c. Acid rain problems can be reduced by lowering SO_2 emissions into the atmosphere. This can occur by pollution control, such as lime slurry scrubbing, or by reducing the amount of electricity produced and used. What responsibility do individuals have, if any, not to waste electricity? Argue both sides of the question in a two-page paper.

4.16 A stream, (Figure 4.25) flowing at 150 L/s and 20 mg/L suspended solids, receives wastewater from three separate sources:

Source	Quantity (L/s)	Solids Concentration (mg/L)
A	100	200
B	300	50
C	50	200

What are the flow and suspended solids concentration downstream at the sampling point?

4.17 An industrial flow of 12 L/min has a solids concentration of 80 mg/L. A solids removal process extracts 960 mg/min of the solids from the flow without affecting the liquid flow. What is the recovery (in percent)?

4.18 A manufacturer of beef sticks produces a wastewater flow of 2000 m^3/d containing 120,000 mg/L salt. It is discharging into a river with a flow of 34,000 m^3/d and a salt concentration of 50 mg/L. The regulatory agency has told the company to maintain a salt level of no greater than 250 mg/L downstream of the discharge.

a. What recovery of salt must be accomplished in the plant before the waste is discharged?
b. What could they do with the recovered salt? Suggest alternatives.

4.19 A community has a persistent air pollution problem called an inversion (discussed in Chapter 12) that creates a mixing depth of only about 500 m above the community. The town's area is about 9 km by 12 km. A constant emission of particulates of 70 $kg/m^2/d$ enters the atmosphere.

a. If there is no wind during one day (24 h) and if the pollutants are not removed in any way, what will the *lowest* concentration level of particles in the air above the town be at the end of that period?
b. If one of the major sources of particulates is wood burning stoves and fireplaces, should the government have the power to prohibit the use of wood stoves and fireplaces? Why or why not?
c. How much should government restrict our lifestyle because of pollution controls? Why does the government seem to be doing more of this now than just a few years ago?

Figure 4.25 Stream receiving three discharges. See Problem 4.16.

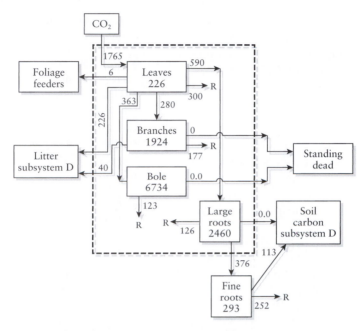

Figure 4.26 A forest ecosystem. See Problem 4.20.

Where do you think it's all headed, and why is it happening? Is there anything that we can do about it? Respond to these questions in a one-page paper.

4.20 Figure 4.26 shows a diagram of organic matter in a forest ecosystem in Tennessee. All compartmented (boxed) numbers are in grams per square meter of the forest floor, and all transfers are in grams per square meter per year. This system is not in equilibrium. Calculate the change ($g/m^2/yr$) in leaves, branches, bole, large roots, and fine roots. *R* is loss by decay.

4.21 The Wrightsville Beach, North Carolina, water treatment plant produces a distilled water of zero salinity. To reduce the cost of water treatment and supply, the city mixes this desalinated water with untreated groundwater that has a salinity of 500 mg/L. If the desired drinking water is to have a salinity not exceeding 200 mg/L, how much of each water—distilled water and untreated groundwater—is needed to produce a finished water flow of 4400 L/s?

4.22 A cyclone used to control the emissions from a furniture manufacturer is expected to be 80% efficient (80% recovery). The hopper under the cyclone holds 1 tonnes of wood dust. If the air flow to the cyclone holds 1.2 tonnes per day of wood dust, how often must the hopper be emptied?

4.23 An industry has been fined heavily by the EPA because it has been polluting a stream with a substance called "gloop." It is presently discharging 200 L/min of effluent at a gloop concentration of 100 mg/L. The stream flowing past the plant into which the effluent is discharging has an upstream flow rate of 3800 L/min and has no gloop in it. The EPA has told the plant that it must reduce its effluent gloop concentration to below 20 mg/L. The plant engineer decides that the cheapest way to meet the EPA requirement is to divert part of the stream into the plant and dilute the effluent so as to lower the gloop concentration to the required level.

a. What is the present gloop concentration downstream from the plant discharge?
b. How much stream water would the plant engineer need to achieve the required gloop concentration in the plant effluent?
c. What will be the gloop concentration in the stream downstream from the plant if the plant engineer's scheme is put into operation?

4.24 A 1000-tonnes/d ore-processing plant processes an ore that contains 20% mineral. The plant recovers 100 tonnes/d of the mineral. This is not a very effective operation, and some of the mineral is still in the waste ore (the tailings). These tailings are sent to a secondary processing plant that is able to recover 50% of the mineral in the tailings. The recovered mineral from the secondary recovery plant is sent to the main plant, where it is combined with the other recovered mineral, and then all of it is shipped to the user.

a. How much mineral is shipped to the user?
b. How much total tailings are to be wasted?
c. How much mineral is wasted in the tailings?

4.25 An aluminum foundry emits particulates that are to be controlled by first a cyclone, followed by a bag filter. Both of these are particulate removal devices and are fully described in Chapter 13. For now, you can consider them as black boxes that remove particulates (solid particles suspended in air). These particulates can be described by their size, measured as their diameter in micrometers (μm). The effectiveness of the cyclone and the bag filter is as follows:

Particle size (μm)	0–10	10–20	20–40	>40
Percentage by weight of each size	15	30	40	15
Percentage recovery of cyclone for each size of particles	20	50	85	95
Percentage recovery of bag filter for each size of particles	90	92	95	100

a. What is the overall percentage of recovery for the cyclone? (*Hint*: Assume 100 tonnes/d are being treated by the cyclone.)
b. What would be the overall percentage of recovery for the bag filter if this was the only treatment device?
c. What is the overall percentage of recovery by this system? Remember that the cyclone is in front of the bag filter.
d. The system has the cyclone in front of the bag filter. Why would you not have placed the bag filter in front of the cyclone?
e. Assuming all the particles were perfect spheres, would a cubic meter of say 5 μm spherical particulates weigh more or less than a cubic meter of perfectly spherical 40 μm particles? Show the calculations on which you base your answer.

4.26 Newsprint appears to be an especially good candidate for recycling. Yet the price of old newspapers has historically not been very high. One reason for the low price of old newsprint is the economics of converting a paper mill from processing trees to processing old newsprint. This is illustrated by the following situation.

The *Star Tribune* and the *St. Paul Pioneer Press and Dispatch* represented 75% of the newspaper circulation in Minnesota. The *Star Tribune* had a circulation of 2,973,100/d and weighed about 0.5 kg/issue while the *St. Paul Pioneer Press and Dispatch* had a circulation of 1,229,500/d and weighed 0.4 kg/issue.

Of course, not all of what the consumer receives is recyclable newsprint; 17% of the weight is nonnewsprint material for glossy advertisements and magazines, and 1% is ink. Also, 5% of what is printed is overissues and never reaches the consumer, while 3% of the newsprint delivered to the press becomes newsroom scrap and is not printed.

a. Find the mass of newspaper that reached consumers in Minnesota each day. How much of this was recyclable newsprint, including ink but excluding glossy pages?

b. If 65% of consumed newsprint and 100% of overissue newsprint and pressroom scrap was collected for recycling, what was the mass rate of recycled material sent to the pulp mills? Recall that overissue is 5% of circulation and pressroom scrap is 3% of the newsprint required for printing.

c. Newsprint must be deinked before it can be reused. Assume that the deinking process removes all of the ink, but in doing so 15% of the paper is lost as waste. Calculate the mass rate of dry recycled pulp produced.

d. If it takes 1.5 cords of wood to create 1 tonnes of virgin pulp, and there are an average of 21.3 cords per km^2 of timberland, how many km^2 of trees are saved daily by this level of recycling?

e. What would the cost of recycled newsprint have to be in order to justify the conversion of a 750-tonnes/d pulp mill to recycled fiber? Base your calculations on a virgin material cost of $90/tonnes, a conversion cost of $100 million and a 10% return on investment. Assume that the cost of producing pulp from the two feedstocks is identical, so that the only economic incentive for converting to recycled newsprint is the difference between feedstock costs.[7]

4.27 According to statements of the Draeger Works in Luebeck, in the gassing of the whole population in a city only 50% of the evaporated poison gas is effective. The atmosphere must be poisoned up to a height of 20 m at a concentration of 45 mg/m^3. How much phosgene is needed to poison a city of 50,000 inhabitants who live in an area of 4 km^2? How much phosgene would the population inhale with the air they breathe in 10 min without protection against gas, if one person uses 30 L of breathing air per minute? Compare this quantity with the quantity of the poison gas used.[8]

END NOTES

1. Based on Kohn, R. E. 1978. *A Linear Programming Model for Air Pollution Control*. Cambridge, MA: MIT Press.

2. Accessed at http://www.massbalance.org/index.php on 1 December, 2008.

3. National Research Council. 2004. *Materials Count: The Case for Material Flows Analyses*. Washington, DC: National Academies Press. Accessed at http://www.nap.edu/catalog-php?record_id=10705 on 1 December, 2008.

4. Luxembourg Office for Official Publications of the European Communities, 2001. Economy-wide Material Flow Accounts and Derived Indicators: A Methodological Guide. Accessed at http://epp.eurostat.ec.europa.eu/cache/ITY_OFF PUB/KS-34-00-536/EN/KS-34-00-536-EN.pdf on 30 December, 2008.

5. Organisation for Economic Co-operation and Development reports on Measuring Material Flows and Resource Productivity. Accessed at http://www.oecd.org/document/47/0,3343,en_ 1571361_40266644_40464047_1_1_1_1,00. html on 30 December, 2009.

6. This example is adapted from Berthourex, B. M., and D. F. Rudd. 1977. *Strategy for Pollution Control*. New York: John Wiley & Sons.

7. This problem appears in Allen, David T., N. Bakshani, and Kirsten Sinclair Rosselot. 1992. *Pollution Prevention: Homework and Design Problems for Engineering Curricula*. New York: American Institute of Chemical Engineers and other societies. Used with permission.

8. This problem first appeared in *Human Behavior in the Concentration Camp* by Elie A. Cohen, translated by M. H. Braaksma. Copyright 1953 by W. W. Norton & Company, Inc., renewed (c) 1981 by M. H. Braaksma. Used by permission of W. W. Norton & Company, Inc.

Reactions

Burra Monster Mine, Australia

Lava flow

Environmental engineers must keep in mind Newton's third law of motion and determine which reactions are possible, which are probable, and whether they can be controlled.

Many reactions, whether in nature or in human-created environments, are predicable. Engineers and scientists first strive to understand these reactions and then create mathematical models to describe the behavior of the phenomena. With these models, they can try to predict what would happen if . . . .

Such models can be simple or highly complex. If a pipeline under pressure discharges a certain flow rate and if the pressure is doubled, it is possible to predict the new flow rate fairly accurately. Such simple hydraulic models are the bread and butter of civil engineering.

Any large and complex model, such as ozone depletion or photochemical smog formation, are more likely to have errors. What has to be evaluated are the consequences of not believing the model contrasted with the degree of uncertainty. While the ozone depletion model cannot be proven conclusively, it is mathematically and chemically reasonable and logical and is, therefore, most probably correct. Ignoring the potential for global disaster predicted by such models is simply foolhardy. [This concept is the basis for the precautionary principle, which basically says that it's better to prevent a problem (be safe) than to clean up a problem (be sorry).]

Because the processes are complex, global models are intricate and highly interconnected. Fortunately, these powerful mathematical models have humble beginnings. It is assumed that some quantity (mass or volume) changes with time and that the quantity of a component can be predicted by using simple rate equations and material balances.

In the previous chapter material flow is analyzed as a steady state operation. Time is not a variable. In this chapter we consider the case wherein material concentrations change with time. For completely mixed-batch systems, a general mathematical expression describing a rate at which the mass or volume of some material A is changing with time t is

$$\frac{dA}{dt} = r$$

where $r =$ reaction rate. *Zero-order reactions* are defined as those in which r is a constant (k) so that

$$\frac{dA}{dt} = k \tag{5.1}$$

Note that the unit for the reaction rate constant k in zero-order reactions is mass/time, such as kg/s.

First-order reactions are defined as those wherein the change of the component A is proportional to the quantity of the component itself so that

$$r = kA$$

and

$$\frac{dA}{dt} = kA \tag{5.2}$$

Note that the unit of the reaction rate constant k in first-order reactions is time^{-1}, such as d^{-1}.

Second-order reactions are ones wherein the change is proportional to the square of the component, or

$$r = kA^2$$

and

$$\frac{dA}{dt} = kA^2 \tag{5.3}$$

Note that in second-order reactions, the reaction rate constant k has the unit of (time $\times$ mass)$^{-1}$.

In environmental engineering applications, Equations 5.1, 5.2, and 5.3 are usually written in terms of the concentration, so for zero-order reactions,

$$\frac{dC}{dt} = k \tag{5.4}$$

for first-order reactions,

$$\frac{dC}{dt} = kC \tag{5.5}$$

and for second-order reactions,

$$\frac{dC}{dt} = kC^2 \tag{5.6}$$

5.1   ZERO-ORDER REACTIONS

Many changes occur in nature at a constant rate. Consider a simple example wherein a bucket is being filled from a garden hose. The volume of water in the bucket is changing with time, and this change is constant (assuming no one is opening or closing the faucet). If at zero time the bucket has 2 L of water in it, at 2 s it has 3 L, at 4 s it has 4 L, and so on, the change in the volume of water in the bucket is constant, at a rate of

$$1 \text{ L/2 s} = 0.5 \text{ L/s}$$

This can be expressed mathematically as a *zero-order reaction*:

$$\frac{dA}{dt} = r = k$$

where A is the volume of water in the bucket and k is the rate constant. Defining $A = A_0$ at time $t = t_0$, and integrating:

$$\int_{A_0}^{A} dA = k \int_{0}^{t} dt$$

$$A - A_0 = -kt \tag{5.7}$$

Figure 5.1   Plot of a zero-order reaction in which the concentration is decreasing.

This equation is, of course, how the water-in-the-bucket problem is solved. Substitute $A_0 = 2$ L at $t = 0$ and $A = 3$ L at $t = 2$ s:

$$3 = 2 + k(2)$$
$$k = 0.5 \text{ L/s}$$

When the constituent of interest, A, is in terms of mass, then the units of the rate constant k are mass/time, such as kg/s. When the constituent of interest is a concentration, C, which it very often is, the rate constant in a zero-order reaction has units of mass/volume/time, or mg/L/s if C is in milligrams per liter and t is in seconds.

The integrated form of the zero-order reaction when the concentration is *increasing* is

$$C = C_0 + kt \qquad\qquad (5.8)$$

and if the concentration is *decreasing*, the equation is

$$C = C_0 - kt \qquad\qquad (5.9)$$

This equation can be plotted as shown in Figure 5.1 if the material is being destroyed or consumed so that the concentration is decreasing. If, on the other hand, C is being produced and is increasing, the slope is positive.

EXAMPLE

5.1

Problem   An anteater finds an anthill and starts eating. The ants are so plentiful that all he has to do is flick out his tongue and gobble them up at a rate of 200/min. How long will it take to have a concentration of 1000 ants in the anteater?

Solution

$$C = \text{concentration of ants in the anteater at any time } t, \text{ ants/anteater}$$
$$C_0 = \text{initial concentration of ants at time } t = 0, \text{ ants/anteater}$$
$$k = \text{reaction rate, the number of ants consumed per minute}$$
$$= 200 \text{ ants/anteater/mm } (\textit{Note}: k \text{ is positive because the concentration}$$
$$\text{is increasing.})$$

According to Equation 5.8:

$$C = C_0 + kt$$

$$1000 = 0 + 200(t)$$

$$t = 5 \text{ min}$$

5.2 FIRST-ORDER REACTIONS

The *first-order reaction* of a material being consumed, or destroyed, can be expressed as

$$\frac{dA}{dt} = r = -kA$$

This equation can be integrated between A_0 and A, and $t = 0$ and t:

$$\int_{A_0}^{A} \frac{dA}{A} = -k \int_{0}^{t} dt$$

$$\ln \frac{A}{A_0} = -kt \tag{5.10}$$

$$\text{or } \frac{A}{A_0} = e^{-kt}$$

$$\text{or } \ln A - \ln A_0 = -kt$$

This logarithm is the so-called natural, or base e, logarithm. It is also sometimes useful to use the common, or base 10, logarithm. The conversion from natural (base e) to common (base 10) logarithms can be accomplished as follows:

$$\frac{\log A}{\ln A} = \frac{k't}{kt} = 0.434 = \frac{k'}{k}$$

For example, suppose $A = 10$; then

$$\frac{\log 10}{\ln 10} = \frac{1}{2.302} = 0.434 = \frac{k'}{k}$$

Thus:

$$k' = (0.434)k$$

$$k_{\text{base } 10} = (0.434)k_{\text{base } e}$$

The advantage of base 10 logs is that they can be conveniently plotted on ordinary semilog paper. Figure 5.2 shows that plotting $\log C$ versus t yields a straight line with a slope 0.434 k. The intercept is $\log C_0$. (Don't be confused between A and C. The mass of some material is A, and in a given volume V, C is the mass A divided by the volume V. If

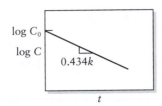

Figure 5.2 Plot of a first-order reaction in which the concentration is decreasing.

the volume is constant, the term drops out. Thus, the preceding argument could have been in terms of C as well as A.)

The slope, of course, on a semilog paper is found by recognizing that

$$\text{Slope} = \frac{\Delta y}{\Delta x}$$

where Δy is the increment on the y axis (the ordinate) corresponding to the increment Δx on the x axis (the abscissa). Choosing Δy as one cycle on the log scale, such as $(\log 10 - \log 1)$, results in the numerator being 1 and the slope being

$$\text{Slope} = \frac{1}{\Delta x_{\text{for one cycle}}}$$

Note again that the mass A and the concentration C are interchangeable in the equations if the volume is constant.

E X A M P L E

5.2

Problem An owl eats frogs as a delicacy, and his intake of frogs is directly dependent on how many frogs are available. There are 200 frogs in the pond, and the rate constant is 0.1 d^{-1}. How many frogs are left at the end of 10 d?

Solution Because the rate is a function of the concentration, this can be described as a first-order reaction:

$$C = \text{number of frogs in the pond at any time } t$$
$$C_0 = \text{initial frog population, 200 frogs per pond}$$
$$k = \text{rate constant, 0.1 } d^{-1}.$$

Using Equation 5.10:

$$\ln \frac{C}{C_0} = -kt$$

$$\ln C - \ln(200) = -0.1(10)$$

$$\ln C = 4.3$$

$$C = 73 \text{ frogs per pond}$$

FOCUS ON

The Exxon Valdez

The 53 million gallons of crude oil were loaded onto the *Exxon Valdez* during the day on March 23, 1989. The tanker left the Alyeska Pipeline Terminal at 9:12 and headed into Prince William Sound under the direction of Pilot William Murphy. Everything seemed normal, and Captain Joseph Hazelwood left the bridge, even though company policy requires two officers to be on the bridge during the passage through the narrows. After the ship cleared the narrows, Pilot Murphy disembarked, leaving Captain Hazelwood as the only officer on the bridge. There apparently were no lookouts posted on the bow as required.

For reasons of convenience, Captain Hazelwood requested permission to travel in the inbound lane as opposed to the outbound lane and

was given permission to do so because there was no inbound traffic. The *Exxon Valdez* started to swing farther to the left. The speed and heading were not reported to the Valdez traffic center as required.

Putting the ship on autopilot, Captain Hazelwood increased the engine speed to normal ocean travel and left the bridge, leaving Third Mate Gregory Cousins in charge. Cousins was supposed to have been relieved at midnight, but he decided to let his friend, who had been busy with the loading during the day, sleep, and he stayed on the bridge, again violating company policy. Later testimony indicated that Cousins had been awake for over 18 hr prior to the accident, and there is evidence that his fatigue contributed directly to the accident.

At about midnight, Cousins plotted the course for turning the *Exxon Valdez* into the inbound lane. About the same time, a lookout reported seeing the lights on Bright Reef off the **starboard** bow (starboard is right, port is left) when the light should have been seen off the **port** bow. The position of the light indicated that the *Exxon Valdez* crossed the inbound lane and was heading directly onto the reef. Cousins gave the right rudder command and took the ship off autopilot, calling Captain Hazelwood to inform him of the change. When the ship did not turn rapidly enough, Cousins ordered a further turn to starboard. But it was too late. At 12:14 A.M., the ship floundered on the rocks and ruptured the hull, eventually spilling 11 million gallons (41.6 million L) of oil.

Captain Hazelwood tried vainly to power the ship off the rocks but to no avail. Stress and stability analysis indicated that the ship had lost enough oil and ballast to make it unstable if it had gotten itself off the rocks. Fortunately their efforts were unsuccessful, and the crew settled in to await assistance, which was slow in coming. Although Exxon eventually sent massive quantities of

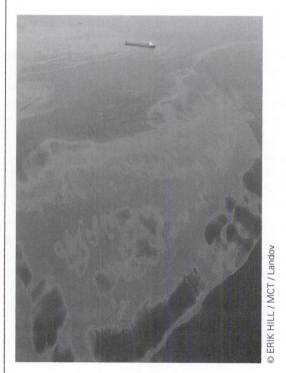

© ERIK HILL / MCT / Landov

Figure 5.3   The *Exxon Valdez* and the spill.

machinery and materials to help in the cleanup, there was little done during the first crucial days.

The effects on the ecosystem, wildlife and plants, were severe and extensive. The oil fouled 1000 miles (1600 km) of beach, and eight years later much of the shoreline in Prince William Sound was still contaminated. Although oil is biodegradable, the low temperature and lack of aquatic biomass in the fragile ecosystem prevented rapid biodegradation from taking place. Fourteen years after the spill, about 50% of the oil was thought to have been either biodegraded or photolysed.

Recovery operations accounted for about 14%, and about 13% had sunk to the sea floor. Over 2% of the oil was still on the beaches, identifiable as either oily slime or oil balls, and still adversely affecting the ecosystem.

Source: Reprinted from *Controlling Environmental Pollution*, Vesilind, P. A. and DiStefano, T. D., 2006. Lancaster, PA: DEStech Publications, Inc. Reproduced with permission of DEStech Publications, Inc.

As described in Chapter 8, owls eating frogs is a part of an ecosystem—frogs eat daphnae, daphnae eat algae, and algae grow in the water. Life in such an ecosystem is in balance, allowing all species to survive. Only people seem to be capable of consciously and purposefully destroying such an ecological balance. These considerations prompted a contemporary Norwegian philosopher, Arne Naess, to try to create a new environmental philosophy that he called "Deep Ecology." He proposed that environmental ethics should be based on two fundamental values: self-realization and biocentric equality. These values are intuited and cannot be rationally justified.

Self-realization is the recognition of oneself as a member of the greater universe and not just as a single individual or even as a member of a restricted community. Self-realization can be achieved, according to Naess, by reflection and contemplation. The tenets of deep ecology are not inherently tied to any particular religious belief, although, like Marxism or any other set of beliefs and values, it may function as a religion for the believer.

> Most people in deep ecology have had the feeling—usually, but not always in nature—that they are connected with something greater than their ego .... Insofar as these deep feelings are religious, deep ecology has a religious component ... [a] fundamental intuition that everyone must cultivate if he or she is to have a life based on value and not function like a computer.[1]

The second tenet of deep ecology is *biocentric equality*. According to Naess, this follows from self-realization, in that once we have understood ourselves to be at one with other creatures and places in the world, we cannot regard ourselves as superior. Everything has an equal right to flourish, and humans are not special. We must eat and use other creatures in nature to survive, of course, but we must not exceed the limits of our "vital" needs. The collecting of material wealth, or goods above the vital necessities, is therefore unethical to a deep ecologist.

If one accepts these two fundamental propositions, then one must agree with the following three corollaries.[2]

1. All life, human and nonhuman, has value in itself, independent of purpose, and humans have no right to reduce its richness and diversity except for vital needs.

2. Humans at present are far too numerous and intrusive with respect to other life forms on earth, with disastrous consequences for all, and must achieve a substantial decrease in population to permit the flourishing of both human and nonhuman life.
3. To achieve this requisite balance, significant change in human economic, technical, and ideological structures must be made, stressing not bigness, growth, and higher standards of living but sustainable societies emphasizing the (nonmaterial) quality of life.

(The latter is, of course, incorporated into sustainable, or green, design.)

Deep ecologists oppose the capitalist/industrial system that seeks to create what they view as unnecessary and detrimental wealth. They disdain the idea of stewardship because it implies human decision making and human intervention in the workings of natural environments. They seek to regain what they view as the sympathetic and supportive relationship between people and nature held by people such as the Native Americans prior to the European invasion. To a deep ecologist, wilderness has a special value and must be preserved for the psychological benefit of humans as well as a place for other species to survive unhindered by human activity. Deep ecologists seek a sense of place in their lives, a part of Earth that they can call home.

There is, of course, no reason why someone who wishes to adopt deep ecology as a personal choice, to adopt it as an ideal, should not do so. Indeed, there may be deep ecologists who consider that the purpose of ethics is to develop a virtuous character. In fact, some philosophers have criticized deep ecology precisely because of what they consider to be its excessive focus on the spiritual development of the individual. But most of the tenets of deep ecology require large-scale changes involving the whole population.

These ideas will, obviously, rattle a lot of cages in the capitalist and political mainstream. The precepts of deep ecology offer a challenge to mainstream ways of doing business. As a result, it has sometimes been characterized as non-humanist because it does not consider humans to be deserving of special benefits and, in fact, indicts humans as the exploiters of nature. Especially galling to critics is the call by deep ecologists to depopulate the world, an objective that has been called fascist[3] and dehumanizing. There is no doubt that, if we are to have a less adverse effect on nature, it would be necessary first to reduce the human population by a significant portion, which would obviously take some time to do through natural attrition. Some deep ecologists have suggested a population reduction of 90% is required. In all likelihood there is little popular support for this view. Critics argue that the acceptance of deep ecology as a global philosophy would revert civilization to the hunter–gatherer societies.

Critics of deep ecology also have argued that it is unacceptably elitist; that is, although its vision of a world largely reverted to wilderness and containing far fewer people may appeal to a small number of the relatively rich and highly educated, it has nothing to offer to the great masses of poor humans trying to survive. And because deep ecology requires voluntary acceptance of a lower material quality of life, it is unlikely that its basic tenets—self-realization and biocentric equality—will be widely and voluntarily accepted.

But back to reactions.

5.3 SECOND-ORDER AND NONINTEGER-ORDER REACTIONS

The *second-order reaction* is defined as

$$\frac{dA}{dt} = r = -kA^2$$

Integrated, it is

$$\int_{A_0}^{A} \frac{dA}{A^2} = -k \int_{0}^{t} dt$$

$$\frac{1}{A} \Big]_{A_0}^{A} = kt$$

$$\frac{1}{A} - \frac{1}{A_0} = kt$$

which plots as a straight line as shown in Figure 5.4.

The *noninteger-order* (any number) *reaction* is defined as

$$\frac{dA}{dt} = r = -kA^n$$

where n is any number. Integrated, it is

$$\left(\frac{A}{A_0}\right)^{1-n} - 1 = \frac{(n-1)\,kt}{A_0^{(1-n)}}$$

These reactions are not as common in environmental engineering.

Figure 5.4 Plot of a second-order reaction in which the concentration is decreasing.

5.4   HALF-LIFE AND DOUBLING TIME

Half-life is defined as the time required to convert one-half of a component. At $t = t_{1/2}$, A is 50% of A_0. By substituting $[A]/[A]_0 = 0.5$ into the preceding expressions, the half-lives for various reaction orders are determined.

$$\text{First order:} \qquad t_{1/2} = \frac{\ln 2}{k} = \frac{0.693}{k} \qquad\qquad (5.11)$$

$$\text{Second order:} \qquad t_{1/2} = \frac{1}{k A_0}$$

$$\text{Noninteger order:} \qquad t_{1/2} = \frac{[(1/2)^{1-n} - 1] \cdot A_0^{1-n}}{(n-1)k}$$

The most common use of half-life is for radioisotopes, which are discussed more in Chapter 15.

E X A M P L E

5.3

Problem   Strontium 90 (Sr^{90}) is a radioactive nuclide found in water and many foods. It has a half-life of 28 years and decays (as do all other radioactive nuclides) as a first-order reaction. Suppose the concentration of Sr^{90} in a water sample is to be reduced by 99.9%. How long would this take?

Solution   Let $A_0 = 1$; then $A = 0.001$ after a 99.9% reduction. Using Equation 5.11, $t_{1/2} = 0.693/k$, and setting $t_{1/2} = 28$ yr, $k = 0.02475$ yr^{-1}. Using Equation 5.10:

$$\ln\left(\frac{0.001}{1}\right) = -0.02475t$$

$$t = \frac{-6.907}{-0.02475} = 279 \text{ yr}$$

Of course, the *doubling time* is the amount of time required to double the amount of a component. At $t = t_2$, A is twice as much as A_0. By substituting $[A]/[A]_0 = 2$ into the expressions, the doubling times for the various reaction orders can be calculated.

5.5   CONSECUTIVE REACTIONS

Finally, some reactions are *consecutive reactions* such that

$$A \rightarrow B \rightarrow C \rightarrow \cdots$$

FOCUS ON

Adverse Effects of Polluted Air on Things

One of the most destructive effects of air pollution is the deterioration of natural and human-made matter from the reactions of the pollutants themselves or the products of their reactions with other airborne substances. Athens, Greece, for example, had one of the highest levels of photochemical smog, given their plethora of sunlight and incredible number of unregulated automobiles. As a result, the buildings of the Acropolis were rapidly deteriorating, as were other remnants of the ancient Greek civilization (Figure 5.5). Ironically, the theft of many of the most valuable pieces by the British around the turn of the century resulted in the saving of these treasures. The clean, controlled air in the British Museum was far better than the putrid atmosphere in Athens. But does this make the theft any less reprehensible?

Do we, in fact, have a duty to preserve things? What is worth preserving? And who decides? The things that may or may not be worth preserving are either constructed by humans, such as buildings, or are natural things and places, such as the Grand Canyon or the Gettysburg Battlefield. We may decide to preserve such landmarks for one of two reasons: to keep the structure or thing in existence purely for its own sake or to protect it against destruction because we enjoy looking at it. There is a distinction, of course, between just enjoying the landmark and using it, as one would a bridge or building. If there is a use for the landmark based on economics or other human desires, then there is no question that the landmark has some value. But some things do not have utility, such as some old buildings and art. Is it possible to develop an argument for the preservation of mere things, irrespective of their utility to humans?

One such argument was advanced by Christopher Stone in his provocative book *Should Trees Have Standing*? Stone argued that, if corporations,

Figure 5.5 The Acropolis.

is important to our heritage, and we want to preserve it for future generations. While it is often municipalities, and ships are considered legal entities, with rights and responsibilities, is it not reasonable to grant similar rights to trees, forests, and mountains? He did not suggest that a tree should have the same rights as humans but that a tree should be able to be represented in court (to have standing), just as corporations are. If this argument is valid, then buildings and other inanimate objects can similarly have legal rights, and their interests can be represented in court. It should then be possible to sue the City of Athens on behalf of the Acropolis.

But the argument by Stone for granting legal rights to trees and other objects is not very strong. Perhaps the only reason we would prefer not to destroy landmarks such as historic buildings and natural wonders is that these things are components of our physical environment and are necessary to a historical grounding of our civilization. The Acropolis in Athens is a bunch of rocks. Not a single roof remains. The land would be worth millions if developed for condominiums. Yet we strive to protect it, and an international project treated the stone to prevent further deterioration. We do this because it

If the first reaction is first-order, then

$$\frac{dA}{dt} = -k_1 A$$

Likewise, if the second reaction is first-order with respect to B, the overall reaction is

$$\frac{dB}{dt} = k_1 A - k_2 B$$

where k_2 is the rate constant for the reaction $B \rightarrow C$. Note that some B is being made at a rate of $k_1 A$ while some is being destroyed at a rate of $-k_2 B$. Integration yields

$$B = \frac{k_1 A_0}{k_2 - k_1}(e^{-k_1 t} - e^{-k_2 t}) + B_0 e^{-k_2 t}$$

This equation is reintroduced in Chapter 8 when the oxygen level in a stream is analyzed; the oxygen deficit is B, being increased as the result of O_2 consumption by the microorganisms and decreased by the O_2 diffusing into the water from the atmosphere. Before this reaction has much meaning, however, the question of stream ecosystems and gas transfer must be addressed.

PROBLEMS

5.1 The die-off of coliform organisms below a wastewater discharge point can be described as a first-order reaction. It has been found that 30% of the coliforms die off in 8 hours and 55% die in 16 h. About how long would it take to have 90% die off? Use semi log paper to solve this problem.

5.2 A radioactive nuclide is reduced by 90% in 12 min. What is its half-life?

5.3 In a first-order process a blue dye reacts to form a purple dye. The amount of blue at the end of 1 h is 480 g and at the end of 3 h is 120 g. Graphically estimate the amount of blue present initially.

5.4 A reaction of great social significance is the fermentation of sugar with yeast. This is a zero-order (in sugar) reaction, where the yeast is a catalyst (it does not enter the reaction itself). If a 0.5-L bottle contains 4 g of sugar, and it takes 30 min to convert 50% of the sugar, what is the rate constant?

5.5   Integrate the differential equation in which A is being made at a rate k_1 and destroyed simultaneously at a rate k_2.

$$\frac{dA}{dt} = k_1 A - k_2 A$$

5.6   A batch reactor is designed to remove gobbledygook by adsorption. The data are as follows:

Time (min)	Concentration of Gobbledygook (mg/L)
0	170
5	160
10	98
20	62
30	40
40	27

What order of reaction does this appear to be? Graphically estimate the rate constant.

5.7   Many everyday processes can be described in terms of reactors.

a. Give one example of a first-order reaction not already described in the text.
b. Describe in words what is going on.
c. Draw a graph describing this reaction.

5.8   An oil storage area, abandoned 19 yr ago, had spilled oil on the ground and saturated the soil at a concentration of perhaps 400 mg/kg of soil. A fast food chain wants to build a restaurant there and samples the soil for contaminants only to discover that the soil still contains oil residues at a concentration of 20 mg/kg. The local engineer concludes that, since the oil must have been destroyed by the soil microorganisms at a rate of 20 mg/kg each year, in one more year the site will be free of all contamination.

a. Is this a good assumption? Why or why not? (*Hint*: Consider kinetics.)
b. How long would *you* figure the soil will take to reach the acceptable contamination of 1 mg/kg?

5.9   A radioactive waste from a clinical laboratory contains $0.2\,\mu\text{Ci}$ (microcuries) of calcium-45 (^{45}Ca) per liter. The half-life of ^{45}Ca is 152 days.

a. How long must this waste be stored before the level of radioactivity falls below the required maximum of $0.01\,\mu\text{Ci/L}$?
b. Radioactive waste can be stored in many ways, including deep well injection and above-ground storage. Deep well injection involves pumping the waste thousands of feet below earth's surface. Above-ground storage is in buildings in isolated areas, which are then guarded to prevent people from getting near the waste. What would be the risk factors associated with each of these storage methods? What could go wrong? Which system of storage would you believe is superior for such a waste?
c. How would a deep ecologist view the use of radioactive substances in the health field, considering especially the problem of storage and disposal of radioactive waste? Try to think like a deep ecologist when responding. Do you agree with the conclusion you have drawn?

END NOTES

1. Sale, Kirkpatrick. May 1988. Deep ecology and its critics. *The Nation* 14: 670–5.
2. See note 1.
3. The term "environmental fascism" was coined by Tom Regan. desJardins, Joseph, 1995. *Environmental Ethics*. Belmont, CA: Wadsworth Publishing Co., p. 142.

Reactors

Sloss Furnaces National Historic Landmark, Alabama

Great Ocean Road Bay of Islands, Australia

Environmental engineers design processes to harness the power of reactions to mimic nature by converting an influent into a beneficial or at least harmless effluent.

Processes can be analyzed by using a black box and writing a material balance (Chapter 4):

$$\begin{bmatrix} \text{Rate of} \\ \text{materials} \\ \text{ACCUMULATED} \end{bmatrix} = \begin{bmatrix} \text{Rate of} \\ \text{materials} \\ \text{IN} \end{bmatrix} - \begin{bmatrix} \text{Rate of} \\ \text{materials} \\ \text{OUT} \end{bmatrix}$$

$$+ \begin{bmatrix} \text{Rate of} \\ \text{materials} \\ \text{PRODUCED} \end{bmatrix} - \begin{bmatrix} \text{Rate of} \\ \text{materials} \\ \text{CONSUMED} \end{bmatrix}$$

In Chapter 4 we assumed the first term [materials accumulated] was zero, invoking the steady state assumption. In Chapter 5 we reviewed reactions, which are, of course, time dependent. If these reactions occur in a black box, this black box becomes a *reactor*, and the first term of the material balance equation can no longer be assumed to be zero.

Many natural processes as well as engineered systems can be conveniently analyzed by using the notion of ideal reactors. A black box can be thought of as a reactor if it has volume and if it is either mixed or materials flow through it.

Three types of ideal reactors are defined on the basis of certain assumptions about their flow and mixing characteristics. The *mixed-batch reactor* is fully mixed and does not have a flow into or out of it. The *plug-flow reactor* is assumed to have no longitudinal mixing but complete latitudinal mixing. The *completely mixed flow reactor* (also known as a continuously or completely stirred tank reactor, CSTR), as the name implies, is a reactor with perfect mixing throughout.

When no reactions occur inside the reactor (in other words, nothing is consumed or produced), the so-called *mixing model* of reactors is adequate. When reactions occur in the reactors, something is produced or consumed. The *reactor model* describes these conditions. We focus first on reactors with no reactions, or the mixing model of reactors.

6.1 MIXING MODEL

To discuss the mixing model, a device known as a *conservative and instantaneous signal* is used. A signal is simply a tracer placed into the flow as the flow enters the reactor. The signal allows for the characterization of the reactor by measuring the signal (typically the tracer concentration) with time. The term "conservative" means that the signal itself does not react. For example, a color dye introduced into water would not react chemically and would neither lose nor gain color. "Instantaneous" means that the signal (again, such as a color dye) is introduced to the reactor all at once, that is, not over time. In real life this can be pictured as dropping a cup of dye into a bucket of water, for example. Conservative and instantaneous signals can be applied to all three basic types of reactors: mixed-batch, plug-flow, and completely mixed-flow reactors.

6.1.1 Mixed-Batch Reactors

The mixed-batch reactor, illustrated in Figure 6.1A, is considered first. Remember that this reactor has no flow in or out. If the conservative instantaneous signal is introduced, it is

Figure 6.1 (A) A mixed-batch reactor and (B) its C-distribution curve. The propeller symbol means that it is perfectly mixed, with no concentration gradients.

assumed that the signal is mixed instantaneously (zero mixing time). The mixing assumption can be illustrated by using a curve shown in Figure 6.1B. The ordinate indicates the concentration at any time t divided by the initial concentration. Before $t = 0$, there is no signal (dye) in the reactor. When the signal is introduced, the signal is immediately and evenly distributed in the reactor vessel because perfect mixing is assumed. Thus, the concentration instantaneously jumps to C_0, the concentration at $t = 0$ after the signal is introduced. The concentration does not change after that time because there is no flow in or out of the reactor and because the dye is not destroyed (it is conservative).

A plot such as Figure 6.1B is commonly called a *C-distribution* and is a useful way of graphically representing the behavior of reactors. A *C*-distribution curve can be plotted as simply C versus t, but it is more often normalized and plotted as C/C_0 versus t. For a mixed-batch reactor, at $t = 0$, after the signal has been introduced, $C = C_0$ and $C/C_0 = 1$.

Although mixed-batch reactors are useful in a number of industrial and pollution-control applications, a far more common reactor is one in which the flow into and out of the reactor is continuous. Such reactors can be described by considering two ideal reactors: the plug-flow and the completely mixed-flow reactor.

6.1.2 Plug-Flow Reactors

Figure 6.2A on the next page illustrates the characteristics of a plug-flow reactor (PFR). Picture a very long tube (such as a garden hose) into which is introduced a continuous flow. Assume that, while the fluid is in the tube, it experiences no longitudinal mixing. If a conservative signal is instantaneously introduced into the reactor at the influent end, any two elements within that signal that enter the reactor together will always exit the effluent end at the same time. All signal elements have exactly equal *retention times*, defined as the time between entering and exiting the reactor and calculated as

$$\bar{t} = \frac{V}{Q}$$

where $\bar{t} =$ hydraulic retention time, s
$V =$ volume of the reactor, m^3
$Q =$ flow rate to the reactor, m^3/s.

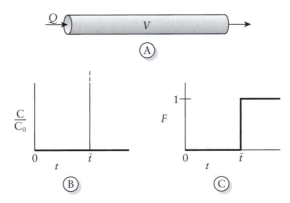

Figure 6.2 (A) A plug-flow reactor, (B) its *C*-distribution curve, and (C) *F*-distribution curve.

Note that the retention time in an ideal PFR is the time *any* particle of water spends in the reactor. Another way of thinking of retention time is the time necessary to *fill* a reactor. If it's plug flow, it's like turning on an empty garden hose and waiting for water to come out the other end. The same definition for retention time holds for *any* type of reactor.

If a conservative instantaneous signal is now introduced into the reactor, the signal moves as a plug (!) through the reactor, exiting at time $\bar{t}$. Before time $\bar{t}$ none of the signal exits the reactor, and the concentration of the signal in the flow is zero ($C = 0$). Immediately after $\bar{t}$, all of the signal has exited, and again $C = 0$. The *C*-distribution curve is, thus, one instantaneous peak shown in Figure 6.2B.

Another convenient means of describing a PFR is to use the *F-distribution* (Figure 6.2C). *F* is defined as the fraction of the signal that has left the reactor at any time *t*:

$$F = \frac{A_0 - A_R}{A_0}$$

where A_0 = amount (usually mass) of tracer added to the reactor
   A_R = amount of tracer remaining in the reactor.

As shown in Figure 6.2C, at $t = \bar{t}$ all the signal exits at once, so $F = 1$.

To recap, perfect plug flow occurs when there is no longitudinal mixing in the reactor. Obviously, this is unrealistic in practice. Equally unrealistic is the third type of ideal reactor, discussed next.

6.1.3 Completely Mixed-Flow Reactors

In a completely mixed-flow (CMF) reactor, perfect mixing is assumed. There are no concentration gradients at any time, and a signal is mixed perfectly and instantaneously. Figure 6.3A illustrates such a reactor.

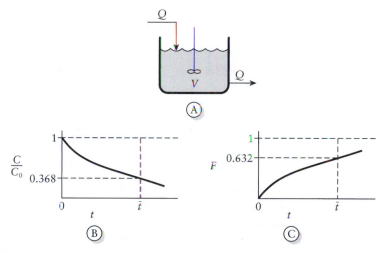

Figure 6.3   (A) A completely mixed-flow reactor, (B) its *C*-distribution curve, and (C) *F*-distribution curve.

A conservative instantaneous signal can be introduced into the CMF reactor feed and the mass balance equation written as

$$
\begin{bmatrix} \text{Rate of} \\ \text{signal} \\ \text{ACCUMULATED} \end{bmatrix} = \begin{bmatrix} \text{Rate of} \\ \text{signal} \\ \text{IN} \end{bmatrix} - \begin{bmatrix} \text{Rate of} \\ \text{signal} \\ \text{OUT} \end{bmatrix}
$$

$$
+ \begin{bmatrix} \text{Rate of} \\ \text{signal} \\ \text{PRODUCED} \end{bmatrix} - \begin{bmatrix} \text{Rate of} \\ \text{signal} \\ \text{CONSUMED} \end{bmatrix}
$$

Because the signal is assumed to be instantaneous, the *rate* at which the signal is introduced is zero. Likewise, the signal is conservative and, therefore, the rate produced and the rate consumed are both zero. Thus,

$$
\begin{bmatrix} \text{Rate of} \\ \text{signal} \\ \text{ACCUMULATED} \end{bmatrix} = 0 - \begin{bmatrix} \text{Rate of} \\ \text{signal} \\ \text{OUT} \end{bmatrix} + 0 - 0
$$

If the amount of signal in the reactor at any time t is A, and A_0 is the amount of signal at $t = 0$, the concentration of the signal in the reactor at $t = 0$ is

$$
C_0 = \frac{A_0}{V}
$$

where C_0 = concentration of signal in the reactor at time zero
A_0 = amount (e.g., mass) of signal in the reactor at time zero
V = reactor volume.

After the signal has been instantaneously introduced, the clear liquid continues to flow into the reactor, so the signal in the reactor is progressively diluted. At any time t the concentration of the signal is

$$C = \frac{A}{V}$$

where C = concentration of signal at any time t
 A = amount of signal at any time t
 V = volume of the reactor

Because the reactor is perfectly mixed, the concentration of the signal in the flow withdrawn from the reactor must also be C.

The *rate* at which the signal is withdrawn from the reactor is equal to the concentration times the flow rate (Chapter 3). Substituting this into the mass balance equation yields

$$\begin{bmatrix} \text{Rate of} \\ \text{signal} \\ \text{ACCUMULATED} \end{bmatrix} = -CQ = -\left(\frac{A}{V}\right)Q$$

where Q = flow rate into and out of the reactor The volume of the reactor, V, is assumed to be constant. The rate that signal A is accumulated is dA/dt, so

$$\frac{dA}{dt} = -\left(\frac{A}{V}\right)Q$$

This equation can be integrated:

$$\int_{A_0}^{A} \frac{dA}{A} = -\int_{0}^{t} \frac{Q}{V}\,dt$$

or

$$\ln A - \ln A_0 = -\frac{Q}{V}t$$

$$\frac{A}{A_0} = e^{-(Qt/V)}$$

The retention time is

$$\bar{t} = \frac{V}{Q}$$

(*Note*: Now the retention time is defined as the average time a particle of water spends in the reactor. This is still numerically equal to the time necessary to fill the reactor volume V with a flow Q, just as for the plug-flow reactor.) Substituting yields

$$\frac{A}{A_0} = e^{-(t/\bar{t})}$$

Both A and A_0 are divided by V to obtain C and C_0, so

$$\frac{C}{C_0} = e^{-(t/\bar{t})}$$

This equation can now be plotted as the C-distribution shown in Figure 6.3B.

At the retention time, $t = \bar{t}$,

$$\frac{C}{C_0} = e^{-1} = 0.368$$

meaning that at the retention time, 36.8% of the signal is still in the reactor, and 63.2% of the signal has exited. This is best illustrated again with the F-distribution, shown in Figure 6.3C.

6.1.4   Completely Mixed-Flow Reactors in Series

The practicality of the mixing model of reactors can be greatly enhanced by considering one further reactor configuration—a series of CMF reactors, as shown in Figure 6.4. Each of the n reactors has a volume of V_0, so that $V = n \times V_0$.

Performing a mass balance on the first reactor and using an instantaneous, conservative signal as before results in

$$\frac{dA_1}{dt} = -\left(\frac{Q}{V_0}\right) A_1$$

where A_1 = amount of the signal in the first reactor at any
                    time t, in units such as kg
           V_0 = volume of reactor 1, m^3
           Q = flow rate, m^3/s

and as before:

$$A_1 = A_0 e^{-(Qt/V_0)} \tag{6.1}$$

where A_0 = amount of signal in the reactor at $t = 0$. For each reactor the retention time is $\bar{t}_0 = V_0/Q$, so that

$$\frac{A_1}{A_0} = e^{-(t/\bar{t}_0)}$$

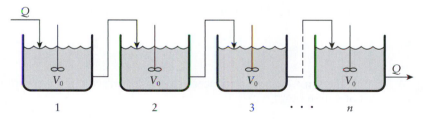

Figure 6.4   A series of CMF reactors.

For the second reactor the mass balance reads as before, and the last two terms are again zero. But now the reactor is *receiving* a signal over time as well as *discharging* it. In addition to the signal accumulation, there is also an inflow and an outflow. In differential form:

$$\frac{dA_2}{dt} = \left(\frac{Q}{V_0}\right) A_1 - \left(\frac{Q}{V_0}\right) A_2$$

Substituting for A_1 (Equation 6.1):

$$\frac{dA_2}{dt} = \left(\frac{Q}{V_0}\right) A_0 e^{-(Qt/V_0)} - \left(\frac{Q}{V_0}\right) A_2$$

and integrating:

$$A_2 = \left(\frac{Qt}{V_0}\right) A_0 e^{-(Qt/V_0)}$$

For three reactors:

$$\frac{A_3}{A_0} = \frac{t}{\bar{t}_0}\left(\frac{e^{-(t/\bar{t}_0)}}{2!}\right)$$

Generally, for i reactors:

$$\frac{A_i}{A_0} = \left(\frac{t}{\bar{t}_0}\right)^{i-1}\left(\frac{e^{-(t/\bar{t}_0)}}{(i-1)!}\right)$$

For a series of n reactors in terms of concentration (obtained by dividing both A_i and A_0 by the reactor volume V_0):

$$\frac{C_n}{C_0} = \left(\frac{t}{\bar{t}_0}\right)^{n-1} e^{-(t/\bar{t}_0)}\left(\frac{1}{(n-1)!}\right) \tag{6.2}$$

Equation 6.2 describes the amount of a conservative, instantaneous signal in any one of a series of n reactors at time t and can be used again to plot the C-distribution. Also, recall that $nV_0 = V$, so that the *total volume* of the entire reactor never changes. The big reactor volume is simply divided into n equal smaller volumes or

$$n\bar{t}_0 = \bar{t} = \frac{V}{Q}$$

Also if a mass M is introduced into a single reactor of volume V_0, the initial concentration is

$$C_{0(1)} = \frac{M}{V_0}$$

If the same mass is introduced into the first reactor of a series of two CMF reactors, each of volume $V_0/2$,

$$C_{0(2)} = \frac{M}{V_0/2} = \frac{2M}{V_0} = 2C_{0(1)}$$

For n reactors,

$$C_{0(n)} = \frac{nM}{V_0} = nC_{0(1)}$$

The F-distribution is a good descriptor of tracer behavior as a CMF reactor is divided into smaller individual CMF volumes.

$$F = \frac{M_0 - M_R}{M_0}$$

The mass of tracer remaining in each of the reactors is

$$M_r = M_1 + M_2 + M_3 + \ldots + M_n = V_0(C_1 + C_2 + C_3 + \ldots + C_n)$$

$$M_0 = V_0 C_0$$

$$F = \frac{C_0 - (C_1 + C_2 + \cdots + C_n)}{C_0} = 1 - \left(\frac{C_1}{C_0} + \frac{C_2}{C_0} + \cdots \frac{C_n}{C_0}\right)$$

Consider now what the concentration would be in the first reactor as the signal is applied. Because the volume of this first reactor of a series of n reactors is $V_0 = V/n$, the concentration at any time must be n times that of only one large reactor (same amount of signal diluted by only one nth of the volume). If it is then necessary to calculate the concentration of the signal in any subsequent reactor, the equation must be

$$\frac{C_n}{C_0} = n \left(\frac{t}{\bar{t}_0}\right)^{(n-1)} \left(\frac{e^{-(t/\bar{t}_0)}}{(n-1)!}\right) = n \left(\frac{nt}{\bar{t}}\right)^{(n-1)} \left(\frac{e^{-(nt/\bar{t})}}{(n-1)!}\right) \qquad (6.3)$$

where $C_0 = $ concentration of signal in the first reactor.

A reactor total volume divided into n volumes must be compared on the basis of the total retention time in the system which is

$$\bar{t} = \frac{V}{Q}$$

So

$$F = 1 - \left[e^{-nt/\bar{t}} + \left(\frac{nt}{\bar{t}}\right)e^{-nt/\bar{t}} + \left(\frac{nt}{\bar{t}}\right)^2\left(\frac{1}{2!}\right)e^{-nt/\bar{t}} + \cdots + \left(\frac{nt}{\bar{t}}\right)^{n-1}\left(\frac{1}{(n-1)!}\right)e^{-nt/\bar{t}}\right]$$

$$= 1 - e^{-nt/\bar{t}}\left[1 + \left(\frac{nt}{\bar{t}}\right) + \left(\frac{nt}{\bar{t}}\right)^2\left(\frac{1}{2!}\right) + \cdots + \left(\frac{nt}{\bar{t}}\right)^{n-1}\left(\frac{1}{(n-1)!}\right)\right]$$

$$= 1 - e^{-nt/\bar{t}}\left[1 + \sum_{i=1}^{n}\left(\frac{nt}{\bar{t}}\right)^{i-1}\frac{1}{(i-1)!}\right]$$

The F-distribution for several reactors in series is shown in Figure 6.5.

Figure 6.5   *F*-distribution curve for a series of CMF reactors.

As the number of reactors increases, the *F*-distribution curve becomes more and more like an *S*-shaped curve. At *n* = infinity it becomes exactly like the *F*-distribution curve for a plug-flow reactor (Figure 6.2C). This can be readily visualized as a great many reactors in series so that a plug is moving rapidly from one very small reactor to another, and because each little reactor has a very short residence time, as soon as it enters one reactor, it gets flushed out into the next one. This, of course, is exactly how the plug moves through the plug-flow reactor.

EXAMPLE
6.1

Problem   It is decided to estimate the effect of dividing a large completely mixed aeration pond (used for wastewater treatment) into 2, 5, 10, and 20 sections so that the flow enters each section in series. Draw the *C*- and *F*-distributions for an instantaneous conservative signal for the single pond and the divided pond.

Solution   For a single pond:

$$\frac{C}{C_0} = e^{-(t/\bar{t})}$$

where C = concentration of signal in the effluent at time t
C_0 = concentration at time t_0
$\bar{t}$ = retention time

Substituting various values of t:

$$
\begin{aligned}
t &= 0.25\bar{t} & C/C_0 &= e^{-(0.25)} & &= 0.779 \\
t &= 0.5\bar{t} & C/C_0 &= e^{-(0.50)} & &= 0.607 \\
t &= 0.75\bar{t} & C/C_0 &= e^{-(0.75)} & &= 0.472 \\
t &= \bar{t} & C/C_0 &= e^{-1} & &= 0.368 \\
t &= 2\bar{t} & C/C_0 &= e^{-2} & &= 0.135
\end{aligned}
$$

These results are plotted in Figure 6.6 and describe a single reactor, $n = 1$.

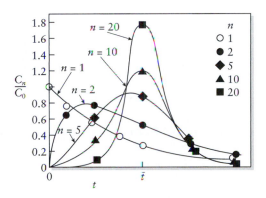

Figure 6.6   Actual C-distribution curves for Example 6.1.

Similar calculations can be performed for a series of reactors. For example, for $n = 10$, remember that $10\bar{t}_0 = \bar{t}$, and using Equation 6.3:

$$t = \bar{t} \qquad \frac{C_{10}}{C_0} = (10)(10)^9 e^{-10} \frac{1}{9!} = 1.25$$

$$t = 0.5\bar{t} \qquad \frac{C_{10}}{C_0} = (10)(5)^9 e^{-5} \frac{1}{9!} = 0.36$$

These data are also plotted in Figure 6.6.

Now for the F-distribution; with one reactor:

$$
\begin{aligned}
t = 0.25\bar{t} \qquad & F = 1 - e^{-0.25} && = 0.221 \\
t = 0.5\bar{t} \qquad & F = 1 - e^{-0.50} && = 0.393 \\
t = \bar{t} \qquad & F = 1 - e^{-1} && = 0.632 \\
t = 2\bar{t} \qquad & F = 1 - e^{-2} && = 0.865
\end{aligned}
$$

and for the 10 reactors in series:

$$F = 1 - e^{-(nt/\bar{t})} \left[1 + \frac{nt}{\bar{t}} + \left(\frac{nt}{\bar{t}}\right)^2 \left(\frac{1}{2!}\right) + \cdots + \left(\frac{nt}{\bar{t}}\right)^{n-1} \frac{1}{(n-1)!} \right]$$

At $t = 0.5\bar{t}$:

$$
\begin{aligned}
F &= 1 - e^{-5}[1 + 5 + (5)^2(1/2!) + (5)^3(1/3!) + \cdots + (5)^9(1/9!)] \\
&= 1 - 0.00674[1 + 5 + 12.5 + 12.5 + 20.83 + 26.04 + 26.04 \\
&\quad + 21.70 + 15.50 + 9.68 + 5.48] \\
&= 1 - 0.969 = 0.031
\end{aligned}
$$

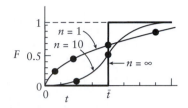

Figure 6.7   Actual *F*-distribution curves for Example 6.1.

Similarly, at $t/\bar{t} = 1$:

$$F = 1 - e^{-10}\left[1 + 10 + (10)^2\left(\frac{1}{2!}\right) + \cdots + (10)^9\left(\frac{1}{9!}\right)\right]$$

and so on. These results are plotted in Figure 6.7.   ∎

Note again, as the C and F curves clearly show, as the number of small reactors (n) increases, the series of completely mixed reactors begin to behave increasingly like an ideal plug-flow reactor.

6.1.5   Mixing Models with Continuous Signals

Thus far the signals used have been instantaneous and conservative. If the instantaneous constraint is removed, the signal can be considered continuous. A continuous conservative signal applied to an ideal PFR simply produces a C-distribution curve with a discontinuity going from $C = 0$ to $C = C_0$ at time t. For a CMF reactor the C curve for when a signal is cut off is exactly like the curve for an instantaneous signal. The reactor is simply being flushed out with clear water, and the concentration of the dye decreases exponentially as before.

If, on the other hand, a continuous signal is introduced to a reactor at $t = 0$ and continued, what does the C curve look like? Start by writing a mass balance equation as before. Remember that the flow of dye *in* is fixed while again the rates of production and consumption are zero. The equation describing a CMF reactor with a continuous (from $t = 0$) conservative signal thus is written in differential form as

$$\frac{dC}{dt} = QC_0 - QC$$

and after integration:

$$C = C_0(1 - e^{Qt})$$

6.1.6   Arbitrary-Flow Reactors

Nothing in this world is ideal, including reactors. In plug-flow reactors there obviously is *some* longitudinal mixing, producing a C-distribution more like that shown in Figure 6.8A

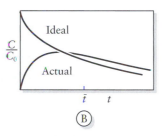

Figure 6.8   C-distribution curves for ideal and actual reactors; (A) plug-flow reactor and (B) completely mixed-flow reactor.

instead of like Figure 6.2B. Likewise, a completely mixed flow reactor cannot be ideally mixed, so that its *C*-distribution curve behaves more like that shown in Figure 6.8B instead of Figure 6.3B. These nonideal reactors are commonly called *arbitrary-flow reactors*.

It should be apparent that the actual *C*-distribution curve in Figure 6.8B for the nonideal CMF (arbitrary-flow) reactor looks suspiciously like the *C* curve for two CMF in series, in Figure 6.6. In fact, looking at Figure 6.6, as the number of CMF in series increases, the *C*-distribution curve approaches the *C* curve for the perfect (ideal) plug-flow reactor. Thus, it seems reasonable to expect that all real-life (arbitrary-flow) reactors really operate in the mixing mode as a series of CMF reactors. This observation allows for a quantitative description of reactor flow properties in terms of *n* CMF reactors in series, where *n* represents the number of CMF in series and defines the type of reactor.

Further analysis of mixing models is beyond the scope of this brief discussion, and the student is directed to any modern text on reactor theory for more advanced study.

6.2   REACTOR MODELS

As noted earlier, reactors can be described in two ways: (1) in terms of their mixing properties only, with no reactions taking place, or (2) as true reactors, in which a reaction occurs. In the previous section the mixing concept was introduced, and now it is time to introduce reactions into these reactors. Stated in another way, the constraint that the signal is conservative is now removed.

Again consider three different ideal reactors: the mixed-batch reactor, the plug-flow reactor, and the completely mixed-flow reactor.

6.2.1   Mixed-Batch Reactors

The same assumptions hold here as in the previous section, namely, perfect mixing is assumed (there are no concentration gradients). The mass balance in terms of the material undergoing some reaction is

The Meuse Valley Episode[1,2]

The Meuse River flows northward on the edge of Belgium. During the 1930s this area was the center of heavy industry in the region. With its coal mines and steel plants and the transportation provided by the Meuse River, towns like Liege became wealthy and economically important. There was no control of air pollution, and the people in the region had learned to live with the dirty atmosphere. Smoke meant jobs and prosperity.

The 20-km river valley from Huy to Liege is narrow (1–2 km) with steep valley walls (60–80 m) (Figure 6.9). Usually a wind sweeps the pollution out of the valley, but at the beginning of December 1930 there was little wind, and a fog had settled over the valley. A radiation inversion "just above the tallest chimneys"[1] occurred and did not break due to the fog – lasting for 5 days. The first illnesses began on December 3 after 2 days; the first deaths occurred on December 4. Eventually 60 people lost their lives, most of whom were elderly and already in poor health. But even healthy younger individuals were coughing and wheezing. In fact some of those who fell ill had stayed indoors during the fog. Domestic and wild animals were also affected with respiratory problems, and cattle, birds, and rats were reported as having died during the episode.

The cause of the disaster was investigated by an "expert commission . . . which included physicians, chemists, and a meteorologist."[1] Although the commission "identified about 30 substances released by 27 factories"[1] in the valley, they determined that the most likely cause was a high SO_2 concentration (and SO_2's oxidation products, particularly H_2SO_4) caused by the emission of sulfur

Figure 6.9 Meuse Valley, Belgium. Based on Nemery, Benoit, Peter H. M. Hoet, and Abderrrahim Nemmar. March 3, 2001. "The Meuse Valley Fog of 1930: An Air Pollution Disaster." Lancet. Vol. 357, Issue 9257, Figure 1, p. 705.

compounds from industrial and domestic coal fires. They also postulated that very fine soot particles that could absorb irritants likely contributed to the problem. Others have implicated fluorine as a major contributing factor in the deaths.

Nothing was done after the episode to assist the people living in the Meuse Valley, although the expert commission warned that "the same accident would happen again" without changes and a subsequent commission appointed to evaluate industrial air pollution legislation recommended changes. "Air pollution was considered an unavoidable consequence of prosperity."[1]

However, the episode received international attention, becoming the first documented air pollution episode involving human fatalities and establishing the effects of temperature inversions and fog in air pollution. The experience in the Meuse Valley helped investigators better understand the cause and effect of air pollution during the Donora episode 18 years later (Chapter 1).

Source: Nemery, Benoit, Peter H. M. Hoet, and Abderrahim Nemmar. March 3, 2001. "The Meuse Valley Fog of 1930: An Air Pollution Disaster." *Lancet*. Vol. 357, Issue 9257, pp. 704–708; Roholm, Kaj. 1937. "The Fog Disaster in the Meuse Valley, 1930: A Fluorine Intoxication." *The Journal of Industrial Hygiene and Toxicology*. Vol. 19, pp. 126–137.

$$\begin{bmatrix} \text{Rate} \\ \text{ACCUMULATED} \end{bmatrix} = \begin{bmatrix} \text{Rate} \\ \text{IN} \end{bmatrix} - \begin{bmatrix} \text{Rate} \\ \text{OUT} \end{bmatrix}$$

$$+ \begin{bmatrix} \text{Rate} \\ \text{PRODUCED} \end{bmatrix} - \begin{bmatrix} \text{Rate} \\ \text{CONSUMED} \end{bmatrix}$$

Because it is a batch reactor, there is no inflow or outflow, and thus

$$\begin{bmatrix} \text{Rate} \\ \text{ACCUMULATED} \end{bmatrix} = \begin{bmatrix} \text{Rate} \\ \text{PRODUCED} \end{bmatrix} - \begin{bmatrix} \text{Rate} \\ \text{CONSUMED} \end{bmatrix}$$

If the material is being produced and there is no consumption, this equation can be written as

$$\frac{dC}{dt} V = r V$$

where C = concentration of material at any time t, mg/L
V = volume of the reactor, L
r = reaction rate, mg/L/s
t = time, s.

The volume term, V, appears because this is a mass balance, and recall that

$$\begin{bmatrix} \text{Mass} \\ \text{flow} \end{bmatrix} = [\text{Concentration}] \times \begin{bmatrix} \text{Volume} \\ \text{flow} \end{bmatrix}$$

The left (accumulation) term units are

$$\frac{(mg/L)}{s} L = \frac{mg}{s}$$

and the right-side units are

$$(mg/L/s) \times (L) = \frac{mg}{s}$$

The volume term, of course, can be canceled, so that

$$\frac{dC}{dt} = r$$

and integrated

$$\int_{C_0}^{C} dC = r \int_{0}^{t} dt$$

For a zero-order reaction, $r = k$, where k = reaction rate constant, and thus

$$C - C_0 = kt \tag{6.4}$$

Equation 6.4 holds when the material in question is being produced. In the situation wherein the reactor destroys the component, the reaction rate is negative, so

$$C - C_0 = -kt \tag{6.5}$$

In both cases:

$$C = \text{concentration of the material at any time } t$$
$$C_0 = \text{concentration of the material at } t = 0$$
$$k = \text{reaction rate constant}$$

If the reaction is first-order:

$$r = kC$$

so

$$\frac{dC}{dt} = kC$$

with the rate constant k having units of time^{-1}. Integrated,

$$\int_{C_0}^{C} \frac{dC}{C} = k \int_{0}^{t} dt$$

$$\ln \frac{C}{C_0} = kt \tag{6.6}$$

$$C = C_0 e^{kt}$$

If the material is being consumed, the reaction rate is negative so

$$\ln \frac{C}{C_0} = -kt \tag{6.7a}$$

$$\ln \frac{C_0}{C} = kt \tag{6.7b}$$

$$C = C_0 e^{-kt}$$

EXAMPLE

6.2

Problem   An industrial wastewater treatment process uses activated carbon to remove color from the water. The color is reduced as a first-order reaction in a batch adsorption system. If the rate constant (k) is 0.35 d^{-1}, how long will it take to remove 90% of the color?

Solution   Let $C_0 =$ initial concentration of the color, and $C =$ concentration of the color at any time t. It is necessary to reach $0.1\ C_0$. Use Equation 6.7b.

$$\ln \left(\frac{C_0}{C} \right) = kt$$

$$\ln \left(\frac{C_0}{0.1C_0} \right) = 0.35t$$

$$\ln \left(\frac{1}{0.1} \right) = 0.35t$$

$$t = \frac{2.30}{0.35} = 6.6 \text{ days}$$

6.2.2   Plug-Flow Reactors

The equations for mixed-batch reactors apply equally well to plug-flow reactors because it is assumed that in perfect plug-flow reactors a plug of reacting materials flows through the reactor and that this plug is itself like a miniature batch reactor. Thus, for a zero-order reaction occurring in a plug-flow reactor in which the material is produced:

$$C = C_0 + k\bar{t} \tag{6.8}$$

where $C =$ effluent concentration
$\quad\quad\quad C_0 =$ influent concentration
$\quad\quad\quad \bar{t} =$ retention time of the reactor $= V/Q$
$\quad\quad\quad V =$ volume of the reactor
$\quad\quad\quad Q =$ flow rate through the reactor

If the material is consumed as a zero-order reaction:

$$C = C_0 - k\bar{t} \tag{6.9}$$

If the material is being produced as a first-order reaction:

$$\ln \frac{C}{C_0} = k\bar{t} \tag{6.10}$$

$$\ln \frac{C}{C_0} = k \left(\frac{V}{Q} \right)$$

$$V = \left(\frac{Q}{k} \right) \ln \frac{C}{C_0}$$

and if it is being consumed:

$$\ln \frac{C}{C_0} = -k\bar{t} \tag{6.11}$$

$$V = \left(-\frac{Q}{k} \right) \ln \frac{C}{C_0}$$

$$V = \frac{Q}{k} \ln \frac{C_0}{C}$$

EXAMPLE
6.3

Problem   An industry wants to use a long drainage ditch to remove odor from their waste. Assume that the ditch acts as a plug-flow reactor. The odor reduction behaves as a first-order reaction, with the rate constant $k = 0.35$ d^{-1}. The flow rate is 1600 L/d. How long must the ditch be if the velocity of the flow is 0.5 m/s and 90% odor reduction is desired?

Solution   Using Equation 6.11:

$$\ln \frac{C}{C_0} = -k\bar{t}$$

$$\ln \frac{0.1C_0}{C_0} = -(0.35)\bar{t}$$

$$\ln(0.1) = -(0.35)\bar{t}$$

$$\bar{t} = 6.58 \text{ d}$$

$$\text{Length of ditch} = 0.5 \ \frac{\text{m}}{\text{s}} \times 6.58 \text{ d} \times 86{,}400 \ \frac{\text{s}}{\text{d}}$$

$$= 2.8 \times 10^5 \text{ m}(!)$$

■

6.2.3 Completely Mixed-Flow Reactors

The mass balance for the completely mixed-flow reactor can be written in terms of the material in question as

$$\begin{bmatrix} \text{Rate} \\ \text{ACCUMULATED} \end{bmatrix} = \begin{bmatrix} \text{Rate} \\ \text{IN} \end{bmatrix} - \begin{bmatrix} \text{Rate} \\ \text{OUT} \end{bmatrix}$$

$$+ \begin{bmatrix} \text{Rate} \\ \text{PRODUCED} \end{bmatrix} - \begin{bmatrix} \text{Rate} \\ \text{CONSUMED} \end{bmatrix}$$

$$\frac{dC}{dt} V = QC_0 - QC + r_1 V - r_2 V$$

where C_0 = concentration of the material in the influent, mg/L
   C = concentration of material in the effluent, and at any place and
      time in the reactor, mg/L
   r = reaction rate
   V = volume of reactor, L

It is now necessary to again assume a steady state operation. Although a reaction is taking place within the reactor, the effluent concentration (and hence the concentration in the reactor) is not changing with time. That is:

$$\frac{dC}{dt} V = 0$$

If the reaction is zero-order and the material is being produced, $r = k$, the consumption term is zero, and

$$0 = QC_0 - QC + kV$$

$$C = C_0 + k\frac{V}{Q}$$

$$C = C_0 + k\bar{t} \tag{6.12}$$

Or, if the material is being consumed by a zero-order reaction:

$$C = C_0 - k\bar{t} \tag{6.13}$$

If the reaction is first-order, and the material is produced, $r = kC$, and

$$0 = QC_0 - QC + kCV$$

$$\frac{C_0}{C} = 1 - k\frac{V}{Q}$$

$$\frac{C_0}{C} = 1 - k\bar{t}$$

$$C = \frac{QC_0}{Q - kV} \tag{6.14}$$

If the reaction is first-order and the material is being destroyed:

$$\frac{C_0}{C} = 1 + k\left(\frac{V}{Q}\right) \tag{6.15a}$$

$$\frac{C_0}{C} = 1 + k\bar{t} \tag{6.15b}$$

$$C = \frac{QC_0}{Q + kV}$$

$$V = \frac{Q}{k}\left[\frac{C_0}{C} - 1\right]$$

Equation 6.15a makes common sense. Suppose it is necessary to maximize the performance of a reactor that destroys a component, that is, it is desired to *increase* (C_0/C), or make C small. This can be accomplished by:

a. increasing the volume of the reactor, V
b. decreasing the flow rate to the reactor, Q
c. increasing the rate constant, k

The rate constant k is dependent on numerous variables such as temperature and the intensity of mixing. The variability of k with temperature is commonly expressed in exponential form as

$$k_T = k_0 e^{\Phi(T - T_0)}$$

where $\Phi =$ a constant

$k_0 =$ the rate constant at temperature T_0

$k_T =$ the rate constant at temperature T

EXAMPLE

6.4

Problem   A new disinfection process destroys coliform (*coli*) organisms in water by using a completely mixed-flow reactor. The reaction is first-order with $k = 1.0$ d^{-1}. The influent concentration is 100 *coli*/mL. The reactor volume is 400 L, and the flow rate 1600 L/d. What is the effluent concentration of coliforms?

Solution

$$\begin{bmatrix} \text{Rate} \\ \text{ACCUMULATED} \end{bmatrix} = \begin{bmatrix} \text{Rate} \\ \text{IN} \end{bmatrix} - \begin{bmatrix} \text{Rate} \\ \text{OUT} \end{bmatrix}$$

$$+ \begin{bmatrix} \text{Rate} \\ \text{PRODUCED} \end{bmatrix} - \begin{bmatrix} \text{Rate} \\ \text{CONSUMED} \end{bmatrix}$$

$$0 = QC_0 - QC + 0 - rV$$

where $r = kC$

$$0 = (1600 \text{ L/d})(100 \ coli/\text{mL}) - (1600 \text{ L/d})C - (1.0 \text{ d}^{-1})(400 \text{ L})C$$

$$C = 80 \ coli/\text{mL}$$

■

6.2.4 Completely Mixed-Flow Reactors in Series

For two CMF reactors in series, the effluent from the first is C_1, and assuming a first-order reaction in which the material is being destroyed:

$$\frac{C_0}{C_1} = 1 + k\left(\frac{V_0}{Q}\right)$$

$$\frac{C_1}{C_2} = 1 + k\left(\frac{V_0}{Q}\right)$$

where V_0 is the volume of the individual reactor. Similarly, for the second reactor, the influent is C_1, and C_2 is its effluent, and for the two reactors,

$$\frac{C_0}{C_1} \cdot \frac{C_1}{C_2} = \frac{C_0}{C_2} = \left[1 + k\left(\frac{V_0}{Q}\right)\right]^2$$

Table 6.1   Performance Characteristics of Completely Mixed-Flow Reactors

Zero-order reaction, material produced	$C = C_0 + k\bar{t}$
Zero-order reaction, material destroyed	$C = C_0 - k\bar{t}$
First-order reaction, material produced	$\dfrac{C_0}{C} = 1 - k\bar{t}$
First-order reaction, material destroyed	$\dfrac{C_0}{C} = 1 + k\bar{t}$
Second-order reaction, material destroyed	$C = \dfrac{-1 + [1 + 4k\bar{t}C_0]^{1/2}}{2k\bar{t}}$
Series of n CMF reactors, zero-order reaction, material destroyed	$\dfrac{C_0}{C_n} = \left(1 + \dfrac{k\bar{t}}{C_0}\right)^n$
Series of n CMF reactors, first-order reaction, material destroyed	$\dfrac{C_0}{C_n} = (1 + k\bar{t}_0)^n$

Note: C = effluent concentration; C_0 = influent concentration; $\bar{t}$ = retention time = V/Q; k = rate constant; C_n = concentration of the nth reactor; $\bar{t}_0$ = retention time in each of n reactors

Table 6.2 Summary of Ideal Reactor Performance

| Reaction Order | CMF | | Plug-Flow Reactor |
	Single Reactor	n Reactors	
Zero	$V = \dfrac{Q}{k}(C_0 - C)$	$V = \dfrac{Q}{k}(C_0 - C_n)$	$V = \dfrac{Q}{k}(C_0 - C)$
First	$V = \dfrac{Q}{k}\left(\dfrac{C_0}{C} - 1\right)$	$V = \dfrac{Qn}{k}\left[\left(\dfrac{C_0}{C}\right)^{1/n} - 1\right]$	$V = \dfrac{Q}{k}\ln\dfrac{C_0}{C}$
Second	$V = \dfrac{Q}{k}\left(\dfrac{C_0}{C} - 1\right)\dfrac{1}{C}$	Complex	$V = \dfrac{Q}{k}\left(\dfrac{1}{C} - \dfrac{1}{C_0}\right)$

Note: Material destroyed; V = reactor volume; Q = flow rate; k = reaction constant; C_0 = influent concentration; C = effluent concentration; n = number of CMF reactors in series.

For any number of reactors in series,

$$\frac{C_0}{C_n} = \left[1 + k\left(\frac{V_0}{Q}\right)\right]^n$$

$$\left(\frac{C_0}{C_n}\right)^{1/n} = 1 + k\frac{V}{nQ}$$

where V = volume of all the reactors, equal to nV_0
  n = number of reactors
  V_0 = volume of each reactor.

The equations for completely mixed-flow reactors are summarized in Table 6.1.

6.2.5 Comparison of Reactor Performance

The efficiency of ideal reactors can be compared by first solving all the descriptive equations in terms of reactor volume, as shown in Table 6.2 and Example 6.5.

EXAMPLE
6.5

Problem   Consider a first-order reaction, requiring 50% reduction in the concentration. Would a plug-flow or a CMF reactor require the least volume?

Solution   From Table 6.2:

$$\frac{V_{\text{CMF}}}{V_{\text{PF}}} = \frac{\dfrac{Q}{k}\left(\dfrac{C_0}{C} - 1\right)}{\dfrac{Q}{k}\left(\ln\dfrac{C_0}{C}\right)}$$

For 50% conversion,

$$\frac{C_0}{C} = 2$$

$$\frac{V_{CMF}}{V_{PF}} = \frac{(2-1)}{\ln 2} = 1.44$$

Conclusion: A CMF reactor would require 44% more volume than a PFR.

◼

The conclusion reached in Example 6.5 is a very important concept used in many environmental engineering systems. Stated generally, for reaction orders of greater than or equal to one, the ideal plug-flow reactor will always outperform the ideal completely mixed-flow reactor. This fact is a powerful tool in the design and operation of treatment systems.

PROBLEMS

6.1 A dye mill has a highly colored wastewater with a flow of 32 million L/d. One suggestion has been to use biological means to treat this wastewater and remove the coloration, and a pilot study is performed. Using a mixed-batch reactor, the following data result:

Time (hours)	Dye Concentration (mg/L)
0	900
10	720
20	570
40	360
80	230

A completely mixed aerated lagoon (batch reactor) is to be used. How large must the lagoon be to achieve an effluent of 50 mg/L?

6.2 A settling tank has an influent rate of 2.4 million L/d. It is 3.6 m deep and has a surface area of 700 m^2. What is the hydraulic retention time?

6.3 An activated sludge tank, $9 \times 9 \times 60$ m, is designed as a plug-flow reactor, with an influent BOD of 200 mg/L and a flow rate of 4 million L/d.

a. If BOD removal is a first-order reaction, and the rate constant is 2.5 d^{-1}, what is the effluent BOD concentration?

b. If the same system operates as a completely mixed flow reactor, what must its volume be (for the same BOD reaction)? How much bigger is this, as a percent of the plug-flow volume?

c. If the plug-flow system was constructed and found to have an effluent concentration of 27.6 mg/L, the system could be characterized as a series of completely mixed-flow reactors. How many? ($n =$?)

6.4 The plant manager has a decision to make. She needs to reduce the concentration of salt in the 32,000 L tank from 30,000 mg/L to 1000 mg/L. She can do it in one of two ways.

1. She can start flushing it out by keeping the tank well mixed while running in a hose with clean water (zero salt) at a flow rate of 240 L/min (with an effluent of 240 L/min, obviously).

2. She can empty out some of the saline water and fill it up again with enough clean water to get 1000 mg/L. The maximum rate at which the tank will empty is 240 L/min, and the maximum flow of clean water is 400 L/min.

a. If she intends to do this job in the shortest time possible, which alternative will she choose?

b. Jeremy Rifkin points out that the most pervasive concept of modern times is *efficiency*. Everything has to be done so as to expend the least energy, effort, and especially time. He notes that we are losing our perspective on time, especially if we think of time in a digital way (as digital numbers on a watch) instead of in an analog way (hands on a conventional watch). With the digits we cannot see where we have been, and we cannot see where we are going, and we lose all perspective of time. Why, indeed, would the plant manager want to empty out the tank in the shortest time? Why is she so hung up with time? Has the issue of having time (and not wasting it) become a pervasive value in our lives, sometimes overwhelming our other values. Write a one-page paper on how you value time in your life and how this value influences your other values.

6.5 A first-order reaction is employed in the destruction of a certain kind of microorganism. Ozone is used as the disinfectant, and the reaction is found to be

$$\frac{dC}{dt} = -kC$$

where C = concentration of microorganisms, microbes/mL
k = rate constant, 0.1 min^{-1}
t = time, min

The present system employs a completelymixed flow reactor, and there is some thought of baffling it to create a series of CMF reactors.

a. If the objective is to increase the percentage microorganism destruction from 80% to 95%, how many CMF reactors are needed in series?

b. We routinely kill microorganisms and think nothing of it. But do microorganisms have the same right to exist as larger organisms, such as whales, for example? Or as people? Should we afford moral protection to microorganisms? Can a microorganism ever become an endangered species? As a part of this assignment, write a letter to the editor of your school newspaper on behalf of Microbe Coliform, a typical microorganism who is fed up with not being given equal protection and consideration within the human society, and who is demanding microbe rights. What philosophical arguments can be mounted to argue for microbe rights? Do not make this a silly letter. Consider the question seriously, because it reflects on the entire problem of environmental ethics.

6.6 A completely mixed-flow reactor, with a volume of V and a flow rate of Q, has a zero-order reaction, $dA/dt = k$, where A is the material being produced and k is the rate constant. Derive an equation that would allow for the direct calculation of the required reactor volume.

6.7 A completely mixed continuous bioreactor used for growing penicillin operates as a zero-order system. The input, glucose, is converted to various organic yeasts. The flow rate to this system is 20 L/min, and the conversion rate constant is 4 mg/(min-L). The influent glucose concentration is 800 mg/L, and the effluent must be less than 100 mg/L. What is the smallest reactor capable of producing this conversion?

6.8 Suppose you are to design a chlorination tank for killing microorganisms in the effluent from a wastewater treatment plant. It is necessary to achieve 99.99% kill in a wastewater flow of 100 m^3/hr. Assume the disinfection is a first-order reaction with a rate constant of 0.2 min^{-1}.

a. Calculate the tank volume if the contact tank is a CMF reactor.

b. Calculate the tank volume if the contact tank is a plug-flow reactor.

c. What is the retention time of both reactors?

6.9 A completely mixed-flow reactor, operating at steady state, has an inflow of 4 L/min and an inflow "gloop" concentration of 400 mg/L. The volume is 60 L; the reaction is zero-order. The gloop concentration in the reactor is 100 mg/L.

a. What is the reaction rate constant?

b. What is the hydraulic retention time?

c. What is the outflow (effluent) gloop concentration?

Energy Flows and Balances

Corn field

Flooded road, Illinois

Environmental engineers need to consider the energy used by designs as well as the energy changes, even if unintentional, that may be produced by designs.

An energetic person is one who is continually in motion, someone who has a lot of energy and is always active. But some energetic people never seem to get anything *done*; they expend a lot of effort but have little to show for it. Obviously, it is not enough to be energetic; one is also expected to be *efficient*. Available energy, in order to be useful, must be funneled efficiently into productive use.

In this chapter we look at quantities of energy, how it flows and is put to use, and the efficiencies of such use.

7.1   UNITS OF MEASURE

The internationally accepted unit of energy is the joule. Other common units for energy are the calorie and kilowatt-hour (kWh)—the former used in natural sciences, the latter in engineering.

7.2   ENERGY BALANCES AND CONVERSION

There are, of course, many forms of energy, such as chemical, heat, and potential energy due to elevation. Often the form of energy available is not the form that is most useful, and one form of energy must be converted to another form. For example, the water in a mountain lake has potential energy and can be run through a turbine to convert this potential to electrical energy that can, in turn, be converted to heat or light, both forms of useful energy. Chemical energy in organic matter, stored in the carbon–carbon and carbon–hydrogen bonds formed by plants, can be severed by a process such as combustion, which liberates heat energy that can, in turn, be used directly or indirectly to produce steam to drive electrical generators. Wind has kinetic energy, and a windmill can convert this to mechanical energy that can, be further converted to electrical energy to produce heat energy, which warms your house. Energy conversion is, thus, an important and ancient engineering process. Unfortunately, energy conversions are *always* less than 100% efficient.

Energy (of whatever kind), when expressed in common units, can be pictured as a quantity that flows, and thus, it is possible to analyze energy flows by using the same concepts used for material flows and balances. As before, a black box is any process or operation into which certain flows enter and from which others leave. If all the flows can be correctly accounted for, then there must be a balance.

Looking at Figure 7.1, note that in a black box the energy in has to equal the energy out (energy wasted in the conversion + useful energy) plus the energy accumulated in the box. This can be expressed as

$$
\begin{bmatrix} \text{Rate of} \\ \text{energy} \\ \text{ACCUMULATED} \end{bmatrix} = \begin{bmatrix} \text{Rate of} \\ \text{energy} \\ \text{IN} \end{bmatrix} - \begin{bmatrix} \text{Rate of} \\ \text{energy} \\ \text{OUT} \end{bmatrix} + \begin{bmatrix} \text{Rate of} \\ \text{energy} \\ \text{PRODUCED} \end{bmatrix} - \begin{bmatrix} \text{Rate of} \\ \text{energy} \\ \text{CONSUMED} \end{bmatrix}
$$

Of course, energy is never produced or consumed in the strict sense; it is simply changed in form. In addition, just as processes involving materials can be studied in their steady state

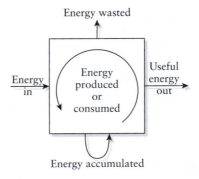

Figure 7.1 Black box for energy flows.

condition, defined as no change occurring over time, energy systems can also be thought of as being in steady state. Obviously, if there is no change over time, there cannot be a continuous accumulation of energy, so the equation must read

$$[\text{Rate of energy IN}] = [\text{Rate of energy OUT}]$$

Remember that energy out has two terms (energy wasted in the conversion and useful energy), so

$$[\text{Rate of energy IN}] = [\text{Rate of useful energy OUT}]$$
$$+ [\text{Rate of wasted energy OUT}]$$

If the input and useful output from a black box are known, the efficiency of the process can be calculated as

$$\text{Efficiency (\%)} = \frac{\text{Useful energy OUT}}{\text{Energy IN}} \times 100$$

EXAMPLE

7.1

Problem   A coal-fired power plant uses 1000 Mg of coal per day. (*Note*: 1 Mg is 1000 kg, commonly called a metric ton or simply a tonne.) The energy value of the coal is 28,000 kJ/kg. The plant produces 2.8×10^6 kWh of electricity each day. What is the efficiency of the power plant?

Solution

$$\text{Energy IN} = (28{,}000 \text{ kJ/kg}) (1000 \text{ Mg/d}) (1000 \text{ kg/Mg})$$
$$= 28 \times 10^9 \text{ kJ/d}$$

$$\text{Useful energy output} = (2.8 \times 10^6 \text{ kWh/d}) (3.6 \times 10^6 \text{ J/kWh}) (10^{-3} \text{ kJ/J})$$
$$= 10.1 \times 10^9 \text{ kJ/d}$$

$$\text{Efficiency (\%)} = [10.1 \times 10^9 \text{ kJ/d}]/[28 \times 10^9 \text{ kJ/d}] \times 100 = 36\%$$

Figure 7.2   Simplified drawing of a bomb calorimeter.

Another example of how one form of energy can be converted to another form is the *calorimeter*, the standard means of measuring the heat energy value of materials when they combust. Figure 7.2 on page 195 shows a schematic sketch of a *bomb calorimeter*. The bomb is a stainless steel ball that screws apart. The ball has an empty space inside into which the sample to be combusted is placed. A sample of known weight, such as a small piece of coal, is placed into the bomb and the two halves screwed shut. Oxygen under high pressure is then injected into the bomb and the bomb is placed in an adiabatic water bath with wires leading from the bomb to a source of electrical current. By means of a spark from the wires, the material in the steel ball combusts, heats the bomb, which in turn heats the water. The temperature rise in the water is measured with a thermometer and recorded as a function of time. Figure 7.3 shows the trace of a typical calorimeter curve.

Figure 7.3   Results of a bomb calorimeter test.

FOCUS ON

Designing a Building to Last—the Flex-Energy Approach

Energy economics of the 21st century are uncertain, but what is certain is that change is coming. If a well constructed building cannot adapt to this change, it may not fall down, but it might get torn down—or at least saddle its owner with higher-than-needed energy bills or higher-than-needed renovation costs.

Even though you can't predict the future, you can design for it. The Flex Energy building design approach can help make this happen, and it might not cost you more than the time it takes your team to think a few things through and come up with a few answers.

The first thing to realize is that not all energy-related design decisions are the same.

- *Invest in the immovable.* If it will be expensive or impossible to upgrade later, look for ways to do it now. Building orientation, daylighting, roof angle, insulation thickness, and high-performance windows—you can't afford to ignore them today because you won't be able to afford to add them tomorrow.

- *Postpone the painless.* If it is easy to add later—especially if it is getting cheaper or better—postpone that investment and use the money for the hard things (in the previous bullet item). Things like state-of-the-art lighting and energy management systems aren't bad ideas, but they are getting better and cheaper—and, thanks to wireless communication technology, easy to add later.

- *Prepare for the probable.* Change is coming. Will your building be ready? Many future energy solutions can be very easy to add, *but only if you plan for them!* It will be hard to add solar energy if your roofs are sloped in the wrong direction; it will be hard to switch from natural gas to electric ground-source heat pumps if your mechanical-electrical room is located in the middle of the building on the second floor.

Where to begin? The Flex Energy 7 are future-energy scenarios that your design team should consider.

1. If a fossil fuel (e.g., natural gas, propane, coal) used on-site becomes economically undesirable or unavailable, what will you do? *Be fossil fuel substitution ready.*
2. If real-time electric rates vary dramatically between peak demand periods and low-demand periods, and this peak pattern varies from day to day, what will you do? *Be demand responsive ready.*
3. If on-site photovoltaic (PV) generation of electricity becomes a low-cost energy source, what will you do? *Be PV ready.*
4. If solar thermal becomes cheaper than natural gas or electricity, what will you do? *Be solar thermal ready.*
5. If electric vehicles (all-electric and plug-in hybrids) are in widespread use, what will you do? *Be electric vehicle ready.*
6. If solid fuels like wood biomass and agricultural biomass become economically attractive, what will you do? *Be biomass ready.*
7. If district energy systems are built near the building and offer hot and chilled water (or steam) at attractive rates, what will you do? *Be district energy ready.*

To get answers that you and your team like, you might need to slightly modify your current design. And that is the value of the Flex Energy approach! Your building will be more adaptable. You and your client will have a better building that is much more likely to be part of the world at the end of the 21st century.

Source: Contributed by Dan Bihn, Low-carbon Energy Consultant.

Note that from time zero the water is heating due to the heat in the room. At $t = 5$ min the switch is closed, and combustion in the bomb occurs. The temperature rise continues to $t = 10$ min, at which time the water starts to cool. The net rise due to the combustion is calculated by extrapolating both the initial heating and cooling lines and determining the difference in temperature.

Now consider the calorimeter as a black box, and assume that the container is well insulated, so no heat energy escapes the system. Because this is a simple batch operation, there is no accumulation. The [Energy IN] is due entirely to the material combusted, and this must equal the [Energy OUT], or the energy expressed as heat and measured as temperature with the thermometer. Remember also that only *heat* energy is considered, so by assuming that no heat is lost to the atmosphere, there is no wasted energy.

The heat energy out is calculated as the temperature increase of the water times the mass of the water plus bomb. Recall that one calorie is defined as the amount of energy necessary to raise the temperature of 1 g of water 1 °C. Knowing the grams of water in the calorimeter, it is possible to calculate the energy accumulation in calories. An equal amount of energy must have been liberated by the combustion of the sample, and knowing the weight of the sample, its energy value can be calculated. (This discussion is substantially simplified, and anyone interested in the details of calorimetry should consult any modern thermodynamics text for a thorough discussion.)

EXAMPLE

7.2

Problem   A calorimeter holds 4 L of water. Ignition of a 10-g sample of a waste-derived fuel of unknown energy value yields a temperature rise of 12.5°C. What is the energy value of this fuel? Ignore the mass of the bomb.

Solution   Making use of a material balance:

$$[\text{Energy IN}] = [\text{Energy OUT}]$$
$$\text{Energy out} = (4.184 \text{ J/g/K}) \, m\Delta T$$
$$= (4.184 \text{ J/g/K})\rho \, V\Delta T$$
$$= (4.184 \text{ J/g/K})(1 \text{ g/mL})(4\text{L})(10^3 \text{ mL/L})(12.5 \text{ °C})$$
$$= 209 \times 10^3 \text{ J}$$
$$\text{Energy IN} = \text{ Energy OUT } = 209 \times 10^3 \text{ J}$$
$$\text{Energy value of the fuel} = [209 \times 10^3 \text{ J}]/[10 \text{ g}] = 20,900 \text{ J/g}$$

Heat energy is easy to analyze by energy balances because the quantity of heat energy in a material is simply its mass times its absolute temperature. (This is true if the heat capacity is independent of temperature, in particular, if phase changes do not occur, as in the conversion of water to steam. Such situations are addressed in a thermodynamics course, highly recommended for all environmental engineers.) An energy balance for heat energy would then be in terms of the quantity of heat, or

$$\begin{bmatrix} \text{Heat} \\ \text{energy} \end{bmatrix} = \begin{bmatrix} \text{Mass of} \\ \text{material} \end{bmatrix} \times \begin{bmatrix} \text{Absolute temperature} \\ \text{of the material} \end{bmatrix}$$

This is analogous to mass flows discussed earlier, except now the flow is energy flow. When two heat energy flows are combined, for example, the temperature of the resulting flow at steady state is calculated by using the black box technique:

$$0 = [\text{Heat energy IN}] - [\text{Heat energy OUT}] + 0 - 0$$

or stated another way:

$$0 = [T_1 Q_1 + T_2 Q_2] - [T_3 Q_3]$$

Solving for the final temperature:

$$T_3 = \frac{T_1 Q_1 + T_2 Q_2}{Q_3} \tag{7.1}$$

where T = absolute temperature
          Q = flow, mass/unit time (or volume/unit time if constant density)
    1 and 2 = input streams
          3 = output stream

The mass/volume balance is

$$Q_3 = Q_1 + Q_2$$

Although strict thermodynamics requires that the temperature be expressed in *absolute* terms, the conversion from celsius (C) to kelvin (K) simply cancels and T can be conveniently expressed in degrees celsius.

E X A M P L E

7.3   **Problem**   A coal-fired power plant discharges 3 m³/s of cooling water at 80°C into a river that has a flow of 15 m³/s and a temperature of 20°C. What will be the temperature in the river immediately below the discharge?

Solution   Using Equation 7.1 and considering the confluence of the river and cooling water as a black box:

$$T_3 = \frac{T_1 Q_1 + T_2 Q_2}{Q_3}$$

$$T_3 = \frac{[(80 + 273) \text{ K } (3 \text{ m}^3/\text{s})] + [(20 + 273) \text{ K } (15 \text{ m}^3/\text{s})]}{(3 + 15) \text{ m}^3/\text{s}} = 303 \text{ K}$$

or 30°C. Note that the use of absolute temperatures is not necessary because the 273 cancels.

7.3   ENERGY SOURCES AND AVAILABILITY

It makes sense for power utilities to use the most efficient fuels possible because these will produce the least ash for disposal and will most likely be the cheapest to use in terms of kilowatt-hours of electricity produced per dollar of fuel cost. But the best fuels, natural gas and oil, are in finite supply. Estimates vary as to how much natural gaseous and liquid fuels remain in earth's crust within our reach, but most experts agree that, if they continue to be used at the present expanding rate, the existing supplies will be depleted within 50 years. Others argue that, as the supplies begin to run low and the price of these fuels increases, other fuels will become less expensive by comparison, so market forces will limit the use of the resources.

If the pessimists are correct and the world runs out of oil and natural gas within 50 years, will the next generation blame us for our unwise use of natural resources? Or should we, indeed, even be worried about the next generation?

There is a strong argument to be made that the most important thing we can do for the coming generations is not to plan for them. The world will be so different, and the state of technology would have changed so markedly, that it is impossible to estimate what future generations would need. For us to deprive ourselves of needed resources today just so some future person would have the benefit of these resources is, by this argument, simply ludicrous.

On the other hand, are there not some things we can be fairly certain of in terms of the future generations? We can fairly well guess that they will appreciate and value many of the same things we do. They will like to have clean air and plentiful and safe water. They will like to have open spaces and wilderness. They will appreciate it if there are no hazardous waste time bombs. And (this is moot) they will appreciate that some of what we regard as natural resources are there for them to use and manage.

But the most important question is why we should care at all about the future generations. What have they ever done for us? Where is the quid pro quo in all this?

If we ignore the problem of reciprocity and decide to conserve energy resources, what are the available sources of renewable energy? The sources in present use include

- hydropower from rivers
- hydropower from tidal estuaries
- solar power
- refuse and other waste materials
- wind
- wood and other biomass, such as sugarcane and rice hulls.

The nonrenewable energy sources include

- nuclear power
- coal, peat, and similar materials
- natural gas
- oil.

The nonrenewable sources are our energy capital, the amount of energy goods that we have to spend. The renewable sources are analogous to our energy income, resources that we can continue to use as long as the sun shines and the wind blows. Previous energy use in the

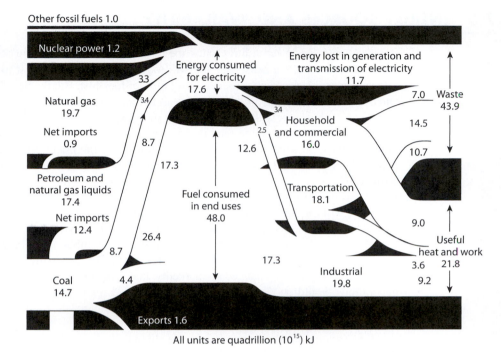

Figure 7.4 Energy flow in the United States. (After Wagner, R.H. 1978. *Environment and Man.* New York: Norton E. Cook.)

United States is illustrated in Figure 7.4. Note that most of the renewable sources were so small that they didn't even make the chart. We are, still, however, rapidly depleting our energy capital and rely almost not at all on the renewable energy income.

7.3.1 Energy Equivalence

There is a big difference between potentially available energy and energy that can be efficiently harnessed. For example, one of the greatest sources of potential energy is tidal energy. The difficulty, however, is how to convert this potential to a useful form, such as electrical energy. With a few notable exceptions, such conversions have not proven cost-effective; that is, the electrical energy from tidal power costs much more than the electrical energy produced by other means.

Further, some types of conversions may not be *energy* efficient, in that it takes more energy to produce the marketable form of energy than the final energy produced. For example, the energy necessary to collect and process household refuse may be more than the energy produced by burning the refuse-derived fuel and producing electricity.

An important distinction must be made between *arithmetic energy equivalence* and *conversion energy equivalence*. The former is calculated simply on the basis of energy while the latter takes into account the energy loss in conversion. Example 7.5 illustrates this point.

EXAMPLE

7.4   **Problem**   What are the arithmetic and conversion energy equivalents between gasoline (20,000 KJ/kg) and refuse-derived fuel (5000 KJ/kg)?

Solution

$$\text{Arithmetic energy equivalence} = \frac{20{,}000 \text{ KJ/kg gasoline}}{5{,}000 \text{ KJ/kg refuse}}$$

$$= 4 \text{ kg refuse/1 kg gasoline}$$

But the processing of refuse to make the fuel also requires energy. This can be estimated at perhaps 50% of the refuse-derived fuel energy, so the actual *net* energy in the refuse is 2,500 KJ/kg. Therefore,

$$\text{Conversion energy equivalence} = \frac{20{,}000 \text{ KJ/kg gasoline}}{2{,}500 \text{ KJ/kg refuse}}$$

$$= 8 \text{ kg refuse/1 kg gasoline}$$

(Of course, we have a large quantity of refuse available.)

Finally, there is the very practical problem of running an automobile on refuse-derived fuel. If this were a *true* equivalence, it would be a simple matter to substitute one fuel for another without penalty. Obviously, this is not possible, and it is necessary to realize that a conversion equivalence does not mean that one fuel can be substituted for another. In addition, the measured energy value of a fuel, such as gasoline, is not the net energy of that fuel. This is the energy value as measured by a calorimeter, but net, or true, value must be calculated by subtracting the energy cost of the surveys, drilling, production, and transport necessary to produce the gasoline. Suffice it to say that energy equivalence calculations are not simple, and thus, grandiose pronouncements by politicians such as "we can save 15 zillion barrels of gasoline a year if we would only start burning all our cow pies" should be treated with proper skepticism.

7.3.2   Electric Power Production

One of the most distressing conversion problems is the production of electricity from fossil fuels. The present power plants are less than 40% efficient. Why is this?

First, consider how a power plant operates. Figure 7.5 shows that the water is heated to steam in a boiler, and the steam is used to run a turbine that, in turn, drives a generator. The waste steam must be condensed to water before it can again be converted to high-pressure steam. This system can be simplified as in Figure 7.6, and the resulting schematic is called a *heat engine*. If the work performed is also expressed as energy and steady state is assumed, an energy balance on this heat engine yields

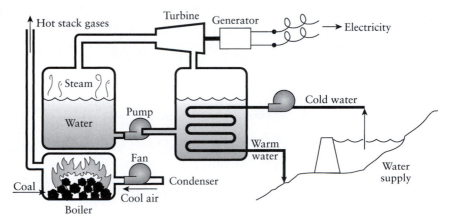

Figure 7.5 Simplified drawing of a coal-fired power plant.

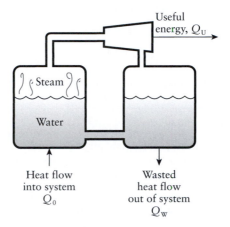

Figure 7.6 The heat engine.

$$\begin{bmatrix} \text{Rate of} \\ \text{energy} \\ \text{ACCUMULATION} \end{bmatrix} = \begin{bmatrix} \text{Rate of} \\ \text{energy} \\ \text{IN} \end{bmatrix} - \begin{bmatrix} \text{Rate of} \\ \text{useful energy} \\ \text{OUT} \end{bmatrix} - \begin{bmatrix} \text{Rate of} \\ \text{wasted energy} \\ \text{OUT} \end{bmatrix}$$

$$0 = Q_0 - Q_U - Q_W$$

where Q_0 = energy flow into the black box
Q_U = useful energy out of the black box
Q_W = wasted energy out of the black box

The efficiency of this system, as previously defined, is

$$\text{Efficiency (\%)} = \frac{Q_U}{Q_0} \times 100$$

FOCUS ON

Beauty and Function

When you think of municipal solid waste incinerators, what is the first thing that comes to mind? Probably not beauty. But there are no rules against function coexisting with beauty in engineering facilities. Values such as beauty are impossible to quantify. Nonetheless, they are important to society and can provide great value to communities.

For example, in Vienna, the famous and controversial Austrian painter, sculptor, and architect Friedensreich Regentag Dunkelbunt Hundertwasser (1928–2000) was solicited to redesign the exterior of the Spittelau Thermal Waste Treatment Plant. Hundertwasser was famous for his revolutionary architectural designs, which incorporate natural features of the landscape and use irregular forms in the building design.

Hundertwasser initially opposed the idea because he had fundamental objections to a garbage incinerating plant as long as all possibilities for avoiding garbage were not exhausted. Over time, Hundertwasser agreed to design the facility; he was promised that:

- The plant would be equipped with the most modern emission purification technology.
- 60,000 apartments would be heated with water heated by the plant, thus avoiding emissions that would otherwise be a further source of pollution.
- The net result would make Vienna's air cleaner. A metropolis such as Vienna would need a garbage incinerating plant despite the greatest efforts to avoid garbage.

The facility even utilizes selective catalytic reduction (SCR) to react the flue gas with vaporized ammonia, water, and oxygen to convert the nitrogen oxides (NOx) to nitrogen and steam. This step also results in dioxin and furan destruction. The exhaust gas is then cooled to 115 °C in the third heat exchanger and finally released into the atmosphere through a 126-m-high stack.

Figure 7.7 Spittelau Thermal Waste Treatment Plant.

The Spittelau Thermal Waste Treatment Plant is striking in its form and function (Figure 7.7). The facility provides heat to a local hospital in addition to private homes. It is located centrally and handles approximately 40% of Vienna's waste. It is significant that the waste is handled locally, rather than being shipped elsewhere. The city is taking responsibility for managing its own waste and turning what could be a polluting eyesore into a high-performance attraction. To some people, however, no matter how you dress up a municipal solid waste incinerator, it is still considered a highly undesirable means of disposing of waste. This is, in part, because incineration destroys material value. It also generates air pollution and an ash that must be managed.

Source: The Spittelau Thermal Waste Treatment Plant, Published by: Thermische Abfallbehandlungsanlage Spittelau, Spittelauer Lände 45, A-1090 Wien; Field trip to WTE facility Spittelau/Fernwaerme Wien GmbH Vienna, Austria, December 22, 2005 by Werner Sunk.

From thermodynamics it is possible to prove that the most efficient engine (least wasted energy) is called the Carnot engine and that its efficiency is determined by the absolute temperature of the surroundings. The efficiency of the Carnot engine is defined as

$$E_C \, (\%) = \frac{T_1 - T_0}{T_1} \times 100$$

where T_1 = absolute temperature of the boiler
T_0 = absolute temperature of the condenser (cooling water)

Because this is the best possible, any real-world system cannot be more efficient:

$$\frac{Q_U}{Q_0} \leq \frac{T_1 - T_0}{T_1}$$

Modern boilers can run at temperatures as high as 600°C. Environmental restrictions limit condenser water temperature to about 20°C. Thus, the best expected efficiency is

$$E_C \, (\%) = \frac{(600 + 273) - (20 + 273)}{(600 + 273)} \times 100 = 66\%$$

A real power plant also has losses in energy due to hot stack gases, evaporation, friction losses, etc. The best plants so far have been hovering around 40% efficiency. When various energy losses are subtracted from nuclear-powered plants, the efficiencies for these plants seem to be even lower than for fossil-fuel plants.

If 60% of the heat energy in coal is not used, it must be wasted, and this energy must somehow be dissipated into the environment. Waste heat energy is emitted from the power plant in two primary ways: stack gases and cooling water. The schematic of the power plant in Figure 7.5 can be reduced to a black box and a heat balance performed, as shown in Figure 7.8. Assume for the sake of simplicity that the heat in the cool air is negligible. So the energy balance at steady state is

$$\begin{bmatrix} \text{Rate of} \\ \text{energy} \\ \text{ACCUMULATED} \end{bmatrix} = \begin{bmatrix} \text{Rate of} \\ \text{energy IN} \\ \text{the coal} \end{bmatrix} - \begin{bmatrix} \text{Rate of} \\ \text{energy OUT} \\ \text{in the} \\ \text{stack gases} \end{bmatrix}$$

$$- \begin{bmatrix} \text{Rate of} \\ \text{energy OUT} \\ \text{in the} \\ \text{cooling water} \end{bmatrix} - \begin{bmatrix} \text{Rate of} \\ \text{energy OUT} \\ \text{as useful} \\ \text{electrical power} \end{bmatrix}$$

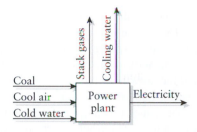

Figure 7.8   The power plant as a black box.

Commonly, the energy lost in the stack gases accounts for 15% of the energy in the coal while the cooling water accounts for the remaining 45%. This large fraction illustrates the problems associated with what is known as *thermal pollution*, the increase in temperature of lakes and rivers due to cooling water discharges.

Most states restrict thermal discharges so that the rise above ambient stream temperature levels will be equal to or less than 1°C. Therefore, some of the heat in the cooling water must be dissipated into the atmosphere before the water is discharged. Various means are used for dissipating this energy, including large shallow ponds and cooling towers. A cutaway drawing of a typical cooling tower is shown in Figure 7.9. Cooling towers represent a substantial additional cost to the generation of electricity. They are estimated to double

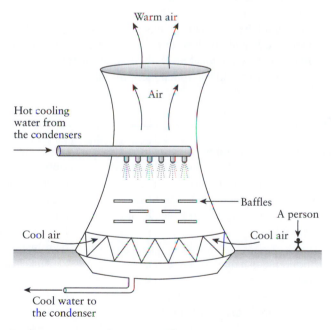

Figure 7.9   Cooling tower used in power plants.

the cost of power production for fossil-fuel plants and may increase the costs by as much as 2.5 times for nuclear power plants.

Even with this expense, watercourses immediately below cooling water discharges are often significantly warmer than normal. This results in the absence of ice during hard winters, and the growth of immense fish. Stories about the size of fish caught in artificially warmed streams and lakes abound, and these places become not only favorite fishing sites for people but winter roosting places for birds. Wild animals similarly use the unfrozen water during winter when other surface waters are frozen. In other parts of the world, cooling water from power generation is used for space heating, piping the hot water through underground lines to homes and businesses.

With this much potential good resulting from the discharge of waste heat, why is it necessary to spend so much money cooling the water before discharge? In the United States, waste heat is not used for space heating because the power plants have been intentionally built as far away from civilization as possible. Our unwillingness to have power plants as neighbors deprives us of the opportunity to obtain "free" heat.

The reasoning for why hot water should not be discharged into watercourses is less obvious. The heat clearly changes the aquatic ecosystem, but some would claim that this change is for the better. Everyone seems to benefit from having the warm water. Yet changes in aquatic ecosystems are often unpredictable and potentially disastrous. Heat can increase the chances of various types of disease in fish, and heat will certainly restrict the types of fish that can exist in the warm water. Many cold-water fish, such as trout, cannot spawn in warmer water, so they will die, their place taken by fish that can survive, such as catfish and carp.

It is unclear what values are involved in governmental restrictions on thermal discharges, such as "no more than 1°C rise in temperature." Is it our intent to protect the trout, or would it be acceptable to have the stream populated by other fish? What about the advantage to other life by having ice-free water during winter? And what about the people who like to fish? On a local level, thermal discharges do not seem to produce lasting effects, so why are we paying so much more for our electricity?

PROBLEMS

7.1   How many pounds of coal must be burned to keep one 100-W light bulb lit for one hour? Assume the efficiency of the power plant is 35%, the transmission losses are 10% of the delivered power, and the heating value of coal is 30,000 KJ/kg.

7.2   Coal is a nonreplenishable resource, and a valuable source of carbon for the manufacture of plastics, tires, etc. Once the stores of coal are depleted, there will not be any more coal to be mined. Is it our responsibility to make sure that there are coal resources left in, say, 200 yr, to be used by people alive then? Should we even worry about this, or should we use up coal as fast as is necessary and prudent for our own needs, figuring that future generations can take care of themselves? Write a one-page argument for either not conserving coal or conserving coal for use by future generations.

7.3 A nuclear-power station with a life of 25 yr produces 750 MW/yr as useful energy. The energy cost is as follows:

150 MW lost in distribution
20 MW needed to mine the fuel
50 MW needed to enrich the fuel
80 MW (spread over 25 yr) to build the plant
290 MW lost as heat

What is the efficiency of the plant? What is the efficiency of the system (including distribution losses)?

7.4 One of the greatest problems with nuclear power is the disposal of radioactive wastes. If we all received electrical energy from a nuclear power plant, our personal contribution to the nuclear waste is about 0.25 L pint, a small milk carton. This does not seem like a lot to worry about. Suppose, however, that the entire New York City area (population 10 million) receives its electricity from only nuclear power plants. How much waste would be generated each year? What should be done with it? Devise a novel (?) method for high-level nuclear waste disposal and defend your selection with a one-page discussion. Consider both present and future human generations as well as environmental quality, ecosystems, and future use of resources.

7.5 A 1-g sample of an unknown fuel is tested in a 2-L (equivalent) calorimeter, with the following results:

Time (min)	Temperature (°C)
0	18.5
5	19.0
6	19.8
7	19.9
8	20.0
9	19.9
10	19.8

What is the heating value of this fuel in kilojoules per kilogram?

7.6 One of the cleanest forms of energy is hydroelectric power. Unfortunately, most of our rivers have already been dammed up as much as is feasible, and there is little likelihood that we will be able to obtain much more hydroelectric power. In Canada, however, the James Bay area was an ideal location for massive new dams that would provide clean and inexpensive electrical power to the northeast. The so-called Hydro-Quebec project was initiated to take advantage of this fact.

The dams, however, created lakes that flooded Native American ancestral lands, and the Native Americans were quite upset by this. Because of these and other environmental concerns, New York State and other possible customers backed out of original purchase arrangements, casting a shadow over the project expansion.

Discuss in a two-page paper the conflict of values as you perceive them. Did the Canadian government have legitimate right to expropriate the lands? Recognize that if this project were not built, other power plants would be constructed. How should the Canadian government have resolved this issue?

7.7 The balance of light hitting the planet earth is shown in Figure 7.10. What fraction of the light is actually useful energy absorbed by Earth's surface?

7.8 Figure 7.4 shows an energy balance for the United States. All values are in quadrillion KJ (10^{15} KJ). The chart shows how various sources of energy were used, and a large part of our energy budget was wasted. Check the numbers on this figure by using an energy balance.

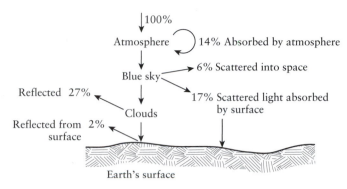

Figure 7.10   Global energy flow. See Problem 7.7.

7.9   The Dickey and Lincoln hydroelectric dams on the St. John River in Maine were planned with careful consideration to the Furbish lousewort (*Pedicularis furbishiae*), an endangered plant species. An engineer writing in *World Oil* (January 1977, p. 5) termed this concern for the "lousy lousewort" to be "total stupidity." Based on only this information, construct an ethical profile of this engineer.

7.10   Disposable diapers, manufactured from paper and petroleum products, are one of the most convenient diapering systems available. Disposable diapers are also considered by many to be anti-environment, but the truth is not so clear-cut.

In this problem three diapering systems are considered: home-laundered cloth diapers, commercially laundered cloth diapers, and disposable diapers containing a superabsorbent gel. Energy and materials balances are used to determine the relative merits of each system.

The energy and waste inventory data for each system are shown in Table 7.1. The data are for 1000 diapers.

An average of 68 cloth diapers are used per week per baby. Because disposable diapers last longer and never need double

Table 7.1

	Disposable Diapers	Cloth Diapers	
		Commercially Laundered	Home Laundered
Energy requirements, 10^6 KJ	1.9	2.1	3.8
Solid waste, m^3	17	2.3	2.3
Atmospheric emissions, kg	8.3	4.5	9.6
Waterborne wastes, kg	1.5	5.8	6.1
Water volume requirements, L	1300	3400	2700

Table 7.2

| | Disposable Diapers | Cloth Diapers | |
		Commercially Laundered	Home Laundered
Energy requirements	0.5	0.55	1.0
Solid waste			1.0
Atmospheric emissions			1.0
Waterborne wastes			1.0
Water volume requirements			1.0

diapering, the number of disposable diapers can be expected to be less.

a. Determine the number of disposable diapers required to match the 68 cloth diapers per week. Assume the following.

- 15.8 billion disposable diapers are sold annually.
- 3,787,000 babies are born each year.
- Children wear diapers for the first 30 months.
- Disposable diapers are worn by only 85% of the babies.

b. Complete Table 7.2, showing the ratio of impact relative to the home-laundered diapers. The first line is already completed.

c. Using the following data, determine the percentage of disposable diapers that would have to be recycled to make the landfill requirements equal for cloth and disposable systems. The table shows the ratios of the impact of solid waste for disposable diapers.

Percentage of Diapers Recycled	Solid Waste per 1000 Diapers (m³)
0	17
25	13
50	9.0
75	4.9
100	0.80

d. Based on all this information, what are the relative adverse environmental impacts of the three different diaper systems?

e. In your opinion, were all of the factors considered fairly in the above exercise? What was not? Was anything left out?*

7.11 You are having lunch with an economist and a lawyer. They are discussing ways to reduce the use of gasoline by private individuals.

Economist: The way to solve this problem is to place such a high tax on gasoline and electric power that people will start to conserve. Imagine if each liter of gas cost $5? You fill up your tank for $300! You'd think twice about driving somewhere when you can walk instead.

*This problem is by David R. Allen, N. Bakshani, and Kirsten Sinclair Rosselot. 1991. *Pollution Prevention: Homework and Design Problems for Engineering Curricula*. New York: American Institute of Chemical Engineers and other societies.

Lawyer: That would be effective, but a law would be better. Imagine a federal law that forbade the use of a car except when at least four people were passengers. Or we could make a law that required cars to be the size and weight of small motorcycles, with only one passenger seat. That way we could still travel wherever we wanted, but we would triple the gas mileage.

Economist: We could also levy a huge tax on spare parts. It would be more likely then that older cars would be junked sooner and that cleaner, more efficient, newer cars would take their place.

Lawyer: I still like the idea of a 72 Km/h speed limit on our interstates. This is the most efficient speed for gasoline consumption, and it would also hold down accidents. Besides, the state could raise a lot of revenue by ticketing everyone exceeding the limit. It would be a money maker!

They continue in this vein until they realize that you have been keeping out of the conversation. They ask you for your opinion about reducing the use of gasoline. How would you respond to their ideas, and what ideas (if any) would you have to offer? Write a one-page response.

7.12   Large consulting firms commonly have many offices, and often communication among the offices is less than efficient. Engineer Stan, in the Atlanta office, is retained by a neighborhood association to write an environmental impact study that concludes that the plans by a private oil company to build a petrochemical complex would harm the habitat of several endangered species. The client, the neighborhood association, has already reviewed draft copies of the report and is planning to hold a press conference when the final report is delivered. Stan is asked to attend the news conference, in his professional capacity, and charge time to the project.

Engineer Bruce, a partner in the firm and working out of the New York office, receives a phone call.

"Bruce, this is J. C. Octane, president of Bigness Oil Company. As you well know, we have retained your firm for all of our business and have been quite satisfied with your work. There is, however, a minor problem. We are intending to build a refinery in the Atlanta area, and hope to use you as the design engineers."

"We would be pleased to work with you again" replies Bruce, already counting the $1 million design fee.

"There is, however, a small problem," continues J. C. Octane, "It seems that one of your engineers in the Atlanta office has conducted a study for a neighborhood group opposing our refinery. I have received a draft copy of the study, and my understanding is that the engineer and leaders of the neighborhood organization are to hold a press conference in a few days and conclude unfavorable environmental impact as a result of the refinery. I need not tell you how disappointed we will be if this occurs."

As soon as Bruce hangs up the phone with J. C. Octane, he calls Stan in Atlanta.

"Stan, you must postpone the press conference at all costs," Bruce yells into the phone.

"Why? It's all ready to go," responds Stan.

"Here's why. You had no way of knowing this, but Bigness Oil is one of the firm's most valued clients. The president of the oil company has found out about your report and threatens to pull all of their business should the report be delivered to the neighborhood association. You have to rewrite the report in such a way as to show that there would be no significant damage to the environment."

"I can't do that!" pleads Stan.

"Let me see if I can make it clear to you, then," replies Bruce. "You either rewrite the report or withdraw from the project and write a letter to the neighborhood association stating that the draft report was in error, and offer to refund all of their money. You have no other choice!"

Is a consulting engineer an employee of a firm and subject to the dictates and orders of his/her superior, or is he/she functioning as an independent professional who happens to be cooperating with other engineers in the firm? What price loyalty?

State all the alternatives Stan might have, and probable ramifications of these courses of action, then recommend what action he should take. Make sure you can justify this action.

Ecosystems

Black bear, South Carolina

Bear-proof trash can in Shenandoah National Park

Environmental engineers need to consider the ecosystems impacted by designs, which means incorporating the expertise of professionals in other disciplines.

Some of the most fascinating reactors imaginable are *ecosystems*. *Ecology*, the topic of this chapter, is the study of plants, animals, and their physical environment; that is, the study of ecosystems and how energy and materials behave in ecosystems.

Specific ecosystems are often difficult to define because all plants and animals are in some way related to each other. Because of its sheer complexity, it is not possible to study Earth as a single ecosystem (except in a very crude way), so it is necessary to select functionally simpler and spatially smaller systems, such as ponds, forests, or even gardens. When the system is narrowed down too far, however, there are too many ongoing external processes that affect the system, so it is not possible to develop a meaningful model. The interaction of squirrels and blue jays at a bird feeder may be fun to watch, but it is not very interesting scientifically to an ecologist because the ecosystem (bird feeder) is too limited in scope. There are many more organisms and environmental factors that become important in the functioning of the bird feeder, and these must be taken into account to make this ecosystem meaningful. The problem is deciding where to stop. Is the backyard large enough to study, or must the entire neighborhood be included in the ecosystem? If this is still too limited, where are the boundaries? There are none, of course, and everything truly is connected to everything else.

With that caveat, what follows is a brief introduction to the study of ecosystems, however they may be defined.

8.1 ENERGY AND MATERIAL FLOWS IN ECOSYSTEMS

Both energy and materials flow inside ecosystems but with a fundamental difference. Energy flow is in only one direction while material flow is cyclical.

All energy on Earth originates from the sun as light energy. Plants trap this energy through a process called *photosynthesis* and, using nutrients and carbon dioxide, convert the light energy to chemical energy by building high-energy molecules of starch, sugar, proteins, fats, and vitamins. In a crude way photosynthesis can be pictured as

$$[Nutrients] + CO_2 \xrightarrow{Sunlight} O_2 + [High\text{-}energy\ molecules]$$

All other organisms must use this energy for nourishment and growth through a process called *respiration*:

$$[High\text{-}energy\ molecules] + O_2 \longrightarrow CO_2 + [Nutrients]$$

This conversion process is highly inefficient, with only about 1.6% of the total energy available being converted into carbohydrates through photosynthesis.

There are three main groups of organisms within an ecosystem. Plants, because they manufacture the high-energy molecules, are called *producers*, and the animals using these molecules as a source of energy are called *consumers*. Both plants and animals produce wastes and eventually die. This material forms a pool of dead organic matter known as *detritus*, which still contains considerable energy. (That's why we need wastewater

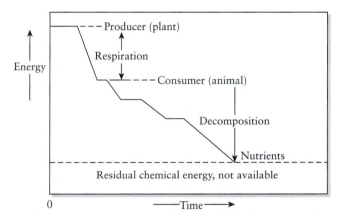

Figure 8.1 Loss of energy through the food chain. (After P. H. McGaughy, 1968. *Engineering Management of Water Quality*. New York: McGraw-Hill.)

treatment plants!) In the ecosystem the organisms that use this detritus are known as *decomposers*.

This one-way flow is illustrated in Figure 8.1. Note that the rate at which this energy is extracted (symbolized by the slope of the line) slows considerably as the energy level decreases—a concept important in wastewater treatment (Chapter 11). Because energy flow is one way (from the sun to the plants, to be used by the consumers and decomposers for making new cellular material and for maintenance), energy is not recycled within an ecosystem, as illustrated by the following argument.

Suppose a plant receives 1000 J of energy from the sun. Of that amount, 760 J is rejected (not absorbed) and only 240 J is absorbed. Most of this is released as heat, and only 12 J are used for production, 7 of which must go for respiration (maintenance) and the remaining 5 J toward building new tissue. If the plant is eaten by a consumer, 90% of the 5 J will go toward the animal's maintenance and only 10% (or 0.5 J) to new tissue. If this animal is, in turn, eaten, then again only 10% (or 0.05 J) will be used for new tissue, and the remaining energy is used for maintenance. If the second animal is a human being, then of the 1000 J coming from the sun, only 0.05 J or 0.005% is used for tissue building—a highly inefficient system.

While energy flow is in one direction only, nutrient flow through an ecosystem is cyclical, as represented by Figure 8.2. Starting with the dead organics, or the detritus, the initial decomposition by microorganisms produces compounds such as ammonia (NH_3), carbon dioxide (CO_2), and hydrogen sulfide (H_2S) for nitrogenous, carbonaceous, and sulfurous matter, respectively. These products are, in turn, decomposed further, until the final stabilized, or fully oxidized, forms are nitrates (NO_3^-), carbon dioxide, sulfates (SO_4^{2-}), and phosphates (PO_4^{3-}). The carbon dioxide is, of course, used by the plants as a source of carbon while the nitrates, phosphates, and sulfates are used as nutrients, or the building blocks for the formation of new plant tissue. The plants die or are used by consumers, which eventually die, returning us to decomposition.

Figure 8.2 Aerobic cycle for phosphorus, nitrogen, carbon, and sulfur. (After P. H. McGaughy, 1968. *Engineering Management of Water Quality.* New York: McGraw-Hill.)

There are three types of microbial decomposers: aerobic, anaerobic, and facultative. The microbes are classified according to whether or not they require molecular oxygen in their metabolic activity, that is, whether or not the microorganisms have the ability to use dissolved oxygen (O_2) as the *electron acceptor* in the decomposition reaction. The general equations for aerobic and anaerobic decomposers are

Aerobic: $\quad$ [Detritus] $+ O_2 \rightarrow CO_2 + H_2O +$ [Nutrients]

Anaerobic: $\quad$ [Detritus] $\rightarrow CO_2 + CH_4 + H_2S + NH_3 + \cdots +$ [Nutrients]

Obligate aerobes are microorganisms that must have dissolved oxygen to survive because they use oxygen as the electron acceptor, so in fairly simple terms, the hydrogen from the organic compounds ends up combining with the reduced oxygen to form water, as in the aerobic equation above. For *obligate anaerobes*, dissolved oxygen is, in fact, toxic, so they must use anaerobic decomposition processes. In anaerobic processes, the electron acceptor is an inorganic oxygen-containing compound, such as nitrates and sulfates. The nitrates are converted to nitrogen or ammonia (NH_3) and the sulfates to hydrogen sulfide (H_2S), as shown in the previous anaerobic equation. The microorganisms find it easier to use nitrates, so this process occurs more often. *Facultative microorganisms* use oxygen when it is available but can use anaerobic reactions if it is not available.

The decomposition carried out by the aerobic organisms is much more complete because some of the end products of anaerobic decomposition (e.g., ammonia nitrogen) are not in their final fully oxidized state. For example, aerobic decomposition is necessary to oxidize ammonia nitrogen to the fully oxidized nitrate nitrogen (Figure 8.2). All three types of microorganisms are used in wastewater treatment as discussed in Chapter 11.

Because nutrient flow in ecosystems is cyclical, it is possible to analyze these flows by using the techniques already introduced for material flow analysis:

$$\begin{bmatrix} \text{Rate of} \\ \text{materials} \\ \text{ACCUMULATED} \end{bmatrix} = \begin{bmatrix} \text{Rate of} \\ \text{materials} \\ \text{IN} \end{bmatrix} - \begin{bmatrix} \text{Rate of} \\ \text{materials} \\ \text{OUT} \end{bmatrix}$$

$$+ \begin{bmatrix} \text{Rate of} \\ \text{materials} \\ \text{PRODUCED} \end{bmatrix} - \begin{bmatrix} \text{Rate of} \\ \text{materials} \\ \text{CONSUMED} \end{bmatrix}$$

EXAMPLE
8.1

Problem   A major concern with the wide use of fertilizers is the leaching of nitrates into the groundwater. Such leaching is difficult to measure, unless it is possible to construct a nitrogen balance for a given ecosystem. Consider, for example, the diagram shown as Figure 8.3 on the next page, which represents nitrogen transfer in a meadow fertilized with 34 g/m^2/yr of ammonia + nitrate nitrogen $(17 + 17)$, both expressed as nitrogen. (Recall that the atomic weight of N is 14 and H is 1.0, so that 17 g/m^2/yr of nitrogen requires the application of $17 \times 17/14 = 20.6$ g/m^2/yr of NH_3.) What is the rate of nitrogen leaching into the soil?

Solution   First, note that organic nitrogen originates from three sources (cows, clover, and the atmosphere). The total output of organic N must equal the input, or $22 + 7 + 1 = 30$ g/m^2/yr. The organic N is converted to ammonia N, and the output of ammonia N must also be equal to the input from the organic N and the inorganic N (fertilizer), so $17 + 30 = 47$ g/m^2/yr. Of this 47 g/m^2/yr, 10 is used by the grass, leaving the difference, 37 g/m^2/yr, to be oxidized to nitrate nitrogen. Conducting a mass balance on nitrate nitrogen in g/m^2/yr and assuming steady state:

$$[\text{IN}] = [\text{OUT}]$$

$$37 + 17 + 1 = 8 + 20 + \text{Leachate}$$

$$\text{Leachate} = 27 \text{ g/m}^2/\text{yr}$$

◼

The process by which an ecosystem remains in a steady-state condition is called *home-ostasis*. There are, of course, fluctuations within a system, but the overall effect is steady state. To illustrate this idea, consider a very simple ecosystem consisting of grass, field mice, and owls, as pictorially represented in Figure 8.4 on page 216.

The grass receives energy from the sun, the mice eat the seeds from the grass, and the owl eats the mice. This progression is known as a *food chain*, and the interaction among the various organisms is a *food web*. Each organism is said to occupy a *trophic level*, depending on its proximity to the producers. Because the mouse eats the plants, it is at trophic level 1; the owl is at trophic level 2. It's also possible that a grasshopper eats the grass (trophic level 1) and a praying mantis eats the grasshopper (trophic level 2) and a shrew eats the praying mantis (trophic level 3). If the owl now gobbles up the shrew, it is performing at

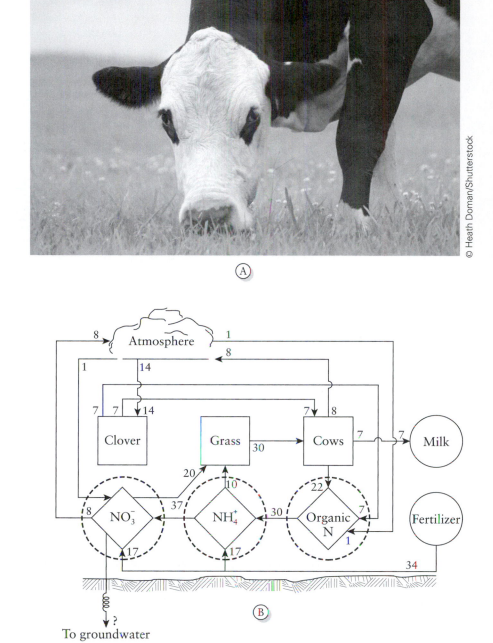

Figure 8.3   (A) An ecosystem showing (B) a nitrogen balance.

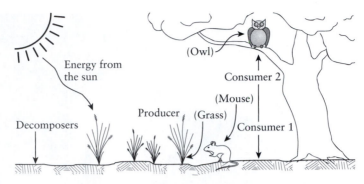

Figure 8.4   A simple ecosystem. The numbers indicate the trophic level.

trophic level 4. Figure 8.5 illustrates such a land-based food web. The arrows show how energy is received from the sun and flows through the system.

If a species is free to grow unconstrained by food, space, or predators, its growth is described as a first-order reaction:

$$\frac{dN}{dt} = kN$$

where N = number of organisms of a species
    k = rate constant
    t = time

Fortunately, populations (except for human populations!) within an ecosystem are constrained by food availability, space, and predators. Considering the first two constraints, the maximum population that can exist can be described in mathematical terms as

$$\frac{dN}{dt} = kN - \frac{k}{K}N^2$$

where K = maximum possible population in the ecosystem. Note that at steady state, $dN/dt = 0$, so

$$0 = kN - \frac{k}{K}N^2$$

and

$$K = N$$

or the population is the maximum population possible in that system. If for whatever reason N is reduced to below K, the population increases to eventually again attain the level K.

Various species can, of course, affect the population level of any other species. Suppose there are M organisms of another species that is in competition with the species of N organisms. Then the growth rate of the original species is expressed as

$$\frac{dN}{dt} = kN - \frac{k}{K}N^2 - sMN$$

where s = growth rate constant for the competitive species
    M = population of the competitive species

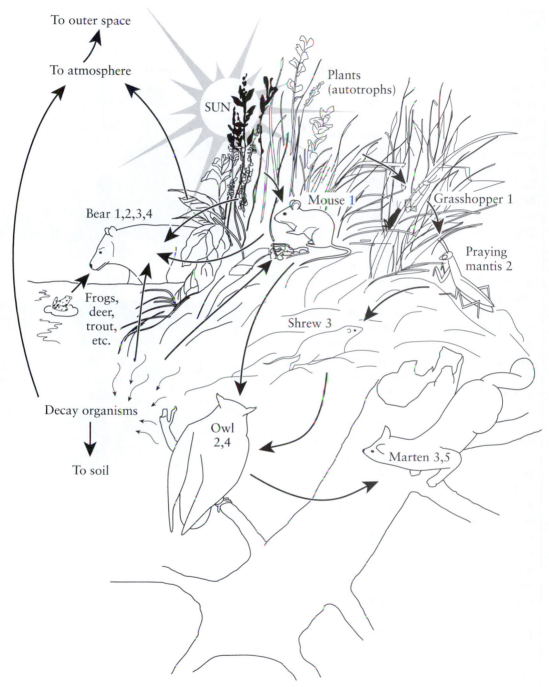

To outer space

To atmosphere

SUN

Plants (autotrophs)

Bear 1,2,3,4

Mouse 1

Grasshopper 1

Praying mantis 2

Frogs, deer, trout, etc.

Shrew 3

Decay organisms

Owl 2,4

Marten 3,5

To soil

Figure 8.5 A terrestrial food web. The numbers show the trophic level. (From Turk. *Ecosystem, Energy, Population*, 1E. © 1975 Brooks/Cole, a part of Cengage Learning, Inc. Reproduced by permission. www.cengage.com/permissions.)

FOCUS ON

The Kelp/Urchin/Sea Otter Ecological Homeostasis

A simplified example of a complex natural interaction is the restoration of the giant kelp forests along the Pacific coasts of Canada, the United States, and Mexico. This valuable seaweed, *Macrocystes*, forms 200-ft-long (61m) streamers fastened on the ocean floor and rising to the surface. Twice each year the kelp dies and regrows, being one of the fastest growing plants. In addition to the wealth of marine life for which the kelp forests provided habitat, kelp is a source of algin, a chemical used in foods, paints, cosmetics, and pharmaceuticals. Commercial harvesting is controlled to ensure the reproduction of the forests.

Some years ago, however, the destruction of these beds seemed inevitable. They were rapidly disappearing, leaving behind a barren ocean floor with no habitat for the marine life that had been so prevalent. The reason for the disappearance was disputed. Was it pollution? Overharvesting? Or some other more subtle cause?

The mystery was finally solved when it was discovered that sea urchins, which feed on the bottom, were eating the lower parts of the kelp plants, weakening their hold on the ocean bottom and allowing them to float away. The sea urchins in turn are the main source of food for the sea otters, and it is by this means that the population of sea urchins is kept in check (Figure 8.6). However, the hunting of sea otters had sufficiently depleted their numbers and allowed for an explosion of the sea urchin population, which in turn resulted in the depletion of the kelp forests. The sea otter is considered a keystone predator, a top predator that has a major influence on community structure. Kelp, on the other hand, is a dominator species because there is so much of it. Its very abundance helps keep things in balance. In each case, however, the species provides a platform on which a complicated food web is built. With the

(A)

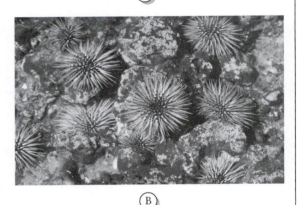

(B)

Figure 8.6 (A) Sea otter (© Heather A. Craig/Shutterstock) and (B) sea urchin (© Karoline Cullen/Shutterstock).

protection of the sea otters, the balance of life was restored, and the kelp forests are again growing in the Pacific.

Source: Reprinted from *Controlling Environmental Pollution*, Vesilind, P. A. and DiStefano, T. D., 2006. Lancaster, PA: DEStech Publications, Inc. Reproduced with permission of DEStech Publications, Inc.

Said in words, this expression reads

$$\begin{bmatrix} \text{Growth} \\ \text{rate} \end{bmatrix} = \begin{bmatrix} \text{Unlimited} \\ \text{growth rate} \end{bmatrix} - \begin{bmatrix} \text{Self-crowding} \\ \text{effects} \end{bmatrix} - \begin{bmatrix} \text{Competitive} \\ \text{effects} \end{bmatrix}$$

If s is small, then both species should be able to exist. If the first species of population N is not influenced by overcrowding, then the competition from the competitive species will be able to keep the population in check.

Note again that at steady state ($dN/dt = 0$):

$$N = K \left[1 - \frac{Ms}{k} \right]$$

so that, if $M = 0$, $N = K$.

Competition in an ecosystem occurs in niches. A *niche* is an animal's or plant's best accommodation with its environment, where it can best exist in the food web. Returning to the simple grass/mouse/owl example, each of the participants occupies a niche in the food web. If there is more than one kind of grass, each may occupy a very similar niche, but there will always be extremely important differences. Two kinds of clover, for example, may seem at first to occupy the same niche, until it is recognized that one species blossoms early and the other late in the summer; thus, they don't compete directly.

The greater the number of organisms available to occupy various niches within the food web, the more stable is the system. If, in the above example, the single species of grass dies due to a drought or disease, the mice would have no food, and they as well as the owls would die of starvation. If, however, there were *two* grasses, each of which the mouse could use as a food source (i.e., they both fill almost the same niche relative to the mouse's needs), the death of one grass would not result in the collapse of the system. This system is, therefore, more stable because it can withstand perturbations without collapsing. Examples of very stable ecosystems are tropical forests and estuaries while unstable ecosystems include the northern tundra and the deep oceans.

Some of the perturbations to ecosystems are natural (witness the destruction caused by the eruption of Mount St. Helens in Washington State) while many more are caused by human activities. Humans have, of course, adversely affected ecosystems on a small scale (streams and lakes) and on a large scale (global). The actions of humans can be thought of as *domination* of nature (as in Genesis I), and this idea has spawned a new philosophical approach to environmental ethics: ecofeminism.

Ecofeminism is "the position that there are important connections—historical, symbolic, theoretical—between the domination of women and domination of nonhuman nature."[1] The basic premise is that it should be possible to construct a better environmental ethic by incorporating into it, by analogy, the problems of domination of women by men.

If science is the dominant means of understanding the environment, ecofeminists argue that science needs to be transformed by feminism. Because women tend to be more nurturing and caring, this experience can be brought to bear on our understanding of the environment. They argue that the whole idea of domination of nature and detachment from nature has to change. Ecofeminism concentrates on the ethics of care, as opposed to the ethics of justice, and by that means avoids many of the problems inherent in the application of classical ethical thinking to the environment. Animal suffering, for example, is not a philosophical problem to ecofeminists because the care of all creatures is what is important.

Ecofeminists often invoke the spiritual dimension, asserting that earth has always been thought of as feminine (Mother Earth and Mother Nature) and that the domination of women by men has a striking parallel with the domination of earth. But such a similarity does not prove a connection. In addition, over the years women have often participated in the destruction of nature with the same vigor as men. Such destruction seems to be a *human* trait and not necessarily a masculine characteristic.

8.2 HUMAN INFLUENCE ON ECOSYSTEMS

Probably the greatest difference between people and all other living and dead parts of the global ecosystem is that people are unpredictable. All other creatures play according to well established rules. Ants build ant colonies, for example. They will always do that. The thought of an ant suddenly deciding that it wants to go to the moon or write a novel is ludicrous. And yet people, creatures who are just as much part of the global ecosystem as ants, can make unpredictable decisions, such as developing chlorinated pesticides, dropping nuclear bombs, or even changing the global climate. From a purely ecological perspective, we are indeed different—and scary.

Following are several of the more evident ways unpredictable humans can affect ecosystems.

8.2.1 Effect of Pesticides on an Ecosystem

The large-scale use of pesticides for the control of undesired organisms began during World War II with the invention and wide use of the first effective organic pesticide, DDT. Before that time, arsenic and other chemicals had been employed as agricultural pesticides, but the high cost of these chemicals and their toxicity to humans limited their use. DDT, however, was cheap, lasting, effective, and it did not seem to harm human beings. Many years later it was discovered that DDT decomposes very slowly, is stored in the fatty tissues of animals, and is readily transferred from one organism to another through the food chain. As it moves through the food chain, it is *biomagnified*, or concentrated as the trophic levels increase. For example, the concentration of DDT in one estuarine food chain is shown in Table 8.1. Note that the concentration factor from the DDT level in water to the larger birds is about 500,000. As the result of these very high concentrations of DDT (and subsequently from other chlorinated hydrocarbon pesticides as well), a number of birds were on the verge of extinction because the DDT affects their calcium metabolism, resulting in the laying of eggs with very thin (and easily broken) shells.

People, of course, occupy the top of the food chain and would be expected to have very high levels of DDT. Because it is impossible to relate DDT directly to any acute human disease, the increase in human DDT levels for many years was not an area of public health concern. Then, the subtle effects of chemicals such as DDT on human reproduction systems came into focus, and public concern finally forced DDT to be banned when it

Table 8.1   DDT Residues in an Estuarine Food Web

	DDT Residues (ppm)
Water	0.00005
Plankton	0.04
Minnows	0.23
Pickerel (predatory fish)	1.33
Heron (feeds on small animals)	3.57
Herring Gull (scavenger)	6.00
Merganser (fish-eating duck)	22.8

Source: Data from Woodell, G. M., et al. 1967. *Science*, 156: 821–824.

was discovered that human milk fed to infants often contained four to five times the DDT content allowable for the interstate shipment of *cows'* milk!

An interesting example of how DDT can cause a perturbation of an ecosystem occurred in a remote village in Borneo. A World Health Organization worker, attempting to enhance the health of the people, sprayed DDT inside of thatch huts in order to kill flies, which he feared would carry disease. The dying flies were easy prey for small lizards that live inside the thatch and feed on the flies. The large dose of DDT, however, made the lizards ill, and they, in turn, fell prey to the village cats. As the cats began to die, the village was invaded by rats, which were suspected of carrying bubonic plague. As a result, live cats were parachuted into the village to try to restore the balance upset by a well-meaning health official.[2]

8.2.2   Effect of Nutrients on a Lake Ecosystem

Lakes represent a second example of an ecosystem affected by people. A model of a lake ecosystem is shown in Figure 8.7. Note that the producers (algae) receive energy from the sun and through the process of photosynthesis produce biomass and oxygen. Because

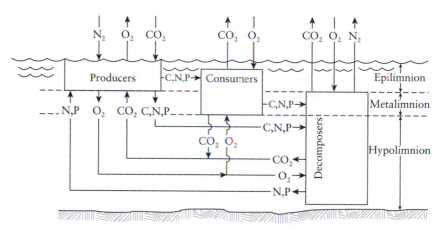

Figure 8.7   Movement of carbon, oxygen, and nutrients in a lake. (After Don Francisco, Department of Environmental Science and Engineering, University of North Carolina, Chapel Hill.)

the producers (the algae) receive energy from the sun, they obviously must be restricted to the surface waters in the lake. Fish and other animals also exist mostly in the surface water because much of the food is there, but some scavengers are on the bottom. The decomposers mostly inhabit the bottom waters because this is the source of their food supply (the detritus).

Through photosynthesis, the algae use nutrients and carbon dioxide to produce high-energy molecules and oxygen. The consumers, including fish, plankton, and many other organisms, all use the oxygen, produce CO_2, and transfer the nutrients to the decomposers in the form of dead organic matter. The decomposers, including scavengers such as worms and various forms of microorganisms, reduce the energy level further by the process of respiration, using oxygen and producing carbon dioxide. The nutrients, nitrogen and phosphorus (as well as other nutrients, often called micronutrients), are then again used by the producers. Some types of algae are able to fix nitrogen from the atmosphere while some decomposers produce ammonia nitrogen that bubbles out. (This discussion is considerably simplified from what actually occurs in an aquatic ecosystem. For a more accurate representation of such systems, see any modern text on aquatic ecology.)

The only element of major importance that does not enter the system from the atmosphere is phosphorus. A given quantity of phosphorus is recycled from the decomposers back to the producers. The fact that only a certain amount of phosphorus is available to the ecosystem limits the rate of metabolic activity. Were it not for the limited quantity of phosphorus, the ecosystem metabolic activity could accelerate and eventually self-destruct because all other chemicals (and energy) are in plentiful supply. For the system to remain at homeostasis (steady state), some key component, in this case phosphorus, must limit the rate of metabolic activity by acting as a brake in the process.

Consider now what would occur if an external source of phosphorus, such as from farm runoff or wastewater treatment effluent, would be introduced. The brake on the system would be released, and the algae would begin to reproduce at a higher rate, resulting in a greater production of food for the consumers, which in turn would grow at a higher rate. All of this activity would produce ever increasing quantities of dead organic matter for the decomposers, which would greatly multiply.

Unfortunately, the dead organic matter is distributed throughout the body of water (as shown in Figure 8.7) while the algae, which produce the necessary oxygen for the decomposition to take place, live only near the surface of the lake, where there is sunlight. The other supply of oxygen for the decomposers is from the atmosphere, also at the water surface. The oxygen, therefore, must travel through the lake to the bottom to supply the aerobic decomposers. If the water is well mixed, this would not provide much difficulty, but unfortunately, most lakes are thermally stratified and not mixed throughout the full depth, so not enough O_2 can be transported quickly enough to the lake bottom.

The reason for such stratification is shown in Figure 8.8. In the winter, with ice on the surface, the temperature of northern lakes is about 4°C, the temperature at which water is the densest. In warmer climates, where ice does not form, the deepest waters are still normally at this temperature. In the spring, the surface water begins to warm, and for a time, the water in the entire lake is about the same temperature. For a brief time, wind can produce mixing that can extend to the bottom. As summer comes, however, the deeper, denser water sits on the bottom as the surface water continues to warm, producing a steep gradient and three distinct sections: *epilimnion*, *metalimnion*, and *hypolimnion*. The

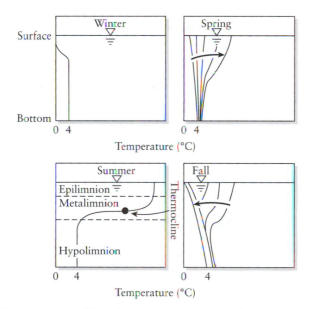

Figure 8.8   Temperature profiles in a temperate climate lake.

inflection point is called the *thermocline*. In this condition the lake is thermally stratified, so there is no mixing between the three strata. During this season, when the metabolic activity can be expected to be the greatest, oxygen gets to the bottom only by diffusion. As winter approaches and the surface water cools, it is possible for the water on top to cool enough to be denser than the lower water, and a *fall turnover* occurs, a thorough mixing. During the winter, when the water on the top is again lighter, the lake is again stratified.

In stratified lakes with an increased food supply, oxygen demand by the decomposers increases and finally outstrips the supply. The aerobic decomposers die and are replaced by anaerobic forms, which produce large quantities of biomass and incomplete decomposition. Eventually, with an ever increasing supply of phosphorus, the entire lake becomes anaerobic, most fish die, and the algae is concentrated on the very surface of the water, forming green slimy algae mats called *algal blooms*. Over time the lake fills with dead and decaying organic matter, and a peat bog results.

This process is called *eutrophication*. It is actually a naturally occurring phenomenon because all lakes receive some additional nutrients from the air and overland flow. Natural eutrophication is, however, usually very slow, measured in thousands of years before major changes occur. What people have accomplished, of course, is to speed up this process with the introduction of large quantities of nutrients, resulting in *accelerated eutrophication*. Thus, an aquatic ecosystem that might have been in essentially steady state (homeostasis) has been disturbed, and an undesirable condition results.

Incidentally, the process of eutrophication can be reversed by reducing nutrient flow into a lake and flushing or dredging as much of the phosphorus out as possible, but this is an expensive proposition.

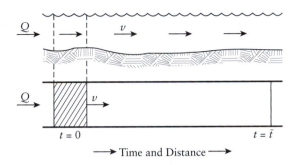

Figure 8.9 A stream acting as a plug-flow reactor.

8.2.3 Effect of Organic Wastes on a Stream Ecosystem

The primary difference between a stream and a lake is that the former is continually flushed out. Streams, unless they are exceptionally lethargic, are therefore seldom highly eutrophied. The stream can, in fact, be hydraulically characterized as a plug-flow reactor, as pictured in Figure 8.9. A plug of water moves downstream at the velocity of the flow, and the reactions that take place in this plug can be analyzed.

Similar to lakes, the reaction of greatest concern in stream pollution is the depletion of oxygen. Thinking of the plug as a black box and a mixed batch reactor, it is possible to write a material balance in terms of oxygen as

$$
\begin{bmatrix} \text{Rate of} \\ \text{oxygen} \\ \text{ACCUMULATED} \end{bmatrix} = \begin{bmatrix} \text{Rate of} \\ \text{oxygen} \\ \text{IN} \end{bmatrix} - \begin{bmatrix} \text{Rate of} \\ \text{oxygen} \\ \text{OUT} \end{bmatrix}
$$
$$
+ \begin{bmatrix} \text{Rate of} \\ \text{oxygen} \\ \text{PRODUCED} \end{bmatrix} - \begin{bmatrix} \text{Rate of} \\ \text{oxygen} \\ \text{CONSUMED} \end{bmatrix}
$$

Because this is not a steady state situation, the first term is not zero. The oxygen IN to the black box (the stream) is called *reoxygenation* and in this case consists of diffusion from the atmosphere. There is no oxygen OUT because the water is less than saturated with oxygen. If the stream is swift and algae do not have time to grow, there is no oxygen PRODUCED. The oxygen used by the microorganisms in respiration is called *deoxygenation* and is the oxygen CONSUMED term. The accumulation, as before, is expressed as a differential equation, so that

$$
\frac{dC}{dt} = \begin{bmatrix} \text{Rate } O_2 \text{ IN} \\ \text{(reoxygenation)} \end{bmatrix} - 0 + 0 - \begin{bmatrix} \text{Rate } O_2 \text{ CONSUMED} \\ \text{(deoxygenation)} \end{bmatrix}
$$

where $C = $ concentration of oxygen, mg/L.

Both reoxygenation and deoxygenation can be described by first-order reactions. The rate of oxygen use, or deoxygenation, can be expressed as

$$
\text{Rate of deoxygenation} = -k_1 C
$$

FOCUS ON

Nutrient Loads

The City of Seattle began discharging raw sewage into Lake Washington in 1900. It wasn't until 1926 that the city began to treat its wastewater and divert the effluent to the Duwamish River that discharges into Puget Sound. By then, development had circled Lake Washington, and 10 treatment plants were constructed around the lake to handle the wastewater discharges. But none of these plants removed nutrients, and the level of phosphorus in the lake water was constantly climbing. In 1955, researchers at the University of Washington discovered a new kind of algae in the lake, the classic telltale blue-green algae that accompanies accelerated eutrophication and water quality deterioration. Using knowledge gained from accelerated eutrophication in lakes in Switzerland, the researchers pinpointed the problem as excessive phosphorus.

As the studies continued, water quality in the lake continued to deteriorate. The *Seattle Times* in one 1963 article referred to the lake as Lake Stinko. But relief was on the way. A trunk sewer was constructed around the lake to intercept all of the wastewater and carry it out of the lake watershed, thus significantly reducing the level of phosphorus entering the lake. By 1971 the lake water quality was beginning to show distinct improvement, and the clarity as measured by the Secchi disc was better than it had been in 1955. The lake became a model of what can be done with concerted public commitment.

On the opposite side of the United States, on the southern shore of Long Island, the duckling industry was booming in the 1940s.[3] New farms were established every year. During this time, the residents began to notice a significant change in the water quality of the Long Island sounds. The oyster and hard-shell crabs disappeared, and huge green algal blooms began to appear along the shores. The clams survived, but the levels of e-coli were so high that the clams were not commercially usable. Studies showed that the algae were not growing in the water but were washed from the bays and tributaries and that the major source of pollution was the duck farms. Subsequent experiments showed that the limiting nutrient in the bays and tributaries was not phosphorus but nitrogen! Ducks produce large quantities of uric acid, which contains nitrogen, and this was what was responsible for the production of excessive algae.

where k_1 = deoxygenation constant, a function of the type of waste material decomposing, temperature, etc., d^{-1}

C = the amount of oxygen per unit volume necessary for decomposition, measured as mg/L

The value of k_1, the deoxygenation constant, is measured in the laboratory using analytical techniques discussed in the next chapter. In Chapter 9, C is defined as the difference between the biochemical oxygen demand (BOD) at any time t and the ultimate BOD. For now consider C as the oxygen necessary by the microorganisms at any time to complete the decomposition. If C is high, then the rate of oxygen use is great.

The reoxygenation of water can be expressed as

$$\text{Rate of reoxygenation} = k_2 D$$

where $D =$ deficit in dissolved oxygen (DO), or the difference between saturation
(maximum dissolved oxygen the water can hold) and the
actual DO, mg/L
$k_2 =$ reoxygenation constant, d^{-1}

Water can hold only a limited amount of a gas; the amount of oxygen that can be dissolved in water depends on the water temperature, atmospheric pressure, and the concentration of dissolved solids. The saturation level of oxygen in deionized water at one atmosphere and at various temperatures is shown in Table 8.2. The value of k_2 is obtained by conducting a study on a stream using tracers, using empirical equations that include stream conditions, or using tables that describe various types of streams (Table 8.3).

The constants can be adjusted for temperature using[4]

$$k_1' = k_1 (1.135)^{T-20}$$
$$k_2' = k_2 (1.024)^{T-20}$$

where $T =$ temperature of the water, °C.

In a stream loaded with organic material, the simultaneous action of deoxygenation and reoxygenation forms what is called the *dissolved oxygen (or DO) sag curve*, first

Table 8.2   Oxygen Solubility in Fresh Water at 1 atm*

Temperature (°C)	Dissolved Oxygen (mg/L)	Temperature (°C)	Dissolved Oxygen (mg/L)
0	14.60	23	8.56
1	14.19	24	8.40
2	13.81	25	8.24
3	13.44	26	8.09
4	13.09	27	7.95
5	12.75	28	7.81
6	12.43	29	7.67
7	12.12	30	7.54
8	11.83	31	7.41
9	11.55	32	7.28
10	11.27	33	7.16
11	11.01	34	7.05
12	10.76	35	6.93
13	10.52	36	6.82
14	10.29	37	6.71
15	10.07	38	6.61
16	9.85	39	6.51
17	9.65	40	6.41
18	9.45	41	6.31
19	9.26	42	6.22
20	9.07	43	6.13
21	8.90	44	6.04
22	8.72	45	5.95

Source: *For atmospheric pressures of less than 1 atm, the dissolved oxygen concentration has to be reduced because there is less pressure to push the oxygen into the water. Published correction factors are available.

Table 8.3 Empirical Reoxygenation Constants

	k_2 at 20°C (d^{-1})
Small backwaters	0.10–0.23
Lethargic streams	0.23–0.35
Large streams, low velocity	0.35–0.46
Large streams, normal velocity	0.46–0.69
Swift streams	0.69–1.15
Rapids	1.15

Source: Reproduces with permission from Mechanism of reaereation in natural streams by O'Connor and Dobbins (ASCE 1958).

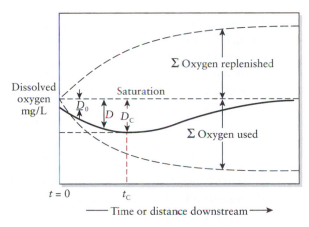

Figure 8.10 The dissolved oxygen sag curve in a stream is the difference between the oxygen used and oxygen supplied.

described by Streeter and Phelps in 1925.[5] The shape of the oxygen sag curve, as shown in Figure 8.10, is the result of adding the rate of oxygen use (consumption) and the rate of supply (reoxygenation). If the rate of use is great, as in the stretch of stream immediately after the introduction of organic pollution, the dissolved oxygen level drops because the supply rate cannot keep up with the use of oxygen, creating a deficit. The deficit (D) is defined as the difference between the oxygen concentration in the streamwater (C) and the total amount the water *could* hold, or saturation (S). That is:

$$D = S - C \tag{8.1}$$

where D = oxygen deficit, mg/L
S = saturation level of oxygen in the water (the most it can ever hold), mg/L
C = concentration of dissolved oxygen in the water, mg/L

After the initial high rate of decomposition when the readily degraded material is used by the microorganisms, the rate of oxygen use decreases because only the less readily decomposable materials remain. Because so much oxygen has been used, the deficit (the difference between oxygen saturation level and actual dissolved oxygen) is great, but the supply of oxygen from the atmosphere is high and eventually begins to keep up with the use, so the deficit begins to level off. Eventually, the dissolved oxygen once again reaches saturation levels, creating the dissolved oxygen sag.

This process can be described in terms of consecutive reactions (Chapter 5), the rate of oxygen use and the rate of oxygen resupply, and can be expressed as:

$$\frac{dD}{dt} = k_1 z - k_2 D$$

where z is the amount of oxygen still required by the microorganisms decomposing the organic material. The rate of change in the deficit (D) depends on the concentration of decomposable organic matter, or the remaining need by the microorganisms for oxygen (z), and the deficit at any time t. As explained more fully in the next chapter, the demand for oxygen at any time t can be expressed as

$$z = L e^{-k_1 t}$$

where $L =$ ultimate oxygen demand, or the maximum oxygen required, mg/L.

Similar to Chapter 5, by substituting this into the above expression and integrating, we obtain the Streeter–Phelps deficit equation:

$$D = \frac{k_1 L_0}{k_2 - k_1} \left(e^{-k_1 t} - e^{-k_2 t} \right) + D_0 e^{-k_2 t} \tag{8.2}$$

where $D =$ oxygen deficit at any time t, mg/L
$D_0 =$ oxygen deficit immediately below the pollutant discharge
   location, mg/L
$L_0 =$ ultimate oxygen demand immediately below the pollutant discharge
   location, mg/L

The most serious concern of water quality is, of course, the point in the stream where the deficit is the greatest, which is the point where the dissolved oxygen concentration is least. By setting $dD/dt = 0$ (where the dissolved oxygen sag curve flattens out and begins to rise), it is possible to solve for the critical time as

$$t_c = \frac{1}{k_2 - k_1} \ln \left[\frac{k_2}{k_1} \left(1 - \frac{D_0(k_2 - k_1)}{k_1 L_0} \right) \right] \tag{8.3}$$

where $t_c =$ time downstream when the dissolved oxygen is at the lowest concentration.

EXAMPLE
8.2

Problem   A large stream has a reoxygenation constant of 0.4 d^{-1} and a velocity of 0.85 m/s. At the point at which an organic pollutant is discharged, it is saturated with oxygen at 10 mg/L ($D_0 = 0$). Below the outfall the ultimate demand for oxygen is found to be 20 mg/L, and the deoxygenation constant is 0.2 d^{-1}. What is the dissolved oxygen 48.3 km downstream?

Solution   Velocity = 0.85 m/s, hence it takes

$$\frac{48.3 \times 10^3 \text{ m}}{0.85 \text{ m/s}} = 56.8 \times 10^3 \text{ s} = 0.66 \text{ day}$$

to travel the 48.3 km.

Using the Streeter–Phelps deficit equation (Equation 8.2):

$$D = \frac{(0.2 \text{ day}^{-1})(20 \text{ mg/L})}{(0.4 \text{ day}^{-1}) - (0.2 \text{ day}^{-1})}(e^{-0.2 \text{ day}^{-1}(0.66 \text{ day})} - e^{-0.4 \text{ day}^{-1}(0.66 \text{ day})}) + 0$$

$$D = 2.2 \text{ mg/L}$$

If the stream at 48.3 km is still saturated with dissolved oxygen at 10 mg/L and there were no additional flows, then the dissolved oxygen content at 48.3 km is (Equation 8.1) $10 - 2.2 = 7.8$ mg/L.

An inherent assumption made in Example 8.2 is not always true—that the volume of the polluting stream is very small compared to the stream flow and, thus, the initial deficit in the stream below the outfall is the same as the deficit above the outfall. It is more accurate to calculate the deficit by performing a material balance on the plug at $t = 0$ under the assumption that within this plug the two incoming streams are well mixed. Figure 8.11 shows that the flow of the stream and its dissolved oxygen concentration are Q_s and C_s, respectively, and the pollution stream has a flow and dissolved oxygen of Q_p and C_p. Thus,

Figure 8.11   The concentration of oxygen in the stream downstream of the source of pollution is a combination of the oxygen concentration in the stream upstream of the discharge (C_s) and the oxygen concentration in the pollutant stream (C_p).

a volume flow balance gives

$$[IN] = [OUT]$$

$$Q_s + Q_p = Q_0$$

where Q_s = upstream flow
  Q_p = flow from the pollution source
  Q_0 = downstream flow

and according to Chapter 3, Mass = Volume × Concentration, so the balance in terms of oxygen mass can be written as

$$Q_sC_s + Q_pC_p = Q_0C_0$$

where C_0 = dissolved oxygen concentration in the stream immediately below the confluence of the pollution. Rearranged:

$$C_0 = \frac{Q_sC_s + Q_pC_p}{Q_s + Q_p} \tag{8.4}$$

Note that this assumes perfect mixing at the pollution source, not a very realistic assumption, especially if the river is broad and lethargic. In this text we ignore such problems.

Once C_0 is determined, the water temperature is calculated by using Equation 7.1 (a material balance on heat energy):

$$T_0 = \frac{Q_sT_s + Q_pT_p}{Q_s + Q_p} \tag{8.5}$$

and the saturation value S_0 at T_0 is found in Table 8.2. The initial deficit is then calculated as

$$D_0 = S_0 - C_0 \tag{8.6}$$

The dissolved oxygen sag curve that incorporates the effect of the pollutant stream on the initial deficit D_0 is shown in Figure 8.12.

The stream may also have a demand for oxygen at the point at which it reaches the outfall. Assuming once again complete mixing, the oxygen demand below the outfall must be calculated as

$$L_0 = \frac{L_sQ_s + L_pQ_p}{Q_s + Q_p} \tag{8.7}$$

where L_0 = ultimate oxygen demand immediately below the outfall, mg/L
  L_s = ultimate oxygen demand of the stream immediately
    above the outfall, mg/L
  L_p = ultimate oxygen demand of the pollutant discharge, mg/L

Figure 8.12   The initial deficit D_0 is influenced by the deficit in the incoming pollutant stream.

EXAMPLE 8.3

Problem   Suppose the waste stream in Example 8.2 has a dissolved oxygen concentration of 1.5 mg/L, a flow of 0.5 m³/s, a temperature of 26°C, and an ultimate biochemical oxygen demand (BOD) of 48 mg/L. The streamwater is running at 2.2 m³/s at a saturated dissolved oxygen concentration, a temperature of 12°C, and an ultimate BOD of 13.6 mg/L. Calculate the dissolved oxygen concentration 48.3 km downstream.

Solution   From Table 8.2, $S = 10.8$ mg/L at 12°C; because the stream is saturated, $S = C_s$, so using Equations 8.4 and 8.5

$$C_0 = \frac{Q_s C_s + Q_p C_p}{Q_s + Q_p} = \frac{2.2 \text{ m}^3/\text{s}(10.8 \text{ mg/L}) + 0.5 \text{ m}^3/\text{s}(1.5 \text{ mg/L})}{(2.2 + 0.5) \text{ m}^3/\text{s}} = 9.1 \text{ mg/L}$$

$$T_0 = \frac{Q_s T_s + Q_p T_p}{Q_s + Q_p} = \frac{2.2 \text{ m}^3/\text{s}(12°\text{C}) + 0.5 \text{ m}^3/\text{s}(26°\text{C})}{(2.2 + 0.5) \text{ m}^3/\text{s}} = 14.6°\text{C}$$

At $T_0 = 14.6°$C, $S_0 = 10.2$ mg/L from Table 8.2, thus using Equation 8.6

$$D_0 = S_0 - C_0 = 10.2 - 9.1 = 1.1 \text{ mg/L}$$

Using Equation 8.7, the ultimate BOD in the stream immediately below the outfall is

$$L_0 = \frac{L_s Q_s + L_p Q_p}{Q_s + Q_p} = \frac{13.6 \text{ mg/L}(2.2 \text{ m}^3/\text{s}) + 48 \text{ mg/L}(0.5 \text{ m}^3/\text{s})}{(2.2 + 0.5) \text{ m}^3/\text{s}} = 20 \text{ mg/L}$$

Using the Streeter–Phelps equation (Equation 8.2), the deficit 48.3 km downstream is then

$$D = \frac{k_1 L_0}{k_2 - k_1}(e^{-k_1 t} - e^{-k_2 t}) + D_0(e^{-k_2 t})$$

calculated as:

$$D = \frac{0.2 \text{ day}^{-1}(20 \text{ mg/L})}{(0.4 - 0.2) \text{ day}^{-1}}(e^{-0.2 \text{ day}^{-1}(0.66 \text{ day})} - e^{-0.4 \text{ day}^{-1}(0.66 \text{ day})})$$
$$+ 1.1 \text{ mg/L}(e^{-0.4 \text{ day}^{-1}(0.66 \text{ day})}) = 3.0 \text{ mg/L}$$

Again, assuming the temperature 48.3 km downstream is still T_0 (14.6°C) and there are no additional flows (Equation 8.1), $C = S - D = 10.2 - 3.0 = 7.2$ mg/L.

The reason, of course, to calculate the dissolved oxygen (DO) is to determine whether organisms will be able to survive in the water. Can a fish survive at a 7.2 mg/L DO? The answer is yes. The optimum DO level depends on the fish species (in particular whether it lives in cold water or warm water) plus a variety of other factors (e.g., fish age, activity level, and size). However, fish usually die when the DO falls to 2 or 3 mg/L. Between 3 and 4 mg/L, they will be severely stressed. If the level stays in this range, the fish will likely die. Typically, 5 mg/L or higher is considered adequate. But it is possible to have too much of a good thing. Fish in water with a very high DO level may die from bubbles of oxygen blocking the blood flow.

Obviously, the impact of organics on a stream is not limited to the effect on dissolved oxygen. As the environment changes, the competition for food and survival results in a change in various species of microorganisms in a stream, and the chemical makeup changes as well. Figure 8.13 illustrates the effect of an organic pollutant load on a stream. Note especially the shift in nitrogen species through organic nitrogen to ammonia to nitrite to nitrate. Compare this to the previous discussion of nitrogen in the nutrient cycle (Figure 8.2). The changes in stream quality as the decomposers reduce the oxygen-demanding material, finally achieving a clean stream, is known as *self-purification*. This process is no different from what occurs in a wastewater treatment plant because in both cases energy is intentionally wasted. The organics contain too much energy, and they must be oxidized to more inert materials.

In the next chapter the means for characterizing water and its pollution are presented, and later the idea of energy wasting in a wastewater treatment plant is discussed in greater detail.

8.2.4   Effect of Design on an Ecosystem

As illustrated by the discharge of organics and nutrients into waterways, engineers' designs often have major impacts on ecosystems. However, with some forethought, engineers can use techniques to minimize the negative impacts. For example, traffic kills many animals every year in addition to fragmenting ecosystems. In Florida, transportation engineers have designed a wildlife barrier wall with culvert underpasses to reduce the number of animals killed on roadways.[6] A 1m high tapered concrete wall with a 15 cm lip on the top, based on walls used in animal exhibits, is located 10 m from the edge of the travel lane to discourage animals from entering the roadway. Approximately every 1 km, culvert underpasses with drop boxes in the median for light allow the animals to pass from one side to the

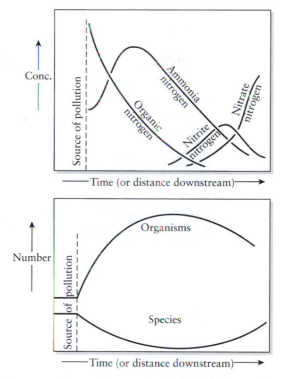

Figure 8.13 Nitrogen and aquatic organisms below a source of organic stream pollution.

other. At another location, a landscaped overpass (or land bridge) is being used to provide a route for animals to safely cross an interstate. Canadian transportation engineers also used underpasses and overpasses but in conjunction with 2.4-m-high wildlife exclusion fencing attached to a 1.5-m chainlink fencing apron buried at a 45-degree angle (to prevent tenacious animals from pulling up or digging beneath the fence). The average construction cost of the overpasses was 60 times more than the least expensive option used (creek pathways within open span culverts), but subsequent research showed that almost half the crossings occurred at the overpasses. The design of fencing in combination with animal crossing structures achieved an impressive 97% reduction in wildlife-vehicle collisions for most species.[7] Obviously, besides reducing the negative impacts on the ecosystems, these designs improve safety and reduce the costs associated with accidents caused by wildlife-vehicle collisions.

The design of new structures and the remodeling of existing structures offer engineers many opportunities to incorporate energy-efficient, water-efficient, and environmentally-friendly materials and devices into their designs, a process known as *green*, or *sustainable*, design. While potentially costing more up front, these options often result in long-term operational savings and healthier living and work spaces. Natural ventilation and day-lighting offer building occupants personal control over their space and can reduce energy costs. Vegetated, or green, roofs can be used to reduce energy costs and stormwater runoff

(A)

(B)

Figure 8.14   Examples of green roofs: (A) G.R.E.E.N.'s Southern Illinois University Edwardsville Engineering Building research site; (B) residential site in Arizona.

(Figure 8.14 on the next page). Stormwater can be kept onsite and allowed to infiltrate or used onsite to water landscaping. Onsite electricity generation with solar systems, wind power, or fuel cells can reduce electrical costs and environmental pollution. Choosing less toxic materials and sustainably-derived materials reduces waste generation and improves indoor air quality. While these types of choices are a (large) step in the right direction, engineers also need to consider what will happen to the structure at the end of its life and design accordingly—for example, designing to allow for easy deconstruction.

Of course, these types of designs mean that engineers have to work in teams, often with nonengineers, such as biologists and landscape architects. And nonengineers rarely speak the same language as engineers, and they typically have different priorities. But they can offer insights and solutions that engineers often miss. So, teamwork, communication skills, and patience are vital. The result, however, can be a vastly improved design.

FOCUS ON

The Beaver Deceiver

Hundreds of years ago, before humans nearly wiped out Europe's indigenous beaver populations for the fur trade, there were dams along virtually every stream in what is now modern Europe. Trapping beavers for their fur was exported to the United States, with similarly disastrous results. Today, in both Europe and North America, beavers are making a comeback. Their return is due in part to the increasing understanding and appreciation for the environmental services they provide. Beaver activities create and sustain wetlands, allow sediments and toxic materials to filter out of watersheds before they can harm aquatic ecosystems, create biodiversity, and provide habitats for both plants and animals. Some biologists have argued that beaver dams throughout a major drainage area—even ones as large as the Mississippi and Missouri basins—will help prevent catastrophic floods.

The first thing to understand about beavers is that the urge to dam is deeply ingrained in them. Damming behavior is instinctive and is stimulated by the sound and feel of flowing water. Beavers dam to surround themselves with a stable body of water that can protect them from predators and provide watery access to food sources in the vicinity. Beavers are expert on where to construct dams and typically select sites where running water is the most vulnerable—at natural or human-made constrictions—including culverts that channel streams under roadways.

In early 2007 a pair of adult beavers arrived in Alhambra Creek in downtown Martinez, California. The city found itself the target of international attention when city officials decided to euthanize the beavers. The public strongly objected. California's Department of Fish and Game offered to relocate the beavers, but relocation of beavers is generally unsuccessful and the public strongly supported keeping the beavers in Martinez. The city was in a jam. They called in wildlife biologist Skip Lisle to install a device called a "beaver deceiver," which Lisle designed to deter beavers from building dams. Lisle used his knowledge of beaver biology and his experience in construction to create the deceiver concept, which is being used in growing numbers in North America and Europe (Figure 8.15).

Lisle determined that by building a fence outward from a road culvert, beavers are forced away from the vulnerable spot in their attempts to dam. The fence, typically made with wood posts and sheets of heavy-gauge steel fencing, blocks off a trapezoidal space in the stream, just outside the culvert opening. The deceivers block access for the beavers but leave the culvert open, allowing fish and other aquatic animals through. The fences push the beaver far enough away from the vulnerable opening that the animal apparently decides to abandon its efforts to dam the stream at that point.

Sometimes the fences are not needed at all, especially when the objective is just to lower the level of water that a dam is holding back. In these instances, a simple pipe system can be constructed. The inlet end of the pipe lies in the beaver-made pond, surrounded by fencing to prevent the animal from blocking the drain. The discharge end of the pipe lies on the other side of the dam. The trick is to access the dam at the desired water level.

Even beyond arguments for humane treatment, the logic for living compatibly with beavers is overwhelming. The challenge is to prevent beavers from doing harm. Note that, as superb as beavers are as engineers, their survival depends on human engineers being even better.

Figure 8.15 (A) The Beaver Deceiver and (B) its target.

Source: Adapted from "U.S. Beaver Management Tool Crosses the Pond," by John Hadidian, November 11, 2003, HSUS. Used by permission.

PROBLEMS

8.1 It has been suggested that the limiting concentration of phosphorus for accelerated eutrophication is between 0.1 and 0.01 mg/L of P. Typical river water might contain 0.2 mg/L P, 50% of which comes from farm and urban runoff, and 50% from domestic and industrial wastes. Synthetic detergents contribute 50% of the P in municipal and industrial waste.

  a. If all phosphorus-based detergents are banned, what level of P would you expect in a typical river?
  b. If this river flows into a lake, would you expect the phosphate detergent ban to have much effect on the eutrophication potential of the lakewater? Why or why not?

8.2 A stream has a dissolved oxygen level of 9 mg/L, an ultimate oxygen demand of 12 mg/L, and an average flow of 0.2 m^3/s. An industrial waste at zero dissolved oxygen with an ultimate oxygen demand of 20,000 mg/L and a flow rate of 0.006 m^3/s is discharged into the stream. What are the ultimate oxygen demand and the dissolved oxygen in the stream immediately below the discharge?

8.3 Below a discharge from a wastewater treatment plant, an 8.6-km stream has a reoxygenation constant of 0.4 d^{-1}, a velocity of 0.15 m/s, a dissolved oxygen concentration of 6 mg/L, and an ultimate oxygen demand (L) of 25 mg/L. The stream is at 15°C. The deoxygenation constant is estimated at 0.25 d^{-1}.

  a. Will there be fish in this stream?
  b. Why should we care if there are fish in the stream? Do the fish deserve moral consideration and protection? What arguments can you muster to support this view?

8.4 A municipal wastewater treatment plant discharges into a stream that, at some times of the year, has no other flow. The characteristics of the waste are

Flow $= 0.1$ m^3/s
Dissolved oxygen $= 6$ mg/L
Temperature $= 18$°C
$k_1 = 0.23$ d^{-1}
Ultimate BOD (L) $= 280$ mg/L

The velocity in the stream is 0.5 m/s, and the reoxygenation constant k_2 is assumed to be 0.45 d^{-1}.

  a. Will the stream maintain a minimum DO of 4 mg/L?
  b. If the flow of streamwater above the outfall has a temperature of 18°C, has no demand for oxygen, and is saturated with DO, how great must the streamflow be to ensure a minimum dissolved oxygen of 4 mg/L downstream of the discharge?

(*Note*: Please do not try this by hand. Use a computer!)

8.5 Ellerbe Creek, a large stream at normal velocity, is the recipient for the wastewater from the 0.44 m^3/s Durham Northside Wastewater Treatment Plant. It has a mean summertime flow of 0.28 m^3/s, a temperature of 24°C, and a velocity of 0.25 m/s. Assume the dissolved oxygen is saturated and the ultimate BOD is O mg/L. The wastewater characteristics are

Temperature $= 28$°C
Ultimate BOD (L) $= 40$ mg/L
$k_1 = 0.23$ d^{-1}
Dissolved oxygen $= 2$ mg/L

The total stream length from outfall to river is 23 km, at which point it empties into the Neuse River. (*Note*: Use Table 8.3 for k_2.)

  a. Should the State of North Carolina be concerned about the effect of this discharge on Ellerbe Creek?
  b. Other than legal considerations, why *should* the state be concerned with the oxygen levels in Ellerbe Creek? It isn't much of a creek, actually, and empties out into the Neuse River without being of much use to

anyone. And yet the state has set dissolved oxygen levels of 4 mg/L for the 10-yr, 7-day low flow. Write a letter to the editor of a fictitious local newspaper decrying the spending of tax revenues for the improvements to the Northside Treatment Plant just so the dissolved oxygen levels in Ellerbe Creek can be maintained above 4 mg/L.

c. Pretend you are a fish in Ellerbe Creek. You have read the letter to the editor in part b. above, and you are royally ticked off. You take pen in fin and respond. Write a letter to the editor from the standpoint of the fish. The quality of your letter will be judged on the basis of the strength of your arguments.

8.6 A large stream with a velocity of 0.85 m/s, saturated with oxygen, has a reoxygenation constant $k_2 = 0.4$ d^{-1} and a temperature of $T = 12°C$, with an ultimate BOD $= 13.6$ mg/L and a flow rate $Q = 2.2$ m^3/s. Into this stream flows a wastewater stream with a flow rate of 0.5 m^3/s, $T = 26°C$, $L = 220$ mg/L. Downstream the deoxygenation constant is $k_1 = 0.2$ d^{-1} and the oxygen level is 1.5 mg/L. What is the dissolved oxygen 48.3 km downstream?

8.7 If algae contain P:N:C, in the proportion of 1:16:100, which of the three elements would be limiting algal growth if the concentration in the water were

0.20 mg/L P
0.32 mg/L N
1.00 mg/L C

Show your calculations.

8.8 If offensive football teams are described in terms of the positions as linemen (L), receivers (R), running backs (B), and quarterbacks (Q) in the ratio of L:R:B:Q of 5:3:2:l, and a squad has the following distribution of players: L = 20, R = 16, B = 6, Q = 12, how many offensive teams can be created with all the positions filled, and what is the "limiting position?" Show your calculations.

8.9 Suppose you and a nonenvironmental engineering friend are walking by a stream in the woods, and your friend remarks, "I wonder if this stream is polluted." How would you answer, and what questions would you have to ask him/her before you can answer the question?

8.10 Why would an environmental engineer possibly want to converse with a wildlife biologist during a project?[8]

8.11 The Environmental Advisory Council of Canada published a booklet entitled "An Environmental Ethic—Its Formulation and Implications"[9] in which they suggest the following as a concise environmental ethic:

> Every person shall strive to protect and enhance the beautiful everywhere his or her impact is felt, and to maintain or increase the functional diversity of the environment in general.

In a one-page essay, critique this formulation of the environmental ethic.

8.12 A discharge from a potential wastewater treatment plant may affect the dissolved oxygen level in a stream. The waste characteristics are expected to be:

Flow $= 0.56$ m^3/s
Ultimate BOD $= 6.5$ mg/L
DO $= 2.0$ mg/L
Temperature $= 25°C$

The water in the stream, upstream from the planned discharge, has the following characteristics:

Flow $= 1.9$ m^3/s
Ultimate BOD $= 2.0$ mg/L
DO $= 9.1$ mg/L
Temperature $= 15°C$

The state DO standard is 4 mg/L. Use:

Deoxygenation constant $= 0.2$ d^{-1}
Reoxygenation constant $= 0.40$ d^{-1}

a. Will the construction of this plant cause the DO to drop below the state standard?

b. If during a hot summer day, the flow in the stream drops to 0.2 m³/s and the temperature in the stream increases to 30°C, will the state standard be met?

c. In the wintertime the stream is ice covered, so that there cannot be any reaeration ($k_2 = 0$). If the temperature of the water in the stream is 4°C, and all other characteristics are the same (as in a. above), will the state standard be met?

d. Suppose your calculation showed that in part c. above the dissolved oxygen dropped to zero during the winter, under the ice. Is this bad? After all, who cares?

8.13 Libby was in good spirits. She loved her job as assistant engineer for the town, and the weather was perfect for working out-of-doors. She had gotten the job of onsite inspector for the new gravity trunk sewer, and this not only gave her significant responsibility, but it allowed her to get out of the office. Not bad for a young engineer only a few months out of school.

The town was doing the work in-house, partly because of Bud, an experienced foreman who would see that the job got done right. Bud was wonderful to work with and was full of stories and practical construction know-how. Libby expected to learn a lot from Bud.

This particular morning the job required the cutting and clearing of a strip of woods on the right of way. When Libby arrived, the crew was already noisily getting prepared for the morning's work. She decided to walk ahead up the right of way to see what the terrain was like.

About 91 m up the right-of-way, she came upon a huge oak tree, somewhat off the center-line, but still on the right-of-way, and therefore destined for cutting down. It was a magnificent oak, perhaps 300 years old, and had somehow survived the clear-cutting that occurred on this land in the mid 1800s. Hardly any trees here were over 150 years old, all having fallen to the tobacco farmer's thirst for more land. But there was this magnificent tree. Awesome!

Libby literally dragged Bud up to the tree and exclaimed, "We cannot cut down this tree. We can run the line around it and still stay in the right-of-way."

"Nope. It has to come down," responded Bud. "First, we are running a gravity sewer. You just don't go changing sewer alignment. We'd have to construct additional manholes and redesign the whole line. And most importantly, you cannot have such a large tree on a sewer line right-of-way. The roots will eventually break into the pipe and cause cracks. In the worst case, the roots will fill up the whole pipe, and this requires a cleaning and possible replacement if the problem is bad enough. We simply cannot allow this tree to remain here"

"But think of this tree as a treasure. It's maybe 300 years old. There probably are no other trees like this in the county," implored Libby.

"A tree is a tree. We're in the business of building a sewer line, and the tree is in the way," insisted Bud.

"Well, I think this tree is special, and I insist that we save it. Since I am the engineer in charge," she gulped inwardly, surprised at her own courage, "I say we do not cut down this tree."

She looked around and saw that some of the crew had walked up to them and were standing around, chainsaws in hand, with wry smiles on their faces. Bud was looking very uncomfortable.

"OK," he said. "You're the boss. The tree stays."

That afternoon in her office, Libby reflected on the confrontation, and tried to understand her strong feelings for the old tree. What caused her to stand up on her hind legs like that? To save a tree? So what if it was special? There were many other trees that were being killed to run the sewer line. What was special about this one? Was it just its age, or was there something more?

The next morning Libby went back to the construction site, and was shocked to find that the old tree was gone. She stormed into Bud's construction trailer and almost screamed,

"Bud! What happened to the old tree?"

"Don't you get your pretty head upset now. I called the Director of Public Works and described to him what we talked about, and he said to cut down the tree. It was the right thing to do. If you don't agree, you have a lot to learn about construction."

Why did Libby feel so attached to the old tree? Why did she want to save it? Was it for herself, or for other humans, or for the sake of the tree itself? What should Libby do now? Does she have any recourse at all? The tree is already dead, regardless of her future actions.

Write a one-page memorandum from Libby to her boss relating the events concerning the tree and expressing her (your) feelings.

8.14   Besides playing important roles in ecosystems, microorganisms play important roles in engineering. What are some of these roles?

8.15   Why have transportation engineers concerned themselves with providing animal crossing structures? (See Section 8.2.4.)

END NOTES

1. Warren, Karen J. 1990. The power and promise of ecological feminism. *Environmental Ethics* 12: No. 5: 125.

2. Smith, G. J. C. et al. 1974. *Our Ecological Crisis*. New York: Macmillan.

3. Law, E. A. 2000. *Aquatic Pollution*. Hoboken, NJ: Wiley.

4. Metcalf & Eddy, Inc. 1991. Wastewater Engineering: *Treatment, Disposal, and Reuse*. New York: McGraw-Hill.

5. Phelps, E. B. 1944. *Stream Sanitation*. New York: John Wiley & Sons.

6. Anonymous. 2000. Structures keep animals off roadways in Florida. *Civil Engineering* 70 No. 7: 29.

7. Morrall, J. F., and T. M. McGuire. 2000. Sustainable highway development in a national park. *Transportation Research Record* 1702: No. 3.

8. James Coolbaugh. 1976. *Environment* 18: No. 6.

9. Morse, Norma H. 1975. *An Environmental Ethic—Its Formulation and Implications*. Ottawa: The Environmental Advisory Council of Canada. Report No. 2, January.

APPLICATIONS

Environmental engineers apply knowledge and judgment to design processes and systems that improve people's lives.

CHAPTER NINE

Water Quality

Courtesy of Corey Green

Mississippi River in St. Louis

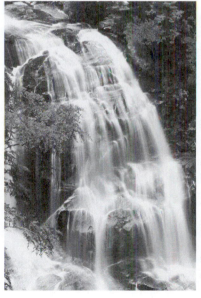

© Jim Hargon/www.hargononline.com/Alamy

Whitewater Falls, North Carolina

Environmental engineers protect water quality. Although we can take any quality of water and improve it, it may not be worth the cost.

When is water dirty? The answer of course depends on what we mean by dirty. For some people, the question is silly, such as the rural judge in a county courthouse who, presiding over a water pollution case, intoned that "Any damn fool knows if water is fit to drink." But what may be pollution to some people may in fact be an absolutely necessary component in the water to others. For example, trace nutrients are necessary for algal growth (and hence for all aquatic life) and fish require organics as a food source in order to survive. These same constituents, however, may be highly detrimental if the water is to be used for industrial cooling. In this chapter, various parameters used to measure water quality are discussed first. Then the question of what is clean and dirty water is considered further.

9.1   MEASURES OF WATER QUALITY

Although there are many water quality parameters, this discussion is restricted to the following.

- Dissolved oxygen is major determinant of water quality in streams, lakes, and other watercourses.
- Oxygen demand, in particular biochemical oxygen demand (which was introduced in Section 8.2.3), is a major parameter indicating the pollution potential of various discharges to watercourses.
- Solids include suspended solids, which are unsightly in natural waters, and total solids, which include dissolved solids, some of which could be detrimental to aquatic life or to people who drink the water.
- Nitrogen is a useful measure of water quality in streams and lakes.
- The potential presence of infectious bacteria and viruses has obvious human health impacts.

9.1.1   Dissolved Oxygen

Dissolved oxygen (DO) is measured with an oxygen probe and meter (Figure 9.1). One of the simplest (and historically oldest) meters operates as a galvanic cell, in which lead and silver electrodes are put in an electrolyte solution with a microammeter between. The reaction at the lead electrode is:

$$Pb + 20H^- \rightarrow PbO + H_2O + 2e^-$$

The electrons that are liberated at the lead electrode travel through the microammeter to the silver electrode, where the following reaction takes place:

$$2e^- + (1/2)O_2 + H_2O \rightarrow 2OH^-$$

Figure 9.1 Dissolved-oxygen meter.

The reaction will not take place unless free dissolved oxygen is available, in which case the microammeter will not register any current.

The trick is to construct and calibrate a meter in such a manner that the electricity recorded is proportional to the concentration of oxygen in the electrolyte solution. In the commercial models the electrodes are insulated from each other with nonconducting plastic and are covered with a permeable membrane with a few drops of an electrolyte between the membrane and electrodes. The amount of oxygen that travels through the membrane is proportional to the DO concentration. A high DO in the water creates a strong driving force to get through the membrane, whereas a low DO forces only limited O_2 through to participate in the reaction to create electrical current. Thus, the current is proportional to the dissolved oxygen level in solution.

As noted in Chapter 8, the saturation of oxygen in water is a function of temperature and pressure. The saturation levels of O_2 in water also depend on the concentration of dissolved solids, with higher solids reducing oxygen solubility. Table 8.2 lists the saturation levels of oxygen in clean water at various temperatures.

The DO meter is calibrated to the maximum (saturated) value in any water by inserting the probe into a sample of the water that has been sufficiently aerated to assure DO saturation. After setting both the zero and saturation values, the meter can be used to read intermediate DO levels in unknown samples. While most meters automatically compensate for temperature change, a variation in dissolved solids requires recalibration.

9.1.2 Oxygen Demand

Perhaps even more important than the determination of dissolved oxygen is the measurement of the rate at which this oxygen is used by microorganisms decomposing organic matter. There are three types of or methods to determine oxygen demand: theoretical oxygen demand, biochemical oxygen demand, and chemical oxygen demand.

Theoretical Oxygen Demand

The oxygen demand for the decomposition of pure materials can be estimated from stoichiometry, assuming that all the organic material completely decomposes. What does "completely decompose" mean? If the compound is a hydrocarbon, which contains only carbon and hydrogen, or an alcohol, which also has oxygen, then the decomposition products are CO_2 and H_2O. If the compound is an amine, which contains carbon, hydrogen, and nitrogen, then the decomposition products are CO_2, H_2O, and NH_3.

Once the chemical reaction is known and balanced, the theoretical oxygen demand (ThOD) can be calculated as:

$$ThOD = C-ThOD + N-ThOD$$

where C-ThOD is due to the decomposition of the carbonaceous (organic) material and N-ThOD is due to the stabilization of the nitrogenous material (NH_3 to NO_3^-, Chapter 8).

EXAMPLE
9.1

Problem   What is the theoretical oxygen demand in mg/L for a 1.67×10^{-3} molar solution of glucose, $C_6H_{12}O_6$, to decompose completely?

Solution   First balance the decomposition reaction (which is an algebra exercise):

$$C_6H_{12}O_6 + aO_2 \rightarrow bCO_2 + cH_2O$$

as

$$C_6H_{12}O_6 + 6O_2 \rightarrow 6CO_2 + 6H_2O$$

That is, for every mole of glucose decomposed, 6 mol of oxygen are required. This gives us a constant to use to change moles per liter of glucose to milligrams per liter of O_2 required, a (relatively) simple unit conversion.

$$\left[\frac{1.67 \times 10^{-3}\,\text{mol glucose}}{L}\right] \times \left[\frac{6\ \text{mol of } O_2}{\text{mol glucose}}\right] \times \left[\frac{32\ \text{g } O_2}{\text{mol } O_2}\right] \times \left[\frac{1000\ \text{mg}}{\text{g}}\right] = 321\,\frac{\text{mg } O_2}{L}$$

EXAMPLE
9.2

Problem   What is the theoretical oxygen demand in liters of air for a 50 mg/L solution of acetone, CH_3COCH_3, to decompose completely?

Solution   The first step is, again, to balance the decomposition reaction:

$$C_3H_6O + aO_2 \rightarrow bCO_2 + cH_2O$$

as

$$C_3H_6O + 4O_2 \rightarrow 3CO_2 + 3H_2O$$

That is, for every mole of acetone decomposed, 4 mol of oxygen are required. Use the ideal gas law and the percent oxygen by volume in air to calculate the liters of air required.

$$\left[\frac{50\ \text{mg acetone}}{L}\right] \times \left[\frac{\text{g}}{1000\ \text{mg}}\right] \times \left[\frac{\text{mol acetone}}{58.08\ \text{g acetone}}\right] \times \left[\frac{4\text{mol of } O_2}{\text{mol acetone}}\right]$$
$$\times \left[\frac{22.4\ \text{L } O_2}{\text{mol } O_2}\right] \times \left[\frac{\text{L air}}{0.21\ \text{L } O_2}\right] \cong 0.4\frac{\text{L air}}{\text{L solution}}$$

Note that in both examples ThOD = C-ThOD. In other words, the N-ThOD = 0 because there was no nitrogen in the organic chemical.

EXAMPLE
9.3

Problem   What is the theoretical oxygen demand in liters of air for a 300 mg/L solution of methylamine, CH_3NH_2, to decompose completely?

Solution   The first step is to balance the decomposition reaction:

$$CH_5N + aO_2 \rightarrow bCO_2 + cH_2O + dNH_3$$

as

$$CH_5N + 1.5O_2 \rightarrow 1CO_2 + 1H_2O + 1NH_3$$

That is, for every mole of methylamine decomposed, 1.5 mol of oxygen are required for the C–ThOD.

$$C-ThOD = \left[\frac{300 \text{ mg } CH_5N}{L}\right] \times \left[\frac{g}{1000 \text{ mg}}\right] \times \left[\frac{mol \ CH_5N}{31.058 \text{ g } CH_5N}\right]$$
$$\times \left[\frac{1.5 \text{ mol of } O_2}{mol \ CH_5N}\right] \times \left[\frac{22.4 \text{ L } O_2}{mol \ O_2}\right] \times \left[\frac{L \text{ air}}{0.21 \ LO_2}\right] \cong 1.55$$

But the NH_3 will also use O_2:

$$NH_3 + 2O_2 \rightarrow HNO_3 + H_2O$$

So there will be a N-ThOD.

$$N-ThOD = \left[\frac{300 \text{ mg } CH_5N}{L}\right] \times \left[\frac{g}{1000 \text{ mg}}\right] \times \left[\frac{mol \ CH_5N}{31.058 \text{ g } CH_5N}\right] \times \left[\frac{1 \text{ mol of } NH_3}{mol \ CH_5N}\right]$$
$$\times \left[\frac{2 mol \ O_2}{mol \ NH_3}\right] \times \left[\frac{22.4 \text{ L } O_2}{mol \ O_2}\right] \times \left[\frac{L \text{ air}}{0.21 \ LO_2}\right] \cong 2.06 \frac{L \text{ air}}{L \text{ solution}}$$

So the total theoretical oxygen demand is:

$$ThOD = C-ThOD + N-ThOD \cong 1.55 + 2.06 \cong 3.6 \text{ L air per L solution.}$$

It is important to realize that to calculate theoretical oxygen demand, you must know the chemical formula of the material decomposing plus the stoichiometric reaction. You may have a good handle on this information in an industrial setting with a limited number of materials used. Unfortunately, municipal wastewaters are seldom pure materials, and it is not possible to calculate the demand for oxygen from stoichiometry. It is, in fact, necessary to conduct a test in which the microorganisms that do the converting are actually employed and the use of oxygen by these microorganisms is measured.

Biochemical Oxygen Demand

The rate of oxygen use by these microorganisms is commonly referred to as *biochemical oxygen demand* (BOD). BOD is not a measure of some specific pollutant, but rather a measure of the amount of oxygen required by aerobic bacteria and other microorganisms to stabilize decomposable organic matter. If the microorganisms are brought into contact with a food supply (such as human waste), oxygen is used by the microorganisms during the decomposition. A very low rate of use would indicate (1) contamination is absent, (2) the available microorganisms are uninterested in consuming the available organics, or (3) the microorganisms are dead or dying. (Nothing decreases oxygen consumption by aquatic microorganisms quite so well as a healthy slug of arsenic.)

The standard BOD test is run in the dark at 20 °C for five days. This is defined as five-day BOD (BOD_5), which is the oxygen used in the first five days. The temperature is specified because the rate of oxygen consumption is temperature dependent. The reaction must occur in the dark because algae may be present and, if light is available, may actually produce oxygen in the bottle. The BOD test is almost universally run using a standard BOD bottle (about 300 mL volume), as shown in Figure 9.2. The bottle is made of special

© Courtesy of P. Aarne Vesilind

Figure 9.2 A BOD bottle, made of a special nonreactive glass and supplied with a ground-glass stopper.

nonreactive glass and has a ground stopper with a lip, which is used to create a water seal so that no oxygen can get into or out of the bottle.

Although the five-day BOD is the standard, it is, of course, also possible to have a two-day, 10-day, or any other day BOD. One form of BOD introduced in Chapter 8 is the ultimate BOD (BOD$_u$, or L), which is the O$_2$ demand after a very long time—when the microorganisms have oxidized as much of the organics as they can. The ultimate BOD is usually run for 30 days, at which point little additional oxygen depletion will occur.

If the dissolved oxygen is measured every day for five days, a curve such as Figure 9.3 is obtained. In this figure, sample A has an initial DO of 8 mg/L, and in five days the DO drops to 2 mg/L. The BOD is the initial DO minus the final DO, which equals the DO used by the microorganisms. In this case, it is $8 - 2 = 6$ mg/L. In equation form:

$$BOD = I - F$$

where I = initial DO, mg/L
F = final DO, mg/L

Referring again to Figure 9.3, sample B has an initial DO of 8 mg/L, but the oxygen was used up so fast that it dropped to zero in two days. If after five days the DO is measured as zero, all we know is that the BOD of sample B is more than $8 - 0 = 8$ mg/L because the organisms might have used more DO if had it been available, but we don't know how much more than 8 mg/L. In general, for samples with an oxygen demand greater than about 8 mg/L, it is not possible to measure BOD directly and dilution of the sample is necessary.

Suppose sample C in the figure is really sample B diluted with distilled water by 1:10. The BOD of sample B is therefore 10 times greater than the measured value, or

$$BOD = (8 - 4)10 = 40 \text{ mg/L}$$

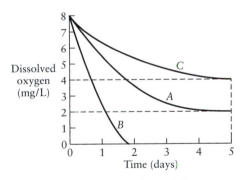

Figure 9.3   Decrease in dissolved oxygen in three different BOD bottles: *A* is a valid test for a 5-day BOD; *B* is invalid because it reaches zero dissolved oxygen before the fifth day; and *C* is the same as *B* but with prior dilution.

With dilution (using water with no BOD of its own), the BOD equation becomes:

$$BOD = (I - F)D \tag{9.1}$$

where D = dilution represented as a fraction and defined as

$$D = \frac{\text{Total volume of bottle}}{\text{Total volume of bottle} - \text{Volume of dilution water}} \tag{9.2a}$$

which is the same as

$$D = \frac{\text{Total volume of bottle}}{\text{Volume of sample}} \tag{9.2b}$$

FOCUS ON

Laboratory Results

In the 1960s how estuaries flushed themselves was being studied, and one of the variables was how the BOD of the water in the estuary was progressively reduced by the flushing. The Delaware Estuary study was a large project to show how the effects of pollutants (and their control) would affect water quality. As a part of this study, a group of young interns were to run up and down the estuary taking samples at specified locations, bring these samples back to the laboratory, and run the BODs.

As the first large batch (well over 100) bottles came in, the intern assigned to the laboratory measured the initial dissolved oxygen in all of the bottles and placed them into an incubator. Five days later, he took the bottles out and measured the final DO. Much to his surprise, the final DO numbers were *higher* than the initial DO values. Suspecting chemical error he remixed all the chemicals very carefully, following exactly the procedures in *Standard Methods*. But the results did not change. The final DO values were still higher, and many bottles were actually supersaturated (higher than saturation).

Swallowing his pride, he finally asked for help. The director of the project decided to be a good mentor and to lend a hand. But after he had remixed the chemicals and run the tests, the results were the same.

After eliminating all other possibilities that would explain the anomalous result, the intern looked into the incubator. When he opened the door, there was a light on. But when he depressed the switch in the door that was to turn off the light, the light remained illuminated. With the door closed, it was impossible to know that the light had stayed on during the five days of incubation and provided the energy for the algae in the bottles to produce oxygen. The light bulb was removed, and the project continued.

Source: Reprinted from *Controlling Environmental Pollution*, Vesilind, P. A. and DiStefano, T. D., 2006. Lancaster, PA: DEStech Publications, Inc. Reproduced with permission of DEStech Publications, Inc.

EXAMPLE
9.4

Problem   Three BOD bottles were prepared with sample and dilution water, as shown in the following table.

Bottle	Sample (mL)	Dilution Water (mL)
1	3	297
2	1.5	298.5
3	0.75	299.25

Calculate the dilution (D) for each.

Solution   Recall that the volume of a standard BOD bottle is 300 mL. For bottle 1, the dilution, D, is (Equation 9.2b):

$$D = \frac{300}{300 - 297} = 100$$

Similarly, for the other two bottles, D is calculated as 200 and 400.

The assumption in the dilution method is that the results from each dilution of a single sample will yield the same BOD value. Sometimes (often) when a series of BOD bottles are run at different dilutions, the calculations do not converge. Consider the following example.

EXAMPLE
9.5

Problem   A series of BOD test were run at three different dilutions. The results were as follows:

Bottle	Dilution	Initial DO (mg/L)	Final DO (mg/L)
1	100	10.0	2.5
2	200	10.0	6.0
3	400	10.0	7.5

What is the BOD?

Solution   First the individual BODs must be calculated. Then the results need to be compared. Recall Equation 9.1:

$$BOD = (I - F)D$$

Therefore, the results of the lab test are

Bottle	BOD (mg/L)
1	750
2	800
3	1000

The results vary by 250 mg/L, a substantial amount. This problem represents what is called a "sliding scale." Ideally, all tests should have resulted in the same (or nearly same) calculated BOD, but clearly they do not. Because this example represents a fairly benign sliding scale, the middle reading might be the best representative of the oxygen use, so one would probably report the BOD as 800 mg/L. A less pragmatic solution would be to clean the bottles, make sure the dilution water is perfectly clean (it does not have a BOD of its own), and try it again, hoping for a more consistent result.

The BOD of the sample is seldom known before the BOD test is conducted, so the dilution must be estimated. The test is not very precise if the drop in DO during the five days of incubation is less than about 2 mg/L or if the DO remaining in the bottle is less than 2 mg/L. Generally, a final BOD is estimated from previous tests or similar waste and at least two dilutions used, with one expected to have at least 2 mg/L DO remaining and one expected to use at least 2 mg/L of DO. The dilution is then calculated as

$$D = \frac{\text{Expected BOD}}{\Delta \text{ DO}} \tag{9.3}$$

EXAMPLE 9.6

Problem   The five-day BOD of an influent to an industrial wastewater treatment plant is expected to be about 800 mg/L based on similar wastewaters. What dilutions should be used in a five-day BOD test?

Solution   Assume that the saturation is about 10 mg/L. If at least 2 mg/L is to remain in the bottle, the drop in BOD should be $10 - 2 = 8$ mg/L, so the dilution would be (Equation 9.3)

$$D = \frac{800}{8} = 100$$

Similarly, if at least 2 mg/L of DO is to be used, the dilution of another bottle should be

$$D = \frac{800}{2} = 400$$

For the test, at least two BOD bottles should be run, one with $D = 100$ and one with $D = 400$. Prudence would suggest that a third bottle be run at a dilution between these two, say $D = 200$.

Usually the waste sample contains sufficient microorganisms to promote decomposition. However, this situation may not always be the case. For example, it is possible to measure the BOD of sugar to estimate its influence on a stream that would have plenty of microorganisms that would love to get at the sugar. Pure sugar, of course, does not contain the necessary microorganisms for decomposition, so to conduct the test, it is necessary to seed the sample. Seeding is a process in which the microorganisms that are responsible for the oxygen uptake are added to the BOD bottle with the sample (sugar) in order for the oxygen uptake to occur.

Suppose the water previously described by the A curve in Figure 9.3 is used as seed water because it obviously contains microorganisms (it has a five-day BOD of 6 mg/L). If we now put 100 mL of a sample with an unknown BOD into a bottle and add 200 mL of seed water, the 300-mL bottle is filled. The dilution, D, is 3. Assuming that the initial DO of this mixture is 8 mg/L and the final DO is 3 mg/L, the total oxygen used is 5 mg/L. But some of this drop in DO is due to the seed water because it also has a BOD, so only a portion of the drop in DO is due to the decomposition of the unknown material. The DO uptake due to the seed water is

$$\text{BOD}_{\text{seed}} = 5 \text{ mg/L} \left(\frac{200 \text{ mL}}{300 \text{ mL}} \right) = 3.3 \text{ mg/L}$$

because only 2/3 of the bottle is the seed water. The remaining oxygen uptake ($5 - 3.3 = 1.7$ mg/L) must be due to the sample, which was diluted 1:3 (or $D = 3$). Its BOD must then be $1.7 \times 3 = 5$ mg/L.

If the seeding and dilution methods are combined, the following general formula is used to calculate the BOD:

$$\text{BOD}_t = \left[(I - F) - (I' - F') \left(\frac{X}{Y} \right) \right] D \tag{9.4}$$

where BOD_t = biochemical oxygen demand measured at some time, t, mg/L
  I = initial DO of bottle with sample and seeded dilution water, mg/L
  F = final DO of bottle with sample and seeded dilution water, mg/L
  I' = initial DO of bottle with only seeded dilution water, mg/L
  F' = final DO of bottle with only seeded dilution water, mg/L
  X = seeded dilution water in sample bottle, mL
  Y = volume of BOD bottle, mL
  D = dilution of sample

EXAMPLE
9.7

Problem   Calculate the BOD$_5$ if the temperature of the sample and seeded dilution water are 20 °C, the initial DOs are saturation, and the sample dilution is 1:30 with seeded dilution water. The final DO of the seeded dilution water is 8 mg/L, and the final DO of the sample and seeded dilution water is 2 mg/L. Recall that the volume of a BOD bottle is 300 mL.

Solution   From Table 8.2 at 20 °C, saturation is 9.07 mg/L; hence, this is the initial DO of the diluted sample and seeded dilution water (I and I'). Using Equation 9.2b,

$$D = \frac{30}{1} = \frac{30 \text{ mL}}{V_s}$$

therefore

$$V_s = 10 \text{ mL} \quad \text{and} \quad X = 300 \text{ mL} - 10 \text{ mL} = 290 \text{ mL}$$

The BOD is then calculated using Equation 9.4:

$$BOD_5 = \left[(9.07 \text{ mg}/L - 2 \text{ mg/L}) - (9.07 \text{ mg/L} - 8 \text{ mg/L}) \left(\frac{290 \text{ mL}}{300 \text{ mL}} \right) \right] 30 = 181 \text{ mg/L}$$

∎

Remember that BOD is a measure of oxygen use or potential use. An effluent with a high BOD can be harmful to a stream if the oxygen consumption is great enough eventually to cause anaerobic conditions (i.e., the DO sag curve approaches zero dissolved oxygen in the stream). Obviously, a small trickle of wastewater going into a great river probably will have negligible effect, regardless of the mg/L of BOD involved. Similarly, a large flow into a small stream can seriously affect the stream even though the BOD might be low.

The BOD of most domestic sewage is about 250 mg/L, although many industrial wastes run as high as 30,000 mg/L. The potential detrimental effect of an untreated dairy waste that might have a BOD of 25,000 mg/L is quite obvious because it represents a 100 times greater effect on the oxygen levels in a stream than raw sewage.

The reactions in a BOD bottle can be described mathematically by first writing a material balance in terms of the dissolved oxygen, starting as always with

$$\left[\begin{array}{c} \text{Rate of DO} \\ \text{ACCUMULATED} \end{array} \right] = \left[\begin{array}{c} \text{Rate of DO} \\ \text{IN} \end{array} \right] - \left[\begin{array}{c} \text{Rate of DO} \\ \text{OUT} \end{array} \right]$$
$$+ \left[\begin{array}{c} \text{Rate of DO} \\ \text{PRODUCED} \end{array} \right] - \left[\begin{array}{c} \text{Rate of DO} \\ \text{CONSUMED} \end{array} \right]$$

Because the BOD bottle is a closed system and because the test is run in the dark so that there is no DO production, the material balance reduces to

$$\left[\begin{array}{c} \text{Rate of DO} \\ \text{ACCUMULATED} \end{array} \right] = - \left[\begin{array}{c} \text{Rate of DO} \\ \text{CONSUMED} \end{array} \right]$$
$$\frac{dz}{dt} V = -rV$$

where z = dissolved oxygen (necessary for the microorganisms to decompose the
           organic matter), mg/L
        t = time
        V = volume of the BOD bottle, mL
        r = reaction rate

This may be assumed to be a first-order reaction (a point of some controversy, incidentally), as initially introduced in Chapter 8:

$$\frac{dz}{dt} = -k_1 z$$

That is, the rate at which the need for oxygen is reduced (dz/dt) is directly proportional to the amount of oxygen necessary for the decomposition to occur (z). Integrated, this expression yields

$$z = z_0 e^{-k_1 t}$$

As the microorganisms use oxygen, at any time, t, the amount of oxygen still to be used is z (Figure 9.4A), and the amount of oxygen already used at any time, t, is y, or the oxygen already demanded by the organisms (Figure 9.4B). The total amount of oxygen that will ever by used by the microorganisms is the sum of what has been used, y, and what is still to be used, z (Figure 9.4):

$$L = z + y$$

where
y = DO already used or demanded at any time, t (i.e., the BOD), mg/L
z = DO still required to satisfy the ultimate demand, mg/L
L = ultimate demand for oxygen, mg/L

Substituting $z = L - y$ into the equation yields

$$L - y = z_0\, e^{-k_1 t}$$

and recognizing from Figure 9.4A that $z_0 = L_0$:

$$y = L - L(e^{-k_1 t})$$

or, equivalently

$$y = L(1 - e^{-k_1 t}) \tag{9.5}$$

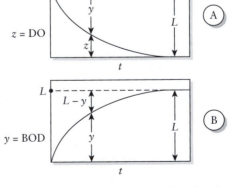

Figure 9.4   BOD definitions. Note that the L used in the dissolved oxygen sag equation is the ultimate carbonaceous BOD.

where $y =$ BOD at any time, t, in days, mg/L
$L =$ ultimate BOD, mg/L
$k_1 =$ deoxygenation constant for base e, d^{-1}

This equation can be used to calculate the BOD at any time, including the ultimate BOD.

EXAMPLE
9.8

Problem   Assuming a deoxygenation constant of 0.25 d^{-1}, calculate the expected BOD_5 if the BOD_3 is 148 mg/L.

Solution   Knowing y_3 and the deoxygenation coefficient, k_1, calculate the ultimate BOD, L, using Equation 9.5.

$$148 \text{ mg/L} = L \left[1 - e^{-(0.25 \ d^{-1})(3 \ d)} \right]$$

$$L = 280 \text{ mg/L}$$

With the ultimate BOD and k_1, the BOD at any time can be calculated, again using Equation 9.5.

$$y_5 = (280 \text{ mg/L}) \left[1 - e^{-(0.25 \ d^{-1})(5 \ d)} \right] = 200 \text{ mg/L}$$

With reference to Figure 9.4B, note that

$$\frac{dy}{dt} = k_1(L - y)$$

so that as y approaches L, $(L - y) \rightarrow$ O and $dy/dt \rightarrow$ O. Integrating this expression again produces Equation 9.5.

It is often necessary, such as when modeling the DO sag curve in a stream (Chapter 8), to know both k_1 and L. The results of the BOD test, however, produce a curve showing the oxygen used over time. How do we calculate the ultimate oxygen use, L, and the deoxygenation rate, k_1, from such a curve?

There are a number of techniques for calculating k_1 and L, one of the simplest being a method devised by Thomas:[1]

$$\left(\frac{t}{y} \right)^{1/3} = \left(\frac{1}{(k_1 L)^{1/3}} \right) + \left(\frac{k_1^{2/3}}{6 \, L^{1/3}} \right) t \tag{9.6}$$

This equation is in the form of a straight line:

$$a = b + mt$$

where $a = (t/y)^{1/3}$
$b = (k_1 L)^{-1/3}$
$m = (1/6)(k_1{}^{2/3} L^{-1/3})$

Thus, plotting a versus t, the slope (m) and intercept (b) can be obtained, and

$$k_1 = 6 \ (m/b)$$

$$L = \frac{1}{6 \, m b^2}$$

**E X A M P L E
9.9**

Problem   The BOD versus time data for the first five days of a BOD test are obtained as follows:

Time, t (days)	BOD, y (mg/L)
2	10
4	16
6	20

Calculate k_1 and L.

Solution   Plot Equation 9.6. The $(t/y)^{1/3}$ values are 0.585, 0.630, and 0.669. These are plotted as shown in Figure 9.5. From the graph, the intercept is $b = 0.545$ and the slope is $m = 0.021$. Thus

$$k_1 = 6 \left(\frac{0.021}{0.545} \right) = 0.23 \ d^{-1}$$

$$L = \frac{1}{6(0.021)(0.545)^2} = 27 \ \text{mg/L}$$

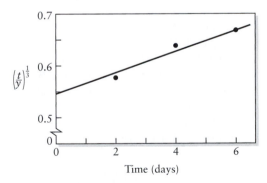

Figure 9.5   Plot used for the calculation of k_1 and L.

But things aren't always that simple. If the BOD of an effluent from a wastewater treatment plant is measured and, instead of stopping the test after five days, the reaction

Figure 9.6   Idealized BOD curves.

is allowed to proceed and the DO measured each day, then a curve like Figure 9.6 might result. Note that some time after five days the curve takes a sudden jump. This jump is due to the exertion of oxygen demand by the microorganisms that decompose nitrogenous organics and convert these to the stable nitrate, NO_3^- (Chapter 8). The oxygen use curve can therefore be divided into two regions: *nitrogenous BOD* (NBOD) and *carbonaceous BOD* (CBOD).

Note the definition of the ultimate BOD on this curve. If no nitrogenous organics are in the sample or if the action of these microorganisms is suppressed, only the carbonaceous curve results. However, for streams and rivers with travel times greater than about five days, the ultimate demand for oxygen must include the nitrogenous demand. It is often assumed that the ultimate BOD can be calculated as

$$L = \text{BOD}_{\text{ult}} = a(\text{BOD}_5) + b(KN)$$

where KN = Kjeldahl nitrogen (organic nitrogen plus ammonia, Section 9.1.4), mg/L
a and b = constants

North Carolina, for example, uses $a = 1.2$ and $b = 4.0$ for calculating the ultimate BOD. This model emphasizes the need for wastewater treatment plants to achieve nitrification (conversion of nitrogen to NO_3^-) or denitrification (conversion to N_2 gas). More on this in Chapter 11.

Chemical Oxygen Demand

The third method for determining the oxygen demand of a water sample, *chemical oxygen demand* (COD), is a laboratory method that essentially determines the ThOD. A sample is mixed with a strong chemical oxidizing agent (potassium dichromate, $K_2Cr_2O_7$) plus a strong acid (sulfuric, H_2SO_4) and then heated. The COD is determined by measuring the consumption of $K_2Cr_2O_7$. The chemicals oxidize all the organic matter—biodegradable and nonbiodegradable. Therefore, the COD for a sample will be greater than the BOD unless the sample contains primarily easily biodegradable wastes. For example, DDT, a refractory organic, would yield a high COD but a zero BOD. The COD should be similar to the ThOD.

The test obviously does not represent stream or lake conditions. So why is it used? Wastewater treatment plant operators will use it for day-to-day operations. Although the BOD test is required because permit levels are based on BOD, the test takes too long to get results to run a plant. COD, on the other hand, can be obtained in about 3 h. It also provides more consistent results because it is chemically, rather than biologically based. Sometimes the COD is also used to estimate BOD_u and to indicate the presence of biologically resistant organics.

9.1.3 Solids

The separation of solids from water is one of the primary objectives of wastewater treatment. Strictly speaking, in wastewater anything other than water or gas is classified as solid, which means that much of the wastewater is actually solids. The usual definition of solids, however, is the residue on evaporation at 103 °C (a temperature slightly higher than the boiling point of water). These solids are known as *total solids*. The test is conducted by placing a known volume of sample in a large evaporating dish (Figure 9.7) and allowing the water to evaporate. The total solids are then calculated as:

$$TS = \frac{W_{ds} - W_d}{V}$$

Figure 9.7 The Gooch crucible, evaporating dish and aluminum boats used for measuring suspended and total solids.

where TS = total solids, mg/L
  W_{ds} = weight of dish plus the dry solids after evaporation, mg
  W_d = weight of the clean dish, mg
  V = volume of sample, L

If the volume of sample is in milliliters, the most common unit of measure, and the weights are in terms of grams, the equation reads:

$$TS = \frac{W_{ds} - W_d}{V} \times 10^6 \tag{9.7}$$

where TS = total solids, mg/L
  W_{ds} = weight of dish plus the dry solids after evaporation, g
  W_d = weight of the clean dish, g
  V = volume of sample, mL

Total solids can be divided into two fractions: *dissolved solids* and *suspended solids*. If a teaspoonful of common table salt is placed in a glass of water, the salt dissolves. The water does not look any different, but the salt remains behind if the water is evaporated. A spoonful of sand, however, does not dissolve and remains as sand grains in the water. The salt is an example of dissolved solids whereas the sand is a suspended solid.

A *Gooch crucible* is used to separate suspended solids from dissolved solids. As shown in Figure 9.7, the Gooch crucible has holes in the bottom on which a glass fiber filter is placed. The sample is then drawn through the crucible with the aid of a vacuum. The suspended material is retained on the filter while the dissolved fraction passes through. If the initial dry weight of the crucible plus filter is known, the subtraction of this weight from the total weight of crucible, filter, and dried solids caught on the filter yields the weight of suspended solids, expressed as mg/L. The equation is

$$SS = \frac{W_{df} - W_d}{V} \times 10^6 \tag{9.8}$$

where SS = suspended solids, mg/L
  W_{df} = weight of dish plus dry filtered solids, g
  W_d = weight of clean crucible and filter, g
  V = volume of sample, mL

Solids can be classified in another way: those that are volatilized at a high temperature and those that are not. The former are known as *volatile solids*, the latter as *fixed solids*. Although volatile solids are considered to be organic, some of the inorganics are decomposed and volatilized as well at the 600 °C used for the test. However, this is not considered a serious drawback. The equation to calculate fixed solids is

$$FS = \frac{W_{du} - W_d}{V} \times 10^6 \tag{9.9}$$

where FS = fixed solids, mg/L
   W_{du} = weight of dish plus unburned solids, g
   W_d = weight of clean crucible and filter, g
   V = volume of sample, mL.

The volatile solids can then be calculated as

$$VS = TS - FS \qquad\qquad (9.10)$$

where VS = volatile solids, mg/L.

EXAMPLE
9.10

Problem   A laboratory runs a solids test. The weight of the crucible = 48.6212 g. A 100-mL sample is placed in the crucible and the water is evaporated. The weight of the crucible and dry solids = 48.6432 g. The crucible is placed in a 600 °C furnace for 24 hr and cooled in a desiccator. The weight of the cooled crucible and residue, or unburned solids, = 48.6300 g. Find the total, volatile, and fixed solids.

Solution   Use Equations 9.7, 9.9, and 9.10.

$$TS = \frac{(48.6432 \text{ g}) - (48.6212 \text{ g})}{100 \text{ mL}} \times 10^6 = 220 \text{ mg/L}$$

$$FS = \frac{(48.6300 \text{ g}) - (48.6212 \text{ g})}{100 \text{ mL}} \times 10^6 = 88 \text{ mg/L}$$

$$VS = 220 - 88 = 132 \text{ mg/L}$$

It is often necessary to measure the volatile fraction of suspended material because this is a quick (if gross) measure of the amount of microorganisms present. The *volatile suspended solids* (VSS) are determined by simply placing the Gooch (filter) crucible in a hot oven (600 °C), allowing the organic fraction to burn off, and weighing the crucible again. The loss in weight is interpreted as volatile suspended solids.

9.1.4 Nitrogen

Recall from Chapter 8 that nitrogen is an important element in biological reactions. Nitrogen can be tied up in high-energy compounds, such as amino acids and amines, and in this form the nitrogen is known as organic nitrogen. One of the intermediate compounds formed during biological metabolism is ammonia nitrogen. Together with organic nitrogen, ammonia is considered an indicator of recent pollution. These two forms of nitrogen are often combined in one measure, known as *Kjeldahl nitrogen*, named after the scientist who first suggested the analytical procedure.

Aerobic decomposition eventually leads to nitrite (NO_2^-) and finally nitrate (NO_3^-) nitrogen. High-nitrate nitrogen with low-ammonia nitrogen suggests that pollution has occurred but quite some time ago.

Light source Filter Sample Photocell Ammeter

Figure 9.8 A photometer used for measuring light penetration through a colored sample, wherein the intensity of the color is proportional to the chemical constituent being measured.

All these forms of nitrogen can be measured analytically by *colorimetric techniques*. The basic idea of colorimetry is that the ion in question combines with some compound and forms a color. The compound is in excess (i.e., there is much more of it than the ion), so the intensity of the color is proportional to the original concentration of the ion being measured. For example, ammonia can be measured by adding a compound called *Nessler reagent* to the unknown sample. This reagent is a solution of potassium mercuric iodide, K_2HgI_4, and reacts with ammonium ions to form a yellow-brown colloid. Since Nessler reagent is in excess, the amount of colloid formed is proportional to the concentration of ammonia ions in the sample.

The color is measured photometrically. The basic workings of a photometer, illustrated in Figure 9.8, consist of a light source, a filter, the sample, and a photocell. The filter allows only certain wavelengths of light to pass through, thus lessening interferences and increasing the sensitivity of the photocell, which converts light energy to electrical current. An intense color allows only a limited amount of light to pass through and creates little current. On the other hand, a sample containing very little of the chemical in question will be clear, allowing almost all of the light to pass through and creating substantial current. If the color intensity and hence light absorbance are directly proportional to the concentration of the unknown ion, the color formed is said to obey *Beer's law*.

A photometer can be used to measure ammonia concentration by measuring the absorbance of light by samples containing known ammonia concentration and comparing the absorbance of the unknown sample to these standards.

EXAMPLE

9.11

Problem Several known samples and an unknown sample containing ammonia nitrogen are treated with Nessler reagent, and the resulting color is measured with a photometer. Find the ammonia concentration of the unknown sample.

Standards (mg/L ammonia)	Absorbance
0 (distilled water)	0
1	0.06
2	0.12
3	0.18
4	0.24
Unknown sample	0.15

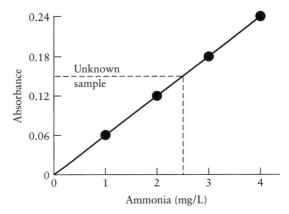

Figure 9.9 A typical calibration curve used for measuring ammonia concentration.

Solution   A calibration curve is plotted (Figure 9.9) using the ammonia standards. The plot results in a straight line, so Beer's law is applicable. The concentration of ammonia in the unknown sample corresponds to 0.15 absorbance and, from the plot, is 2.5 mg/L.

9.1.5 Bacteriological Measurements

From a public health standpoint, the bacteriological quality of water is as important as the chemical quality. A number of diseases can be transmitted by water, among them typhoid and cholera. However, it is one thing to declare that water must not be contaminated by *pathogens* (disease-causing organisms) and another to discover the existence of these organisms.

Seeking the presence of pathogens presents several problems. First, there are many kinds of pathogens. Each pathogen has a specific detection procedure and must be screened individually. Second, the concentration of these organisms can be so small as to make their detection impossible. Looking for pathogens in most surface waters is a perfect example of the proverbial needle in a haystack. And yet only one or two organisms in the water might be sufficient to cause an infection if the water is consumed.

Water-related illnesses are the leading cause of human sickness and death worldwide, far outstripping deaths due to armed conflicts. The relatively few deaths in the United States attributed to drinking water are usually because the water was improperly treated, over half the time being improperly or not disinfected.[2] Pathogens of importance include *Salmonella*, *Shigella*, the hepatitis virus, *Entamoeba histolytica*, *Giardia lamblia*, *Escherichia coli*, and *Cryptosporidium*.

Salmonellosis is caused by various species of *Salmonella*, and the symptoms include acute gastroenteritis, septicemia (blood poisoning), and fever. Gastroenteritis usually consists of severe stomach cramps, diarrhea, and fever, and although it makes for a horrible few days, it is seldom fatal. Typhoid fever is caused by the *Salmonella typhi*, and this disease is much more serious, lasting for weeks, and can be fatal if treated improperly. About

3% of the victims become carriers of *Salmonella typhi*, and, although they exhibit no further symptoms, they pass on the bacteria to others mainly through the contamination of water. Shigellosis, also called bacillary dysentery, is another gastrointestinal disease and has symptoms similar to salmonellosis.

Infectious hepatitis, caused by the hepatitis virus, has been known to be transmitted through poorly treated water supplies. Symptoms include headache, back pains, fever, and eventually jaundiced skin color. Although rarely fatal, it can cause severe debilitation. The hepatitis virus can escape dirty sand filters in water plants and can survive for a long time outside the human body.

Amoebic dysentery, or amebiasis, is also a gastrointestinal disease, resulting in severe cramps and diarrhea. Although its normal habitat is in the large intestine, it can produce cysts that pass to other people through contaminated water and cause gastrointestinal infections. The cysts are resistant to disinfection and can survive for many days outside the intestine.

Originally known as beaver disease in the north country, giardiasis is caused by *Giardia lamblia*, a flagellated protozoan that usually resides in the small intestine. Out of the intestine, its cysts can inflict severe gastrointestinal problems, often lasting two to three months. Giardiasis was known as beaver disease because beavers can act as hosts, greatly magnifying the concentration of cysts in fresh water. Giardia cysts are not destroyed by usual chlorination levels but are effectively removed by sand filtration. Backpackers should take care by purifying drinking water with halogen tablets or small hand-held filters. Giardia has ruined more than one summer for unwary campers.

Escherichia coli (pronounced "ess-shur-eek'-ee-uh ko-lie" and commonly referred to as *E. coli*) are "almost universal inhabitants of the intestinal tract of humans and warm-blooded animals"[3] and are the most common aerobic bacteria in the gut that do not require special environmental or nutritional conditions for growth.[4] Although most *E. coli* are not infectious, some are pathogenic, causing diarrhea, fever, and/or nausea. In fact, up to 70% of Montezuma's revenge (or travelers' diarrhea) cases are due to a particular strain of *E. coli*, and 50 to 80% of urinary tract infections are caused by *E. coli*.[4]

Due to the prevalence of *E. coli*, outbreaks are in the news regularly. However, while a contaminated water supply in Walkerton, Ontario, killed nine and made more than 700 ill[5] and swimming in sewage-contaminated water has been linked to infections,[6] most infections occur from eating undercooked contaminated ground beef. In particular, *E. coli* O157:H7, which lives in the intestines of healthy cattle, can be problematic for humans, often causing severe bloody diarrhea and abdominal cramps and occasionally causing kidney failure.[6] (The letters and numbers indicating the particular strain of *E. coli* refer to specific markers on its surface that distinguish it from other strains.) One federal study found that 28% of cattle entering large slaughterhouses have the organism, which can be spread throughout ground meat during processing.[7] There are, however, other sources of contamination, including unpasteurized milk and juice, produce (such as sprouts and lettuce) that has been irrigated with contaminated water, and person-to-person contact in families and child care centers. Although engineers can design water treatment systems to prevent the spread of pathogenic *E. coli* via drinking water, prevention primarily has to be dealt with through public education and behavior modification (e.g., cooking ground meat to at least 344 K, washing produce, and practicing good hygiene).

FOCUS ON

The Dreaded E. Coli

At birth, a baby's intestine is sterile, but it soon begins to fill with bacteria so that, when the baby is but a few days old, it is hosting millions of bacteria that have found their way into the intestine through the mouth and other body cavities. By the time the baby is an adult the intestine, now fully 9m long, has hundreds of different species and perhaps as many as a trillion organisms. The most important microbe in the intestines of all warm-blooded animals, including humans, is known as *Escherichia coli* (named for the German pediatrician and bacteriologist Theodor Escherich, who discovered them in 1885, Figure 9.10) These bacteria also inhabit other parts of the body, including the nose and ears. They are essentially harmless, unless an anomalous growth occurs. Only one of the strains of *E. coli* is known to be toxic—the dreaded O157:H7, which can occur in undercooked meat and can also be found in improperly treated drinking water.

How can something so toxic as the O157:H7 microbe be classified as an *E. coli*? The answer is in the methods used by microbiologists to make sense of the microbial world. The only way that microbes can be classified and identified is by how they behave in the laboratory. Some, for example, take up the gram stain and become gram-positive, whereas those that do not are gram-negative. The great majority of microorganisms in the warm-blooded intestine ferment lactose in a test tube, causing it to become cloudy and produce gas. The organisms that do so then have become known as *E. coli*. But not all organisms that ferment lactose are common residents of the intestine, the chief one being O157:H7. It is an *E. coli* because it ferments lactose, but it has no other similarities to the common intestinal bacteria. This is the reason we can say that *E. coli* are harmless and at the same time admit that *E. coli* (specifically one strain) can cause great harm and even death to humans.

Figure 9.10   (A) *Escherichia coli*; O157:H7 (B) Theodor Escherich

Another public health problem has been the incidence of cryptosporidiosis, caused by *Cryptosporidium*, an enteric protozoan. The disease is often debilitating, with severe diarrhea and abdominal pain, and can last two weeks. There is no reliable curative treatment for cryptosporidiosis. In Milwaukee in 1993 an outbreak of criptosporidiosis has the dubious distinction of being the largest recorded outbreak of waterborne disease in U.S. history. It affected about 400,000 people and resulted in at least 50 deaths.[2] A possible cause of the outbreak was a decline in a treatment plant's performance that occurred when treatment chemicals were being changed and when the source water may have been contaminated by stormwater runoff.[2] The usual source of the pathogen seems to be agricultural runoff contaminating water supplies. Because the cysts are resistant to the common methods of drinking water disinfection, filtration provides the best barrier against contamination of the drinking water supply.

Our knowledge of the presence of pathogenic microorganisms in water is actually fairly recent. Water-borne diseases, such as cholera, typhoid, and dysentery, were highly prevalent even in the mid-19th century. Even though microscopes were able to show living organisms in water, it was not at all obvious that these little critters were able to cause such dreaded diseases. In retrospect, the connection between contaminated water and disease should have been obvious based on empirical evidence. For example, during the mid-1800s, the Thames River below London was grossly contaminated with human waste, and one Sunday afternoon a large pleasure craft capsized, throwing everyone into the drink. Although nobody drowned, most of the passengers died of cholera a few weeks later!

However, it took a shrewd public health physician named John Snow to make the connection between contaminated water and infectious disease. In the mid-19th century, water supply to the citizens of London was delivered by a number of private firms, each pumping water out of the Thames and selling it at pumps in the city. One such company took its supply downstream of waste discharges and provided water by means of a pump on Broad Street. As would be expected, cholera epidemics in London were common, and during one particularly virulent episode, John Snow noticed that the cases of cholera seemed to be concentrated around the Broad Street pump. He carefully recorded all the cases and marked them on a map, which clearly showed that the center of the epidemic was the pump. He convinced the city fathers to order the pump handle to be removed, and the epidemic subsided. The connection between contaminated water and infectious disease was proven. John Snow's contribution is immortalized by the naming of the John Snow Pub on what is now the renamed Broadwick Street in London. A plaque is on the wall of the pub showing the location of the famous pump (Figure 9.11).

Of course, many pathogenic organisms can be carried by water. How then is it possible to measure for bacteriological quality? The answer lies in the concept of *indicator organisms*. The indicator most often used is a group of microbes called coliforms (which includes the 150 strains of *E. coli*). These critters have five important attributes. They are:

1. normal inhabitants of the digestive tracts of warm-blooded animals
2. plentiful and hence not difficult to find
3. easily detected with a simple test
4. generally harmless except in unusual circumstances
5. hardy, surviving longer than most known pathogens.

Courtesy of P. Aarne Vesilind

Figure 9.11   The John Snow Pub and a marker indicating the location of the famous pump.

Because of these five attributes, coliforms have become universal indicator organisms. But the presence of coliforms does not prove the presence of pathogens! If a large number of coliforms are present, there is a good chance of recent pollution by wastes from warm-blooded animals; therefore, the water may contain pathogenic organisms, but this is not proof of the presence of such pathogens. It simply means that pathogens might be present. However, a high coliform count is suspicious, and although the water may in fact be perfectly safe to drink, it should not be consumed.

The opposite is also true. The absence of coliforms does not prove that there are no pathogens in the water. However, it is taken as an indication that the water is safe to drink.

There are two principal ways of measuring for coliforms. The simplest is *membrane filtration*—running a sample through a sterile filter, thus capturing any coliforms. The filter is then placed in a petri dish containing a sterile agar that soaks into the filter and promotes the growth of coliforms while inhibiting other organisms. After 24 or 48 hours of incubation, the shiny dark blue-green dots, which indicate coliform colonies, are counted. If it is known how many milliliters of water were poured through the filter, the concentration of coliforms can be expressed as coliforms per 100 mL.

The second method of measuring for coliforms is called the *most probable number* (MPN), a test based on the fact that in a lactose broth coliforms produce gas and make the broth cloudy. The production of gas is detected by placing a small tube upside down inside a larger tube (Figure 9.12) so as not to have air bubbles in the smaller tube. After incubation, if gas is produced, some of it will become trapped in the smaller tube and this, along with a cloudy broth, indicates that the tube has been inoculated with at least one coliform.

And that is the trouble. Theoretically, one coliform can cause a positive tube just as easily as a million coliforms can. Hence it is not possible to ascertain the concentration of coliforms in the water sample from just one tube. The solution is to inoculate a series of tubes with various quantities of the sample, the reasoning being that a 10-mL sample would be 10 times more likely to contain a coliform than a 1-mL sample.

For example, using three different inoculation amounts (10 mL, 1 mL, and 0.1 mL of sample), three tubes are inoculated with each amount. After incubation, an array, as shown in the following table, is compiled to show the results. A plus sign indicates a positive test (cloudy broth with gas formation) and a minus sign represents tubes where no coliforms were found. Based on these data one would suspect that there is at least 1 coliform per 10 mL, but there still is no firm number.

Amount of Sample, mL Put in Test Tube	Tube Number		
	1	2	3
10	+	+	+
1	−	+	−
0.1	−	−	+

Figure 9.12 The capture of gas in a tube where lactose is fermented by coliform organisms.

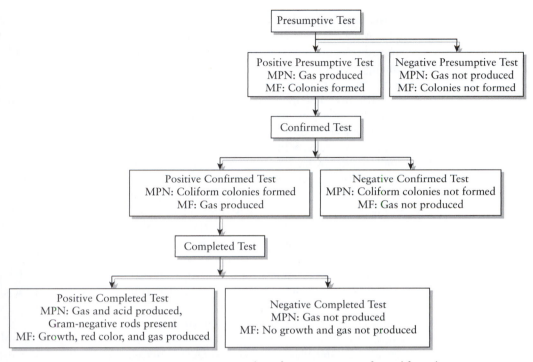

Figure 9.13 Presumptive and confirmatory testing for coliform bacteria, using MPN and membrane filtration (MF) methods.

The solution to this dilemma lies in statistics. It can be proven statistically that such an array of positive and negative results will occur most probably if the coliform concentration is 75 coli/100 mL. A higher concentration would most probably result in more positive tubes, and a lower concentration would most probably result in more negative tubes. Thus, 75 coli/100 mL is the MPN.

Both analysis methods involve three stages of testing to determine the presence or absence of coliforms (Figure 9.13). The first stage of analysis is a presumptive test, which infers the presence or absence of coliform bacteria. However, it does not give absolute proof. For example, although coliforms produce gas, some noncoliform bacteria also produce gas. The second stage of analysis is a confirmatory test, which verifies the presence or absence of total coliforms. The third stage of analysis is a completed test, which verifies the presence or absence of *E. coli*.

9.2 ASSESSING WATER QUALITY

Treated drinking water in the United States is tested for microbes and 80 chemicals a specified number of times each year to ensure its safety (see Section 9.3). To determine proper design and treatment of drinking water, analyses are also commonly conducted for

Table 9.1   Use of Laboratory Analyses in Wastewater Treatment

Analysis	Use
Physical Characteristics	
Color	Condition of wastewater (fresh or septic)
Odor	Treatment requirements, indication of anaerobic conditions
Solids	
TS, VS, SS	Design and operation of treatment process
DS	Effluent reuse potential
Temperature	Design and operation of biological processes
Turbidity	Effluent quality
Transmittance	Suitability of UV disinfection
Inorganic Characteristics	
Alkalinity	Design and operation of treatment process
Chloride	Effluent reuse potential
Hydrogen sulfide	Operation of treatment process, odor control requirements
Metals	Effluent and sludge reuse potential, design and operation of treatment process
Nitrogen	Design and operation of treatment process, effluent and sludge reuse potential
Oxygen	Design and operation of treatment process
pH	Design and operation of treatment process
Phosphorus	Design and operation of treatment process, effluent and sludge reuse potential
Sulfate	Odor potential, treatability of sludge
Organic Characteristics	
BOD_5	Design and operation of treatment process
COD	Operation of treatment process
Methane	Operation of treatment process, energy recovery potential
NOD	Design and operation of treatment process (nitrification/denitrification)
Specific organics	Design and operation of treatment system
Biological Characteristics	
Coliforms	Design and operation of disinfection system
Specific microorganisms	Operation of treatment system

Source: SMALL AND DECENTRALIZED WASTEWATER MANAGEMENT SYSTEMS by Crites, Ronald W., Copyright 1998 by McGraw-Hill Companies, Inc.-Books. Reproduced with permission of McGraw-Hill Companies, Inc.-Books in the format Textbook via Copyright Clearance Center.

parameters such as pH, alkalinity, and hardness (discussed in Chapter 10). Similarly, a range of wastewater characteristics can be analyzed to provide information pertinent to the design and operation of wastewater treatment plants (Table 9.1). However, the seven principal components are SS, BOD, pathogens, TDS, heavy metals, nutrients, and priority organic pollutants[8]. The first three components drive the design of most wastewater treatment systems. In addition to impacting the aesthetic qualities of the effluent, SS impact the amount of sludge produced and the potential for anaerobic conditions to develop. Biodegradable organics (as measured by BOD and COD) require dissolved oxygen for treatment when aerobic processes are used. And, of course, pathogens cause communicable diseases. Dissolved inorganic substances (i.e., TDS) increase through repeated use of water and therefore have implications in the reuse of treated wastewater. Heavy metals (which are the cations with atomic weights above 23 and which are contributed from households as well as industry) can upset biological treatment processes and can reduce sludge management options. Nutrients (i.e., phosphorus and nitrogen) can cause oxygen depletion and eutrophication when discharged to natural water bodies (discussed in Chapter 8). However, they are desirable in sludge used in land application and in effluent used for irrigation, but excessive loadings can contaminate surface water and groundwater. Priority organic pollutants are hazardous and often resist conventional treatment methods.

Over 500 methods for measuring these and other water quality parameters have been standardized in a volume entitled *Standard Methods for the Examination of Water and Wastewater*. The title is commonly shortened to *Standard Methods*. To determine which of these tests to use when confronted with a problem, it is necessary first to define the problem. For example, if the problem is "The water stinks!" it is necessary to decide what measurements will determine just how much it stinks (numerically) and next what is the odor-causing agent in the water. Quantitative measurement of odor is difficult and imprecise simply because it depends on human olfactory senses, and the measurement of the concentration of the causative agent is difficult because it is often not known what to measure. In fortunate circumstances, it is possible to find the specific culprit, but most often an indirect measure, such as volatile solids or ammonia, must be used to measure odor.

If, on the other hand, the problem is "I drank this water and it made me sick!" it is necessary to ask how sick and what was the nature of the illness. This information will give clues as to what may have been in the water that caused the health problem. If the complaint was an upset digestive system (the notorious green-apple-quick-step), the suspicion is that bacterial or viral contaminants were present. The coliform test can then be used to get an indication of such contamination.

If, instead of digestive upset, the health problem is that people in a community are developing mottled teeth, the immediate suspicion is that the drinking water contains excessive fluoride. The fluoride in this case would be a pollutant, even though most communities add fluoride to drinking water to prevent dental caries in children and teenagers.

A third example of a water quality problem may be that "The fish are all dead!" It is then necessary to seek solutions as to what caused the fish kill. Some constituent was present at a high enough concentration to kill the fish, or a constituent necessary for higher aquatic life was absent. For example, a waste pesticide could have caused the kill, and a chemical screening must be done to identify the pesticide. The fish may also have died due to a lack of oxygen (which is the most common cause of fish kills), and it is useful to estimate the BOD and measure the DO.

The entire objective of such measurements is to obtain a quantitative handle on how clean or dirty the water is. Only then is it possible to proceed to solving pollution problems and avoid the problem of judges and lawyers telling us that "Any damn fool knows if the water is fit to drink."

9.3   WATER QUALITY STANDARDS

But what use is the quantitative analysis of water quality unless there exist some standards that describe the desired quality for various beneficial uses of the water? There are, in fact, three main types of water quality standards: drinking water standards, effluent standards, and surface water quality standards.

9.3.1   Drinking Water Standards

Based on public health and epidemiological evidence and tempered by a healthy dose of expediency, the national drinking water standards for many physical, chemical, and bacteriological contaminants have been established by the EPA under the Safe Drinking Water Act (SDWA). Table 9.2 lists some of these standards. The complete standards can be found in Title 40 Part 141 of the *Code of Federal Regulations* (40CFR141).

The list of chemical standards is quite long and includes the usual inorganics (lead, arsenic, chromium, etc.) as well as some organics (e.g., DDT). Bacteriological standards for drinking water are written in terms of the coliform indicators. The normal standard is presently less than 1 coliform per 100 mL of treated drinking water. One example of a physical standard is *turbidity*, or the interference with the passage of light. A water that has high turbidity is cloudy, a condition caused by the presence of colloidal solids. Turbidity does not in itself cause a health problem, but the colloidal solids may prove to be convenient vehicles for pathogenic organisms.

The standards listed in Table 9.2 are examples of *primary standards*. Primary standards specify maximum contaminant levels (MCLs) or treatment techniques. They are set to protect public health and so are enforceable.

Secondary standards, on the other hand, are set to make the water more palatable and usable, reducing, for example, unpleasant tastes and corrosivity (Table 9.3). These standards are not enforceable. For example, the secondary standard for chloride is 250 mg/L, a point at which water has a distinct salty taste. There is no primary standard for chloride because before the salt can become harmful it will taste so bad that nobody would drink the water. Iron likewise is not a health problem (and, in fact, is important for healthy blood), but high iron concentrations make water appear red and discolor laundry. Manganese gives water a blue color and similarly can discolor laundry and ceramic surfaces, such as bathtubs.

Maximum contaminant level goals (MCLGs) are also not enforceable but apply to the primary contaminants. These goals are set at levels that present no known or anticipated health effects. Therefore, they may be lower than MCLs due to technological or economic issues. In other words, it may be so expensive to reduce a contaminant to a concentration at which there are no known effects that the decision is made to accept an increased health risk in order to reduce the cost of compliance.

Because science and technology evolve, the EPA continues to review data to determine whether substances known or anticipated to occur in public water systems need to be regulated under the SDWA. This review process results in a *contaminant candidate list* (CCL). CCL 1, published in 1998, had 60 contaminants. CCL 2, published in 2005, had 51 contaminants that were not removed from CCL 1 through the regulatory determination process. The draft CCL 3, published in 2008, had 104 contaminants: 93 chemicals or chemical groups and 11 microbiological contaminants out of the approximately 7500 chemicals and

Table 9.2 Primary Drinking Water Standards

Contaminant	Concentration (mg/L, unless stated otherwise)		
	MCL	MCLG *	BAT**
Inorganic			
Asbestos	7 million fibers/L (longer than 10 μm)	2, 3, 8	
Chromium	0.1		2, 5, 7, 6 for Cr^{3+}
Cyanide	0.2 (as free cyanide)		5, 7, 10
Fluoride	4.0		
Nitrate	10 (as nitrogen)		5, 7, 9
Nitrite	1 (as nitrogen)		5, 7
Thallium	0.002	0.0005	1, 5
Organic			
Alachlor	0.002	0	4
Carbofuran	0.04		4
Toluene	1		4, 12
Trichloroethylene	0.005	0	4, 12
Xylenes (total)	10		4, 12
2,3,7,8-TCDD (dioxin)	0.00000003	0	4
Disinfection by-product			
Bromate	0.010	0	17
Bromoform	–	0	
Chlorite	1.0	0.8	17
Haloacetic acids (HAA5)	0.060	–	2, 4, 6
Total trihalomethanes	0.080	–	2, 4, 6
Biological			
Total coliforms		0	2, 13, 14, 15, 16
≥ 40 samples/month	≤ 5% positive		
< 40 samples/month	≤ 1 sample positive		
	Negative repeat sample after fecal coliform or *E. coli* positive sample		
Cryptosporidium	–	0	
Viruses	–	0	
Radioactive			
Combined radium-226 & radium-228	5 pCi/L	0	5, 6, 7
Uranium	30 mg/L	0	2, 5, 6, 7
Physical			
Turbidity	1 NTU		

* Only those MCLGs that are different from the MCL are specified.
** Best available technology (BAT) or other method for achieving compliance:

1 = activated alumina	7 = reverse osmosis	13 = oxidation
2 = coagulation and filtration	8 = corrosion control	14 = well placement and construction
3 = direct and diatomite filtration	9 = electrodialysis	15 = disinfectant residual
4 = granular activated carbon	10 = chlorine	16 = distribution system maintenance
5 = ion exchange	11 = ultraviolet ligh	17 = control of treatment process
6 = lime softening	12 = packed tower aeration	

Table 9.3 Secondary Drinking Water Standards

Contaminant	Standard (mg/L, unless stated otherwise)
Cation	
Aluminum	0.05–0.2
Copper	1.0
Iron	0.3
Manganese	0.05
Silver	0.10
Zinc	5
Anion	
Chloride	250
Fluoride	2.0
Sulfate	250
Other	
TDS	500
Foaming agents	0.5
Color	15 color units
Odor	3 threshold odor number
Corrosivity	Noncorrosive
pH	6.5–8.5

microbes evaluated. Contaminants on the CCL are studied to determine how to detect them, whether they occur in drinking water, what health effects they have, and how to remove them. As of July 2008 the EPA had determined that no regulatory action is appropriate or necessary for 9 contaminants on CCL 1 and 11 on CCL 2.

9.3.2 Effluent Standards

The Clean Water Act (CWA) is used to reduce the flow of pollutants into natural watercourses. All point source dischargers to natural watercourses are required to obtain a National Pollution Discharge Elimination System (NPDES) permit (40CFR122). (Businesses discharging to a sewer system rather than a natural watercourse are not required to obtain an NPDES permit; however, they must obtain permits from the municipal treatment plants receiving the waste.) Although some detractors have labeled these "permits to continue polluting," the permitting system has nevertheless had a major beneficial effect on the quality of surface waters. Typical effluent standards for a domestic wastewater treatment plant range from 5 to 20 mg/L BOD, for example. The intent is to tighten these limits as required to enhance water quality.

9.3.3 Surface Water Quality Standards

Tied to the effluent standards are surface water standards, often called "stream standards." All surface waters in the United States are classified according to a system of standards based on their greatest beneficial use. The highest classification is usually reserved for pristine waters, which are often used as sources of drinking water. The next highest classification includes waters that have had wastes discharged into them but that nevertheless exhibit high levels of quality. The categories continue in order of decreasing quality, with the lowest water quality useful only for irrigation and transport.

The CWA Section 305(b) established the National Water Quality Inventory, highlighting which waterbodies are meeting water quality standards and which are not. The CWA Section 303(d) established a prioritized list of impaired waters that need total maximum daily loads (TMDLs). A TMDL is the maximum amount of a pollutant a waterbody can receive from all sources, point and nonpoint, to still meet water quality standards. TMDLs used at the watershed level can result in increased focus on prevention and treatment of nonpoint source pollution, including trading among point and nonpoint sources and regulating nonpoint sources, as well as a focus on reducing or eliminating combined sewer overflows (CSOs). CSOs occur after rain and snowmelt events in older urban areas that have one sewer system to handle both stormwater and sewage.

The objective is to attempt to establish the highest possible classification for all surface waters and then to use the NPDES permits to turn the screws on polluters, enhance the water quality, and increase the classification of the watercourse. Once at a higher

---

FOCUS ON

The Origin of U.S. National Drinking Water Quality Standards

In the early 1900s the death rate from typhoid, cholera, and other waterborne diseases was high in many American cities, partly from poor sanitation and contaminated milk and other food supplies but almost certainly from contaminated drinking water. The city of Pittsburgh, for example, took water from the Allegheny River at locations that were immediately downstream from wastewater discharges. More wealthy people used bottled water and avoided this hazard, but the poor had little choice but to drink contaminated water.

At this time, the Constitution of the United States was interpreted very narrowly, and any prospect of the federal government getting involved in setting and enforcing national drinking water standards was highly unlikely. Public health leaders within the U.S. Public Health Service (PHS), at that time an agency of the U.S. Department of Treasury, decided to try to improve the drinking water quality by implementing the interstate commerce clause, which gives the federal government the right to facilitate interstate commerce. At that time the major interstate transportation was by trains and having trains stop at a community was a major advantage. The scientists, engineers, and physicians at the PHS decided to develop drinking water

quality standards that would be applicable only to communities where the trains stopped and would take on water. The argument was that such water should be safe for interstate travelers. The public position of the PHS was that these standards, which for the first time used coliform organisms as a bacteriological indicator, were not intended to be national standards. The only club the PHS had was that, if a community water supply did not meet these standards, the trains would not stop. But this was a very big club. As one chemist from the Iowa State Board of Health noted, "the effect of the posting of notices in (railroad) stations and the condemnation of local waters has been to force the standard upon the local plants through the action of governmental prestige and public opinion." Indeed, even communities where the trains did not stop had an incentive to clean up their drinking water supplies. The effect nationally was to reduce the death rate from typhoid and cholera to almost zero by 1940.

Source: Patrick Gurian and Joel A. Tarr, The First Federal Drinking Water Quality Standards and their Evolution, in Improving Regulation, ed. Paul S. Fischbeck and R. Scott Farrow, Resources for the Future, Washington, DC, 2001.

classification, no discharge would be allowed that would degrade the water to a lower quality level. The objective is to attain eventually pure water in all surface watercourses. As Pollyannaish as this may sound, it is an honorable goal, and the thousands of engineers and scientists who devote their professional careers to achieving that end understand the joy of small victories and share in the ultimate dream of pollution-free water.

PROBLEMS

9.1 Given the following BOD_5 test results:

Initial DO	8 mg/L
Final DO	0 mg/L
Dilution	1:10

What can you say about

a. BOD_5?
b. ultimate BOD?

9.2 If you have two bottles full of lake water and keep one in the dark and the other in daylight, which one would have a higher DO after a few days? Why?

9.3 The following data are obtained for a wastewater sample:

Total solids	4000 mg/L
Suspended solids	5000 mg/L
Volatile suspended solids	2000 mg/L
Fixed suspended solids	1000 mg/L

Which of these numbers is questionable (wrong?) Why?

9.4 A water has a BOD_5 of 10 mg/L. The initial DO in the BOD bottle is 8 mg/L and the dilution is 1:10. What is the final DO in the BOD bottle?

9.5 If the BOD_5 of a waste is 100 mg/L, draw a curve showing the effect on the BOD_5 of adding progressively higher doses of chromium (a toxic chemical).

9.6 Some years ago an industrial plant in New Jersey was having trouble with its downstream neighbors. It seems that the plant was discharging apparently harmless dyes into the water and making the stream turn all sorts of colors. The dye did not seem to harm the aquatic life, and it did not soil boats or docks. It was, in short, an aesthetic nuisance.

The plant wastewater treatment engineer was asked to come up with solutions to the problem. She found that the expansion of the plant, adding activated carbon columns, would cost about $500,000, but that there was a simpler solution. They could build a holding basin and hold the plant effluents in this basin during the day and release it at night, or hold it until they had enough blue and green color to make the resulting effluent appear to be blue/green. The basin would cost only $100,000 to construct. The plant would not be violating any standard or regulation, so the operation would be legal. In effect, the plant would discharge wastes so as to reduce public complaints but not actually treat the wastewater to remove the dyes. The total discharge of dye waste would be unchanged.

You are the president of the company and must make a decision to either spend $500,000 and treat the waste or to spend $100,000 and eliminate the complaints from the public. Write a memo to the engineer advising her how to proceed. Include in the memo your rationale for making the decision.

9.7 Consider the following data from a BOD test:

Day	DO (mg/L)	Day	DO (mg/L)
0	9	5	6
1	9	6	6
2	9	7	4
3	8	8	3
4	7	9	3

If there is no dilution (i.e., the dilution factor is 1), what is the:

a. BOD_5?
b. ultimate carbonaceous BOD?
c. ultimate nitrogenous BOD?
d. Why do you think no oxygen is used until the third day?

9.8   A chemical engineer working for a private corporation is asked to develop a means for disinfecting its industrial sludge. He decides to use high doses of chlorine to do the job because it is available at the plant. Laboratory studies show that this method is highly efficient and inexpensive. The plant is constructed. After years of operation, it is discovered that the effluent from the plant contains very high concentrations of trihalomethane (e.g., chloroform). This chemical is carcinogenic, and people downstream have been drinking this water.

a. It is possible that the company engineer knew about the formation of trihalomethane and of its health effect but decided to construct the facility anyway.
b. It is also possible that the company engineer did not know that the chlorine would cause potential health problems, even though the effect of chlorine and high organic materials such as wastewater sludge have been known for a long time to competent environmental engineers.

Discuss the engineer's responsibility in both of these cases.

9.9   An industry discharges 40 million L/d of a waste that has a BOD_5 of 2000 mg/L. How many pounds of BOD_5 are discharged?

9.10 An industry applies to the state for a discharge permit into a highly polluted stream (zero DO, a stench, oil slicks on the surface, black in color). The state denies the permit. The engineer working for the industry is told to write a letter to the state appealing the permit denial based on the premise that the planned discharge is actually *cleaner* than the present

stream water and would actually *dilute* the pollutants in the stream. He is, however, a lousy writer, and asks you to compose a one-page letter for him to send to the state.

a. Write a letter from the engineer to the state arguing his case.
b. After you have written the letter arguing for the permit, write a letter back from the state to the industry justifying the state's decision not to allow the discharge.
c. If the case went to court and a judge had both letters to read as the primary arguments, what would be the outcome? Write an opinion from the judge deciding the case. What elements of environmental ethics might the judge employ to make a decision?

9.11 If you dumped two Liter of milk everyday into a stream, what would be your discharge in kg BOD_5/d? Milk has a BOD_5 of about 20,000 mg/L.

9.12 Given the same standard ammonia samples as in Example 9.11, if your unknown sample measured 20% absorbance, what is the ammonia concentration?

9.13 Suppose you ran a multiple tube coliform test and got the following results: 10 mL samples, all 5 positive; 1 mL samples, all 5 positive; 0.1 samples, all 5 negative. Use the table in *Standard Methods* to estimate the concentration of coliforms.

9.14 If coliform bacteria are to be used as an indicator of viral pollution as well as an indicator of bacterial pollution, what attributes must the coliform organisms have relative to viruses?

9.15 Draw a typical DO curve for a BOD run at the following conditions.

a. Stream water, 20 °C, dark
b. Unseeded sugar water, 20 °C, dark
c. Stream water, 20 °, with light
d. Stream water, 40 °C, dark

9.16 Consider the following data for a BOD test.

Day	DO mg/L
0	9
1	8
2	7
3	6
4	5
5	4.5
6	4

Assume there is no dilution.

a. Calculate BOD_5.
b. Plot the BOD versus time.
c. Suppose you took the sample and after six days aerated it, put it into the incubator, and measured the DO every day for five days. Draw this curve on the graph as a dotted line.

9.17 Suppose two water samples have the following forms of nitrogen at day zero.

	Sample A (mg/L)	Sample B (mg/L)
Organic	40	0
Ammonia	20	0
Nitrite	0	0
Nitrate	2	10

For each sample, draw the curves for the four forms of nitrogen as they might exist in a BOD bottle during 10 days of incubation.

9.18 A wastewater sample has $k_1 = 0.2$ d^{-1} and an ultimate BOD $(L) = 200$ mg/L. What is the final dissolved oxygen at five days in a BOD bottle in which the sample is diluted 1:20 and where the initial DO is 10.2 mg/L?

9.19 Calculate the concentration of oxygen in milligrams per liter required to completely oxidize the following organic compounds. Also calculate the volume of air in liters per liter of solution treated. Use 500 mg/L as the concentration of each compound.

a. methyl tertiary butyl ether (MTBE), $C_5H_{12}O$
b. benzylmorphine, $C_{24}H_{25}NO_3$
c. bepridil, $C_{24}H_{34}N_2O$
d. toluene, $H_5C_6H_2C - H$

9.20 A student places two BOD bottles in an incubator, having measured the initial DO of both as 9.0 mg/L. In bottle A, she has 100% sample, and in bottle B she puts 50% sample and 50% unseeded dilution water. The final DO at the end of five days is 3 mg/L in bottle A and 4 mg/L in bottle B.

a. What was the BOD_5 of the sample as measured in each bottle?
b. What might have happened to make these values different?
c. Do you think the BOD measure included:
   i. only carbonaceous BOD?
   ii. only nitrogenous BOD?
   iii. both carbonaceous and nitrogenous BOD?

Why so you think so?

9.21 The BOD_5 of an industrial waste after pretreatment is 220 mg/L, and the ultimate BOD is 320 mg/L.

a. What is the deoxygenation constant k_1' (base 10)?
b. What is the deoxygenation constant k_1 (base e)?

9.22 The ultimate BOD of each of two wastes is 280 mg/L. For the first, the deoxygenation constant k_1 (base e) $= 0.08$ d^{-1}, and for the second, $k_1 = 0.12$ d^{-1}. What is the BOD_5 of each? Show graphically how this can be so.

9.23 You are the chief environmental engineer for a large industry and routinely receive the test results of the wastewater treatment plant effluent quality. One day you are shocked to discover that the level of cadmium is about 1000 times higher than the effluent permit. You call the laboratory technician, and he tells you that he also thought that was strange, so he ran the test several times to be sure.

You have no idea where the cadmium came from or whether it will ever show up again. You are sure that if you report this peak to the state, they may shut down the entire industrial operation because the treated effluent flows into a stream that is used as a water supply, and they will insist on knowing where the source was so that it cannot happen again. Such a shutdown would kill the company, which is already tottering on the verge of bankruptcy. Many people would lose their jobs and the community would suffer.

You have several options:

a. Erase the offending data entry and forget the whole thing.

b. Delay reporting the data to the state and start a massive search for the source, even though you have doubts it will ever be found.

c. Bring this to attention of your superiors, hoping that they will make a decision and you will be off the hook.

d. Report the data to the state and accept the consequences.

e. Other?

What would you decide, and how would you decide it if:

a. you are 24 years old, two years out of school, not married.

b. you are 48 years old, married, with two children in college.

How would your decisions differ in these circumstances?

9.24 Community water systems are required to provide annually a consumer confidence report (CCR).

a. Chris and Kelly recently moved and realized that they no longer receive a CCR. Why might that be?

b. Compare your community's CCR, if available, to another community's. If yours is unavailable, compare two available reports. (Some reports are available on the EPA's website at http://www.epa.gov/safewater/dwinfo/index.html.)

END NOTES

1. Thomas, H. A., Jr. 1950. Graphical determination of BOD curve constants. *Water and Sewer Works* 97:123.

2. Symons, J. M. 1997. *Plain Talk About Drinking Water: Questions and Answers About the Water You Drink*. Denver, CO: American Water Works Association.

3. Madigan, Michael T., John M. Martinko, and Jack Parker. 1997. *Brock Biology of Microorganisms*. Prentice Hall: Upper Saddle River, NJ.

4. Talaro, Kathleen Park, and Arthur Talaro. 1999. *Foundations in Microbiology*. Boston: WCB McGraw-Hill.

5. Kondro, Wayne. 2000. *E. coli* Outbreak deaths spark judicial inquiry in Canada. *Lancet*, June 10. Accessed at http://www.findarticles.com on December 8, 2008.

6. Centers for Disease Control and Prevention. 2000. *Disease Information: Escherichia coli O157:H7*. May 30. Accessed at http://www.cdc.gov/ncidod/dbmd/diseaseinfo/escherichiacoli_g.htm on December 8, 2008.

7. Raloff, J. Toxic bugs taint Large Numbers of Cattle. *Science News*. March 25, 2000. Located at http://www.findarticles.com.

8. Crites, Ron, and George Tchobanoglous. 1998. *Small and Decentralized Wastewater Management Systems*. WCB McGraw-Hill: Boston.

Water Supply and Treatment

Campground and reservoir, Illinois

Aboveground storage tanks, Illinois

Environmental engineers create and protect drinking water sources and provide safe drinking water.

281

As long as population densities are sufficiently low, the ready availability of water for drinking and other uses and the effective disposal of waterborne wastes may not pose a serious problem. For example, in colonial America wells and surface streams provided adequate water, and wastes were disposed of into other nearby watercourses without fuss or bother. Even today, much of rural America has no need for water and wastewater systems more sophisticated than a well and a septic tank. But people are social and commercial animals and, in the process of congregating in cities, have created a problem of adequate water supply and disposal.

In this chapter the availability of water for public use by communities is considered first, followed by a discussion of how this water is treated and then distributed to individual users. In Chapter 11 the collection and treatment of used water, or wastewater, is described.

10.1 THE HYDROLOGIC CYCLE AND WATER AVAILABILITY

The concept of the *hydrologic cycle*, already presented in Chapter 4, is a useful starting point for the study of water supply. Illustrated in Figure 10.1, this cycle includes the precipitation of water from clouds, infiltration into the ground or runoff into surface watercourses, followed by evaporation and transpiration of the water back into the atmosphere.

Precipitation is the term applied to all forms of moisture originating in the atmosphere and falling to the ground (e.g., rain, sleet, and snow). Precipitation is measured with gauges that record in inches of water. The depth of precipitation over a given region is often useful in estimating the availability of water.

Evaporation and *transpiration* are the two ways water reenters the atmosphere. Evaporation is loss from free water surfaces while transpiration is loss by plants. The same meteorological factors that influence evaporation are at work in the transpiration process—solar radiation, ambient air temperature, humidity, and wind speed, as well as the amount of soil moisture available to the plants—all impact the rate of transpiration. Because

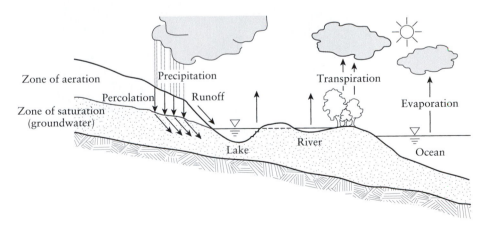

Figure 10.1 The hydrologic cycle in diagram form.

evaporation and transpiration are so difficult to measure separately, they are often combined into a single term, *evapotranspiration*, or the total water loss to the atmosphere by both evaporation and transpiration.

Water on the surface of earth that is exposed to the atmosphere is called *surface water*. Surface waters include rivers, lakes, oceans, etc. Through the process of *percolation*, some surface water (especially during a precipitation event) seeps into the ground and becomes *groundwater*. Both groundwater and surface water can be used as sources of water for communities.

10.1.1 Groundwater Supplies

Groundwater is both an important direct source of water supply and a significant indirect source of supply as a large portion of the flow to streams is derived from subsurface water. Water exists both near and far below the soil surface.

Near the surface of earth, soil pore spaces contain both air and water. This zone is known as the *zone of aeration*, or vadose zone. It may have zero thickness in swamplands and be several hundred feet thick in arid regions. Moisture from the zone of aeration cannot be tapped as a water supply source because this water is held to the soil particles by capillary forces and is not readily released.

Below the zone of aeration is the *zone of saturation*, in which the pores are filled with water. Water within the zone of saturation is what is often referred to as *groundwater*. A stratum that contains a substantial amount of groundwater is called an *aquifer*, and the surface of this saturated layer is known as the *water table*. If the aquifer is underlain by an impervious stratum, it is called an *unconfined aquifer*. If the stratum containing water is trapped between two impervious layers, it is known as a *confined aquifer*. Confined aquifers can sometimes be under pressure, just like pipes, and if a well is tapped into a confined aquifer under pressure, an *artesian well* results. Sometimes the pressure is sufficient to allow these artesian wells to flow freely without the necessity of pumping.

The amount of water that can be stored in the aquifer is equal to the volume of the void spaces between the soil grains. The fraction of voids volume to total volume of the soil is termed *porosity*, defined as

$$\text{Porosity} = \frac{\text{Volume of voids}}{\text{Total volume}}$$

But not all of this water is available for extraction and use because it is so tightly tied to the soil particles. The amount of water that can be extracted is known as *specific yield*, defined as

$$\text{Specific yield} = \frac{\text{Volume of water that will drain freely from a soil}}{\text{Total volume of water in the soil}}$$

The flow of water out of a soil can be illustrated by using Figure 10.2. The flow rate must be proportional to the area through which flow occurs times the velocity:

$$Q = Av$$

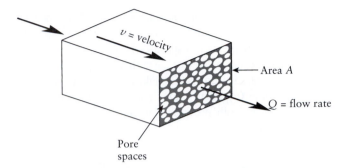

Figure 10.2 Flow from a porous medium such as soil.

where Q = flow rate, m³/s
$\quad A$ = area of porous material through which flow occurs, m²
$\quad v$ = superficial velocity, m/s

The superficial velocity is, of course, not the actual velocity of the water in the soil because the volume occupied by the soil solid particles greatly reduces the available area for flow. If a is the area available for flow, then

$$Q = Av = av'$$

where v' = the actual velocity of water flowing through the soil
$\quad a$ = the area available for flow

Solving for v' yields

$$v' = \frac{Av}{a}$$

If a sample of soil is of some length L, then

$$v' = \frac{Av}{a} = \frac{AvL}{aL} = \frac{v}{\text{Porosity}}$$

because the total volume of the soil sample is AL and the volume occupied by the water is aL.

Water flowing through the soil at a velocity v' loses energy, just as water flowing through a pipeline or an open channel loses energy. This energy loss per distance traveled is defined as

$$\frac{dh}{dL}$$

where h = energy, measured as elevation of the water table in an unconfined aquifer or as pressure in a confined aquifer, m
$\quad L$ = horizontal distance in the direction of flow, m

Table 10.1   Typical Aquifer Parameters

Aquifer Material	Porosity (%)	Specific Yield (%)	Coefficient of Permeability (m/s)
Clay	55	3	1×10^{-6}
Loam	35	5	5×10^{-6}
Fine sand	45	10	3×10^{-5}
Medium sand	37	25	1×10^{-4}
Coarse sand	30	25	8×10^{-4}
Sand and gravel	20	16	6×10^{-4}
Gravel	25	22	6×10^{-3}

Source: Adapted from Davis, M., and D. Cornwell. 1991. *Introduction to Environmental Engineering.* New York: McGraw-Hill. The McGraw-Hill Companies, Inc.

In an unconfined aquifer, the drop in the elevation of the water table with distance is the slope of the water table, dh/dL, in the direction of flow. The elevation of the water surface is the potential energy of the water, and water flows from a higher elevation to a lower elevation, losing energy along the way. Flow through a porous medium, such as soil, is related to the energy loss using Darcy's equation:

$$Q = KA \frac{dh}{dL} \qquad (10.1)$$

where K = coefficient of permeability, m/s
    A = cross-sectional area, m^2

Table 10.1 shows some typical values of porosity, specific yield, and coefficient of permeability.

Darcy's equation makes intuitive sense, in that the flow rate (Q) increases with increasing area through which the flow occurs (A) and with increasing elevation or pressure differences (dh/dL). The greater the driving force (e.g., the difference in upstream and downstream pressures), the greater the flow. The fudge factor K is the *coefficient of permeability*, an indirect measure of the ability of a soil sample to transmit water. It varies dramatically for different soils, ranging from about 0.05 m/d for clay to over 5000 m/d for gravel. The coefficient of permeability is commonly measured in the laboratory by using *permeameters*, which consist of a soil sample through which a fluid, such as water, is forced. The flow rate is measured for a given driving force (difference in pressures) through a known area of soil sample, and the permeability is calculated.

EXAMPLE
10.1

Problem   A soil sample is placed in a permeameter as shown in Figure 10.3 on the next page. The length of the sample is 0.1 m, and it has a cross-sectional area of 0.05 m^2. The water pressure on the upflow side is 2.5 m, and on the downstream side the water pressure is 0.5 m. A flow rate of 2.0 m^3/d is observed. What is the coefficient of permeability?

Figure 10.3 Permeameter used for measuring coefficient of permeability, using the Darcy equation.

Solution Use Darcy's equation (Equation 10.1). The pressure drop is the difference between the upstream and downstream pressures, or $h = 2.5 - 0.5 = 2.0$ m. Solving for K:

$$K = \frac{Q}{A \dfrac{dh}{dL}} = \frac{2.0 \text{ m}^3/\text{day}}{0.05 \text{ m}^2 \times \dfrac{2 \text{ m}}{0.1 \text{ m}}} = 2 \text{ m/d} = 2 \times 10^{-5} \text{ m/s}$$

From Table 10.1, the sample appears to contain fine sand.

If a well is sunk into an unconfined aquifer and water is pumped out, the water in the aquifer will begin to flow toward the well (Figure 10.4). As the water approaches the well, the area through which it flows gets progressively smaller, so a higher superficial (and actual) velocity is required. The higher velocity results, of course, in an increasing loss of energy, so the energy gradient must increase, forming a *cone of depression*. The reduction in the water table is known in groundwater terms as a *drawdown*. If the rate of water flowing toward the well is equal to the rate of water being pumped out of the well, the condition is at equilibrium, and the drawdown remains constant. If, however, the rate of water pumping is increased, the radial flow toward the well has to compensate, resulting in a deeper cone or drawdown.

Consider a cylinder shown in Figure 10.5 through which water flows toward the center. Using Darcy's equation:

$$Q = KA \frac{dh}{dL} = K(2\pi rw) \frac{dh}{dr}$$

where r = radius of the cylinder
$A = 2\pi rw$, the cross-sectional surface area of the cylinder

If water is pumped out of the center of the cylinder at the same rate as water is moving in through the cylinder surface area, the depth of the cylinder through which water flows

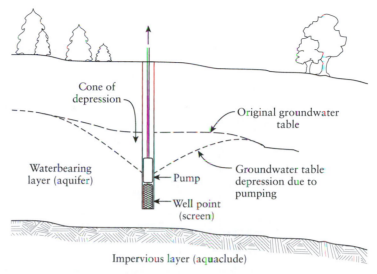

Cone of
depression

Original groundwater
table

Waterbearing
layer (aquifer)

Pump

Groundwater table
depression due to
pumping

Well point
(screen)

Impervious layer (aquaclude)

Figure 10.4   Drawdown in water table due to pumping from a well.

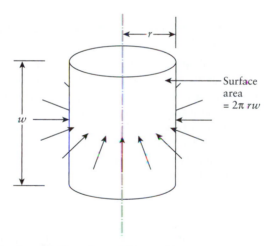

Surface
area
$= 2\pi \, rw$

w

Figure 10.5   Cylinder with flow through the surface.

into the well, w, can be replaced by the height of the water above the impermeable layer, h. This equation can then be integrated as

$$\int_{r_2}^{r_1} Q \, \frac{dr}{r} = 2\pi K \int_{h_2}^{h_1} h \, dh$$

$$Q \ln \frac{r_1}{r_2} = \pi K (h_1^2 - h_2^2); \qquad Q = \frac{\pi K (h_1^2 - h_2^2)}{\ln \dfrac{r_1}{r_2}} \qquad (10.2)$$

Note that the integration is between any two arbitrary values of r and h.

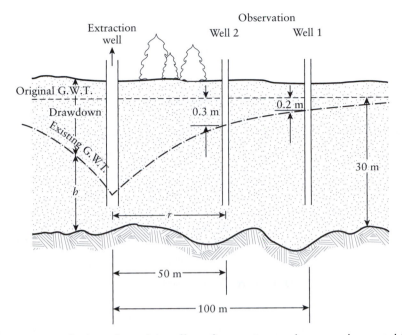

Figure 10.6 Multiple wells and the effect of extraction on the groundwater table.

This equation can be used to estimate the pumping rate for a given drawdown any distance away from a well in an unconfined aquifer using the water level measurements in two observation wells, as shown in Figure 10.6. Also, knowing the diameter of a well, it is possible to estimate the drawdown at the well, the critical point in the cone of depression. If the drawdown is depressed all the way to the bottom of the aquifer, the well goes dry; it cannot pump water at the desired rate. Although the derivation of the above equation is for an unconfined aquifer, the same situation would occur for a confined aquifer, where the pressure would be measured by observation wells.

EXAMPLE

10.2 **Problem** A well is 0.2 m in diameter and pumps from an unconfined aquifer 30 m deep at an equilibrium (steady-state) rate of 1000 m³/d. Two observation wells are located at distances 50 m and 100 m from the well, and they have been drawn down by 0.3 m and 0.2 m, respectively. What are the coefficient of permeability and estimated drawdown at the well? (See Figure 10.6.)

Solution Use Equation 10.2 with $h_1 = 30$ m $- 0.2$ m $= 29.8$ m and $h_2 = 30$ m $- 0.3$ m $= 29.7$ m.

$$K = \frac{Q \ln \dfrac{r_1}{r_2}}{\pi(h_1^2 - h_2^2)} = \frac{(1000 \text{ m}^3/\text{d}) \ln\left(\dfrac{100 \text{ m}}{50 \text{ m}}\right)}{\pi[(29.8 \text{ m})^2 - (29.7 \text{ m})^2)]} = 37.1 \text{ m/d}$$

If the radius of the well is 0.2 m/2 = 0.1 m, this can be plugged into the same equation as

$$Q = \frac{\pi K (h_1^2 - h_2^2)}{\ln \dfrac{r_1}{r_2}} = \frac{\pi (37.1 \text{ m/d})[(29.7 \text{ m})^2 - h_2^2]}{\ln \dfrac{50 \text{ m}}{0.1 \text{ m}}} = 1000 \text{ m}^3/\text{d}$$

Solving for h_2:

$$h_2 = 28.8 \text{ m}$$

Because the aquifer is 30 m deep, the drawdown at the well is $30 - 28.8 = 1.2$ m.

FOCUS ON

Arsenic in Groundwater in Bangladesh

Environmental engineers need to consider local cultural, geological, and logistical issues before implementing a project. A case in point is Bangladesh, where groundwater from tubewells, introduced by humanitarian agencies to prevent waterborne diseases, puts over 70 million people at risk of severe arsenic poisoning. Rural poverty and social stigma against arsenic poisoning sufferers exacerbate the problem.[1]

Until the 1960s, most Bangladeshis obtained their drinking and irrigation water from bacterially contaminated ponds, rivers, and shallow wells and consequentially experienced high rates of infant mortality and diarrheal disease.[2] In an effort to reduce deaths from bacterial disease, the World Bank and UNICEF backed a project to dig 4.5 million tubewells to provide safe drinking water to the country's dense and largely rural population, a technique that had been effective in other countries (Figure 10.7).[3]

Although in other countries concentrations of arsenic are often the result of human activities (such as burning coal, installing arsenic-treated telephone poles, and fertilizer runoff) arsenic in the alluvial soils of Bangladesh is natural in origin. Arsenopyrite (FeAsS), a ferric arsenic compound, composes up to 7.5% of the geologic substrates in the country. In the presence of air, arsenopyrite is oxidized by oxygen, and the arsenic adsorbed with iron sulfide becomes mobilized. Individuals who drink groundwater are at risk of arsenicosis, a slow disease that leads to disfigurement, neurological damage, and cancer of the internal organs.[4]

Though the well-drilling project did succeed in reducing infant diarrheal disease mortality from 250,000 in 1983 to 110,000 in 1996, it also led to millions of new cases of debilitating arsenic poisoning (Figure 10.8).[2] Furthermore, the widespread use of well water in irrigation meant vegetables and agricultural products contained elevated levels of arsenic. Individuals marked by the black skin lesions that characterize arsenic poisoning are often socially outcast and forced out of their home communities. However, most rural Bangladeshis do not associate the lesions, which may take years to appear, with contaminated well water.[1]

Despite the disadvantages of tubewell water, returning to the use of arsenic-free but bacteriologically contaminated surface water was unfeasible. The Bangladeshi population continues to grow at 1.9% annually, and as human impact increases, surface water quality declines.[2] In addition, people who own the use-rights to a well are often respected and influential and see filtration and boiling of surface water as fit only for the poor.

In an attempt to address what the World Health Organization called a "disaster," in 2007 the National Academy of Engineering in the United States sponsored a competition for the Grainger Challenge Prize.[3] The competition's goal was to promote innovative solutions to arsenic removal from tubewell water that would be "affordable, robust, reliable, easy to maintain, and environmentally friendly."[5]

Figure 10.7   A well painted green (water safe).

The first prize of $1 million went to Abul Hussam of George Mason University, whose own home in Bangladesh had contaminated tube wells. Hussam's SONO filter was a more sustainable and less expensive point-of-use filtration method than the other entries—able to supply arsenic-free water to two to three families using locally found or manufactured materials.[5] Water first passes through a composite iron matrix that removes arsenic through surface complexation reactions. It then goes into a second bucket where it passes through coarse river sand and wood charcoal to remove organics. A third bucket contains wet brick chips to remove fine particulates. The resulting water contains less than 0.01 mg/L As(V) and almost no inorganic As(III), one-fifth the WHO-established limit. SONO filters cost only $40. Importantly, the waste product is nontoxic iron-arsenate cement that does not leach into groundwater.[6]

Figure 10.8   (A) Arsenic lesions on feet; (B) arsenic lesions on hand, cancer; (C) keratoses on hand.

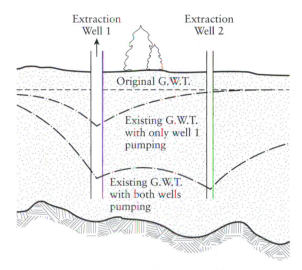

Figure 10.9 Effect of two extraction wells on the groundwater table.

Multiple wells in an aquifer can interfere with each other and cause excessive draw-down. Consider the situation in Figure 10.9, where first a single well creates a cone of depression. If a second production well is installed, the cones will overlap, causing greater drawdown at each well. If many wells are sunk into an aquifer, the combined effect of the wells could deplete the groundwater resources, and all wells would go dry.

The reverse is also true, of course. Suppose one of the wells is used as an injection well; then the injected water flows from this well into the others, building up the groundwater table and reducing the drawdown. The judicious use of extraction and injection wells is one way that the flow of contaminants from hazardous waste or refuse dumps can be controlled, as discussed further in Chapter 15.

Finally, a lot of assumptions are made in the above discussion. First, it is assumed that the aquifer is homogeneous and infinite, meaning that it sits on a level aquaclude and that the permeability of the soil is the same at all places for an infinite distance in all directions. Second, steady state and uniform radial flow are assumed. The well is assumed to penetrate the entire aquifer and be open for the entire depth of the aquifer. Finally, the pumping rate is assumed to be constant. Clearly, any of these conditions may cause the analysis to be faulty, and this model of aquifer behavior is only the beginning of the story. Modeling the behavior of groundwater is a complex and sophisticated science.

10.1.2 Surface Water Supplies

Surface water supplies are not as reliable as groundwater sources because quantities often fluctuate widely during the course of a year or even a week, and the quality of surface water is easily degraded by various sources of pollution. The variation in the river or stream flow can be so great that even a small demand cannot be met during dry periods, so storage facilities must be constructed to hold the water during wet periods so it can be saved for the dry ones. The objective is to build these reservoirs sufficiently large to have dependable supplies.

One method of arriving at the proper reservoir size is by constructing a *mass curve*. In this analysis the total flow in a stream at the point of a proposed reservoir is summed and plotted against time. On the same curve the water demand is plotted, and the difference between the total water flowing in and the water demanded is the quantity that the reservoir must hold if the demand is to be met. The method is illustrated by the following example.

EXAMPLE

10.3

Problem   A reservoir is needed to provide a constant flow of 425 L/s. The monthly stream flow records, in total cubic feet of water for each month, are

Month	J	F	M	A	M	J	J	A	S	O	N	D
Water ($\times 10^8$ L)	14	17	20	11	8	6	14	22	3	14	17	22

Calculate the reservoir storage necessary to provide the constant 425 L/s demand.

Solution   The storage requirement is calculated by plotting the cumulative water flows (Figure 10.10). For example, for January, 14×10^8 L is plotted while for February 17 is added to that and 31×10^8 L is plotted. The demand for water is constant at 425 L/s, $\times$ 60 s/min $\times$ 60 min/h $\times$ 24 h/d $\times$ 30 days/month $= 11 \times 10^8$ L/month. This can be plotted as a sloped line on the curved supply line. Notice that the stream flow in May is lower than the demand, and this is the start of a drought lasting into June. Another way of looking at it is that the demand slope is greater than the supply slope, and thus, the reservoir has to make up the deficit. In July the rains come, and the supply increases until the reservoir can be filled up again, late in August. The reservoir capacity needed to get through that particular drought is 17×10^8 L. A second drought, starting in September, lasts into November and requires 10×10^8 L of capacity. If the municipality has a reservoir with at least 17×10^8 L capacity, it can draw water from it throughout the year.

■

A mass curve, such as Figure 10.10, is actually of little use if only limited stream flow data are available. One year's data yield very little information about long-term variations. For example, was the drought in the above example the worst drought in 20 years, or was the year shown actually a fairly wet year? To get around this problem, it is necessary to predict statistically the recurrence of events such as droughts and then to design the structures according to a known risk. This procedure is discussed in Chapter 3.

Water supplies are often designed to meet demands 19 out of 20 years. In other words, once in 20 years the drought will be so severe that the reservoir capacity will not be adequate to meet the demand for water. If running out of water once every 20 years is unacceptable, the community can choose to build a bigger reservoir and expect to be dry only once every 50 years, or whatever time frame is chosen. The question is one of an increasing investment of capital for a steadily smaller added benefit.

Using a *frequency analysis* of recurring natural events, such as droughts, as described in Chapter 3, a 100-year drought or a 10-year drought can be calculated. Although this 10-year drought occurs on the average once every 10 years, there is no guarantee that it

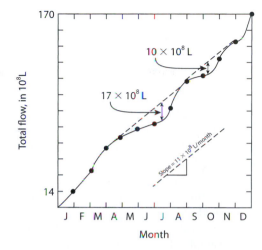

Figure 10.10   Mass curve showing required storage volumes.

would indeed occur once every 10 years. In fact, it could, for example, happen three years in a row, and then not again for 50 years.

When a particularly severe and unanticipated drought occurs, it is assumed that people will cooperate and reduce their use of water. If they don't, the community can impose sanctions on those that use more than their share. Many communities in southern California have imposed severe fines and other penalties for excessive use.

A far better way of attaining the same end is to have everyone cooperate voluntarily. Such community voluntary cooperation is often difficult to achieve, however, and nowhere is this more evident than in matters of environmental degradation or fair use of resources. The essence of this problem can be illustrated by what has become known as the "prisoner dilemma."

Suppose there are two prisoners, A and B, both of whom are being separately interrogated. Each prisoner is told that, if neither of them confess, they will both receive a two-year sentence. If one prisoner accuses the other, the accuser is set free and the one who was accused receives a 20-year sentence. If, however, they both accuse each other, each one gets a 10-year sentence. Should prisoner A trust prisoner B not to accuse him and do so likewise? If prisoner B does not accuse A and A does not accuse B, they both get only two years. But all A has to do to go free is to accuse B and hope that B does not accuse him. Prisoner B, of course, has the same option. Graphically, the options look like this matrix:

		B Accuses A	
		NO	**YES**
	NO	2, 2	20, 0
A Accuses B			
	YES	0, 20	10, 10

The same dilemma might occur in the use of a scarce resource, such as a water supply. During a severe drought, all industries are asked to cooperate by voluntarily restricting water use. Industry A agrees to do so, and if the other industries cooperate, they would all experience a loss of production, but all the companies would survive. If Industry A decides not to cooperate, on the other hand, and if everyone else does, Industry A wins big because it has full use of the water and its production and profits would be large. By using all of the water, Industry A would have deprived others of this resource, and the other industries may have been forced to shut down. If, on the other hand, *everyone* behaved like Industry A and decided not to cooperate, then they would all run out of water and all of them would be driven into bankruptcy.

Many theorists hold that human nature prevents us from choosing the cooperation alternative and that the essence of human nature is competition and not cooperation. In that sense, humans are no different from animals and plants that survive in a stable ecosystem—not by cooperation but by competition.* Within any such system, there is continual competition for resources, and the only sharing that occurs is instinctive sharing, such as a mother feeding her young. What might be seen as cooperation in an ecosystem is actually parasitic behavior: the use of another creature for one's own ends. A maple tree, for example, does not exist to have a honeysuckle vine grow on it. If a maple could make a conscious choice, it would likely try to prevent the honeysuckle from twisting itself around its trunk and branches and reducing the amount of sunlight the maple tree receives.

If competition is the essence of human (and other living creature) conduct, can we ever then expect humans to act in a cooperative mode?

The greatest single threat to the global ecosystem is the cancerous growth of the human population. This trend somehow has to be reversed if the human species has any hope of long-term survival. Yet out of altruistic and cooperative motives, we help other humans in need. We mount massive food-supply efforts for starving people and use international pressures to prevent wars of attrition. If we are truly competitive, then such actions would make no sense, no more than one loblolly pine seedling helping its neighbor. Both seedlings "know" that only one will survive, and each tries its best to be the taller and stronger of the two.

For years it was thought that competition and aggression (wars) would be the undoing of humans on this planet. It is ironic that cooperation may actually be the single most species-destructive trait in *homo sapiens*.

10.2 WATER TREATMENT

Many aquifers and isolated surface waters are of high water quality and may be pumped from the supply and transmission network directly to any number of end uses, including human consumption, irrigation, industrial processes, and fire control. However, such clean water sources are the exception to the rule, particularly in regions with dense populations or regions that are heavily agricultural. Here, the water supply must receive varying degrees of treatment prior to distribution.

---

*This is an exaggeration, of course. In some human societies, such as the Native Americans and the Mexican Americans, the perception of the individual within the community is different than in most Western cultures. Similarly, some animals, such as whales, have been suspected of occasionally behaving in a cooperative manner.

Figure 10.11   A typical water treatment plant.

A typical water treatment plant is diagramed in Figure 10.11. Such plants are made up of a series of reactors or unit operations, with the water flowing from one to the next to achieve a desired end product. Each operation is designed to perform a specific function, and the order of these operations is important. Described below are a number of the most important of these processes.

10.2.1   Softening

Some waters (both surface waters and groundwaters) need hardness removed to use them as a potable water source. Hardness is caused by multivalent cations (or minerals)—such as calcium, magnesium, and iron—that dissolve from soil and rocks (particularly limestone). While hardness does not cause health problems, it does reduce the effectiveness of soaps and cause scale formation.

The reaction of hardness ions with soaps causes bathtub ring and reduces lather. Soaps are typically made up of long, chain-like molecules with two distinct ends. The hydrophilic end interacts with water while the hydrophobic end interacts with oil and grease. When the hydrophobic end interacts with hardness ions instead, the soap bunches together, forming a soap scum, or film. Besides leaving a bathtub ring, soap scum can cause skin irritation by disrupting the skin's pH and can make hair dull. The same effect happens to laundry detergents, and the scum formed can make clothes look drab and feel stiff and can cause fabric to wear out faster. Scale, which forms when calcium carbonate precipitates from heated water, is a more serious problem because it reduces heat transfer efficiency by coating water heaters, boilers, heat exchangers, tea pots—anything in which water is heated—and can eventually clog pipes. Hardness also sometimes causes objectionable tastes.

Total hardness (TH) is defined as the sum of the multivalent cations in the water. Calcium (Ca^{2+}) and magnesium (Mg^{2+}) tend to be the largest components of hardness, so TH is typically approximated as the sum of these two components. However, iron (Fe^{2+}

and Fe^{3+}), manganese (Mn^{2+}), strontium (Sr^{2+}), and aluminum (Al^{3+}) may also be present in water supplies.

$$TH = \sum(\text{Multivalent cations}) \cong Ca^{2+} + Mg^{2+} \tag{10.3}$$

Hardness can be calculated by analyzing for all cations present in a sample. Cations can be measured by sophisticated instruments, such as atomic absorption and ion-specific electrodes. Alternatively, hardness can be determined through titration. Ethylenediaminete-traacetic acid (EDTA) is used as the titrant, and Eriochrome Black T, which turns from blue to red when metal ions are present, is used as the indicator.

Typical units for hardness are mg/L as $CaCO_3$ and meq/L. By using these units, the contributions of different substances (e.g., calcium and magnesium) can be added directly. (The units of mg/L of a particular substance, such as 10 mg/L of Ca^{2+}, cannot be added directly to the mg/L of a different substance, such as 5 mg/L of Mg^{2+}—similar to the problem of adding 25 cm to 1.5 m).

To convert a concentration in mg/L to meq/L, divide the concentration by the substance's equivalent weight (EW):

$$C_q = \frac{C}{EW} \tag{10.4}$$

where C_q = concentration in meq/L
C = concentration in mg/L
EW = equivalent weight in g/eq or mg/meq

A substance's equivalent weight is calculated by dividing its atomic weight (AW) or molecular weight (MW) by its valence or ionic charge (n, which is always positive):

$$EW = \frac{AW \text{ or } MW}{n} \tag{10.5}$$

where AW or MW has units of g/mole and n has units of equivalents/ mole (eq/mol).

To convert to the standard unit mg/L as $CaCO_3$, the meq/L concentration is multiplied by the equivalent weight of $CaCO_3$, which is 50.0 mg/meq (100 g/mol/ 2 eq/mol):

$$C_{CaCO_3} = C_q \times 50.0 \tag{10.6}$$

where C_{CaCO_3} = concentration in mg/L as $CaCO_3$
C_q = concentration in meq/L

EXAMPLE

10.4   **Problem**   The concentration of calcium in a water sample is 100 mg/L. What is the concentration in (a) meq/L and (b) mg/L as $CaCO_3$?

Solution   The valence or ionic charge of calcium is +2, so n is 2 eq/mol. Calcium's atomic weight is 40.1 g/mol. Therefore, its equivalent weight is (Equation 10.5):

$$EW = \frac{AW}{n} = \frac{40.1 \text{ g/mol}}{2 \text{ eq/mol}} = 20.0 \text{ g/eq} = 20.0 \text{ mg/meq}$$

Note that the equivalent weight for a substance is constant because its atomic or molecular weight and ionic change are constant.

a. The concentration in meq/L is then simply obtained through unit conversion (Equation 10.4):

$$C_q = \frac{C}{EW} = \frac{100 \text{ mg/L}}{20.0 \text{ mg/meq}} = 5.0 \text{ meq/L}$$

b. Again, the concentration in mg/L as $CaCO_3$ is simply obtained through unit conversion (Equation 10.6):

$$C_{CaCO_3} = C_q \times 50$$

$$= (5.0 \text{ meq/L})(50 \text{ mg/meq})$$

$$= 250 \text{ mg/L as } CaCO_3$$

Note that the *correct* unit includes "as $CaCO_3$."

EXAMPLE

10.5

Problem   A water sample contains 60 mg/L of calcium, 60 mg/L of magnesium, and 25 mg/L of sodium. What is the total hardness in (a) meq/L and (b) mg/L as $CaCO_3$?

Solution   The first step when calculating total hardness is to determine which species are relevant. Remember that only multivalent cations contribute to hardness. Therefore, we can ignore sodium. The second step is to put the concentrations of the relevant species into units that can be summed—either meq/L or mg/L as $CaCO_3$.

a. In units of meq/L (Equations 10.4 and 10.5 followed by Equation 10.3):

$$Ca^{2+} = \frac{60 \text{ mg/L}}{20.0 \text{ mg/meq}} = 3.0 \text{ meq/L}$$

$$Mg^{2+} = \frac{60 \text{ mg/L}}{\dfrac{24.3 \text{ mg/mmol}}{2 \text{ meq/mmol}}} = 4.9 \text{ meq/L}$$

$$TH = Ca^{2+} + Mg^{2+} = (3.0 \text{ meq/L}) + (4.9 \text{ meq/L}) = 7.9 \text{ meq/L}$$

b. To obtain mg/L as $CaCO_3$, multiply meq/L by 50 mg/meq (Equation 10.6). Either multiply each component of hardness by 50 mg/meq and then sum them, or sum them and then multiply by 50 mg/meq. The answer is the same.

$$TH = (7.9 \text{ meq/L})(50 \text{ mg/meq}) = 395 \text{ mg/L as } CaCO_3$$

Table 10.2 Water Hardness Classifications

	Hardness	
Classification	meq/L	mg/L as CaCO$_3$
Extremely soft to soft	0–0.9	0–45
Soft to moderately hard	0.9–1.8	46–90
Moderately hard to hard	1.8–2.6	91–130
Hard to very hard	2.6–3.4	131–170
Very hard to excessively hard	3.4–5	171–250
Too hard for ordinary domestic use	>5	>250

Source: L.A. Lipe and M. D. Curry, "Ion Exchange Water Softening," a discussion for water treatment plant operators, 1974–75 seminar series sponsored by Illinois Environmental Protection Agency.

Water is classified as soft or hard depending on the amount of hardness ions present. The water in the above example would be classified as very hard (Table 10.2), which is typical of a groundwater source. Surface water is generally soft because fewer minerals dissolve in it. However, surface water can be hard. Water treatment plants typically distribute moderately hard water, in the range of 80 to 90 mg/L as CaCO$_3$. (It is difficult to rinse off soap if the water is too soft, and some hardness can protect the distribution system from corrosion.)

Total hardness can be divided into two components—carbonate hardness (CH), also known as temporary hardness, and noncarbonate hardness (NCH), also known as permanent hardness:

$$TH = CH + NCH \qquad (10.7)$$

Carbonate hardness is the component of total hardness associated with the anions carbonate (CO$_3^{2-}$) and bicarbonate (HCO$_3^-$); it is that portion of hardness that forms scale. Noncarbonate hardness is the component of total hardness associated with all other anions. The form of the hardness impacts the amount and types of chemicals required to remove the hardness.

Carbonate hardness is equal to the smaller of alkalinity or total hardness. Noncarbonate hardness is equal to the difference between total hardness and carbonate hardness. If the alkalinity is equal to or greater than the total hardness, then the noncarbonate hardness is zero because all the hardness ions are associated with alkalinity. A good check on calculations is to remember that the sum of carbonate and noncarbonate hardness *cannot* be greater than total hardness.

Alkalinity is a measure of the buffering capacity of water (or the capacity of the water to neutralize acid, or H$^+$). It is not the same as pH; water does not have to be basic to have high alkalinity. The natural sources of alkalinity in water are the atmosphere and limestone. Atmospheric carbon dioxide dissolved in water forms carbonic acid (H$_2$CO$_3$), which can dissociate into bicarbonate (HCO$_3^-$) and carbonate (CO$_3^{2-}$). Limestone formations (CaCO$_3$) can also dissolve in water and produce HCO$_3^-$ and CO$_3^{2-}$. Both HCO$_3^-$ and CO$_3^{2-}$ can "grab" (or neutralize) H$^+$ that is added to the water. Notice that carbonate can grab two H$^+$ while bicarbonate can grab only one. Water (H$_2$O) dissociates into hydrogen (H$^+$)

and hydroxide (OH^-). Hydroxide can grab one H^+ to reform water; however, H^+ already in the water cannot neutralize any added H^+. Based on this chemistry, alkalinity can be calculated as

$$
\text{Alkalinity (mol/L)} = \left(\frac{\text{mol HCO}_3^-}{L}\right)\left(\frac{1 \text{ mol ALK}}{\text{mol HCO}_3^-}\right)
$$

$$
+ \left(\frac{\text{mol CO}_3^{2-}}{L}\right)\left(\frac{2 \text{ mol ALK}}{\text{mol CO}_3^{2-}}\right)
$$

$$
+ \left(\frac{\text{mol OH}^-}{L}\right)\left(\frac{1 \text{ mol ALK}}{\text{mol OH}^-}\right)
$$

$$
- \left(\frac{\text{mol H}^+}{L}\right)\left(\frac{1 \text{ mol ALK}}{\text{mol H}^+}\right)
$$

or simply

$$
\text{Alkalinity (mol/L)} = [\text{HCO}_3^-] + 2[\text{CO}_3^{2-}] + [\text{OH}^-] - [\text{H}^+]
$$

When the units moles per liter are used, carbonate is multiplied by 2 (because it can neutralize 2 mol of H^+). However, when the units meq/L or mg/L as $CaCO_3$ are used, the 2 is already taken into account,

$$
\text{CO}_3^{2-} \text{ (eq/L)} = \left(\frac{\text{mol CO}_3^{2-}}{L}\right)\left(\frac{2 \text{ eq}}{\text{mol}}\right)
$$

so the equation becomes

$$
\text{Alkalinity (meq/L)} = \left(\frac{1 \text{ meq HCO}_3^-}{L}\right)\left(\frac{1 \text{ meq ALK}}{\text{meq HCO}_3^-}\right)
$$

$$
+ \left(\frac{1 \text{ meq CO}_3^{2-}}{L}\right)\left(\frac{1 \text{ meq ALK}}{\text{meq CO}_3^{2-}}\right)
$$

$$
+ \left(\frac{1 \text{ meq OH}^-}{L}\right)\left(\frac{1 \text{ meq ALK}}{\text{meq OH}^-}\right)
$$

$$
- \left(\frac{1 \text{ meq H}^+}{L}\right)\left(\frac{1 \text{ meq ALK}}{\text{meq H}^+}\right)
$$

or simply

$$
\text{Alkalinity (meq/L)} = (\text{HCO}_3^-) + (\text{CO}_3^{2-}) + (\text{OH}^-) - (\text{H}^+)
$$

At pH less than 8.3, the carbonate concentration is much less than the bicarbonate concentration, and around neutral pH, $[H^+] \approx [OH^-]$. Therefore, for most potable water sources:

$$
\text{Alkalinity} \cong (\text{HCO}_3^-)
$$

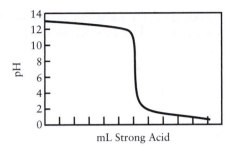

Figure 10.12 Titration curve.

A two-step titration is used for the laboratory measurement of alkalinity. The first step measures carbonate alkalinity; the second step measures bicarbonate alkalinity. In the first step sulfuric acid or hydrochloric acid is used to drop the sample pH to 8.3. Phenolphthalein, which changes from pink to colorless, is typically used as the indicator for this step. In the second step the pH is dropped to 4.5 using bromocresol green as the indicator. Figure 10.12 shows a typical titration curve of a basic solution with a strong acid. The buffering capacity of the solution is high in the beginning of the titration, so the pH change is gradual. Once the buffering capacity is exceeded, however, the pH change becomes quite rapid, highlighting the need for close pH monitoring and control in reactors in which alkalinity is being consumed.

EXAMPLE
10.6

Problem   From the following water analysis, determine the total hardness, carbonate hardness, and noncarbonate hardness in (a) milliequivalents per liter (meq/L) and (b) milligrams per liter (mg/L) as $CaCO_3$.

Component	Concentration (mg/L)
CO_2	6.0
Ca^{2+}	50.0
Mg^{2+}	20.0
Na^+	5.0
Alkalinity	120 as $CaCO_3$
SO_4^{2-}	94.0
pH	7.3

Solution   The first step is to convert the relevant concentrations to meq/L or mg/L as $CaCO_3$ as shown in the following table. This step is the same as shown in Examples 10.4 and 10.5. Note that, because alkalinity is given in mg/L as $CaCO_3$, the equivalent weight used to calculate the meq/L is the equivalent weight of $CaCO_3$, not of bicarbonate. (Because the pH is near neutral, the alkalinity is approximately all bicarbonate.)

Component	EW (mg/meq)	Concentration		
		mg/L	meq/L	mg/L as CaCO$_3$
Ca^{2+}	20.0	50	2.5	125
Mg^{2+}	12.2	20	1.6	82
Alkalinity	50.0	120	2.4	120

Total hardness is the sum of the multivalent cations (Equation 10.3), in this case Ca^{2+} and Mg^{2+}:

$$TH = (2.5 \text{ meq/L}) + (1.6 \text{ meq/L}) = 4.1 \text{ meq/L}$$

or

$$TH = (125 \text{ mg/L as CaCO}_3) + (82 \text{ mg/L as CaCO}_3) = 207 \text{ mg/L as CaCO}_3$$

This water is considered very hard to excessively hard (Table 10.2).

To determine carbonate hardness, compare the alkalinity to the total hardness. In this case the alkalinity (2.4 meq/L) is less than the total hardness (4.1 meq/L). Therefore, the carbonate hardness equals the alkalinity, 2.4 meq/L or 120 mg/L as CaCO$_3$. (If the alkalinity would have been greater than the total hardness, then the carbonate hardness would have equaled the total hardness. Remember, carbonate hardness cannot be greater than total hardness.)

To determine noncarbonate hardness, subtract the carbonate hardness from the total hardness (Equation 10.7):

$$NCH = TH - CH = (4.1 \text{ meq/L}) - (2.4 \text{ meq/L}) = 1.7 \text{ meq/L}$$

or

$$NCH = (207 \text{ mg/L as CaCO}_3) - (120 \text{ mg/L as CaCO}_3) = 87 \text{ mg/L as CaCO}_3$$

In this case the carbonate hardness was less than the total hardness; therefore, there was noncarbonate hardness. If the carbonate hardness had equaled the total hardness, then the noncarbonate hardness would have been zero.

■

Bar charts (Figure 10.13 on the next page) are useful for visualizing hardness speciation. The speciation is important to know when using lime-soda softening to remove the hardness. When constructing a bar chart, calcium is placed first and magnesium second because magnesium is more expensive to remove. These are followed by other hardness ions and then other cations. Bicarbonate is placed first on the anion bar and is followed by other anions present because carbonate hardness requires fewer chemicals to remove. Units that can be summed, i.e., meq/L or mg/L as CaCO$_3$, must be used, and the chart must be drawn to a consistent scale. Note that carbonate hardness and noncarbonate hardness must sum to total hardness, and they cannot be greater than total hardness.

Figure 10.13 Generalized bar chart.

Problem   From the previous water analysis, construct a bar chart to determine the speciation of the hardness.

Component	Concentration (mg/L)
CO_2	6.0
Ca^{2+}	50.0
Mg^{2+}	20.0
Na^+	5.0
Alkalinity	120 as $CaCO_3$
SO_4^{2-}	94.0
pH	7.3

Solution   The first step is to convert the concentrations of all ions to meq/L or mg/L as $CaCO_3$ as shown in the following table.

Component	EW (mg/meq)	Concentration mg/L	meq/L	mg/L as $CaCO_3$
Ca^{2+}	20.0	50	2.5	125
Mg^{2+}	12.2	20	1.6	82
Na^+	23.0	5	0.2	11
Alkalinity	50.0	120	2.4	120
SO_4^{2-}	48.0	94	2.0	98

The next step is to construct the bar chart, placing cations on top and anions on bottom in the specified order and to scale (Figure 10.14).

The first thing to check with the bar chart is whether the sum of the cations is approximately equal to the sum of the anions. In this case they are (4.3 meq/L versus 4.4 meq/L, which is close enough). This indicates that the water analysis appears to be analytically correct and relatively complete; no major cations or anions are missing.

Next, determine the speciation. Almost all the calcium (2.4 of the 2.5 meq/L) is associated with bicarbonate, but none of the magnesium is associated with bicarbonate.

Figure 10.14   Speciation of hardness bar chart.

The remaining calcium (0.1 meq/L) and all the magnesium (1.6 meq/L) are associated with sulfate. Therefore, the speciation can be reported as

calcium carbonate hardness (CCH) = 2.4 meq/L
calcium noncarbonate hardness (CNCH) = 0.1 meq/L
magnesium carbonate hardness (MCH) = 0 meq/L
magnesium noncarbonate hardness (MNCH) = 1.6 meq/L.

The general steps for solving hardness problems can be summarized as follows.

Step 1.   Calculate total hardness (TH) as the sum of the multivalent cations.
Step 2.   Calculate alkalinity (ALK), which is typically the bicarbonate concentration.
Step 3.   Calculate carbonate hardness (CH) and noncarbonate hardness (NCH).
$$CH = ALK \text{ if } TH > ALK$$
$$CH = TH \text{ if } TH < ALK$$
$$NCH = TH - CH.$$
Step 4.   Determine the hardness speciation.
$$CCH = Ca^{2+} \text{ if } Ca^{2+} < CH$$
$$CCH = CH \text{ if } Ca^{2+} > CH$$
$$CNCH = Ca^{2+} - CCH.$$
$$MCH = CH - CCH.$$
$$MNCH = Mg^{2+} - MCH.$$
Step 5.   Check your calculations.
$$CH = CCH + MCH$$
$$NCH = CNCH + MNCH$$
$$TH = CCH + CNCH + MCH + MNCH$$
$$Ca^{2+} = CCH + CNCH$$
$$Mg^{2+} = MCH + MNCH$$

The comparisons can be done visually with bar charts or mathematically. Note that we do not need to know what the noncarbonate anions are; we simply need to know how much of the hardness is not associated with alkalinity.

Ion Exchange and Reverse Osmosis

Softening is the process of removing hardness. Ion exchange and precipitation are the typical methods used, but reverse osmosis (RO) is also used. Point-of-use RO units are used in homes (Figure 10.15) while large-scale units are used in locations with severely limited freshwater supplies but copious quantities of saltwater. RO uses

© Susan Morgan

Figure 10.15 (A) A typical point-of-use reverse osmosis unit; (B) schematic.

high pressure to push water molecules through a membrane, resulting in treated water on one side and a concentrated wastewater on the other. (Remember that osmosis is the opposite process; water will migrate through a membrane to dilute the concentrated solution.) The process produces water with low dissolved minerals and removes some bacteria. However, RO is slow and generates large quantities of wastewater; a typical unit will produce about 4 L of potable water in 6 hr while generating 8 to 24 L of wastewater. It is also expensive to operate. Home units often come with an activated carbon filter to remove chlorine and improve the taste of the water. For waters with high hardness, they will typically follow a whole-house ion exchange softener.

Ion exchange, or zeolite, softening is most applicable to waters that are high in noncarbonate hardness (because it can be removed without chemical addition, unlike precipitation) and that have less than 350 mg/L as $CaCO_3$ total hardness.[8] Ion exchange softeners are often used in residences that have wells (Figure 10.16 on the next page). The hard water passes through a column containing resin. The resin adsorbs the hardness ions, exchanging them for sodium typically. (This is why softened water often tastes salty and why people on low-sodium and sodium-free diets should avoid drinking it.) Once the resin no longer removes the amount of hardness desired, a concentrated salt, or brine, solution (NaCl) is used to regenerate the resin (remove the hardness ions) so that it can be reused. Hydrochloric acid can be used instead of salt, so the exchanged ion is hydrogen (H^+) instead of sodium. However, in addition to being hazardous and reducing pH, hydrochloric acid is more expensive than salt and is therefore not widely used. Usually, regeneration requires 2.1 to 3.5 kg salt/kg hardness removed, and regeneration rates are 2.2 to 4.4 L/s/m^3 of resin for 55 min followed by 6.6 to 11 L/s/m^3 of resin for 5 min for municipal plants.[8] A short backwash is performed before regeneration to expand the bed 75% to 100% and remove particulates.

As long as the resin is relatively fresh (i.e., it has plenty of sodium remaining), essentially 100% of the hardness will be removed. Because not all the hardness needs to be removed, part of the water can bypass the system so that, when the treated and untreated water mix, the desired hardness is obtained. This scenario, of course, is a classical material balance.

EXAMPLE

10.8   **Problem**   A water softener has 0.07 m^3 of ion exchange resin with an exchange capacity of 46 kg/m^3. The water use is 1500 L/day. If the incoming water contains 245 mg/L of hardness as $CaCO_3$ and the softened water needs to be 100 mg/L as $CaCO_3$, how much water should bypass the softener, and what is the time between regeneration cycles?

Solution   The amount of water that should bypass the softener is a function of the desired hardness and the initial hardness. A schematic of the process is shown in Figure 10.16. Writing a material balance equation around the mixing point (the circle in the diagram) provides the solution at a particular point in time.

$$Q_{IX}C_{IX} + Q_{BP}C_{BP} = Q_0C_f$$

Figure 10.16 (A–B) Typical ion exchange water softener; (C) schematic.

For the case of a fresh resin, the effluent hardness is zero, so

$$Q_{IX}(0) + Q_{BP}C_{BP} = Q_0C_f$$

$$\frac{Q_{BP}}{Q_0} = \frac{C_f}{C_{BP}}$$

or

$$\% \text{ Bypass} = \frac{\text{Desired hardness}}{\text{Initial hardness}}(100)$$

In this case the amount that should be bypassed is

$$\% \text{ Bypass} = \frac{100 \text{ mg/L as CaCO}_3}{245 \text{ mg/L as CaCO}_3}(100) \cong 41\%$$

$$\text{Amount to bypass} = 0.41(1500 \text{ L/day}) \cong 610 \text{ L/d}$$

The length of the cycle, or time to breakthrough, is a function of the exchange capacity of the resin. If we assume complete saturation of the resin before regenerating, then

$$\text{Breakthrough} = \frac{(\text{Capacity})(V_{\text{resin}})}{Q_{\text{IX}}C_0} = \frac{(\text{Capacity})(V_{\text{resin}})}{(1 - \text{Bypass})Q_0 C_0}$$

$$\text{Breakthrough} = \frac{(46 \text{ kg/m}^3)(0.07 \text{ m}^3)(10^6 \text{ mg/kg})}{(1 - 0.41)(1500 \text{ L/day})(245 \text{ mg/L as CaCO}_3)} \cong 15 \text{ d}$$

Therefore, salt will have to be added approximately every two weeks.

Of course, residential units are typically less sophisticated and do not have bypass lines. The problem is actually more complicated than the solution given because the amount to bypass would fluctuate as the resin capacity was used. Figure 10.17 shows a general breakthrough curve for ion exchange columns. Breakthrough may be considered as occurring once the effluent concentration is approximately equal to the influent concentration. However, it is more typical at municipal plants for the breakthrough criterion to be set much lower, for example, at 5% to 10% of the influent concentration. Residences typically have one ion exchange column. Municipal water treatment plants will have multiple columns, with the effluent from one column flowing into the next column. This arrangement allows more effective use of the resin before regeneration is required,

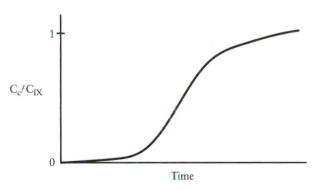

Figure 10.17   Ion exchange breakthrough curve.

reducing operating costs. Disposal of the spent regeneration chemical can be a significant problem for municipal plants because it can be corrosive and toxic (due to the high concentration of chloride salts and the large volume).

Lime Soda Softening

Although some municipal water treatment plants use ion exchange, most use chemical precipitation. The pH of the water is increased, often through the addition of lime. Either quicklime (CaO, unslaked lime) or hydrated lime ($Ca(OH)_2$), slaked lime) is used. (Although lime is a calcium species, it is very effective at softening water. Sodium hydroxide can be used, but it is more expensive.) As the pH increases to approximately 10.3, carbonate becomes the dominant species of alkalinity, and $CaCO_3$ (scale) precipitates. As the pH increases to approximately 11, magnesium precipitates as magnesium hydroxide ($Mg(OH)_2$). Noncarbonate hardness is more expensive to precipitate because a carbonate (typically, soda ash, Na_2CO_3) must be added. Therefore, calcium carbonate hardness (CCH) is targeted for removal first, then magnesium carbonate hardness (MCH), and finally calcium noncarbonate hardness (CNCH) and magnesium noncarbonate hardness (MNCH) (Figure 10.18). Carbon dioxide in water forms carbonic acid, which must be neutralized (by lime or caustic addition) or removed (through air stripping) before the pH will rise. Due to solubility constraints, precipitation can reduce total hardness to as low as 40 mg/L as $CaCO_3$ (the practical solubility limit). Due to time constraints, excess lime (lime over the stoichiometric amount) is typically added.

Carbon dioxide
$$CO_2 + Ca(OH)_2 \rightleftharpoons CaCO_3 \text{ (s)} + H_2O$$

Calcium carbonate hardness (CCH)
$$Ca(HCO_3)_2 + Ca(OH)_2 \rightleftharpoons 2CaCO_3 \text{ (s)} + 2H_2O$$

Calcium noncarbonate hardness (CNCH)
$$CaSO_4 + Na_2CO_3 \rightleftharpoons CaCO_3 \text{ (s)} + Na_2SO_4$$

Magnesium carbonate hardness (MCH)
$$Mg(HCO_3)_2 + Ca(OH)_2 \rightleftharpoons CaCO_3 \text{ (s)} + MgCO_3 + 2H_2O$$
$$MgCO_3 + Ca(OH)_2 \rightleftharpoons Mg(OH)_2 \text{ (s)} + CaCO_3 \text{ (s)}$$

Magnesium noncarbonate hardness (MNCH)
$$MgSO_4 + Na_2CO_3 \rightleftharpoons MgCO_3 + Na_2SO_4$$
$$MgCO_3 + Ca(OH)_2 \rightleftharpoons Mg(OH)_2 \text{ (s)} + CaCO_3 \text{ (s)}$$

Species	Ratio of Chemical to Species (meq/meq)	
	Lime	Soda Ash
CO_2	1	0
CCH	1	0
CNCH	0	1
MCH	1	0
MNCH	1	1

Figure 10.18 Lime–soda softening reactions.

Figure 10.19 Chemical precipitation water softening process.

Figure 10.19 shows a general treatment train for softening. The precipitates are removed through settling. Recarbonation (adding carbon dioxide to the water) is used to lower the pH to ensure that any fine particles not removed in the settling tank resolubilize and that the distributed water has a pH near neutral. If excess lime is used to remove magnesium carbonate hardness and only enough carbon dioxide is added to drop the pH to about 10.4 before a second settling tank, then additional precipitation and solids removal can be achieved. Recarbonation may not be necessary if part of the water bypasses the softening process, known as split treatment. The treated water can be softened to solubility limits and then mixed with the bypassed water to obtain the desired hardness level (typically in the moderately hard range).

EXAMPLE

10.9

Problem   Using the data from the Example 10.7 and Figure 10.20, determine the mass rate of chemicals required to soften the water to the practical solubility limit if the flow rate is 8 million L/d and 93% pure quicklime (CaO) and 98% pure soda ash are used.

Component	EW (mg/meq)	Concentration		
		mg/L	meq/L	mg/L as CaCO$_3$
CO$_2$	22.0	6.0	0.27	13
Ca^{2+}	20.0	50	2.5	125
Mg^{2+}	12.2	20	1.6	82
Na$^+$	23.0	5	0.2	11
Alkalinity	50.0	120	2.4	120
SO$_4^{2-}$	48.0	94	2.0	98

Solution   [Note that the equivalent weight of carbon dioxide is (44 g/mol) / (2 eq/mol) = 22 mg/meq.] To determine the quantity of chemicals required, use the fact that each meq/L of hardness will require 1 meq/L of chemical to remove it (Figure 10.18). The meq/L of lime and soda ash are shown in the following table. Note that the removal of carbon dioxide and CCH require only lime, the removal of CNCH requires only soda ash, and

Figure 10.20   Speciation of hardness bar chart.

the removal of MNCH requires both lime and soda ash. To soften to the practical solubility limit, 1.25 meq/L excess CaO is needed. [For hydrated lime, $Ca(OH)_2$, 0.4 meq/L is needed.]

Component	Component	Lime	Soda Ash
		Concentration (meq/L)	
CO_2	0.27	0.27	0
CCH	2.4	2.4	0
CNCH	0.1	0	0.1
MCH	0	0	0
MNCH	1.6	1.6	1.6
Excess		1.25	
Total		5.52	1.7

Note that the following equations can also be used to calculate the chemical amounts.

$$Lime = (CO_2) + CH + MNCH + Excess$$

$$Soda\ ash = NCH$$

To determine the mass rate of chemicals required, the meq/L must be converted to mg/L by using the EW of the chemicals.

Chemical	EW (mg/meq)
CaO	28
$Ca(OH)_2$	37
$Na_2(CO_3)$	53

From Chapter 3, conversion from concentration to mass flow rate is done by using

$$Q_M = M = C Q_V$$

The purity of the chemicals used is less than 100%, so the mass rate must be divided by the purity.

For this example 2 mgd of water are being treated with 93% pure CaO and 98% pure soda ash, so the chemical amounts required are

$$M_{CaO} = \frac{(5.52 \text{ meq/L})(28 \text{ mg/meq})(8 \text{ million L/d})}{0.93}$$

$$= 1330 \text{ Kg/d}$$

$$M_{Na_2CO_3} = \frac{(1.7 \text{ meq/L})(53 \text{ mg/meq})(8 \text{ million L/d})}{0.98}$$

$$= 736 \text{ kg/d}$$

EXAMPLE
10.10

Problem   Using the data from Example 10.9, calculate the flow rate of sludge that has to be pumped from the settling tank if the sludge thickens to 10% while in the tank.

Solution   To determine the amount of sludge, use material balances on calcium and magnesium. When softening to the practical solubility limits, 0.6 meq/L of calcium and 0.2 meq/L of magnesium will remain in the water.

Component	Concentration (meq/L)
Original Ca^{2+}	2.5
Original Mg^{2+}	1.6
Residual Ca^{2+}	0.6
Residual Mg^{2+}	0.2
Lime	5.52

$$Mg^{2+}-\text{sludge} = (\text{Original } Mg^{2+}) - (\text{Residual } Mg^{2+}) = (1.6 \text{ meq/L}) - (0.2 \text{ meq/L})$$

$$= 1.4 \text{ meq/L}$$

$$Ca^{2+} - \text{sludge} = (\text{Original } Ca^{2+}) + (\text{Lime}) - (\text{Residual } Ca^{2+})$$

$$= (2.5 \text{ meq/L}) + (5.52 \text{ meq/L}) - (0.6 \text{ meq/L}) = 7.42 \text{ meq/L}$$

Again, use the mass flow rate equation from Chapter 3, $M = CQ$. Calcium will precipitate as $CaCO_3$ while magnesium will precipitate as $Mg(OH)_2$. Note that the sludge is 100% pure sludge.

Chemical	EW (mg/meq)
$CaCO_3$	50
$Mg(OH)_2$	29.2

$$M_{\text{sludge}} = \sum (QC)_i$$
$$= \left[(1.4 \text{ meq/L}) (29.2 \text{ mg/meq}) + (7.42 \text{ meq/L}) (50 \text{ mg/meq}) \right]$$
$$\times (8 \text{ million L/d})$$
$$= 3295 \text{ kg/d}$$

To calculate the flow rate of sludge that will have to be pumped, rearrange the mass flow rate equation and use the thickened concentration. Remember that $1\% = 10,000$ mg/L (Chapter 3).

$$Q = \frac{M}{C} = \frac{(3295 \text{ kg/d})}{(100,000 \text{ mg/L})}$$
$$= 3295 \text{ L/d}$$

It is unnecessary to soften all water to the practical solubility limit (and would result in numerous complaints about slick water), so determining the amount of chemical to add is slightly more complicated and is determined based on how much Mg^{2+} is in the raw water compared to how much can be left in the water (40 mg/L as $CaCO_3$). Another option is to use split treatment (Figure 10.21). Part of the water is treated to the practical solubility limit and part is left untreated. When the streams are recombined, the desired hardness is obtained. Split treatment can be used to tailor the hardness, decrease capital cost for tanks, and lower operating costs.

The amount of water to bypass is based on a material balance on Mg^{2+} around the mixing point of the treated and untreated water flows (Figure 10.22). To reduce operating costs, the balance is on magnesium instead of the total hardness that was used in ion exchange. Assuming steady state, the material balance is

$$Q_b Mg_r + (Q - Q_b) Mg_1 = Q Mg_f$$

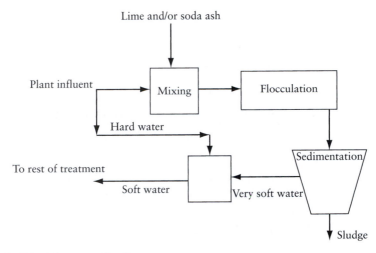

Figure 10.21 Diagram of split treatment.

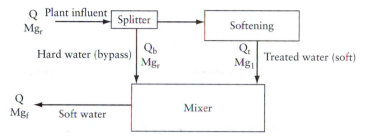

Figure 10.22   Mass balance on magnesium in split treatment.

$$\frac{Q_b}{Q} = X = \frac{Mg_f - Mg_1}{Mg_r - Mg_1} \tag{10.8}$$

When the softening process is to the practical solubility limit and the recombined water leaves the maximum amount of magnesium in the water (again to reduce operating costs),

$$\frac{Q_b}{Q} = X = \frac{(40 \text{ mg/L as } CaCO_3) - (10 \text{ mg/L as } CaCO_3)}{Mg_r - (10 \text{ mg/L as } CaCO_3)} = \frac{30 \text{ mg/L as } CaCO_3}{Mg_r - (10 \text{ mg/L as } CaCO_3)} \tag{10.9}$$

EXAMPLE
10.11

Problem   A water has a total hardness of 255 mg/L as $CaCO_3$ and a Mg^{2+} concentration of 55 mg/L as $CaCO_3$. How much water needs to be treated in split treatment if the water will be treated to the practical solubility limits? What will be the final blended hardness?

Solution   Use Equation 10.9.

$$\frac{Q_b}{Q} = X = \frac{30 \text{ mg/L as } CaCO_3}{Mg_r - (10 \text{ mg/L as } CaCO_3)} = \frac{30 \text{ mg/L as } CaCO_3}{(55 \text{ mg/L as } CaCO_3) - (10 \text{ mg/L as } CaCO_3)}$$
$$= 0.67$$

Therefore, 67% of the water can be bypassed, so 33% is softened, or treated.

The final blended hardness is determined by a material balance on total hardness.

$$Q_b(TH_r) + (Q - Q_b)(TH_1) = Q(TH_f)$$
$$TH_f = \frac{Q_b}{Q}(TH_r) + \frac{Q - Q_b}{Q}(TH_1)$$
$$= (0.67)(255 \text{ mg/L as } CaCO_3) + (0.33)(40 \text{ mg/L as } CaCO_3)$$
$$= 184 \text{ mg/L as } CaCO_3$$

If the final hardness needs to be less than this amount, then a second-stage softening process will be used. Because the maximum amount of magnesium is already in the water (40 mg/L as $CaCO_3$), selective calcium removal will be used in the second stage.

EXAMPLE

10.12   **Problem**   Can selective calcium removal be used for the following waters?

Water Sample	TH (mg/L as $CaCO_3$)	Ca^{2+} (mg/L as $CaCO_3$)
A	330	210
B	330	300

Solution   Assuming that the total hardness is composed of only Ca^{2+} and Mg^{2+}, then

Water Sample	Mg^{2+} (mg/L as $CaCO_3$)
A	$120 > 40$ mg/L as $CaCO_3$
B	$30 < 40$ mg/L as $CaCO_3$

Therefore, selective calcium removal can be used on water B but not A.

■

EXAMPLE

10.13   **Problem**   The alkalinity for Water Sample B in the previous example is 200 mg/L as $CaCO_3$ and the CO_2 is 12 mg/L as $CaCO_3$. Calculate the amount of hydrated lime and soda ash in meq/L required (1) to remove the magnesium to the practical solubility limit and (2) to leave the magnesium. Also calculate the final hardness if the magnesium is left.

Solution   First, determine the speciation of the hardness. Note that 50 mg/meq, the equivalent weight of $CaCO_3$, is used to calculate all the concentrations in meq/L.

Component	C(mg/L)	EW(mg/meq)	C(meq/L)
CO_2	12	50	0.24
Ca^{2+}	300	50	6.0
Mg^{2+}	30	50	0.6
ALK	200	50	4.0
CCH		4.0	
CNCH			2.0
MCH			0
MNCH			0.6
Excess			0.4

Part 1 is calculated the same as previously.

$$\text{Lime} = CO_2 + CCH + MNCH + Excess$$

$$= (0.24 \text{ meq/L}) + (4.0 \text{ meq/L}) + (0.6 \text{ meq/L}) + (0.4 \text{ meq/L})$$

$$= 5.24 \text{ meq/L}$$

$$\text{Soda ash} = CNCH + MNCH = (2.0 \text{ meq/L}) + (0.6 \text{ meq/L}) = 2.6 \text{ meq/L}$$

For part 2, all of the magnesium is left and no excess is needed because the magnesium is being left.

$$Lime = CO_2 + CCH = (0.24 \text{ meq/L}) + (4.0 \text{ meq/L}) = 4.24 \text{ meq/L}$$

$$Soda\ ash = CNCH = 2.0 \text{ meq/L}$$

Therefore, the amount of lime is reduced by about 19%, and the amount of soda ash is reduced by about 23%.

The final hardness is calculated as

$$TH_f = (\text{Residual Ca}^{2+}) + \text{Mg}^{2+}$$

$$\approx (30 \text{ mg/L as CaCO}_3) + (30 \text{ mg/L as CaCO}_3) = 60 \text{ mg/L as CaCO}_3$$

10.2.2   Coagulation and Flocculation

Raw surface water entering a water treatment plant usually has significant turbidity caused by tiny (colloidal) clay and silt particles. These particles have a natural electrostatic charge that keeps them continually in motion and prevents them from colliding and sticking together. Chemicals known as coagulants, such as alum (aluminum sulfate), and coagulant aids, such as lime and polymers, are added to the water (Stage 1 in Figure 10.11), first to neutralize the charge on the particles and then to aid in making the tiny particles "sticky" so they can coalesce and form large, quick-settling particles (Stage 2 in Figure 10.11). The purpose is to clear the water of the suspended colloidal solids by building larger particles that readily settle. *Coagulation* is the chemical alteration of the colloidal particles to make them stick together to form larger particles called *flocs*.

Two mechanisms are thought to be important in the process of coagulation–charge neutralization and bridging. *Charge neutralization* occurs when the coagulant (e.g., aluminum ions) is used to counter the charges on the colloidal particles, pictured in Figure 10.23 on the next page. The colloidal particles in natural waters are commonly negatively charged and, when suspended in water, repel each other due to their like charges. This causes the suspension to be stable and prevents the particles from settling. Positive ions added to the water are drawn to the negatively charged particles, compressing the net negative charge on the particles and making them less stable in terms of their charges. Such an increase in colloidal instability makes the particles more likely to collide with each other and form larger particles.

The second mechanism is *bridging*, in which the colloidal particles stick together by virtue of the macromolecules formed by the coagulant, as illustrated in Figure 10.24 on the next page. The macromolecules (or polymers) have positive charge sites, by which they attach themselves to colloids, bridging the gap between adjacent particles and creating larger particles.

Both of these mechanisms are important in coagulation with alum. When aluminum sulfate, $Al_2(SO_4)_3$, is added to water, the alum initially dissolves to form

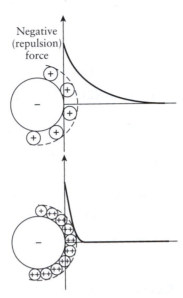

Figure 10.23 Effect of multivalent cations on the negative (repulsion) force of colloidal particles, resulting in charge neutralization.

aluminum ions, Al^{3+}, and sulfate ions, SO_4^{2-}. But the aluminum ion is unstable and forms various charged aluminum oxides and hydroxides. The specific combination of these compounds is dependent on the pH of the water, the temperature, and the method of mixing. Because many of the desirable macromolecular forms of aluminum hydroxide dissolve at low pH, lime $[Ca(OH)_2]$ is often added to raise the pH. Some of the calcium precipitates as calcium carbonate $[CaCO_3]$, assisting in the settling.

Jar tests are used to choose the best coagulant and estimate the minimum dose. Typically, the test is run with six containers holding samples of the water to be treated. Each container receives a different chemical or chemical dose, and after being mixed, the settling characteristics of the solids are observed. The chemical and minimum dose that provide adequate solids removal are chosen. The alkalinity of the water is also measured if metallic salts, such as alum, are used because these salts react with the alkalinity in the water, reducing its buffering capacity.

Figure 10.24 Effect of macromolecules (polymers) in flocculating colloidal particles by the mechanism of bridging.

EXAMPLE

10.14

Problem   Given the following jar test results, which polymer dose should be used?

Container No.	1	2	3	4	5	6
Alum (mg/L)	6	6	6	6	6	6
Polymer (mg/L)	0.25	0.5	1.0	2.0	3.0	4.0
Turbidity (NTU)	0.9	0.7	0.4	0.3	0.7	1.0

Solution   Graph the dose versus the turbidity (Figure 10.25).

While the lowest turbidity is obtained at a dose of 2 mg/L, this turbidity is not much lower than the turbidity obtained at half that dose. In addition, the minimum point on the curve may be between 1 and 2 mg/L. If time permits, it may be useful to run another test. However, since the dose chosen is an estimate, the operator may try 1 mg/L in the plant and adjust the dosage as needed.

Figure 10.25   Polymer dose versus turbidity.

Why does the turbidity increase at higher polymer doses in the above example? The net effect of coagulation is to destabilize the colloidal particles so they have the propensity to grow into larger particles. But too high of a coagulant dose or a coagulant aid dose will restabilize the particles by creating positive (rather than negative) particles or creating particles with large surface areas and low densities.

As mentioned previously, a metal salt used as a coagulant can affect the alkalinity of the water. The relationship is

1 meq/L coagulant consumes 1 meq/L of alkalinity and generates 1 meq/L of chemical sludge   (10.10)

The total sludge generated can be calculated as

$$\text{Total sludge} = \text{Chemical sludge} + \text{SS removed} \qquad (10.11)$$

E X A M P L E

10.15   **Problem**   Based on jar test results, 51 mg/L of alum is used to coagulate a sample containing 114 mg/L suspended solids. How many mg/L as $CaCO_3$ of natural alkalinity are consumed? What is the mass rate of sludge generated if suspended solids are reduced to 10 mg/L? Assume $Al(OH)_3$ is precipitated (EW = 26.0 mg/meq). The plant treats 400,000 L/d.

Solution   To use Equation 10.10, the concentration of alum must be in meq/L. The EW of alum is 100 mg/meq, so

$$C = \frac{51 \text{ mg/L}}{100 \text{ mg/meq}} = 0.51 \text{ meq/L}$$

Therefore, the amount of alkalinity consumed is

$$C = (0.51 \text{ meq/L alum}) \left(\frac{1 \text{ meq/L alkalinity}}{\text{meq/L alum}} \right) (50 \text{ mg/meq}) = 26 \text{ mg/L as } CaCO_3$$

Use Equations 10.10 and 10.11 to determine the amount of sludge generated.

$$\text{Chemical sludge} = (0.51 \text{ meq/L alum}) \left(\frac{1 \text{ meq/L sludge}}{\text{meq/L alum}} \right) (26.0 \text{ mg/meq}) = 13 \text{ mg/L}$$

$$\text{SS removed} = (114 \text{ mg/L}) - (10 \text{ mg/L}) = 104 \text{ mg/L}$$

So

$$\text{Total sludge} = (0.4 \times 10^6 \text{L/d})(13 \text{ mg/L} + 104 \text{ mg/L}) = 47 \text{ kg/d}$$

The assistance in the growth of larger particles is a physical process known as *flocculation*. For particles to come together and stick to each other, either through charge neutralization or bridging, they have to move at different velocities. Consider for a moment the movement of cars on a highway. If all the cars moved at exactly the same velocity, it would be impossible to have car-to-car collisions. Only if the cars have different velocities (speed and direction), with some cars catching up to others, are accidents possible. The intent of the process of flocculation is to produce differential velocities within the water so that the particles can come into contact. Commonly, this is accomplished in a water treatment plant by simply using a large slow-speed paddle that gently stirs the chemically treated water. This is depicted as Stage 2 in Figure 10.11 and illustrated in Figure 10.26. Once these larger particles are formed, the next step is to remove them by the process of settling.

10.2.3 Settling

When the flocs have been formed, they must be separated from the water. This is invariably done in *gravity settling tanks* that simply allow the heavier-than-water particles to settle to the bottom. Settling tanks are designed so as to approximate a plug-flow reactor. That is,

Figure 10.26 Typical flocculator used in water treatment.

the intent is to minimize all turbulence. The two critical design elements of a settling tank are the entrance and exit configurations because this is where plug flow can be severely compromised. Figure 10.27 on the next page shows one type of entrance and exit configuration used for distributing the flow entering and leaving the water treatment settling tank.

The sludge in water treatment plants is composed of aluminum hydroxides, calcium carbonates, and clays, so it is not highly biodegradable and will not decompose at the bottom of the tank. Typically, the sludge is removed every few weeks through a *mud valve* on the tank bottom and is wasted into a sewer or into a sludge holding/drying pond.

Settling tanks work because the density of the solids exceeds that of the liquid. The movement of a solid particle through a fluid under the pull of gravity is governed by a number of variables, including

- particle size (volume)
- particle shape
- particle density
- fluid density
- fluid viscosity.

The latter term may be unfamiliar, but it refers simply to the ability of the fluid to flow. Pancake syrup, for example, has a high viscosity, whereas water has a relatively low viscosity.

© Courtesy of P. Aarne Vesilind

Figure 10.27 Typical settling tank used in water treatment plants.

In settling tanks it is advantageous to get particles to settle at the highest velocity. This requires large particle volumes, compact shapes (low drag), high particle and low fluid densities, and low fluid viscosity. In practical terms it is not feasible to control the last two variables, but coagulation and flocculation certainly result in the growth of particles and changes to their density and shape.

The reason why coagulation/flocculation is so important in preparing particles for the settling tank is shown by some typical settling rates in Table 10.3. Although the settling velocity of particles in a fluid is also dependent on the particle shape and density, these

Table 10.3 Typical Settling Rates

Particle Diameter (mm)	Typical Particle	Settling Velocity (m/s)
1.0	Sand	2×10^{-1}
0.1	Fine sand	1×10^{-2}
0.01	Silt	1×10^{-4}
0.001	Clay	1×10^{-6}

Figure 10.28   An ideal settling tank.

numbers illustrate that even small changes in particle size for typical flocculated solids in water treatment can dramatically affect the efficiency of removal by settling.

Settling tanks can be analyzed by assuming an ideal tank (very much like we analyzed ideal reactors). Such ideal settling tanks can be visualized hydraulically as perfect plug-flow reactors: a plug (column) of water enters the tank and moves through the tank without intermixing (Figure 10.28). If a solid particle enters the tank at the top of the column and settles at a velocity of v_0, it should have settled to the bottom as the imaginary column of water exits the tank, having moved through the tank at a horizontal velocity v_h.

Several assumptions are required in the analysis of ideal settling tanks.

- Uniform flow occurs within the settling tank. (This is the same as saying that there is ideal plug flow because uniform flow is defined as a condition wherein all water flows horizontally at the same velocity.)
- All particles settling to the bottom are removed. That is, as the particles drop to the bottom of the column as depicted in Figure 10.28, they are removed from the flow.
- Particles are evenly distributed in the flow as they enter the settling tank.
- All particles still suspended in the water when the column of water reaches the far side of the tank are not removed and escape the tank.

Consider now a particle entering the settling tank at the water surface. This particle has a settling velocity of v_0 and a horizontal velocity v_h such that the resultant vector defines a trajectory as shown in Figure 10.29. In other words, the particle is just barely removed; it hits the bottom at the last instant. Note that, if the same particle enters the settling tank at any other height, such as height h, its trajectory always carries it to the bottom. Particles having this velocity are termed *critical particles* in that particles with lower settling velocities are not all removed and particles with higher settling velocities are all removed. For example, the particle having velocity v_s, entering the settling tank at the

Figure 10.29   Particle trajectories in an ideal settling tank.

Addressing the Negative Life Cycle Impacts of Chlorine

Drinking water chlorination has contributed to a dramatic decline in waterborne disease rates and increased life expectancy in the United States. Largely because of this success and with few exceptions, many Americans take it for granted that their tap water will be free of disease-causing organisms. It is not always the case, however. A stark example of the continuing public health threat from waterborne disease outbreaks occurred in Peru in 1991, where a major causative factor was inadequate drinking water disinfection. The result: a five-year epidemic of cholera, the disease's first appearance in the Americas in the 20th century. The epidemic spread to 19 Latin American countries, causing more than one million illnesses and 12,000 deaths. After the outbreak, U.S. and international health officials criticized Peruvian water officials for not chlorinating the entire water supply.

When considering the sustainability of chemicals and chemical processes, it is important to consider impacts across the life cycle. Some people argue about trade-offs as if they were unavoidable,

saying that the risks justify the benefits. But green chemists and green engineers approach such trade-off arguments with skepticism, knowing that hazards can be addressed through design to eliminate the need for the negative trade-offs. The same chemical may be made via different processes, resulting in very different hazard profiles, different levels of purity (or contamination), and different transportation requirements. While water chlorination has undoubtedly saved millions of lives, serious unintended negative consequences have been associated with the production, transportation, and use of chlorine. By addressing the life cycle impacts, the hazard footprint of chlorine as a disinfectant can be improved. In a series of focus boxes that follow in this chapter, we present some of the life cycle issues associated with chlorine from its manufacture, transportation, and use. We also describe a growing technology with a reduced hazard footprint called "onsite hypochlorite generation" and present an emerging green chemistry option.

surface, will not hit the bottom and escape the tank. However, if this same particle enters at some height h, it should just barely hit the bottom and be removed. Any of these particles that happen to enter the settling tank at height h or lower would thus be removed, and those entering above h would not. Because the particles entering the settling tank are assumed to be equally distributed, the proportion of those particles with a velocity of v_s removed is equal to h/H, where H is the height of the settling tank.

From Chapter 3 the hydraulic retention, or detention, time is defined as

$$\bar{t} = \frac{V}{Q} \tag{10.12}$$

The volume of rectangular tanks is calculated as $V = HLW$, where $W =$ width of the tank and $L =$ length of the settling zone (which will be less than the length of the basin because the inlet and outlet zones experience turbulence). Using the continuity equation, the flow rate is $Q = Av$, where A is the area through which the flow occurs and v is the velocity. In the case of a settling tank acting as a plug-flow reactor, the flow occurs through $A = HW$.

$$\bar{t} = \frac{V}{Q} = \frac{HWL}{(HW)v} = \frac{L}{v}$$

Using similar triangles in Figure 10.29 and rearranging yields

$$v_0 = \frac{H}{\bar{t}} \tag{10.13}$$

As noted above, $\bar{t}$ is V/Q, where $V = HWL$, or if the surface area of the settling tank is defined as $A_s = WL$, then $V = A_s H$. Substituting yields

$$v_0 = \frac{H}{\bar{t}} = \frac{H}{A_s H/Q} = \frac{Q}{A_s} \tag{10.14}$$

This equation represents an important design parameter for settling tanks called the *overflow rate*. Note the units:

$$v_0 = \frac{m}{s} = \frac{Q}{A_s} = \frac{m^3/s}{m^2}$$

Overflow rate has the same units as velocity. Commonly, overflow rate is expressed as $m^3/d/m^2$, but it actually is a velocity term and is, in fact, equal to the velocity of the critical particle. When the design of a clarifier is specified by overflow rate, what is really defined is the critical particle because its velocity is specified.

It should also be noted that when any two of the following—overflow rate, retention time, or depth—are defined, the remaining parameter is also fixed, as shown in Example 10.16.

EXAMPLE
10.16

Problem   A water treatment plant settling tank has an overflow rate of 24 $m^3/d/m^2$ and a depth of 2 m. What is its retention time?

Solution   Use Equation 10.13.

$$v_0 = (24\ m^3/d/m^2) = 24\ m/d$$

$$\bar{t} = \frac{H}{v_0} = \frac{2\ m}{24\ m/d} = 0.0833\ d \cong 2\ hr$$

∎

Overflow rate is interesting in that a better understanding of settling can be obtained by looking at individual variables. For example, increasing the flow rate, Q, in a given tank increases v_0; that is, the critical velocity increases and, thus, fewer particles are totally removed because fewer particles have a settling velocity greater than v_0.

It would, of course, be advantageous to decrease v_0 so more particles can be removed. Because v_0 is not a function of the particle, this is done by either reducing Q or changing the tank geometry by increasing A_s. The latter term may be increased by changing the dimensions of the tank so that the depth is shallow and the length and width are very large. For example, the area can be doubled by taking a 3-m-deep tank, slicing it in half (two 1.5-m slices), and placing them alongside each other. The new shallow tank has the same horizontal velocity because it has the same area through which the flow enters ($A = WH$) but double the surface area ($A_s = WL$); hence, v_0 is half the original value.

FOCUS ON

Mercury from Chlor-Alkali Production

A major industry for many years was the production of chlorine gas (Cl_2) and high quality sodium hydroxide (NaOH) from brine using a process that relied on mercury electrodes. An unwanted effect of this was the discharge of some of the mercury in the wastewater. Chlor-alkali plants along Lake St. Claire (between Lake Huron and Lake Erie) discharged enough mercury into the small lake to raise the level by a factor of 100 over several decades. A Dow Chemical chlor-alkali plant along the St. Claire River discharged an estimated 91 tons of mercury during only 20 years of operation. The fish in St. Claire were contaminated to mercury levels of 3 to 7 ppm, significantly above the 0.5 ppm that was considered relatively safe.

Probably the most notorious mercury pollution problem in the U.S. occurred in southwestern Virginia. The Mathieson Chemical Company's waste from their chlor-alkali plant in Saltville was discharged into the Holston River, which feeds into the Tennessee River system. The plant was eventually shut down, partly due to the requirement to cease the discharge of mercury, but mercury has continued to leak from the plant site into the river. One estimate is the soil within the plant grounds contains about 100 tons of mercury, and two large "muck ponds" containing high mercury levels continue to leak into the groundwater. Some estimates are that the plume from the ponds is now more than 25 m deep. There appears to be no reasonable way to remove the muck from the ponds, nor does there appear to be a way of cleaning up the groundwater. The states of Virginia and Tennessee both established bans on eating fish taken from the contaminated rivers. Fish tested 30 years after the plant closure still have mercury concentrations at twice the permitted levels. Unfortunately there are still 4 or 5 chlorine plants left in the U.S. that use mercury cells, even though safer technologies are well established.

Source: *Aquatic Pollution* by Edward Laws, John Wiley & Sons, New York, 2000.

Why not then make *very* shallow tanks? The problem is first a practical one of hydraulics and the even distribution of flow as well as the great expense in concrete and steel. Secondly, as particles settle, they can flocculate, or bump into slower-moving particles and stick together, creating higher settling velocities and enhancing solids removal. Settling tank depth then is an important practical consideration, and typically tanks are built 3 to 4 m deep to take advantage of the natural flocculation that occurs during settling.

EXAMPLE
10.17

Problem   A small water plant has a raw water inflow rate of 0.6 m³/s. Laboratory studies have shown that the flocculated slurry can be expected to have a uniform particle size (only one size), and it has been found through experimentation that all the particles settle at a rate of $v_s = 0.004$ m/s. (This is unrealistic, of course.) A proposed rectangular settling tank has an effective settling zone of $L = 20$ m, $H = 3$ m, and $W = 6$ m. Could 100% removal be expected?

Solution   Remember that the overflow rate is actually the critical particle settling velocity. What is the critical particle settling velocity for the tank (Equation 10.14)?

$$v_0 = \frac{Q}{A_s} = \frac{0.6 \text{ m}^3/\text{s}}{(20 \text{ m})(6 \text{ m})} = 0.005 \text{ m/s}$$

The critical particle settling velocity is greater than the settling velocity of the particle to be settled; hence, not all the incoming particles will be removed.

The same conclusion can be reached by using the particle trajectory. The velocity v through the tank is

$$v = \frac{Q}{HW} = \frac{0.6 \text{ m}^3\text{s}}{(3 \text{ m})(6 \text{ m})} = 0.033 \text{ m/s}$$

Using similar triangles,

$$\frac{v_s}{v} = \frac{H}{L'}$$

where L' is the horizontal distance the particle would need to travel to reach the bottom of the tank.

$$\frac{0.004 \text{ m/s}}{0.033 \text{ m/s}} = \frac{3}{L'}$$

$$L' = 25 \text{ m}$$

Hence, the particles would need 25 m to be totally removed, but only 20 m is available, so 100% removal cannot be expected.

E X A M P L E

10.18   **Problem**   In Example 10.17, what fraction of the particles will be removed?

Solution   Assume that the particles entering the tank are uniformly distributed vertically. If the length of the tank is 20 m, the settling trajectory from the far bottom corner would intersect the front of the tank at height 4/5 (3 m), as shown in Figure 10.30. All those particles entering the tank below this point would be removed and those above would not. The fraction of particles that will be removed is then 4/5, or 80%.

Alternatively, because the critical settling velocity is 0.005 m/s and the actual settling velocity is only 0.004 m/s, the expected effectiveness of the tank is $0.004/0.005 = 0.8$, or 80%.

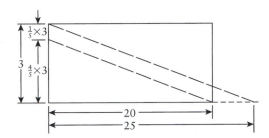

Figure 10.30   Ideal settling tank. See Example 10.18.

There is an important similarity between ideal settling tanks and the ideal world—neither one exists. Yet engineers are continually trying to idealize the world. There is nothing wrong with this, of course, because simplification is a necessary step in solving engineering problems as demonstrated in Chapter 3. The danger of idealizing is that it is easy to forget all the assumptions used in the process. A settling tank, for example, *never* has uniform flow. Wind, density, and temperature currents, as well as inadequate baffling at the tank entrance can all cause nonuniform flow. So, don't expect a settling tank to behave ideally but design them with a large safety factor.

If the (overdesigned) settling tank works well, the water leaving it is essentially clear. However, it is not yet acceptable for domestic consumption; one more polishing step is necessary, usually using a rapid sand filter.

10.2.4   Filtration

In the discussion of groundwater quality, it was noted that the movement of water through soil removes many of the contaminants in water. Environmental engineers have learned to apply this natural process to water treatment systems and developed what is now known as the *rapid sand filter*. The operation of a rapid sand filter involves two phases: filtration and washing.

© Courtesy of P. Aarne Vesilind

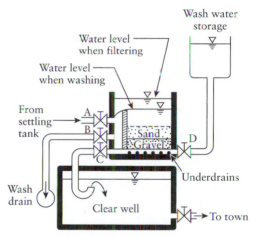

Figure 10.31   A rapid sand filter used in water treatment.

A slightly simplified version of the rapid sand filter is illustrated in a cut-away drawing in Figure 10.31. Water from the settling basins enters the filter and seeps through the sand and gravel bed, through a false floor, and out into a clear well that stores the finished water. During filtration, valves A and C are open. Sometimes anthracite, a type of carbon, is also used in the filter bed; it can remove dissolved organic materials.

The suspended solids that escape the flocculation and settling steps are caught on the filter sand particles and eventually the rapid sand filter becomes clogged, resulting in greater head loss through the filter, so it must be cleaned. This cleaning is performed hydraulically by a process called *backwashing*. The operator first shuts off the flow of water to the filter (closing valves A and C) and then opens valves D and B, which allow wash water (clean water stored in an elevated tank or pumped from the clear well) to enter below the filter bed. This rush of water forces the sand and gravel bed to expand (fluidize) and jolts individual sand particles into motion, rubbing them against their neighbors. The suspended solids trapped within the filter are released and escape with the wash water. After at least 15 min, the wash water is shut off and filtration is resumed. Treatment plants want to minimize the frequency of backwashing because it uses energy and a significant amount of water, the product of the treatment plant. In addition, this water may require treatment prior to discharge.

FOCUS ON

The New York City Chlorine Gas Accident

Chlorine gas is extremely hazardous and its transportation creates serious risks. In the first week of June 1944, a truck carrying a cargo of compressed chlorine passed over the Manhattan Bridge into jam-packed downtown Brooklyn. Stopped at a light, the driver detected the unmistakable odor of his cargo. He pulled over and, on checking, found that one tank was leaking. He removed the tank, sat it on the ground, and then hurried off to call his dispatcher to ask for instructions. Inadvertently, he had placed the tank on a sidewalk grating that covered the ventilation intake of the Brooklyn-Manhattan Transit subway station at Myrtle Avenue and Flatbush Avenue Extension, one of the busiest stations of the subway system.

Chlorine is heaver than air, and it went through the grating as it seeped out of the tank. Down below, the subway trains rushing through the tunnel acted much like the pistons of a vacuum pump, sucking green clouds of chlorine into the station mezzanine. People were knocked over like tenpins. Passengers coming up the stairs from the platform beneath found themselves walking into a gas chamber. Blinded, many stumbled down the stairs, colliding with passengers who were coming up. Meanwhile, the chlorine began to roll down the stairs as well, and it engulfed the passengers at the platform level as they awaited incoming trains.

Within minutes panic ensued. Scores of people groped about, trying to avoid falling off the narrow platform yet unable to see or breathe. Many fainted on the stairways, to be trampled on by the choking hordes that collided on the stairs while trying to escape the gas from above and below. As soon as the alarm was sounded, the city put into operation its wartime procedure for poison-gas attacks, and police and emergency services rushed to the scene. Over a hundred people required hospital care, although amazingly nobody died in the accident.

Source: Reprinted from *Controlling Environmental Pollution*, Vesilind, P. A. and DiStefano, T. D., 2006. Lancaster, PA: DEStech Publications, Inc. Reproduced with permission of DEStech Publications, Inc.

Filters are a very important process in meeting turbidity limits. A common design and operating parameter is the filtration rate (or filter loading), which is the rate of water applied to the surface area of the filter. The calculation and units, such as, $m^3/hr/m^2$, are similar to those for overflow rate. This rate can vary from about 5 to 24 $m^3/hr/m^2$ but may be limited to 5 to 8 m³/hr/m² by state regulations. Backwash rates generally range from 24 to 62 m³/hr/ft².

EXAMPLE

10.19   **Problem**   What is the filtration rate for a 8 m by 6 m filter if it receives 8 million L/d?

Solution

$$\text{Filtration rate} = \frac{Q}{A_s} = \frac{(8 \times 10^6 \text{ L/d})(\text{d}/1440 \text{ min})}{(8 \text{ m})(6 \text{ m})} = 116 \text{ LPM/m}^2 = 8 \text{ m}^3/\text{hr/m}^2$$

EXAMPLE

10.20   **Problem**   How much backwash water is required to clean a 8 m by 6 m filter?

Solution   Assume 48 m³/hr/m² will be used as the backwash rate and the filters will be cleaned for 15 minutes. Then

$$V = (\text{Backwash rate})(A_s)(t) = (48 \text{ m}^3\text{hr/m}^2)(8 \text{ m})(6 \text{ m})(15 \text{ min}) = 576 \text{ m}^3$$

10.2.5   Disinfection

The water is *disinfected* to destroy whatever pathogenic organisms might remain. Prechlorination may be done before filtration to help keep the filters free of growth and provide adequate *contact time* with the disinfectant. Adequate disinfection is a balance between the concentration of the disinfectant (C) and the contact time (T), an analysis known as the CT-concept.

Commonly, disinfection is accomplished by using chlorine, which is purchased as a liquid under pressure and released into the water as chlorine gas using a chlorine feeder system. The dissolved chlorine oxidizes organic material, including pathogenic organisms. The presence of a residual of active chlorine in the water is an indication that no further organics remain to be oxidized and that the water can be assumed to be free of disease-causing organisms. Water pumped into distribution systems in the United States is required to contain a residual of chlorine to guard against any contamination in the distribution system (Table 10.4). It is for this reason that water from drinking fountains or faucets has a slight taste of chlorine.

When chlorine is added to water, it forms hypochlorous acid (HOCl), which is a weak acid that dissociates to the hypochlorite ion (OCl^-) above pH 6. These two species are defined as free available chlorine. When ammonia or organic nitrogen compounds are present, HOCl will react with them to form chloramines. Chloramines are defined as combined available chlorine. They are weaker disinfectants than free available chlorine but are more stable.

Table 10.4 Chlorine Residual Limits

Disinfectant	Minimum Residual* (mg/L)		MRDL** (mg/L)	MRDLG*** (mg/L)
	Leaving Plant	In Distribution System		
Chloramines (as Cl_2)	2.0		4.0	4.0
Combined residual		0.5		
Chlorine (as Cl_2)	2.0		4.0	4.0
Chlorine dioxide (as ClO_2)	2.0		0.8	0.8
Free chlorine residual		0.2		

 * From Illinois and Missouri regulations.
** MRDL = maximum residual disinfectant level.
** MRDLG = maximum residual disinfectant level goal.

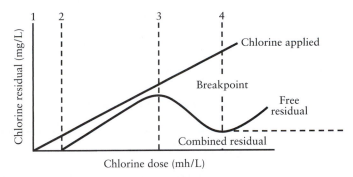

Figure 10.32 Chlorine breakpoint dosage.

To obtain free available residual chlorine in waters containing chemicals that react with chlorine (such as manganese, iron, nitrite, ammonia, and organics), chlorine must be added beyond the breakpoint dose (Figure 10.32). Between points 1 and 2 in Figure 10.32, the chlorine is reacting with reducing compounds (such as manganese, iron, and nitrite). No disinfection is occurring and no residual is formed. Between points 2 and 3 the additional chlorine is reacting with the organics and ammonia (forming chlororganics and chloramines), resulting in combined residual. Between points 3 and 4 the chlororganics and chloramines are partly destroyed. Past point 4 chlorine additions provide free available residual chlorine. Chlorine additions beyond the breakpoint dose result in a residual that is directly proportional to the amount of additional chlorine.

EXAMPLE

10.21

Problem A 18 million L/d water treatment plant uses 10 kg/d of chlorine for disinfection. If the daily chlorine demand is 0.5 mg/L, what is the daily chlorine residual?

Solution Chlorine demand is the dose required to reach the desired residual level. Therefore, the residual is the difference between the chlorine applied and the chlorine demand.

Using the equation $M = QC$:

$$\text{Chlorine applied} = \frac{M}{Q} = \left[\frac{10 \text{ kg/d}}{18 \text{ million L/d}} \right]$$

$$= 0.56 \text{ mg/L}$$

$$\text{Chlorine residual} = (0.56 \text{ mg/L}) - (0.5 \text{ mg/L}) = 0.06 \text{ mg/L}$$

This residual is below the minimum recommended free residual of 0.2 to 0.5 mg/L.

FOCUS ON

Chlorine Toxicity and DBPs

Chlorine, particularly as hypochlorite, is a strong oxidizer. Chlorinated effluents can be lethal to aquatic life. When residual chlorine levels exceed 0.1 mg/L, laboratory tests show that the effluents are lethal to fish and invertebrate species, which may result in changes in aquatic ecosystems, such as reduction in diversity and shifts in species distribution. These effects, which may persist for up to 0.5 km downstream, can be caused by chlorine at levels as low as 0.02 mg/L.

When chlorine reacts with organic matter, it forms disinfection byproducts (DBPs). In recent years, regulators and the general public have focused greater attention on potential health risks from DBPs. The predominant DBPs are trihalomethanes (THMs) (chloroform, bromoform, bromodichloromethane, and dibromochloromethane) and haloacetic acids (HAAs—monochloro-, dichloro-, trichloro-, monobromo-, and dibromo-). THMs and HAAs form when chlorine reacts with organic material in source water (which comes from decomposing plant material, pesticides, etc.). The amount of THMs and HAAs in drinking water can change from day to day, depending on the season, water temperature,

amount of chlorine added, the amount of plant material in the water, and a variety of other factors. Concerned that these chemicals may be carcinogenic to humans, the USEPA set the first regulatory limits for THMs in 1979. Since that time, a wealth of research has improved our understanding of how DBPs are formed, their potential health risks, and how they can be controlled.

It is now recognized that all chemical disinfectants form some potentially harmful byproducts. The byproducts of chlorine disinfection are by far the most thoroughly studied. Most water systems are meeting the DBPs standard by controlling the amount of natural organic matter prior to disinfection.

Chlorine dioxide (ClO_2) is obviously a chlorine-based chemical, but it is not as strong an oxidizer as hypochlorite. It is often used as a preoxidant because, unlike chlorine, it will not chlorinate organic compounds and therefore will not react with organic matter in the water to form THMs. However, it will effectively oxidize reduced iron, manganese, sulfur compounds, and certain odor-causing organic substances in raw water.

A major concern with the use of chlorine gas and hypochlorite compounds to disinfect water is the formation of *disinfection byproducts* (DBPs). DBPs are unwanted compounds.

For example, trihalomethanes (THMs), which are formed by the reaction of chlorine with organic matter, are a DBP and are suspected carcinogens. Chlorine dioxide does not react with organics, so it will not form THMs. In addition, it negates, the need to store and transport chlorine gas, improving safety. So it is being adopted more frequently. Other options for disinfecting water are using ultraviolet light and ozonation. DBP formation is avoided, but these options do not provide a residual to protect the water in the distribution system.

10.2.6 Other Treatment Processes

In addition to the treatment processes already discussed, there are other steps that may or may not have to be taken before the water is distributed. For example, water must be *stable* before it enters the distribution system. Stable water is in chemical balance and will not cause corrosion or scaling in the distribution system. Two tests are used to determine the stability of water—the Marble Test and the Langelier Index. Both tests indicate the calcium carbonate saturation level of water. Water is considered stable when it is saturated with calcium carbonate. Depending on the cause of the stability problem, unstable water can be stabilized by using recarbonation, acid addition, phosphate addition, alkali addition, or aeration. Unstable water from reactions in the distribution system (for example, bacterial decomposition of organic matter and reduction of sulfates to sulfides) can be prevented by providing a suitable chlorine residual throughout the distribution system.

FOCUS ON

Onsite Hypochlorite Generation

The use of chlorine to treat water and wastewater poses both environmental and safety risks, depending on its form and manufacture. An ideal replacement technology would provide the same efficacy of a chlorine gas system while reducing cost and risks by reducing the inherent hazard of the technology.

Onsite hypochlorite generation, as its name implies, can generate hypochlorite onsite, removing the dangers associated with transporting Cl_2. It uses only three materials: water, salt, and electricity. While system designs vary depending on the manufacturer, the basic procedure is the same. The salt and water are combined to make a brine solution. The brine is then passed through a brine proportioner that dilutes the solution to approximately 3.5%, after which it is injected into electrolytic cells.

While in the cells, the brine is converted into sodium hypochlorite (NaOCl).

Anode	**Cathode**
$2Cl^- \rightarrow Cl_2 + 2e^-$	$2Na^+ + 2e^- \rightarrow Na$
$Cl_2 + H_2O \rightarrow HOCl + HCl$	$2Na + 2H_2O \rightarrow H_2$
(rapid and exothermic)	$+ 2\,NaOH$

The acids, hypochlorous and hydrochloric, created at the anode mix with the base, sodium hydroxide, generated at the cathode to form salt and sodium hypochlorite.

$$HCl + NaOH \rightarrow NaCl + H_2O$$

$$HOCl + NaOH \rightarrow NaOCl + H_2O$$

The overall reaction in the cells is $NaCl + H_2O \rightarrow NaOCl + H_2$.

The resulting 0.8% solution of NaOCl is injected into a day tank for storage. When it is needed, it is injected into the water by a metering pump over a specified period of time, providing the same treatment as the more highly concentrated chlorine gas. The 0.8% solution is much weaker than commercial NaOCl (12% NaOCl) and even household bleach (5% NaOCl), so more NaOCl solution is needed. But it provides the same disinfection and performance benefits, including providing a chlorine residual.

The reigning advantage of using onsite hypochlorite generation technology is safety. Not only is it risky to transport, handle, and store hazardous materials, such as chlorine gas and industrial-strength sodium hypochlorite, but complying with the regulations that govern those activities is expensive. That's why more and more innovative municipalities are searching for alternative means of disinfecting their drinking water supply. The use of onsite hypochlorite generation eliminates concerns about transport, storage, and regulation. The only chemical brought onsite is salt. Additionally, hydrogen gas is produced as a byproduct. Typically it is vented to the atmosphere, but it could be collected and used as a valuable fuel.

In addition to their use in drinking water and wastewater disinfection, onsite sodium hypochlorite generating systems are suitable for use in

- odor and corrosion control
- food and beverage processing
- cooling towers, inland power stations, and industry
- irrigation systems
- agriculture/farming
- industrial and institutional cleaning.

Because the system has been shown to be cost-effective, reliable, and safe, more and more owners of in-ground swimming pools are also switching to onsite hypochlorite generation. The pools are commonly referred to as saltwater pools because they are filled with saltwater. While some people mistakenly think they do not contain chlorine at all, saltwater pools generate hypochlorite by ionizing the circulating saltwater. The downside of onsite generation is the need for additional electricity and the need to occasionally replace electrodes as they wear out. But as the technology gains in popularity, engineers know that the design will continue to improve. Maybe even a new disinfectant will be found.

To improve technologies, engineers should consider the full life cycle of chemical products and processes. Onsite hypochlorite generation illustrates that improvements in how a chemical is made and applied can have tremendous impacts on its hazard footprint from the life cycle perspective. Engineering decisions should be made not only on performance and cost but on safety and environmental impacts across the full life cycle.

Taste, odor, and color complaints are the most common types of complaints received by water utilities.[8] Nobody wants to drink smelly, bad-tasting, colored water. There are a variety of causes for taste and odor (T&O) problems. Natural causes include bacterial and algal growth and stratification of the source water (Chapter 8). Human causes include inadequately treated municipal and industrial wastewater, inadequate or incomplete treatment plant and distribution system maintenance, and household plumbing. Prevention of T&O problems is the key to their control. Prevention takes the form of source water management and plant and distribution system maintenance. In addition, some treatment processes are useful for removing some T&Os, including aeration (which will remove gases and volatile organic compounds) and coagulation/flocculation/sedimentation.

FOCUS ON

Emerging Green Chemical Alternatives to Chlorine for Water Disinfection

The use of chlorine for domestic water treatment has unquestionably been effective as a mainstay in reducing waterborne diseases worldwide. However, chlorine and chlorine byproducts from manufacturing and other uses make their way into aquatic and other ecosystems. Once present in the environment, chlorine compounds react with other compounds, leading to the formation of more byproducts, which may bioaccumulate within the food chain. Alternative oxidation technologies, such as those based on the catalytic activation of hydrogen peroxide, for example, may provide effective and more benign oxidation for water disinfection.

In 1999, Dr. Terry Collins of Carnegie Mellon University's Department of Chemistry was awarded the Presidential Green Chemistry Challenge Award (Figure 10.33) for his development of tetra-amido macrocyclic ligands (TAMLs®). TAMLs are catalysts with a host of applications for cleaning up pollutants. The oxidation catalysts are the first highly effective mimics of peroxidase enzymes. When partnered with hydrogen peroxide, they are able to convert harmful pollutants into less toxic substances. Made from the common elements of carbon, hydrogen, nitrogen, and oxygen around a reactive iron core, Fe-TAMLs have low toxicity and are usable at extremely low concentrations. Additionally, their composition results in very strong chemical bonds that keep the molecule stable during the reaction with hydrogen peroxide, allowing them to be used in multiple cycles.

Research by Collins and his colleagues has shown that Fe-TAMLs have enormous potential to provide clean and safe alternatives to existing industrial practices and to provide ways to remediate other pressing environmental problems that currently lack solutions. For example, the catalysts have proven effective in degrading estrogenic compounds, cleaning wastewater containing dyes from textile manufacturing, reducing fuel pollutants, treating pulp and paper processing byproducts, and decontaminating anthrax. TAML technology has the potential to become a platform for drinking water disinfection and water purification, particularly as pollutants such as pharmaceuticals and personal care products in water and wastewater become more of a concern.

Photographer: Matt Mendelsohn, Source: USEPA

Figure 10.33 Presidential Green Chemistry Challenge Award.

Fluoridation of the water or fluoride removal may also be necessary. Fluoride prevents tooth decay, but at concentrations above approximately 1.5 mg/L, it also stains teeth brown. Because most people don't want brown teeth, source waters that naturally have high concentrations of fluoride (e.g., waters in volcanic regions) must be treated to remove some of the fluoride or blended with other source water to reduce the concentration. If the public agrees, source waters that are low in fluoride may have fluoride added, using chemicals such as sodium fluoride and fluorosilicic acid.

10.3   DISTRIBUTION OF WATER

Water is typically stored in a *clear well* following treatment. From the clear well the finished water is pumped into the distribution system. Such systems are under pressure so that contaminants are kept out (Chapter 1) so any tap into a pipe, whether it be a fire hydrant or domestic service, will yield water.

Because the demand for finished water varies with the day of the week and the hour of the day, storage facilities must be used in the distribution system. Most communities have an elevated storage tank, which is filled during periods of low water demand and then supplies water to the distribution system during periods of high demand. Figure 10.34 shows how such a storage tank can assist in providing water during peak demand periods and emergencies.

The calculation of the required elevated storage capacity requires both a frequency analysis as well as a material balance, as illustrated in Example 10.22.

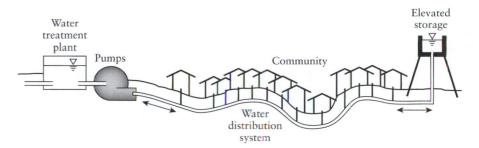

Figure 10.34   During periods of high water demand, the water flows from both the water plant and the elevated storage tanks to meet the demand. During the low-demand periods, the pumps fill the elevated storage tanks.

EXAMPLE
10.22

Problem   It has been determined that a community requires a maximum flow of 40 million L/d of water during 10 h in a peak day, beginning at 8 A.M. and ending at 6 P.M. During the remaining 14 h, it needs a flow of 8 million L/d. During the entire 24 h, the water treatment plant is able to provide a constant flow of 24 million L/d, which is pumped into the distribution system. How large must the elevated storage tank be to meet this peak demand?

Solution   Assume the tank is full at 8 A.M. and run a material balance on the community over the next 10 hours:

$$\begin{bmatrix} \text{Rate of water} \\ \text{ACCUMULATED} \end{bmatrix} = \begin{bmatrix} \text{Rate of} \\ \text{water IN} \end{bmatrix} - \begin{bmatrix} \text{Rate of} \\ \text{water OUT} \end{bmatrix}$$
$$+ \begin{bmatrix} \text{Rate of water} \\ \text{PRODUCED} \end{bmatrix} - \begin{bmatrix} \text{Rate of water} \\ \text{CONSUMED} \end{bmatrix}$$

The flow to the community comes from the tower (Q_1) and the plant (6 mgd).

$$0 = [Q_1 + 24 \text{ million L/d}] - [40 \text{ million L/d}] + 0 - 0$$

Solving,

$$Q_1 = 16 \text{ million L/d}$$

This flow must be provided from the tank over a 10-hr period. The required volume of the tank is

$$\frac{(16 \text{ million L/d})(10 \text{ hr})}{24 \text{ hr/day}} = 6.67 \text{ million L}$$

If the tank holds 6.67 million liter, the community can get the water it needs.

But can the tank be full at 8 A.M.? A material (water) balance on the community between 6 P.M. and 8 A.M. yields

$$0 = [24 \text{ million L/d}] - [8 \text{ million L/d} + Q_2] + 0 - 0$$

The flow coming into the community is 6 mgd, while the flow out is 2 mgd (the water used) plus Q_2, which is water needed to fill the tank.

$$Q_2 = 16 \text{ million L/d}$$

This is spread over 14 hr, so that

$$\frac{(16 \text{ million L/d})(14 \text{ h})}{24 \text{ hr/day}} = 9.33 \text{ million L}$$

will be supplied to the tower. So there is no problem filling the tank.

PROBLEMS

10.1 Suppose you are asked to recommend a series of laboratory tests to be run on a small drinking water treatment plant in a developing county. The plant consists of alum flocculation, settling, rapid sand filtration, and chlorination. What tests would you suggest they run, and at what frequency? Justify your answers, considering cost, human health, and environmental protection.

10.2 One day Farmer Brown drilled a well 60 m deep into an aquifer that is 300 m ft deep and has a groundwater table 9 m below the ground surface. On the same day, a neighbor of Farmer Brown, Farmer Jones, drilled a well only 15 m deep. They both pumped the same quantity of water on that first day.

  a. Explain, using sketches, why it was worth the extra money for Farmer Brown to drill a deep well even though they both obtained the same yield from their wells the first day.

  b. Does Farmer Brown *own* the water he pumps from the ground? Is such ownership of natural resources possible? What would happen to a resource such as groundwater in a capitalistic market economy in the absence of governmental controls?

  c. Suppose the government believes that it owns the water, and then sells it to Farmer Brown. Would that necessarily result in a higher level of conservation of natural resources, or would it encourage the rapid depletion of resources?

  d. What would the deep ecologist say about governmental or private ownership of natural resources such as groundwater?

  e. If you were Farmer Jones, and discovered what Farmer Brown had done, what would you do, and why?

10.3 During the years following the Civil War, the New Orleans Water Company installed sand filters to filter the water out of the Mississippi River and sell it to the folks in New Orleans. The filters resembled the rapid sand filters in use today, and water from the river was pumped directly into the filters. Unfortunately, after the plant was built, the facility failed to produce the expected quantity of water and the company went bankrupt. Why did this happen? What would you have done as company engineer to save the operation?

10.4 A typical colloidal clay particle suspended in water has a diameter of 1.0 μm. If coagulation and flocculation with other particles manage to increase its size 100 times its initial diameter (at the same shape and density), how much shorter will be the settling time in 3 m of water, such as in a settling tank?

10.5 It is necessary to maintain a constant flow of 60 million L/mo to a power plant cooling system. The runoff records for a stream are as follows:

Month	Total Flow During That Month (million L)
1	3760
2	488
3	180
4	20
5	20
6	8
7	0
8	8
9	64
10	28
11	288
12	368
13	84
14	220
15	132

If a reservoir is to be constructed, what would be the storage requirement?

10.6 A storage tank at an oil refinery receives a constant flow into the tank at 0.1 m³/s. It is used to distribute the oil for processing only during the 8-hr working day. What must be the flow out of the tank, and how big must the tank be?

10.7 An unconfined aquifer is 10 m thick and is being pumped so that one observation well placed at a distance of 76 m shows a drawdown of 0.5 m. On the opposite side of the extraction well is another observation well, 100 m from the extraction well, and this well shows a drawdown of 0.3 m. Assume the coefficient of permeability is 50 m/d.

a. What is the discharge of the extraction well?
b. Suppose the well at 100 m from the extraction well is now pumped. Show with a sketch what this will do to the drawdown.
c. Suppose the aquifer sits on an aquaclude that has a slope of 1/100. Show with a sketch how this would change the drawdown.

10.8 A settling tank in a water treatment plant has an inflow of 2 m³/min and a solids concentration of 2100 mg/L. The effluent from this settling tank goes to sand filters. The concentration of sludge coming out of the bottom (the underflow) is 18,000 mg/L, and the flow to the filters is 1.8 m³/min.

a. What is the underflow flow rate?
b. What is the solids concentration in the effluent?
c. How large must the sand filters be (in m²)?

10.9 A settling tank is 20 m long, 10 m deep, and 10 m wide. The flow rate to the tank is 10 m³/minute. The particles to be removed all have a settling velocity of 0.1 m/min.

a. What is the hydraulic retention time?
b. Will all the particles be removed?

10.10 The settling basins for a 200 million L/d wastewater treatment plant are operated in parallel with flow split evenly to 10 settling tanks, each of which is 3 m deep and 25 m wide, with a length of 32 m.

a. What is the expected theoretical percentage removal for particles of 0.1 mm diameter that settle at 1×10^{-2} m/s?
b. What theoretical percentage removal is expected for particles of 0.01 mm diameter that settle at 1×10^{-4} m/s?

10.11 A 0.1-m diameter well fully penetrates an unconfined aquifer 20 m deep. The permeability is 2×10^{-3} m/s. How much can it pump for the drawdown at the well to reach 20 m and the well to start sucking air?

10.12 The settling velocity of a particle is 0.002 m/s, and the overflow rate of a settling tank is 0.008 m/s.

a. What percentage of the particles does the settling tank capture?
b. If the particles are flocculated so that their settling rate is 0.05 m/s, what fraction of the particles is captured?
c. If the particles are not changed, and another settling tank is constructed to run in parallel with the original settling tank, will all of the particles be captured?

10.13 A water treatment plant is being designed for a flow of 1.6 m³/s.

a. How many rapid sand filters, using only sand as the medium, are needed for this plant if each filter is 10 m × 20 m? What assumption do you have to make to solve this problem?
b. How can you reduce the number of filters?

10.14 Engineer Jaan, fresh out of school, is busily at work at a local consulting firm. As one of his first jobs, he is asked to oversee the operation of a hazardous waste cleanup job at a Superfund site. The remediation plan is to drill a series of interception wells and capture the contaminated groundwater. The depths of the wells are based on a series of borings, and it is intended that the wells reach to bedrock, an impermeable layer.

One day, Jaan is on the job and the foreman calls him over to the well being drilled.

"Strange. We were supposed to hit bedrock at 70 m, and we are already at 80 m and haven't hit anything yet. You want me to keep going?" the foreman asks Jaan.

Not knowing exactly how to respond, Jaan calls the office and talks to his immediate superior, engineer Robert.

"What do you mean, we haven't hit rock?" inquires Robert, "We have all the borings to show it's at 70 m."

"Well, the foreman says he hasn't hit anything yet," replies Jaan. "What do you want me to do?"

"We are working on a contract, and we were expecting to hit rock here. Maybe the drill has bent, or we are in some kind of seam in the rock. Whatever, we can't afford to keep on drilling. Let's just stop here, show on the drilling log that we hit rock at 80 m. Nobody will ever know."

"We can't do that! Suppose there *is* a seam down there? The hazardous waste could seep out of the containment."

"So what? It'll be years, maybe decades before anyone will know. And besides, it probably will be so diluted by the groundwater that nobody will even be able to detect it. Just say on the drilling log that you hit rock, and move to the next site."

This is a case of falsification of data, but it could be totally harmless, and nobody will ever know. Should Jaan obey the direct order, or should he take some other action? What would happen if he simply told the foreman to keep drilling, and never told Robert that this has happened? He could mark the drilling log with the false information, and Robert would never know. Discuss the wisdom of such a course of action.

10.15 Why might the actual coagulant dose a treatment plant operator uses be different from the minimum dose found from a jar test?

10.16 Given the following jar test results, choose the coagulant dose and determine the chemical feed rate (in kg/d) at a 10 million L/d plant.

Alum Dose (mg/L)	pH	Turbidity (NTU)	Alkalinity (mg/L as $CaCO_3$)
0	7.7	6.2	200
20	7.4	5.5	200
40	7.2	4.6	177
60	7.1	4.5	180
80	6.9	4.2	189
100	6.8	4.0	146

10.17 A water treatment plant has a sedimentation basin receiving 8 million L/d that has a diameter of 20 m and an average water depth of 3 m. What are the retention time and overflow rate in the basin?

10.18 A 16 million L/day water treatment plant is being designed for an overflow rate of $1.2 \ m^3/hr/m^2$. If two circular sedimentation basins will be used at all times, what should be the diameter of each basin?

10.19 Calculate the alkalinity, total hardness, carbonate hardness, and noncarbonate hardness for the following water in mg/L as $CaCO_3$.

Cations	mg/L	Anions	mg/L
Ca^{2+}	94	HCO_3^-	135
Mg^{2+}	28	SO_4^{2-}	134
Na^+	14	Cl^-	92
K^+	31	pH	7.8

10.20 Calculate the alkalinity, total hardness, carbonate hardness, and noncarbonate hardness for the following water in mg/L as $CaCO_3$.

Cations	mg/L	Anions	mg/L
Ca^{2+}	12	HCO_3^-	75
Mg^{2+}	15	SO_4^{2-}	41
Sr^{2+}	3	Cl^-	25
Na^+	15	NO_3^-	10
K^+	15	pH	7.8

10.21 Calculate the alkalinity, total hardness, carbonate hardness, and noncarbonate hardness for the following water in mg/L as $CaCO_3$.

Cations	mg/L	Anions	mg/L
Ca^{2+}	15	HCO_3^-	165
Mg^{2+}	10	SO_4^{2-}	10
Sr^{2+}	2	Cl^-	6
Na^+	20	NO_3^-	3
K^+	10	pH	6.9

10.22 Determine the quantity of quicklime and soda ash required to soften 4 million L/d of the following water to the practical solubility limit. Also calculate the flow rate of sludge if it thickens to 7% solids.

Component	mg/L
Ca^{2+}	53.0
Mg^{2+}	12.1
HCO_3^-	285.0
CO_2	7.2
pH	8.1

10.23 Determine the quantity of quicklime and soda ash required to soften 20,00,000 L/d of the following water to the practical solubility limit. Also calculate the flow rate of sludge if it thickens to 11% solids.

Component	mg/L
Ca^{2+}	53.0
Mg^{2+}	12.1
HCO_3^-	134.0
CO_2	6.8
pH	7.2

10.24 Determine the quantity of hydrated lime and soda ash required to soften 1,200,000 L/d of the following water to the practical solubility limit. Also calculate the flow rate of sludge if it thickens to 9% solids.

Component	mg/L
CO_2	12
Ca^{2+}	83
Mg^{2+}	42
HCO_3^-	240
pH	7.1

10.25 A water sample has a composition as shown. It will be treated in a 15 million L/d plant.

Constituent	Average Concentration (meq/L)
CO_2	0.23
Ca^{2+}	4.1
Mg^{2+}	1.8
HCO_3^-	3.8

a. If the sample will be softened with lime-soda softening to the minimum level using excess quicklime, how much lime and how much soda ash will be required in kilograms per day?

b. If the sample will be softened using the selective calcium removal process, how much lime and how much soda ash will be required in kilograms per day?

c. What will be the approximate final hardness of the softened water in part b?

10.26 Based on jar test results, 10 mg/L of alum is used to coagulate a sample containing 75 mg/L suspended solids.

a. How many milligrams per liter as $CaCO_3$ of natural alkalinity are consumed?

b. What is the mass rate of sludge generated if suspended solids are reduced to 3 mg/L?

Assume $Al(OH)_3$ is precipitated. The plant treats 900,000 L/d.

10.27 Based on jar test results, 10 mg/L of $FeCl_3$ is used to coagulate a sample containing 75 mg/L suspended solids.

a. How many milligrams per liter as $CaCO_3$ of natural alkalinity are consumed?

b. What is the mass rate of sludge generated if suspended solids are reduced to 3 mg/L?

Assume $Fe(OH)_3$ is precipitated. The plant treats 900,000 L/d.

10.28 Nearly all water on earth contains naturally occurring fluoride. Investigation of the decay-preventing effects of naturally occurring fluoride in water led to the start of community water fluoridation in 1945. Fluorides are effective in dramatically reducing tooth decay, particularly for teenagers and young adults.

a. Are there negative effects of fluoridation? What else may the fluorides added to drinking water do?
b. Are there alternatives?

END NOTES

1. Hassan, M. Manzurul, et al. 2005. Social implications of arsenic poisoning in Bangladesh. *Social Science & Medicine* 61, No. 10: 2201–11.
2. Mukherjee, Arun B., and Prosun Bhattacharya. 2001. Arsenic in groundwater in the Bengal Delta Plain: Slow poisoning in Bangladesh. *Environmental Review* 189–220.
3. World Health Organization. 2008. An interview with Mahmuder Rahman. *Bulletin of the World Health Organization* January 2008. 11–12.
4. Centeno, Jose A. et al. 2007. Global impacts of geogenic arsenic: A medical geology research case. *Royal Swedish Academy of Sciences*. February. 78–81.
5. Atkins, Randy. 2007. National academy of engineering announces $1 million challenge to provide safe drinking water. *National Academy of Engineering* February 1. Accessed at http://nationalacademies.org on 8 December, 2008.
6. Hussam, Abul, and Abul K. M. Munir. 2007. A simple and effective arsenic filter based on composite iron matrix: Development and deployment studies for groundwater of Bangladesh. *Journal of Environmental Science and Health* October.1869–78.
7. Gibson, Don, and Marty Reynolds. 2000. Softening. *Water Treatment Plant Operation: A Field Study Training Program*. California Department of Health Services and USEPA. Sacramento: California State University.
8. Bowen, Russ. 1999. Taste and odor control. *Water Treatment Plant Operation: A Field Study Training Program*. California Department of Health Services and USEPA. Sacramento: California State University.

CHAPTER ELEVEN

Wastewater Treatment

© Susan Morgan

Restrooms at park, Illinois

© Doug Plummes/Photographer's Choice/Getty Images

Salmon, Pacific Northwest

Environmental engineers clean used water to allow its discharge into the environment—protecting the inhabitants and users of waterways.

Water has many uses, including drinking, commercial navigation, recreation, fish propagation, and waste disposal (!). It is easy to forget that a major use of water is simply as a vehicle for transporting wastes. In isolated areas where water is scarce, waste disposal becomes a luxury use, and other methods of waste carriage are employed, such as pneumatic pipes or containers. But in most of the Western world, a beneficial use of water for waste transport is almost universal, and this, of course, results in large quantities of contaminated water.

11.1 WASTEWATER

11.1.1 Transport

Wastewater is discharged from homes, commercial establishments, and industrial plants by means of *sanitary sewers*, usually large pipes flowing partially full (not under pressure). Sewers flow by gravity drainage downhill, and the system of sewers has to be so designed that the *collecting sewers*, which collect the wastewater from homes and industries, all converge to a central point where the waste flows by *trunk sewers* to the wastewater treatment plant. Sometimes it is impossible or impractical to install all gravity sewers, so the waste has to be pumped by pumping stations through *force mains*, or pressurized pipes.

Design and operation of sewers is complicated by the inflow of stormwater, which is supposed to flow off in separate *storm sewers* in newer communities but often seeps into the wastewater sewers through loose manhole covers and broken lines. Such an additional flow to the wastewater sewers is called *inflow*. (Older communities often have *combined sewers*; they were designed to collect and transport both sanitary wastewater and stormwater. They frequently have problems with combined sewer overflows (CSOs) during rain events.) Further, sewers often have to be installed below the groundwater table, so any breaks or cracks in the sewer (such as from the roots of trees seeking water) can result in water seeping into the sewers. This additional flow is known as *infiltration*. Local communities often spend considerable time and expense in rehabilitating sewerage systems to prevent such inflow and infiltration (I/I) because every gallon that enters the sewerage system has to be treated at the wastewater treatment plant.

The wastewater, diluted by I/I, flows downhill and eventually to the edge of the community that the sewerage system serves. In years past, this wastewater simply entered a convenient natural watercourse and was forgotten by the community. The growth of our population and the awareness of public health problems created by the raw sewage makes such a discharge untenable and illegal, and wastewater treatment is required.

11.1.2 Components

Although water can be polluted by many materials, the most common contaminants found in domestic wastewater that can cause damage to natural watercourses or create human health problems are

- organic materials, as measured by the demand for oxygen (BOD)
- nitrogen (N)
- phosphorus (P)
- suspended solids (SS)
- pathogenic organisms (as estimated by coliforms).

FOCUS ON

The Destruction of the Rhine River

In 1986, a fire at the Sandoz Ltd. chemical plant in Switzerland dumped 30 tons of highly toxic organophosphate pesticides called disulfoton and parathion into the Rhine River. The river flows north through Germany and the Netherlands and is both an important waterway and a source of drinking water for many German and Dutch cities (Figure 11.1). The concentrated plug of poison flowed down the river. For hundreds of kilometers, the riverbanks were clogged with dead fish and aquatic plants. The river reeked of decay and seemed to be totally dead. It took several years for the effects of the poisons to wash through the river into the sea, and eventually life returned to the Rhine.

This event was the most notorious destruction of water quality in Europe and presented interesting political problems because the spill was in Switzerland, but the devastation was in Germany and the Netherlands. Sandoz Ltd. "cleaned up" their act by moving its pesticide chemical production to Brazil.

The Rhine has historically been the recipient of both human and industrial waste, and its water quality has been very poor. Samuel Taylor Coleridge, as he watched the river flowing through Cologne (Köln) penned these verses.

> In Köln, a town of monks and bones,
> And pavements fanged with murderous stones,
> And rags, and bags, and hideous wenches;
> I counted two and seventy stenches,
> All well defined, and several stinks.
>
> Ye Nymphs that reign o'er sewers and sinks,
> The river Rhine, it is well known,
> Doth wash your city of Köln;
> But tell me Nymphs: What power divine
> Shall henceforth wash the River Rhine?

Figure 11.1   Map of the Rhine from Switzerland through the Netherlands.

Source: Reprinted from *Controlling Environmental Pollution*, Vesilind, P. A. and DiStefano, T. D., 2006. Lancaster, PA: DEStech Publications, Inc. Reproduced with permission of DEStech Publications, Inc.

Municipal wastewater treatment plants are designed to remove these objectionable characteristics from the influent. The designs vary considerably but often take a general form as shown in Figure 11.2 on the next page.

The typical wastewater treatment plant is divided into five main areas:

- preliminary treatment—removal of large solids to prevent damage to the remainder of the unit operations
- primary treatment—removal of suspended solids by settling
- secondary treatment—removal of the demand for oxygen
- tertiary (or advanced) treatment—a name applied to any number of polishing or cleanup processes, one of which is the removal of nutrients such as phosphorus
- solids treatment and disposal—the collection, stabilization, and subsequent disposal of the solids removed by other processes.

Primary treatment systems are usually physical processes. Secondary treatment processes are commonly biological. Tertiary treatment systems can be physical (e.g., filtration to remove solids), biological (e.g., constructed wetlands to remove BOD), or chemical (e.g., precipitation to remove phosphorus).

A range of wastewater characteristics can be analyzed to provide information pertinent to the design and operation of treatment plants (Table 11.1). Seven principal components, however, are of concern in the design and operation of treatment systems—total suspended solids (TSS), BOD, pathogens, total dissolved solids (TDS), heavy metals, nutrients, and priority organic pollutants.[1] TSS impact the amount of sludge produced and the potential for anaerobic conditions to develop; in addition, TSS affect the aesthetic qualities of the effluent. Biodegradable organics (as measured by BOD) require dissolved oxygen for treatment when aerobic processes are used. Of course, pathogens cause communicable diseases. These three constituents drive the design of most wastewater treatment systems. Dissolved inorganic substances (i.e., TDS, which is made up of substances such as calcium, sodium, and sulfate) increase through repeated use of water and, therefore, have implications in the reuse of treated wastewater. Heavy metals (which are the cations with atomic weights above 23 and which are contributed from households as well as industry) can upset biological treatment processes and can reduce sludge management options if present in sufficient quantity. Nutrients (i.e., phosphorus and nitrogen) can cause oxygen depletion and eutrophication when discharged to natural water bodies. However, they are desirable in sludge used in land application and in effluent used for irrigation, although excessive loadings can contaminate surface water and groundwater. Priority organic pollutants are hazardous and often resist conventional treatment methods.

There is a large range between the typical concentrations of weak and strong sanitary wastewater and between sanitary wastewater and septage (which is the substance remaining in septic tanks after anaerobic treatment) (Table 11.2). While these concentrations are typical in the United States, they may not be applicable in other countries. For example, engineers in Thailand found that wastewater in Bangkok has a BOD between 50 and 70 mg/L and a suspended solids concentration between 90 and 110 mg/L.[2] Obviously, it is important to obtain current local information on wastewater composition and flow rates when designing a new treatment plant or upgrades to an existing plant.

© Courtesy of P. Aarne Vesilind

KEY
① Bar screen
② Grit chamber
③ Primary clarifier
④ Aeration tank
⑤ Final clarifier
⑥ Chlorine contact tank
⑦ Digester
⑧ Dewatering

Figure 11.2 A typical wastewater treatment plant, showing preliminary treatment, primary treatment, secondary treatment, tertiary treatment, and solids treatment. Depending on the need for effluent water quality, the plant can discharge to a watercourse after either secondary or tertiary treatment.

Table 11.1   Use of Laboratory Analyses in Wastewater Treatment Plant Design and Operation

Analysis	Use
Physical Characteristics	
Color	Condition of wastewater (fresh or septic)
Odor	Treatment requirements, indication of anaerobic conditions
Solids	
TS, VS, SS	Design and operation of treatment process
DS	Effluent reuse potential
Temperature	Design and operation of biological processes
Turbidity	Effluent quality
Transmittance	Suitability of UV disinfection
Inorganic Characteristics	
Alkalinity	Design and operation of treatment process
Chloride	Effluent reuse potential
Hydrogen sulfide (H_2S)	Operation of treatment process, odor control requirements
Metals	Effluent and sludge reuse potential, design and operation of treatment process
Nitrogen	Design and operation of treatment process, effluent and sludge reuse potential
Oxygen	Design and operation of treatment process
pH	Design and operation of treatment process
Phosphorus	Design and operation of treatment process, effluent and sludge reuse potential
Sulfate	Odor potential, treatability of sludge
Organic Characteristics	
BOD_5	Design and operation of treatment process
COD	Operation of treatment process
Methane (CH_4)	Operation of treatment process, energy recovery potential
NOD	Design and operation of treatment process (nitrification/denitrification)
Specific organics	Design and operation of treatment system
Biological Characteristics	
Coliforms	Design and operation of disinfection system
Specific microorganisms	Operation of treatment system

Source: SMALL AND DECENTRALIZED WASTEWATER MANAGEMENT SYSTEMS by Crites, Ronald W., Copyright 1998 by McGraw-Hill Companies, Inc.-Books. Reproduced with permission of McGraw-Hill Companies, Inc.-Books in the format Textbook via Copyright Clearance Center.

11.2   PRELIMINARY AND PRIMARY TREATMENT

11.2.1   Preliminary Treatment

The most objectionable aspect of discharging raw sewage into watercourses is the presence of floating material. It is only logical, therefore, that *screens* were the first form of wastewater treatment used by communities, and even today, screens are used as the first step in treatment plants. Typical screens, shown in Figure 11.3 on the next page, consist of

Table 11.2 Typical Wastewater Concentrations in the United States

Component	Concentration (mg/L)			
	Weak Sanitary	Medium Sanitary	Strong Sanitary	Septage
TS	350	720	1200	40,000
SS	100	220	350	15,000
BOD$_5$	110	220	400	6,000
N (as N)	20	40	85	700
P (as P)	4	8	15	250

Source: Metcalf and Eddy, Inc. (Revised by George Tchobanogous and Franklin L. Burton.) 1991. *Wastewater Engineering: Treatment, Disposal, and Reuse*. New York: McGraw-Hill.

© Courtesy of P. Aarne Vesilind

Figure 11.3 A typical bar screen.

a series of steel bars, which might be about 2.5 cm (1 in) apart. The purpose of a screen in modern treatment plants is the removal of larger materials that might damage equipment or hinder further treatment. In some older treatment plants screens are cleaned by hand, but mechanical cleaning equipment is used in almost all new plants. The cleaning rakes are automatically activated when the screens become sufficiently clogged to raise the water level in front of the bars.

In many plants the next treatment step is a *comminutor*, a circular grinder designed to grind the solids coming through the screen into pieces about 0.3 cm or smaller. Many designs are in use; one common design is shown in Figure 11.4.

The third common preliminary treatment step involves the removal of grit or sand (Figure 11.5 on the next page). This is necessary because grit can wear out and damage

© Courtesy of P. Aarne Vesilind

Figure 11.4 A typical comminutor.

© Courtesy of P. Aarne Vesilind

Figure 11.5 A grit chamber used in wastewater treatment.

equipment, such as pumps and flow meters. The most common *grit chamber* is simply a wide place in the channel where the flow is slowed sufficiently to allow the heavy grit to settle. Sand is about 2.5 times as heavy as most organic solids and, thus, settles much faster than the light solids. The objective of a grit chamber is to remove this inorganic grit without removing the organic material. The latter must be further treated in the plant, but the grit can be dumped as fill without undue odor or other problems. One way of ensuring that the light biological solids do not settle is to aerate the grit chamber, allowing the sand and other heavy particles to sink but keeping everything else afloat. Aeration has the additional advantage of driving some oxygen into the sewage, which may have become devoid of oxygen in the sewerage system.

FOCUS ON

The Pigeon River

© Karl Musser

Figure 11.6 The Pigeon River.

The Pigeon River flows from the Great Smoky Mountains of North Carolina into Tennessee, passing through Canton, North Carolina, which boasted a large paper mill (Figure 11.6). Upstream of Canton the Pigeon River was clear and clean, but below the paper mill the river turned into a dark brown color and emitted a sulfuric odor. The State of North Carolina for many years tried to get the paper mill to clean up its discharges, but the importance of the mill to the economically depressed area made firm legal action politically impossible. Thus the Pigeon River remained an open industrial sewer emptying into the State of Tennessee.

This did not sit well with the folks in Tennessee, but they could do nothing about it. The water was polluted out of their jurisdiction, and setting high stream standards on the Tennessee portion of the Pigeon River made no sense because they had to be enforced in North Carolina. Finally, the Tennessee water pollution agency pushed the USEPA to take steps. The controversy escalated with the discovery of dioxin in the stream water, and the State of Tennessee put up warning signs along the river prohibiting the eating of fish caught from the river.

In 1992, the mill was sold to a consortium of the workers and was renamed Blue Ridge Paper Products. With assistance from numerous sources, the new company was able to fund extensive pollution control measures to the tune of $330 million. With this treatment, the dioxin levels dropped to below the toxic limit, and the water color returned to normal in the Tennessee portion of the Pigeon.

Similar to the Rhine River, the dilemma illustrates one of the problems in water pollution control. If each country or state is allowed to set its own water quality standards, they may be incompatible as a river flows across borders. The standards set by the downstream country or state must be enforced by the upstream country or state, which would not attain any direct benefit from the money spent for such pollution control.

Source: Reprinted from *Controlling Environmental Pollution*, Vesilind, P. A. and DiStefano, T. D., 2006. Lancaster, PA: DEStech Publications, Inc. Reproduced with permission of DEStech Publications, Inc.

11.2.2 Primary Treatment

Following the grit chamber most wastewater treatment plants have a *settling tank* to settle as much of the solid matter as possible. These tanks, in principle, are no different from the settling tanks introduced in the previous chapter. It is again desired to operate the tanks as plug-flow reactors, and turbulence is, therefore, kept to a minimum. The solids settle to the bottom and are removed through a pipe while the clarified liquid escapes over a *V-notch weir*, a notched steel plate over which the water flows, promoting equal distribution of the liquid discharge all the way around a tank. Settling tanks can be circular (Figure 11.7) or rectangular (Figure 11.8).

Settling tanks are also known as *sedimentation tanks* and *clarifiers*. The settling tank that follows preliminary treatment, such as screening and grit removal, is known as the *primary clarifier*. The solids that drop to the bottom of a primary clarifier are removed as *raw sludge*, a name that doesn't do justice to the undesirable nature of this stuff.

© Courtesy of P. Aarne Vesilind

Figure 11.7 Circular settling tank (primary clarifier) used in wastewater treatment.

Figure 11.8   Rectangular settling tank (primary clarifier) used in wastewater treatment.

Raw sludge is generally odoriferous, can contain pathogenic organisms, and is full of water—three characteristics that make its disposal difficult. It must be both stabilized, to reduce its possible public health impact and to retard further decomposition, and dewatered for ease of disposal. In addition to the solids from the primary clarifier, solids from other processes must similarly be treated and disposed of. The treatment and disposal of wastewater solids (sludge) is an important part of wastewater treatment and is discussed further in a subsequent section.

Primary treatment, in addition to removing about 60% of the solids, removes about 30% of the demand for oxygen and perhaps 20% of the phosphorus (both as a consequence of the removal of raw sludge). If this removal is adequate and the dilution factor in the watercourse is such that the adverse effects are acceptable, then a primary treatment plant is sufficient wastewater treatment. Governmental regulations, however, have forced all primary plants to add secondary treatment, whether needed or not.

When primary treatment is judged to be inadequate, solids, BOD, and phosphorus removal can be enhanced by the addition of chemicals, such as aluminum sulfate (alum) or calcium hydroxide (lime), to the primary clarifier influent. With such addition, the effluent BOD can be reduced to about 50 mg/L, and this BOD level may be able to meet required effluent standards. Chemical addition to primary treatment is especially attractive for large coastal cities that can achieve high dilution in the dispersion of the plant effluent.

In a more typical wastewater treatment plant, primary treatment without chemical addition is followed by secondary treatment, which is designed specifically to remove the demand for oxygen.

11.3 SECONDARY TREATMENT

The water leaving the primary clarifier has lost much of the suspended organic matter but still contains a high demand for oxygen due to the dissolved biodegradable organics. This demand for oxygen must be reduced (energy expended) if the discharge is to avoid creating unacceptable conditions in the watercourse. The objective of secondary treatment is to remove BOD while, by contrast, the objective of primary treatment is to remove solids. Except in rare circumstances, almost all secondary treatment methods use microbial action to reduce the energy level (BOD) of the waste (a process advocated in the late 1800s by Dibdin and Dupré, as described in Chapter 1). The basic differences among all these alternatives are how the waste is brought into contact with the microorganisms.

11.3.1 Fixed Film Reactors

Although there are many ways the microorganisms can be put to work, the first really successful modern method of secondary treatment was the *trickling filter*. The trickling filter, shown in Figure 11.9, consists of a bed of media (such as fist-sized rocks or various plastic shapes) over which the waste is trickled. An active biological growth forms on the media, and the organisms obtain their food from the waste stream dripping over the bed. Air is either forced through the media or, more commonly, air circulation is obtained automatically by a temperature difference between the air in the bed and ambient temperature. In older filters the waste is sprayed onto the rocks from fixed nozzles; newer designs use a rotating arm that moves under its own power, distributing the waste evenly over the entire bed, like a lawn sprinkler. Often the flow is recirculated, obtaining a higher degree of treatment. The name trickling filter is obviously a misnomer because no filtration takes place.

A modern modification of the trickling filter is the *rotating biological contactor*, or rotating disc, pictured in Figure 11.10 on page 356. The microbial growth occurs on rotating discs that are slowly dipped into the wastewater, which provides their food. By bringing the discs out into the open air, the microbes are able to obtain the necessary oxygen to keep the growth aerobic.

11.3.2 Suspended Growth Reactors

Around 1900, when trickling filtration was already firmly established, some researchers began musing about the wasted space in a filter taken up by the rocks. Could the microorganisms be allowed to float free, and could they be fed oxygen by bubbling in air? Although this concept was quite attractive, it was not until 1914 that the first workable pilot plant was

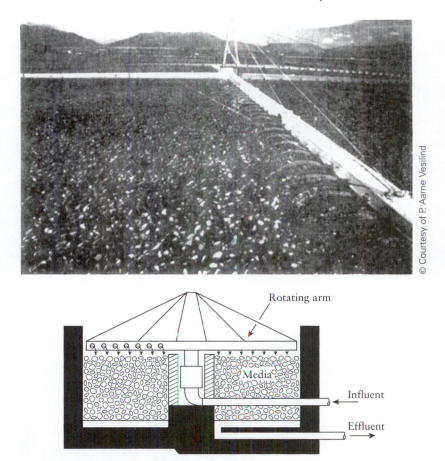

© Courtesy of P. Aarne Vesilind

Figure 11.9   A trickling filter.

constructed. It took some time before this process became established as what we now call the *activated sludge system.*

The key to the activated sludge system is the reuse of microorganisms. The system, illustrated in Figure 11.11 on page 357, consists of a tank full of waste liquid (from the primary clarifier) and a mass of microorganisms. Air is bubbled into this tank (called the *aeration tank*) to provide the necessary oxygen for the survival of the aerobic organisms. The microorganisms come into contact with the dissolved organics and rapidly adsorb these organics on their surface. In time, the microorganisms use the energy and carbon by decomposing this material to CO_2, H_2O, and some stable compounds and in the process produce more microorganisms. The production of new organisms is relatively slow, and most of the aeration tank volume is used for this purpose.

Once most of the food has been used, the microorganisms are separated from the liquid in a settling tank, called a *secondary* or *final clarifier*. The liquid escapes over a V-notch weir.

The separated microorganisms exist on the bottom of the final clarifier without additional food and become hungry waiting for more dissolved organic matter. These microorganisms are said to be activated—hence, the term *activated sludge*.

Figure 11.10 Rotating disc fixed-film biological reactor.

When these settled and hungry microorganisms are pumped to the head of the aeration tank, they find more food (organics in the effluent from the primary clarifier), and the process starts all over again. The sludge pumped from the bottom of the final clarifier to the aeration tank is known as *return activated sludge*.

The activated sludge process is a continuous operation, with continuous sludge pumping and clean water discharge. Unfortunately, one of the end products of this process is excess microorganisms. If the microorganisms are not removed, their concentration eventually increases to the point where the system is clogged with solids. It is, therefore, necessary to waste some of the microorganisms, and this *waste activated sludge* must be processed and disposed of. Its disposal is one of the most difficult aspects of wastewater treatment.

Activated sludge systems are designed on the basis of loading, or the amount of organic matter (food) added relative to the microorganisms available. This ratio is known as the food-to-microorganisms ratio (F/M) and is a major design parameter. Unfortunately, it is difficult to measure either F or M accurately, and engineers have approximated these by BOD and the suspended solids in the aeration tank, respectively. The combination of the

Figure 11.11 A schematic diagram of the activated sludge system.

liquid and microorganisms undergoing aeration is known (for some unknown reason) as *mixed liquor*, and the suspended solids are called *mixed liquor suspended solids* (MLSS). The ratio of incoming BOD to MLSS, the *F/M* ratio, is also known as the *loading* on the system and is calculated as kilograms of BOD/d per kilogram of MLSS in the aeration tank.

If this ratio is low (little food for lots of microorganisms) and the aeration period (detention time in the aeration tank) is long, the microorganisms make maximum use of available food, resulting in a high degree of treatment. Such systems are known as *extended aeration* and are widely used for isolated sources (e.g., motels and small developments). Added advantages of extended aeration are that the ecology within the aeration tank is quite diverse and little excess biomass is created, resulting in little or no waste activated sludge to be disposed of—a significant saving in operating costs and headaches. At the other extreme is the *high-rate* system, in which the aeration periods are very short (thus saving money by building smaller tanks) and the treatment efficiency is lower.

11.3.3 Design of Activated Sludge Systems Using Biological Process Dynamics

The objective of an activated sludge system is to degrade the organics in the influent and oxidize them to CO_2 and H_2O, recognizing that some of this energy must also be used to build new microorganisms. These influent organics provide the food for the microorganisms, and in biological process dynamics are known as *substrate*. As noted above, substrate is usually measured indirectly as the BOD, with the decrease in oxygen indicating microbial degradation of the substrate. Although other methods, such as organic carbon, may be more accurate measures of substrate concentration, BOD already has to be measured for regulatory compliance.

FOCUS ON

The Development of the Activated Sludge System

Probably no other single wastewater treatment process has had as much success as the activated sludge process. This system had an interesting beginning.

The first experiments with aeration were performed at the Lawrence Experiment Station in Massachusetts in 1912 using 1-gal glass jugs. At the end of each run when nitrification had been achieved (which was the primary measure of treatment in those days), they would empty out the jugs, fill them up with new wastewater, and resume aerating. They were astonished to find that the time necessary to achieve nitrification decreased with every test. They did not understand that, by not washing out the jugs at the conclusion of a run, they were leaving a film of microorganisms attached to the jugs' walls and that these microorganisms were the active culture that then seeded the next test. They had invented the activated sludge system but did not know it.

Dr. Gilbert Fowler from the University of Manchester in England happened to be visiting the experiment station and heard about the experiments. He apparently understood the significance of what had occurred but did not reveal his ideas to the Americans. Instead, he went back to England and had two of his associates, Edward Arden and W. T. Lockett, set up a series of tests in which they would aerate a sample of waste for some days, settle out the sludge solids, decant the liquid off the top, and refill the container with new wastewater. Using this procedure, the time to nitrification was reduced from five weeks to 24 hours! It was a simple matter to go from such a fill-and-draw operation to a continuous system in which the sludge is settled in a clarifier and then pumped back to the aeration system. Arden and Lockett are given credit for inventing the activated sludge system (and for giving it the name "activated sludge" because they thought it had been altered similarly to activated carbon), but the original work was done in Massachusetts. It just took a clever researcher like Fowler to recognize what was going on.

Source: MOHLMAN, F. W. Twenty-five years of activated sludge. Modern Sewage Disposal Anniversary Book of the Federation of Sewage Works Association, USA 1938.

As with most living organisms, microorganisms' growth is affected by the availability of food (substrate) and environmental conditions (e.g., pH, temperature, and salinity). The microbial growth curve shown in Figure 11.12 is typical of a closed, batch system for a single type of microorganism (i.e., a pure culture). (Of course, a wastewater treatment plant has a diverse assortment of microbes, but this model is a useful place to start.) While the microorganisms adjust to the environment and substrate during the lag phase, they have limited growth and use little of the substrate. However, once they are adjusted, they undergo rapid, exponential growth. This phase cannot continue indefinitely, though. If it did, a single bacterial cell weighing about 10^{-12} g and doubling every 20 min would produce a population weighing about 4000 times the weight of earth after 48 h of exponential growth![3] Instead, byproduct and waste accumulation and/or restrictions on substrate or nutrient levels limit the maximum population size that can be supported (K). During this period of maximum population (the stationary phase), little or no growth occurs, and there is no net increase or decrease in cell number. Eventually, due to increased waste, by-products, and/or a lack of substrate or nutrients, the death rate becomes greater

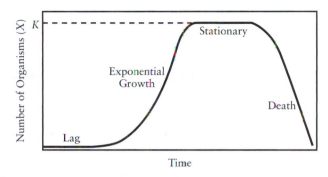

Figure 11.12   Microbial growth curve of a pure culture in a closed system.

than the growth rate, and the microbial population declines (the death, or endogenous, phase).

In this situation the only limitation on growth during the exponential growth phase is the rate at which the microorganisms can reproduce. Therefore, the number of microorganisms is proportional to the growth rate, or in other words, the growth is similar to a first-order reaction (Chapter 5):

$$\frac{dX}{dt} = \mu X \tag{11.1}$$

where X = number of microorganisms
   μ = instantaneous, specific growth rate

Integrating this equation allows us to predict future population density when we know the original population and the instantaneous growth rate:

$$X = X_0 e^{\mu t}$$

The doubling (or generation) time can be determined by substituting $X = 2X_0$ into this equation, as in Chapter 5:

$$t_D = \frac{\ln 2}{\mu}$$

Engineers often make use of continuous culture devices (chemostats) rather than batch culture devices. A conventional activated sludge wastewater treatment plant is an example of a large-scale chemostat. Chemostats maintain cell populations in the exponential growth phase by controlling the dilution rate and the concentration of a limiting nutrient, such as the carbon or nitrogen source. Both the cell density (or population size) and growth rate can be controlled. Chemostat operation is a saturation process, which can be described by

$$\mu = \hat{\mu} \frac{S}{K_S + S} \tag{11.2}$$

where $\hat{\mu}$ = maximum specific growth rate (at nutrient saturation)
   S = substrate or nutrient concentration
   K_S = saturation, or half-velocity, constant

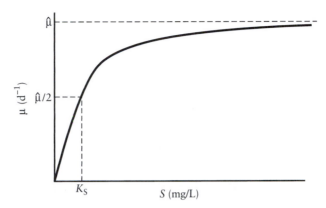

Figure 11.13 Specific growth rate of microbes.

The saturation constant is the nutrient concentration when the growth rate is half the maximum growth rate (Figure 11.13). Estimates of $\hat{\mu}$ and K_S are obtained by plotting $1/S$ versus $1/\mu$. The y intercept is $1/\hat{\mu}$, and the slope is $K_S/\hat{\mu}$.

Of course, in nature the situation is different. Pure cultures do not exist. Exponential growth is limited due to the availability of nutrients, competition among microorganisms, and predator–prey relationships. In addition, because environmental conditions are rarely optimal for maximum growth, the actual maximum growth rate is typically well below the laboratory rate; for example, the doubling time for *Escherichia coli* in the laboratory can be 20 minutes, whereas in the intestinal tract it is 12 hours.[3]

Microbial growth can be measured by counting or weighing cells. The direct microscopic count can be used to determine the total cell count; however, one of the problems with this method is that both living and dead cells are counted. The plate, or colony, count method (in which dilutions of sample are incubated on agar plates) measures only viable (i.e., living) cells. An indirect measure of cell growth is measuring the cell mass by either centrifuging and weighing the cells or by measuring the sample turbidity with a colorimeter or spectrophotometer. Although turbidity is less sensitive than viable counting, it is quick, easy, and does not change the sample.

While treatment plant operators control the microbial concentration in the aeration basin, they are interested in doing so only to reduce the BOD. The microorganisms, expressed as suspended solids, biodegrade and use the BOD (substrate) at a rate r_s. As the food is used, new organisms are produced. The rate of new cell mass (microorganisms) production as a result of the destruction of the substrate is

$$r_x = -Y r_s \tag{11.3}$$

where $Y =$ the yield, or mass of microorganisms produced per mass of substrate used, commonly expressed as kg SS produced per kg BOD used. Y will always be less than one due to inefficiencies in energy conversion processes as discussed in Chapter 8.

Combining the Equations 11.1, 11.2, and 11.3, the expression for substrate utilization commonly employed is derived:

$$r_s = \frac{-X}{Y}(\mu) = \frac{-X}{Y}\left(\frac{\hat{\mu} S}{K_S + S}\right) \tag{11.4}$$

Figure 11.14   A suspended growth reactor with no recycle.

This is known as the Monod model. (The argument for the model validity is beyond the scope of this text. Suffice it to say that the model is empirical but reasonable. If you are interested in the development of this model, see any number of modern textbooks on wastewater processing.)

Because this expression is an empirical model based on experimental work with pure cultures, the two constants, $\hat{\mu}$ and K_S, must be evaluated for each substrate and microorganism culture. However, they remain constant for a given system as S and X are varied. Note that they are a function of the substrate and microbial mass, not of the reactor.

The application of biological process dynamics to the activated sludge process is best illustrated by considering a system shown in Figure 11.14, which is not an activated sludge system because there is no solids recycle but serves to introduce the notation and terminology. This is a simple continuous biological reactor of volume V with a flow rate of Q. The reactor is completely mixed. Recall that this means that the influent is dispersed within the tank immediately upon introduction; thus, there are no concentration gradients in the tank, and the quality of the effluent is exactly the same as that of the tank contents.

In such a reactor it is possible to develop two types of material balances—in terms of solids (microorganisms) and BOD (substrate). There are also two retention, or detention, times—liquid and solids. The liquid, or hydraulic, retention time was introduced in Chapter 3 and is expressed as

$$\bar{t} = \frac{V}{Q} \tag{11.5}$$

Recall that $\bar{t}$ can also be defined as the average time the liquid remains in the reactor.

The *solids retention time* is analogous to the hydraulic retention time and represents the average time *solids* stay in the system (which is longer than the hydraulic retention time when the solids are recycled). The solids retention time is also known as *sludge age* and the *mean cell residence (or detention) time* (MCRT). All three names represent the same parameter, which can be calculated as

$$\theta_C = \frac{\text{Mass of solids (microorganisms) in the system}}{\text{Mass of solids wasted/Time}} = \text{Time}$$

The numerator, the amount of solids in the simple reactor, is expressed as VX (Volume × Solids concentration), and the denominator, the rate of the solids wasted, is equal to QX (Flow rate × Solids concentration). Thus, the mean cell retention time is

$$\theta_C = \frac{VX}{QX} = \frac{V}{Q} \tag{11.6}$$

which, of course, is the same as the hydraulic retention time (Equation 11.5) because the solids are not recycled.

The amount of solids wasted must also be equal to the rate they are produced, or dX/dt. Substituting $dX/dt \times V$ for QX and using Equation 11.3 yields

$$\theta_C = \frac{VX}{r_x V} = \frac{VX}{(-Yr_s)V} = \frac{-X}{Yr_s}$$

Note that the concentration terms have to be multiplied by volume to obtain mass.

Remember that these two equations are for a reactor *without recycle* (Figure 11.14).

Using Figure 11.14 we can write a mass balance in terms of the microorganisms:

$$\begin{bmatrix} \text{Rate of} \\ \text{ACCUMULATION} \end{bmatrix} = \begin{bmatrix} \text{Rate} \\ \text{IN} \end{bmatrix} - \begin{bmatrix} \text{Rate} \\ \text{OUT} \end{bmatrix}$$

$$+ \begin{bmatrix} \text{Rate of} \\ \text{microorganism} \\ \text{GROWTH} \end{bmatrix} - \begin{bmatrix} \text{Rate of} \\ \text{microorganism} \\ \text{DEATH} \end{bmatrix}$$

If the growth and death rates are combined as *net* growth, this equation reads

$$V \frac{dX}{dt} = QX_0 - QX + r_x V$$

or

$$V \frac{dX}{dt} = QX_0 - QX - Yr_s V$$

In a steady-state system $dX/dt = 0$, and assuming there are no cells in the inflow, $X_0 = 0$. Using this information and making use of the Monod substrate utilization model (Equation 11.4) yields

$$r_s = \frac{-X}{Y} \left(\frac{Q}{V} \right) = \frac{-X}{Y} \left(\frac{1}{\theta_C} \right) = \frac{-X}{Y} \left(\frac{\hat{\mu} S}{K_S + S} \right)$$

Therefore:

$$\frac{1}{\theta_C} = \frac{\hat{\mu} S}{K_S + S} = \mu \tag{11.7}$$

or

$$S = \frac{K_S}{\hat{\mu}\theta_C - 1} \tag{11.8}$$

This is an important expression because we can infer from it that the substrate concentration, S, is a function of the kinetic constants (which are beyond our control for a given substrate) and the mean cell retention time. The value of S, which in real life would be the effluent BOD, is influenced then by the mean cell retention time (or the sludge age as previously defined). If the mean cell retention time is increased, the effluent concentration should decrease.

EXAMPLE

11.1   **Problem**   A biological reactor such as the one pictured in Figure 11.14 (with no solids recycle) must be operated so that an influent BOD of 600 mg/L is reduced to 10 mg/L. The kinetic constants have been found to be $K_S = 500$ mg/L and $\hat{\mu} = 4$ d^{-1}. If the flow is 3 m^3/day, how large should the reactor be?

Solution   Remember that the substrate concentration, S, in the effluent is exactly the same as S in the reactor if the reactor is assumed to be perfectly mixed.

Using Equations 11.7 and then 11.6

$$\theta_C = \frac{K_S + S}{S\hat{\mu}} = \frac{500 \text{ mg/L} + 10 \text{ mg/L}}{(10 \text{ mg/L})(4 \text{ day}^{-1})} = 12.75 \text{ days}$$

$$V = \theta_C Q = (12.75 \text{ day})(3 \text{ m}^3/\text{day}) = 38.25 \text{ m}^3 \cong 38 \text{ m}^3$$

EXAMPLE

11.2   **Problem**   Given the conditions in Example 11.1, suppose the only reactor available has a volume of 24 m^3. What would be the percent reduction in substrate (substrate removal efficiency)?

Solution   Using Equations 11.6 and 11.8

$$\theta_C = \frac{V}{Q} = \frac{24 \text{ m}^3}{3 \text{ m}^3/\text{day}} = 8 \text{ days}$$

$$S = \frac{K_S}{\hat{\mu}\theta_C - 1} = \frac{500 \text{ mg/L}}{(4 \text{ day}^{-1})(8 \text{ day}) - 1} = 16 \text{ mg/L}$$

$$\text{Recovery} = \frac{(600 - 16)}{600} \times 100 = 97\%$$

The system pictured in Figure 11.14 is not very efficient because long hydraulic retention times are necessary to prevent the microorganisms from being flushed out of the tank. Their growth rate has to be faster than the rate of being flushed out, or the system will fail. The success of the activated sludge system for wastewater treatment is based on a significant modification: the recycle of the microorganisms. Such a system is shown in Figure 11.15.

Some simplifying assumptions are necessary before this model can be used. First, again assume $X_0 = 0$. Also, once again assume steady-state conditions and perfect mixing. The excess microorganisms (waste activated sludge) are removed from the system at flow rate Q_w and a solids concentration X_r, which is the settler underflow concentration (as well as the concentration of the solids being recycled to the aeration tank).

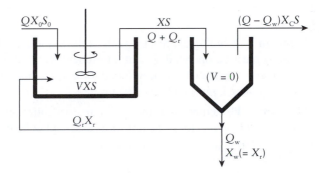

Figure 11.15   A suspended growth reactor with recycle (the activated sludge system).

And lastly, assume that there is no substrate removal in the settling tank and that the settling tank has no volume so that all of the microorganisms in the system are in the reactor (aeration tank). The volume of the aeration tank is, thus, the only active volume, and the settling (microorganism separation) is assumed to take place magically in a zero-volume tank, an obviously incorrect assumption. The mean cell retention time in this case is

$$\theta_C = \frac{\text{Microorganisms in the system}}{\text{Microorganisms wasted/Time}}$$

$$= \frac{XV}{Q_w X_r + (Q - Q_w)X_C} \tag{11.9}$$

If we assume that the microorganism separator (the final settling tank) is a perfect device so that there are no microorganisms in the effluent ($X_C = 0$) or that the effluent concentration is unknown:

$$\theta_C \cong \frac{XV}{Q_w X_r} \tag{11.10}$$

If the sludge is wasted from the aeration basin instead of the recycle line:

$$\theta_C = \frac{XV}{Q_w X + (Q - Q_w)X_C} \approx \frac{V}{Q_w} \tag{11.11}$$

Because MLSS is much less concentrated than the sludge in the bottom of a secondary clarifier, this type of wasting is not used frequently.

EXAMPLE

11.3

Problem   An activated sludge wastewater treatment system uses a 8-million-L aeration basin. The mean cell retention time is 12 days. The MLSS is kept at 3100 mg/L, and the recycled activated sludge (RAS) is 11,000 mg/L. What is the wasted activated sludge (WAS) rate if the sludge is wasted from (a) the aeration basin and (b) the recycle line?

Solution

a. When wasting occurs from the aeration basin, the wasting concentration is the same as the concentration in the aeration basin, which is the MLSS concentration. Assuming the effluent solids concentration (X_C) is negligible, the wasting rate is (Equation 11.11)

$$Q_w \approx \frac{VX}{\theta_C X} = \frac{V}{\theta_C} = \frac{8 \times 10^6 \text{ L}}{12 \text{ day}} = 0.67 \text{ million L/d}$$

b. When wasting occurs from the recycle line, the wasting concentration is the same as the concentration in the recycle line, which is the RAS concentration. Assuming the effluent solids concentration (X_C) is negligible, the wasting rate is (Equation 11.10)

$$Q_w \approx \frac{VX}{\theta_C X_r} = \frac{(8 \times 10^6 \text{ L})(3100 \text{ mg/L})}{(12 \text{ day})(11,000 \text{ mg/L})} = 0.19 \text{ million L/d}$$

In this case, wasting from the recycle line reduces the WAS rate by 75%. Therefore, less pumping is required.

The removal of substrate is often expressed in terms of a *substrate removal velocity* (q), defined as

$$q = \frac{\text{Mass substrate removed/Time}}{\text{Mass microorganisms under aeration}}$$

Using the previous notation:

$$q = \frac{\left(\dfrac{(S_0 - S)}{\bar{t}}\right) V}{XV}$$

$$q = \frac{S_0 - S}{X\bar{t}} \tag{11.12}$$

The substrate removal velocity is a rational measure of the substrate removal activity, or the mass of BOD removed in a given time per mass of microorganisms doing the work. This is also called, in some texts, the *process loading factor*, with the clear implication that it is a useful operation and design tool because, as you recall, the basic design parameter for an activated sludge system is its loading. Remember that the food-to-microorganism (F/M) ratio is the ratio of incoming (rather than removed) BOD to MLSS. It is calculated as

$$F/M = \frac{QS_0}{VX} = \frac{S_0}{\bar{t}X} \tag{11.13}$$

The substrate removal velocity can also be derived by conducting a mass balance in terms of substrate on a continuous system with microorganism recycle (Figure 11.13).

$$\begin{bmatrix} \text{Rate of} \\ \text{ACCUMULATION} \end{bmatrix} = [\text{Rate IN}] - [\text{Rate OUT}]$$
$$+ \begin{bmatrix} \text{Rate of} \\ \text{substrate} \\ \text{PRODUCTION} \end{bmatrix} - \begin{bmatrix} \text{Rate of} \\ \text{substrate} \\ \text{CONSUMPTION} \end{bmatrix}$$

The production term is, of course, zero. The rate of consumption is the substrate removal velocity multiplied by the solids concentration and the reactor volume (to obtain the correct units), so the equation reads

$$\frac{dS}{dt} V = QS_0 - QS + 0 - qXV$$

Assuming steady state conditions—that is, $(dS/dt)V = 0$, and solving for q yields

$$q = \frac{S_0 - S}{X\bar{t}}$$

the same equation as before.

The substrate removal velocity can also be expressed as

$$q = \frac{\mu}{Y} = \begin{bmatrix} \dfrac{\text{Mass microorganisms}}{\text{produced/Time}} \\ \dfrac{}{\text{Mass microorganisms}} \\ \text{in the reactor} \end{bmatrix} \times \begin{bmatrix} \dfrac{\text{Mass substrate}}{\text{removed}} \\ \dfrac{}{\text{Mass microorganisms}} \\ \text{produced} \end{bmatrix}$$

and substituting Equations 11.2 and 11.7 yields

$$q = \frac{\mu}{Y} = \frac{\hat{\mu}S}{Y(K_S + S)} = \frac{1}{\theta_C Y} \tag{11.14}$$

Equating these two expressions for substrate removal velocity and solving for $(S_0 - S)$ gives the substrate removal (reduction in BOD):

$$S_0 - S = \frac{\hat{\mu}SX\bar{t}}{Y(K_S + S)} \tag{11.15}$$

Solving for X in Equation 11.12 yields the concentration of microorganisms in the reactor (mixed liquor suspended solids):

$$X = \frac{S_0 - S}{\bar{t}q} \tag{11.16}$$

EXAMPLE

11.4

Problem   An activated sludge system operates at a flow rate (Q) of 400 m³/d with an incoming BOD (S_0) of 300 mg/L. Through pilot plant work, the kinetic constants for this system are determined to be $Y = 0.5$ kg SS/kg BOD, $K_S = 200$ mg/L, $\hat{\mu} = 2$ d^{-1}. A solids concentration of 4000 mg/L in the aeration tank is considered appropriate. A treatment system must be designed that will produce an effluent BOD of 30 mg/L (90% removal). Determine

a. the volume of the aeration tank
b. the sludge age (or mean cell residence time, MCRT)
c. the quantity of sludge wasted daily
d. the *F/M* ratio.

Solution   (The mixed-liquor suspended solids concentration is usually limited by the ability to keep an aeration tank mixed and to transfer sufficient oxygen to the microorganisms. A reasonable value for the solids under aeration would be $X = 4000$ mg/L as stated in the problem.)

a. The volume of a basin is typically calculated from the hydraulic retention time (HRT), which is currently unknown. Therefore, we need another equation to calculate HRT. Because we know the kinetic constants, we can use Equation 11.15

$$S_0 - S = \frac{\hat{\mu} S X \bar{t}}{Y(K_S + S)}$$

rearranged:

$$\bar{t} = \frac{Y(S_0 - S)(K_S + S)}{\hat{\mu} S X}$$

$$= \frac{(0.5 \text{ kg/kg})(300 \text{ mg/L} - 30 \text{ mg/L})(200 \text{ mg/L} + 30 \text{ mg/L})}{(2 \text{ day}^{-1})(30 \text{ mg/L})(4000 \text{ mg/L})}$$

$$= 0.129 \text{ d} = 3.1 \text{ hr}$$

The volume of the tank is then (Equation 11.5) $V = \bar{t} Q = (0.129 \text{ d})(400 \text{ m}^3/\text{d}) = 51.6 \text{ m}^3 \cong 52 \text{ m}^3$.

b. The sludge age is obtained from Equation 11.7:

$$\theta_C = \frac{1}{\mu} = \frac{K_s + S}{\hat{\mu} S} = \frac{200 \text{ mg/L} + 30 \text{ mg/L}}{(2 \text{ d}^{-1})(30 \text{ mg/L})} = 3.8 \text{ days}$$

Alternately, the sludge age is obtained from Equation 11.14

$$q = \frac{\hat{\mu} S}{Y(K_S + S)} = \frac{\mu}{Y} = \frac{1}{\theta_C Y}$$

as

$$\theta_C = \frac{1}{qY}$$

First q must be calculated. Using the kinetic constants to calculate q (Equation 11.14) yields

$$q = \frac{\hat{\mu} S}{Y(K_S + S)} = \frac{(2 \text{ d}^{-1})(30 \text{ mg/L})}{(0.5 \text{ kg/kg})(200 \text{ mg/L} + 30 \text{ mg/L})} = 0.522 \text{ d}^{-1}$$

or equivalently (Equation 11.12)

$$q = \frac{S_0 - S}{X\bar{t}} = \frac{300 \text{ mg/L} - 30 \text{ mg/L}}{(4000 \text{ mg/L})(0.129 \text{ d})}$$

$$= 0.523 \frac{\text{kg BOD removed/d}}{\text{kg SS in the reactor}} = 0.523 \text{ d}^{-1}$$

So

$$\theta_C = \frac{1}{qY} = \frac{1}{(0.522 \text{ d}^{-1})(0.5 \text{ kg/kg})} = 3.8 \text{ days}$$

c. Now that we have the MCRT, we can calculate the sludge wasting rate (Equations 11.9 and 11.10):

$$\theta_C = \frac{XV}{X_r Q_w - (Q - Q_w)X_C} \cong \frac{XV}{X_r Q_w}$$

$$X_r Q_w \cong \frac{XV}{\theta_C} = \frac{(4000 \text{ mg/L})(51.6 \text{ m}^3)(10^3 \text{ L/m}^3)}{(3.8 \text{ day})(10^6 \text{ mg/kg})} \cong 54 \text{ kg/d}$$

d. Using Equation 11.13,

$$F/M = \frac{S_0}{\bar{t}X} = \frac{300 \text{ mg/L}}{(0.129 \text{ day})(4000 \text{ mg/L})} = 0.58 \frac{\text{kg BOD/d}}{\text{kg SS}}$$

■

EXAMPLE
11.5

Problem   Using the same data as in the previous example, what mixed-liquor solids concentration is necessary to attain a 95% BOD removal (i.e., $S = 15$ mg/L)?

Solution   In this case we can use the two equations for the substrate removal velocity, Equation 11.14

$$q = \frac{\hat{\mu} S}{Y(K_S + S)} = \frac{(2 \text{ day}^{-1})(15 \text{ mg/L})}{(0.5 \text{ kg/kg})(200 \text{ mg/L} + 15 \text{ mg/L})} = 0.28 \text{ d}^{-1}$$

and Equation 11.16

$$X = \frac{S_0 - S}{\bar{t}q} = \frac{300 \text{ mg/L} - 15 \text{ mg/L}}{(0.129 \text{ day})(0.28 \text{ d}^{-1})} = 7890 \text{ mg/L}$$

Notice that the MLSS concentration is almost twice as much to halve the effluent concentration. The mean cell residence time would also now be almost twice as long (Equation 11.14)

$$\theta_C = \frac{1}{qY} = \frac{1}{(0.28 \text{ d}^{-1})(0.5 \text{ kg/kg})} = 7.1 \text{ days}$$

While more microorganisms are required in the aeration tank if higher removal efficiencies are to be attained, their concentration depends on the settling efficiency in the final clarifier. If the sludge does not settle well, the return sludge solids concentration is low, and there is no way to increase the solids concentration in the aeration tank. We'll address this more under solids separation.

11.3.4 Gas Transfer

The two principal means of introducing sufficient oxygen into the aeration tank are by bubbling compressed air through porous diffusers (Figure 11.16 on the next page) or by beating air in mechanically (Figure 11.17 on page 371). In both cases the intent is to transfer one gas (oxygen) from the air into the liquid and simultaneously transfer another gas (carbon dioxide) out of the liquid. These processes are commonly called *gas transfer*.

Gas transfer means simply the process of allowing any gas to dissolve in a fluid or the opposite, promoting the release of a dissolved gas from a fluid. One of the critical aspects of the activated sludge system is the supply of oxygen to the suspension of microorganisms in the aeration basin.

Figure 11.18 on page 371 shows a system where air is forced through a tube and a porous diffuser, creating very small bubbles that rise through clean water. Assume that the water in this system does not contain microorganisms and that nothing uses the dissolved oxygen. Of interest here is only how the aeration system performs. What happens to the oxygen once it is dissolved in the water is not important at this time.

The transfer of oxygen takes place through the bubble gas/liquid interface, as shown in Figure 11.19 on page 372. If the gas inside the bubble is air and an oxygen deficit exists in the water, the oxygen transfers from the bubble into the water. In some cases, depending on the concentration of gases already dissolved in the water, transfer of gases, such as CO_2, from solution and into the bubble can also occur. Before discussing gas transfer further, however, it is necessary to briefly review some concepts of gas solubility.

Most gases are only slightly soluble in water; among these are hydrogen, oxygen, and nitrogen. Other gases are very soluble, including sulfur dioxide (SO_2), chlorine (Cl_2), and carbon dioxide (CO_2). Most of these readily soluble gases dissolve and then ionize in the water. For example, when CO_2 dissolves in water, the following reactions occur:

$$CO_2 \text{ (gas)} \leftrightharpoons CO_2 \text{ (dissolved)} + H_2O \leftrightharpoons H_2CO_3$$

$$H_2CO_3 \leftrightharpoons H^+ + HCO_3^- \leftrightharpoons H^+ + CO_3^{2-}$$

All these reactions are in equilibrium, so as more CO_2 is introduced, more CO_2 is dissolved, and eventually more carbonate ion (CO_3^{2-}) is produced. (The equations are driven

Air header

Diffuser

Figure 11.16 Diffused aeration used in the activated sludge system.

to the right.) At any given pH and temperature, the quantity of a given gas dissolved in a liquid is governed by *Henry's law*, which states that

$$S = KP \tag{11.17}$$

where S = solubility of a gas (the maximum amount that can be dissolved),
          mg gas/liter water
       P = partial pressure of the gas, as measured in pounds
          per square inch (psi), kilopascals (kPa), atmospheres,
          or other pressure terms
       K = solubility constant

Figure 11.17   Mechanical aeration used in the activated sludge system.

Figure 11.18   Gas transfer experiment wherein air is bubbled into the tank through a diffuser and the dissolved oxygen is measured by using a probe and dissolved oxygen meter.

Because the units of K are a function of the units of S and P, tabulated values of K vary. Note that K is different from the solubility *coefficient*. The solubility constant defines the equilibrium condition for various species at a constant pressure.

Figure 11.19   Gas transfer into and out of an air bubble in water.

Henry's law states that the solubility is a direct function of the partial pressure of the gas being considered. In other words, if the partial pressure P is doubled, the solubility of the gas S is likewise doubled, etc. However, solubility is influenced by many variables, such as the presence of impurities and the temperature. The effect of temperature on the solubility of oxygen in water was tabulated in Table 8.2.

The partial pressure is defined as the pressure exerted by the gas of interest. For example, at atmospheric pressure (1 atm = 101 kPa), a gas that is 60% O_2 and 40% N_2 has a partial pressure of oxygen of $0.60 \times 101 = 60.6$ kPa and a partial pressure of nitrogen of $0.4 \times 101 = 40.4$ kPa. The total pressure is always the sum of the partial pressures of the individual gases. This is known as *Dalton's law*.

EXAMPLE

11.6   **Problem**   At one atmosphere the solubility of pure oxygen in water is 46 mg/L if the water contains no dissolved solids. What would be the quantity of dissolved oxygen in pure water if the gas above the water is oxygen (Figure 11.20A)? What would be the solubility if the oxygen is replaced by nitrogen (Figure 11.20B) and then by air (Figure 11.20C)? Assume that in all three cases the total pressure of the gases above the water is one atmosphere and the temperature is 20°C.

Solution   Henry's law is (Equation 11.17)

$$S = KP$$

$$46 \text{ mg/L} = K \times 1 \text{ atm} \quad \text{or} \quad K = 46 \text{ mg/L-atm}$$

The water in Figure 11.20A would therefore contain 46 mg/L of dissolved oxygen because the entire pressure is due to oxygen. When the oxygen is replaced by nitrogen, the

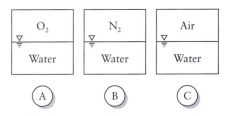

Figure 11.20   Three gases at atmospheric pressure above water. See Example 11.6.

partial pressure of oxygen is zero, and $S = 0$ mg/L. In the third case, since air is 20% oxygen:

$$S = KP = (46 \text{ mg/L-atm})(0.20) = 9.2 \text{ mg/L}$$

Referring to Table 8.2, at 20°C, the solubility of oxygen from air into pure water is indeed 9.2 mg/L.

Henry's law defines the solubility of gas at equilibrium. That is, it is assumed that the gas in contact with the water has had sufficient time to come to a dissolved gas concentration that, over time, will not change. Consider next the changes in dissolved gas concentrations with time.

Using oxygen in air as an example of a gas and water as the liquid, the dissolved oxygen concentration at any time may be visualized as in Figure 11.21. Above the water surface it is assumed that the air is well mixed so that there are no concentration gradients in the air. If the system is allowed to come to equilibrium, the concentration of dissolved oxygen in the water will eventually attain saturation, S. Before equilibrium occurs, however, at some time t, the concentration of dissolved oxygen in the water is C, some value less than S. The difference between the saturation value S and the concentration C is the deficit D, so that $D = (S - C)$ (Equation 8.1). As time passes, the value of C increases until it becomes S, producing a saturated solution and reducing the deficit D to zero.

Below the air/water interface, it is visualized that there exists a diffusion layer through which the oxygen has to pass. The concentration of the gas decreases with depth until the concentration C is reached, and again it is assumed that the water is well mixed so that, except for the thin layer at the air interface, the concentration of dissolved oxygen in the water is everywhere C mg/L. At the interface the concentration increases at a rate of

$$\frac{dC}{dx}$$

where $x =$ thickness of diffusion layer. If this slope is large (dC is big compared to dx), the rate at which oxygen is driven into the water is large. Conversely, when

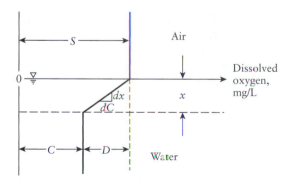

Figure 11.21   Definition sketch for describing gas transfer.

dC/dx is small, C is approaching S, so the rate of change is small. This idea can be expressed as

$$\frac{dC}{dx} \alpha (S - C)$$

Note that as $(S - C)$ approaches zero, $dC/dx \to 0$, and when $(S - C)$ is large, dC/dx is large. It can also be argued that the rate of change in concentration with time must be large when the slope is large (i.e., dC/dx is large). Conversely, as dC/dx approaches zero, the rate of change should also approach zero. The proportionality is

$$\frac{dC}{dt} \alpha (S - C)$$

The proportionality constant is symbolically written as $K_L a$ and given the name *gas transfer coefficient*, and the equation is written as

$$\frac{dC}{dt} = K_L a (S - C)$$

Note that the rate of which the oxygen is driven into the water is high when the driving force, $(S - C)$, is high, and conversely, as C approaches S, the rate decreases. This driving force can be expressed equally well by using the deficit, or the difference between the concentration and saturation (how much oxygen could still be driven into the water). Written in terms of the deficit:

$$\frac{dD}{dt} = -K_L a D$$

Since the deficit is *decreasing* with time as aeration occurs, $K_L a$ is negative.

This equation can be integrated to yield

$$\ln \frac{D}{D_0} = -K_L a t$$

where D_0 is the initial deficit.

$K_L a$ is determined by aeration tests as illustrated in Figure 11.18. The water is first stripped of oxygen, usually by chemical means, so C approaches zero. The air is then turned on and the dissolved oxygen concentration measured with time using a dissolved oxygen meter. If, for example, different types of diffusers are to be tested, the $K_L a$ is calculated for all types using identical test conditions. A higher $K_L a$ implies that the diffuser is more effective in driving oxygen into the water and, thus, presumably the least expensive to operate in a wastewater treatment plant.

$K_L a$ is a function of, among other factors, type of aerator, temperature, size of bubbles, volume of water, path taken by the bubbles, and presence of surface active agents. The explanation of how $K_L a$ is influenced by these variables is beyond the scope of this text.

EXAMPLE

11.7

Problem   Two diffusers are to be tested for their oxygen transfer capability. Tests were conducted at 20°C, using a system as shown in Figure 11.18, with the following results:

Time (min)	Dissolved Oxygen, C (mg/L) Air-Max Diffuser	Wonder Diffuser
0	2.0	3.5
1	4.0	4.8
2	4.8	6.0
3	5.7	6.7

Note that the test does not need to start at $C = 0$ at $t = 0$.

Solution   With $S = 9.2$ mg/L (saturation at 20°C),

t	Air-Max Diffuser $(S - C)$	Wonder Diffuser $(S - C)$
0	7.2	5.7
1	5.2	4.4
2	4.4	3.2
3	3.5	2.5

These numbers are now plotted by first calculating $\ln(S - C)$ and plotting against the time t (Figure 11.22). The slope of the plot is the proportionality factor, or in this case, the gas transfer coefficient $K_L a$. Calculating the slopes, the $K_L a$ for the Air-Max diffuser is found to be 0.27 min^{-1} while the Wonder diffuser has a $K_L a$ of 0.25 min^{-1}. The latter seems to be the better diffuser based on oxygen transfer capability.

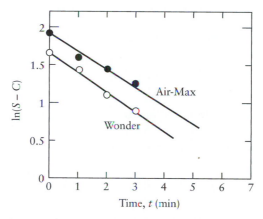

Figure 11.22   Experimental gas transfer data. See Example 10.7.

In Chapter 8 the transfer of oxygen to streamwater is described by the equation

$$\frac{dD}{dt} = -k_2 D$$

where k_2 is the *reaeration constant* and is identical to $K_L a$ as just defined. The reaeration constant is used to describe how oxygen is transferred from the atmosphere into a stream in order to provide sufficient dissolved oxygen for the aerobic aquatic microorganisms. In this chapter the same mechanism is used to describe how oxygen is driven into the liquid in an activated sludge aeration tank.

11.3.5 Solids Separation

The success or failure of an activated sludge system often depends on the performance of the final clarifier, the usual method of separating the solids grown in the aeration tank from the liquid. If the final settling tank is not able to achieve the required return sludge solids, the solid concentration returned to the aeration tank will be low, the MLSS in the aeration tank will drop, and, of course, the treatment efficiency will be reduced because there will be fewer microorganisms to do the work. It is useful to think of the microorganisms in the aeration tank as workers in an industrial plant. If the total number of workers is decreased, the production is cut. Similarly, if fewer microorganisms are available, less work is done.

The MLSS concentration is a combination of the return solids diluted by the influent:

$$X = \frac{Q_r X_r + Q X_0 + r_x V}{Q_r + Q}$$

or

$$X = \frac{\alpha X_r + X_0 + r_x \bar{t}}{\alpha + 1}$$

where α = recycle, or recirculation, ratio $\left(\dfrac{Q_r}{Q}\right)$.

Again assume no solids in the influent ($X_0 = 0$):

$$X = \frac{Q_r X_r + r_x V}{Q_r + Q} = \frac{\alpha X_r + r_x \bar{t}}{\alpha + 1}$$

Results from settling the sludge in a liter cylinder can be used to estimate the return sludge concentration. After 30 min of settling, the solids in the cylinder are at a SS concentration that would be equal to the expected return sludge solids or

$$X_r = \frac{H}{H_S}(X)$$

where X_r = the expected return suspended solids concentration, mg/L
        X = mixed liquor suspended solids, mg/L
        H = height of cylinder, m
        H_S = height of settled sludge, m

Figure 11.23 Filamentous bacteria: (A) *Sphaerotilus natans* (false branching); (B) *Nocardia form* (true branching).

As indicated, when the sludge does not settle well (H_S is larger), the return activated sludge (X_r) becomes thin (low suspended solids concentration), and thus, the concentration of microorganisms in the aeration tank (X) drops. This results in a higher F/M ratio (same food input but fewer microorganisms) and a reduced BOD removal efficiency.

When the microorganisms in the system are very difficult to settle, the sludge is said to be a *bulking sludge*. Often this condition is characterized by a biomass comprised almost totally of filamentous organisms (Figure 11.23), which form a kind of lattice structure with the filaments and refuse to settle. Treatment plant operators must keep a close watch on settling characteristics because a trend toward poor settling can be the forerunner of a badly upset (and hence ineffective) plant.

The settleability of activated sludge is most often described by the sludge volume index (SVI), which is determined by measuring the volume occupied by a sludge after settling for 30 min in a 1-L cylinder (a quick and easy test). It is calculated as

$$\text{SVI} = \frac{(\text{Volume of sludge after 30 min settling, mL/L}) \times 1000 \text{ mg/g}}{\text{mg/L suspended solids}}$$

SVI has units of mL/g, but they are usually left off.

EXAMPLE 11.8

Problem   A sample of mixed liquor was found to have SS = 4000 mg/L, and after settling for 30 min in a 1-L cylinder, it occupied 400 mL. Calculate the SVI.

Solution

$$\text{SVI} = \frac{(400 \text{ mL/L})(1000 \text{ mg/g})}{4000 \text{ mg/L}} = 100$$

SVI values below 100 are usually considered acceptable, but sludges with SVI greater than 200 are badly bulking and will be difficult to settle in the final clarifier. Sludges with SVI values between 100 and 200 may or may not settle well.

The causes of poor settling (high SVI) are not always known, and hence, the solutions are elusive. Wrong or variable F/M ratios, fluctuations in temperature, high concentrations of heavy metals, and deficiencies in nutrients in the incoming wastewater have all been blamed for bulking. Cures include reducing the F/M ratio, changing the dissolved oxygen level in the aeration tank, and dosing with hydrogen peroxide (H_2O_2) to kill the filamentous microorganisms.

11.3.6 Effluent

Effluent from secondary treatment often has a BOD of about 15 mg/L and a suspended solids concentration of about 20 mg/L. This is most often quite adequate for disposal into watercourses, because the BOD and SS of natural streamwater can vary considerably. For example, the BOD can vary from about 2 mg/L to far greater than 15 mg/L. In addition, the effluent from wastewater treatment plants is often diluting the stream.

Before discharge, however, modern wastewater treatment plants are required to disinfect the effluent to reduce further the possibility of disease transmission. Often chlorine is used for the disinfection because it is fairly inexpensive. Chlorination occurs in simple holding basins designed to act as plug-flow reactors (Figure 11.24). The chlorine is injected

© Courtesy of P. Aarne Vesilind

Figure 11.24 Plug-flow reactor for chlorination.

at the beginning of the tank, and it is assumed that all the flow is in contact with the chlorine for 30 minutes. Prior to discharge, excess chlorine, which is toxic to many aquatic organisms, must be removed through *dechlorination*. The most common dechlorination method is bubbling in sulfur dioxide; the chlorine is reduced while the SO_2 is oxidized to sulfate. At this point the flow can be discharged into a receiving stream or other watercourse.

Chlorination does not seem to make much sense from the ecological standpoint because the effluent must be assimilated into the aquatic ecology, and dosing wastewater treatment plant effluents with chlorine results in the production of chlorinated organic compounds, such as chloroform, a carcinogen. In addition, there is no epidemiological evidence that unchlorinated treatment plant effluents cause any public health problems. So why are treatment plants required to chlorinate?

Simple. Few in the regulatory bureaucracy are willing to reduce one layer of protection for humans, even though the chlorination of effluents is an expensive and potentially environmentally harmful practice. It is much simpler to continue to enforce a regulation that may be expensive and cumbersome than to eliminate it and have any chance at all of bearing the brunt of some lawsuit or administrative retribution. The decision is a case of private harm versus public good, and because most people are not utilitarians, they will not decide on the basis of greatest good. The chance of small damage to themselves will overwhelm the real damage to many. A compromise is, of course, to eliminate chlorination and to introduce other methods of disinfection, such as ultraviolet radiation and ozone. In some sensitive areas these techniques are already being used and with time may eventually eliminate chlorination of wastewater plant effluents.

11.4   TERTIARY TREATMENT

There are situations when secondary treatment (even with disinfection) is inadequate to protect the watercourse from harm due to a wastewater discharge. One concern is that nutrients, such as nitrogen and phosphorus, may still cause a problem if the effluent is discharged into a still water body. In addition, if the water downstream of a discharge is used for recreational purposes, a high degree of treatment is necessary, especially in solids and pathogen removal. When upgrading the secondary treatment process will not meet the more stringent discharge limits, the effluent from secondary treatment is treated further to achieve whatever quality is required. Such processes are collectively called tertiary, or advanced, treatment. Some of these processes are discussed below.

11.4.1   Nutrient Removal

Nitrogen removal is accomplished by first treating the waste thoroughly enough in secondary treatment to oxidize all the nitrogen to nitrate. This usually involves longer detention times in secondary treatment, during which bacteria, such as *Nitrobacter* and *Nitrosomonas*, convert ammonia nitrogen to NO_3^-, a process called *nitrification*. These reactions are

$$2NH_4^+ + 3O_2 \xrightarrow{\text{Nitrosomonas}} 2NO_2^- + 2H_2O + 4H^+$$

$$2NO_2^- + O_2 \xrightarrow{\text{Nitrobacter}} 2NO_3^-$$

Alternatives to Phosphate Mining[4]

Phosphorus (P), usually in the form of phosphate, is both a nutrient and a pollutant. It is considered a nutrient when it is found where it is needed—facilitating desired plant growth. It is a pollutant when it is found where it is not needed—facilitating unwanted plant growth that leads to eutrophication in lakes and other waterways.

Phosphate has negative life cycle impacts when it is mined. One of the unwanted byproducts of its mining is hydrogen fluoride. When the Rocky Mountain Phosphate Company in Garrison, Montana, began operations in the early 1960s, it had great difficulty with pollution control and emitted high concentrations of sulfur oxides and fluorides. By the summer of 1965 the vegetation in and around Garrison began to die. The land was barren of wildlife, with most animals dying off because of their inability to walk. Cattle in the farms around Garrison similarly were unable to stand up and soon died. Excess fluoride causes cattle (and people) to have fluorosis, a disease in which their joints swell and they are unable to walk. In effect, the fluoride etches the bones and makes them brittle. The same effect occurs when people drink water that has too much fluoride in it and their teeth become soft and mottled. Finally, in 1967, the federal government intervened and forced the company to cease operations until it installed emission equipment that removed 99.9% of the fluoride.[5]

Additional negative consequences of mining phosphate rock are the release of cadmium and uranium. Phosphate rock may be contaminated with cadmium, a highly toxic metal. Uranium radionuclides may be released from mining, in the processing effluents, and from the use of phosphate rock products (such as fertilizer). Erosion of agricultural soils in areas with heavy fertilizer usage may input the ^{238}U decay radionuclides into drinking water supplies.

Needless to say, getting phosphate to where it is wanted and keeping it from where it is not wanted is a challenge. One very promising technology that is consistent with sustainable materials management principles is the mining of phosphorus from human sewage.[6] The City of Edmonton (pop. approx. 700,000) in Alberta, Canada, is home to the Gold Bar Wastewater Treatment Plant that is successfully operating the world's first industrial-sized nutrient treatment facility to remove phosphorus and other nutrients from municipal sludge and to recycle them into environmentally safe commercial fertilizer. Currently, the technology extracts more than 80% of the phosphorus on average and 10 to 15% of the ammonia from a flow of 500,000 L/d, approximately 20% of the plant's liquid sludge stream.

While many wastewater treatment plants remove phosphorus to reduce their nutrient loading on receiving waters, the nutrients end up in the sludge, which may be dewatered and either land applied or composted to take advantage of the nutrients. But the liquid fraction from dewatering is also nutrient rich and actually adds costs to a system by clogging pipes with a concrete-like scale called "struvite." Struvite is formed from phosphorus and ammonia (nitrogen) combining with magnesium. Rather than fight struvite production, the innovative nutrient recovery process promotes it, processing the liquid fraction from sludge dewatering to recover phosphorus and other nutrients and then converting them into 99.9% pure struvite—a high-quality commercial fertilizer that can generate revenue for the municipality.

The fertilizer contains no heavy metals and has a commercially desirable formulation of nitrogen, phosphorus, and magnesium. Because it dissolves slowly over a nine-month period, it does not leach into water, causing eutrophication. It is useful for turf (golf course) markets, container nurseries, agriculture, and other markets that value slow-release fertilizers. The reactor at Gold Bar produces approximately 500 kg/d. The product is sorted, dried, and bagged onsite and is immediately ready for commercial sale.

In contrast to the negative impacts of mining phosphorus for fertilizer, the struvite recovery process recovers nutrients that would otherwise be released into the environment, helps sewage treatment plants reduce operating costs and meet environmental regulations, and provides municipalities with revenue from the sale of the valuable fertilizer.

In the policy arena, Germany and Sweden have set national targets to obtain phosphorus from internal recycling rather than from mining phosphate rock. On the global scale, if phosphorus mining from sewage were expanded to include manures from agriculture, it is postulated that phosphate mining could be completely displaced.

Both reactions are slow and require sufficient oxygen and long detention times in the aeration tank. The rate of microorganism growth is also low, resulting in low net sludge production, making washout a constant danger. This process removes the oxygen demand caused by the nitrogen.

To remove the nutrient properties of nitrogen, the nitrate must be converted to nitrogen gas. Once the ammonia has been converted to nitrate, it can be reduced to nitrogen gas by a broad range of facultative and anaerobic bacteria, such as *Pseudomonas*. This reduction, called *denitrification*, requires a source of carbon, and methanol (CH_3OH) is often used for that purpose. The sludge containing NO_3^- is placed in an *anoxic* condition, in which the microorganisms use the nitrogen as the electron acceptor, as discussed in Chapter 8. Using the methanol as the source of carbon, the facultative microorganisms convert the nitrate to nitrogen gas, N_2, which then bubbles out of the sludge into the atmosphere. (Sometimes anoxic conditions are not desirable, such as in a primary clarifier. When the sludge in the primary clarifier is not pumped out and all oxygen is depleted in the bottom sludge, the sludge begins to denitrify, creating bubbles that carry some of the solids from the sludge zone.)

Phosphorus removal is accomplished by either chemical or biological means. In wastewater, phosphorus exists as orthophosphate (PO_4^{3-}), polyphosphate (P_2O_7), and organically bound phosphorus. Polyphosphate and organic phosphate may be as much as 70% of the incoming phosphorus load. In the metabolic process, microorganisms use the poly- and organo-phosphates and produce the oxidized form of phosphorus, orthophosphate.

Chemical phosphorus removal requires that the phosphorus be fully oxidized to orthophosphate, and hence, the most effective chemical removal occurs at the end of the secondary biological treatment system. The most popular chemicals used for phosphorus removal are lime, $Ca(OH)_2$, and alum, $Al_2(SO_4)_3$. The calcium ion at high pH will combine with phosphate to form a white, insoluble precipitate called calcium hydroxyapatite that is settled and removed. Insoluble calcium carbonate is also formed and removed and can be recycled by burning in a furnace.

The aluminum ion from alum precipitates as poorly soluble aluminum phosphate, $AlPO_4$, and aluminum hydroxide, $Al(OH)_3$. The hydroxide precipitate forms sticky flocs and helps settle the phosphates. The most common point of alum dosing is in the final clarifier.

The amount of lime or alum required to achieve a given level of phosphorus removal depends on the amount of phosphorus as well as other constituents in the water. The sludge produced can be calculated by using stoichiometric relationships.

Biological methods of phosphorus removal have the advantage of producing fewer solids for disposal. Most biological phosphorus removal systems rely on the fact that microorganisms can be stressed by cutting off their supply of oxygen (anoxic condition) and fooling them into thinking that all is lost and they will surely die! If this anoxic condition is then followed by a sudden reintroduction of oxygen, the cells will start to store the phosphorus in their cellular material and do so at levels far exceeding their normal requirement. This *luxury uptake* of phosphorus is followed by the removal of the cells from the liquid stream, thereby removing much of the phosphorus. Several proprietary processes that use microorganisms to store excess phosphorus can produce effluents that challenge the chemical precipitation techniques.

11.4.2 Further Solids and Organic Removal

Rapid sand filters similar to those in drinking water treatment plants can be used to remove residual suspended solids and to polish the water. The sand filters are typically located between the secondary clarifier and disinfection.

Oxidation ponds are commonly used for BOD removal. The oxidation, or polishing, pond is essentially a hole in the ground, a large pond used to confine the plant effluent before it is discharged into the natural watercourse. Such ponds are designed to be aerobic, and because light penetration for algal growth is important, a large surface area is needed. The reactions occurring within an oxidation pond are depicted in Figure 11.25. (Oxidation ponds are sometimes used as the only treatment step if the waste flow is small; a large pond area is required, however.)

Activated carbon adsorption is another method of BOD removal, but this process has the added advantage that inorganics as well as organics are removed. The mechanism of adsorption on activated carbon is both chemical and physical, with tiny crevices catching and holding colloidal and smaller particles. An activated carbon column is similar to an ion exchange column (as discussed in Chapter 10). It is a completely enclosed tube with dirty water pumped up from the bottom and the clear water exiting at the top. As the carbon becomes saturated with various materials, the dirty carbon must be removed from the column to be regenerated, or cleaned. Removal is often continuous, with clean carbon being added at the top of the column. The regeneration is usually done by heating the carbon in the absence of oxygen, driving off the organic matter. A slight loss in efficiency is noted with regeneration, and some virgin carbon must always be added to ensure effective performance.

The tertiary wastewater treatment processes described are not only complex, they are also expensive, so alternative wastewater management strategies have been sought. One such alternative is to spray secondary effluent on land and allow the soil microorganisms to degrade the remaining organics. Such systems, known as *land treatment*, have a longer history in Europe than in North America. They represent a reasonable alternative to complex and expensive systems, especially for smaller communities where land is plentiful.

There are three major types of land treatment: slow rate infiltration, rapid infiltration, and overland flow. Probably the most promising land treatment method, though, is irrigation. But again the amount of land area required is substantial, and disease transmission is possible because the waste carries pathogenic organisms. Commonly, from 1000 to 2000 hectares (10,000 m^2 to 20,000 m^2) of land are required for every 1 m^3/s of wastewater

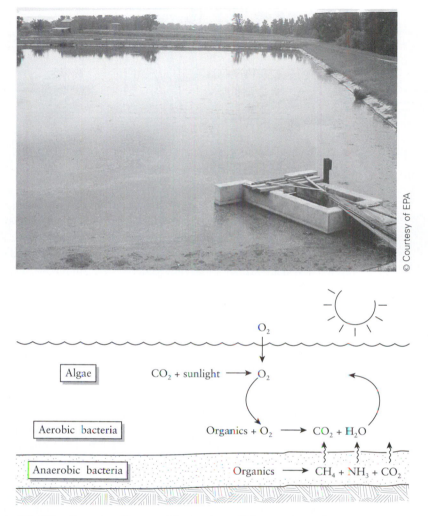

Figure 11.25 Reactions in an oxidation (or stabilization) pond.

flow, depending on the crop and soil. Nutrients such as N and P remaining in the secondary effluent are, of course, beneficial to the crops.

11.4.3 Wetlands

Another option is a constructed wetland. (While existing wetlands have been used for wastewater treatment, the use of constructed wetlands is more common because they can be controlled to perform more reliably and Clean Water Act requirements for discharges to U.S. waterways are avoided.) Constructed wetlands are designed based on natural wetland ecosystems and make use of physical, chemical, and biological processes to remove contaminants. While all wastewater treatment systems rely to some extent on natural processes, such as gravity and biodegradation, constructed wetlands rely primarily on natural

Figure 11.26 Surface flow wetland.

components to maintain the major treatment operations and use mechanical equipment sparingly. This design contrasts with conventional treatment systems, which maintain natural processes with energy-intensive mechanical equipment. As a result, constructed wetlands have lower operation and maintenance requirements and use less energy than conventional systems. In addition, they generate less sludge and provide habitat for wildlife. They do, however, require more land area. Wetlands have been used to treat stormwater runoff, landfill leachate, and wastewater from residences, small communities, businesses (such as truck stops), and rest areas.

The two main categories of wetlands are surface flow and subsurface flow. Surface flow wetlands, which resemble natural wetlands, are more common in wastewater treatment (Figure 11.26). They are also known as free water surface wetlands and open water wetlands. A low-permeability material (such as clay, bentonite, or a synthetic liner) is used on the bottom to avoid groundwater contamination. Subsurface flow wetlands are also known as vegetated submerged bed, gravel bed, reed bed, and root zone wetlands (Figure 11.27). These systems are used to replace septic systems. Because the wastewater is kept below the surface of the medium (which ranges from coarse gravel to sand), these systems reduce mosquito and odor problems. A hybrid system, with both subsurface and surface flow, can also be used.

Constructed wetlands are considered attached-growth biological reactors. The major components of constructed wetland systems are the plants, soils, and microorganisms. The plants serve as support media for microorganisms, provide shade (which reduces algal growth), insulate the water from heat loss, filter solids and pathogens, and provide dissolved oxygen.[1,7] The plants most commonly used are cattails, reeds, rushes, bulrushes, arrowhead, and sedges; the depth of water will dictate which plants will grow well.[1,8] (Due

Figure 11.27 Subsurface flow wetland.

to the high nutrient levels, these plants will typically dominate a system, which is why constructed wetlands do not have the plant diversity of natural wetlands.[8])

Typical design criteria include a detention time of 7 days (longer for cold regions) and a hydraulic loading of 200 m³/ha-d.[8] BOD loadings up to 220 kg/ha-d, depths of 1.5 m, and length-to-width ratios of 3:1 have been used successfully.[7] Water depths in surface flow wetlands have been 100 to 450 mm while bed depths of subsurface flow wetlands have been 0.45 to 1 m.[1] Wetland systems can achieve an impressive effluent of 5 to 10 mg/L BOD and total nitrogen of 5 to 15 mg/L TSS.[8]

11.5   SLUDGE TREATMENT AND DISPOSAL

The slurries produced as underflows from the settling tanks, from both primary treatment and secondary treatment, must be treated and eventually disposed of. Generally speaking, two types of sludges are produced in conventional wastewater treatment plants—*raw primary sludge* and *biological, or secondary, sludge.* The raw primary sludge comes from the bottom of the primary clarifier, and the biological sludge is either solids that have grown on the fixed-film reactor surfaces and sloughed off the media or waste activated sludge grown in the activated sludge system.

The quantity of sludge produced in a treatment plant can be analyzed by using the mass flow technique. Figure 11.28 is a schematic representation of a typical municipal wastewater treatment plant. The symbols in this figure are as follows:

S_0 = influent BOD, kg/h
X_0 = influent suspended solids, kg/h
h = fraction of BOD not removed in the primary clarifier
i = fraction of BOD not removed in activated sludge system
X_e = plant effluent suspended solids, kg/h
k = fraction of influent solids removed in the primary clarifier
ΔX = net solids produced by biological action, kg/h
Y = yield, or the mass of biological solids produced in the aeration tank per mass of BOD destroyed, or

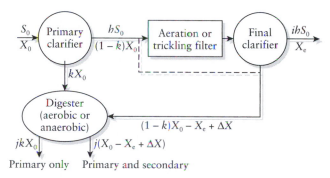

Figure 11.28   Typical wastewater treatment plant used for calculating sludge quantities.

$$\frac{\Delta X}{\Delta S}$$

where $\Delta S = h S_0 - i h S_0$

EXAMPLE
11.9

Problem   A wastewater enters the 24 million L/d treatment plant with a BOD of 200 mg/L and suspended solids of 180 mg/L. The primary clarifier is expected to be 60% effective in removing the solids while it also removes 30% of the BOD. The activated sludge system removes 95% of the BOD that it receives, produces an effluent with a suspended solids concentration of 20 mg/L, and is expected to yield 0.5 kg solids per kilogram of BOD destroyed. The plant is shown schematically in Figure 11.29A. Find the quantity of both raw primary sludge and waste activated sludge produced in this plant.

Solution   The raw primary sludge from the primary clarifier is simply the fraction of solids removed, $k = 0.60$, times the influent solids. The influent solids flow is

$$X_0 = (180 \text{ mg/L}) \, (24 \text{ million L/d}) = 4320 \text{ kg/d}$$

so the production of raw primary sludge is $k X_0 = 0.60(4320 \text{ kg/d}) = 2592 \text{ kg/d}$.

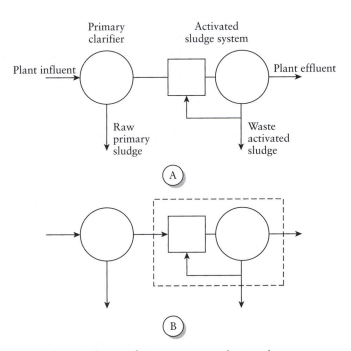

Figure 11.29   Sludge production from primary and secondary treatment.

Drawing a dashed line around and setting up the solids mass balance for the activated sludge system (Figure 11.29B) yields

$$\begin{bmatrix} \text{Rate of} \\ \text{solids} \\ \text{ACCUMULATED} \end{bmatrix} = \begin{bmatrix} \text{Rate of} \\ \text{solids} \\ \text{IN} \end{bmatrix} - \begin{bmatrix} \text{Rate of} \\ \text{solids} \\ \text{OUT} \end{bmatrix}$$

$$+ \begin{bmatrix} \text{Rate of} \\ \text{solids} \\ \text{PRODUCED} \end{bmatrix} - \begin{bmatrix} \text{Rate of} \\ \text{solids} \\ \text{CONSUMED} \end{bmatrix}$$

Assuming steady state and no consumption of solids:

$$0 = [\text{Rate IN}] - [\text{Rate OUT}] + [\text{Rate PRODUCED}] - 0$$

The solids into the activated sludge system are from the solids not captured in the primary clarifier, or $(1 - k)X_0 = 1728$ kg/d. The solids out of the system are of two kinds, the effluent solids and waste activated sludge. The effluent solids are

$$X_e = (20 \text{ mg/L}) \, (24 \text{ million L/d}) = 480 \text{ kg/d}.$$

The waste activated sludge is unknown.

The biological sludge is produced as the BOD is used. The amount of BOD entering the activated sludge system is

$$h S_0 = (0.7)(200 \text{ mg/L}) \, (24 \text{ million L/d}) = 3360 \text{ kg/d}.$$

The activated sludge system is 95% effective in removing this BOD, or $1 - i = 0.95$, so the amount of BOD destroyed within the system is $(1 - i) \times h S_0 = 3192$ kg/d. The yield is assumed to be 0.5 kg solids produced per kg of BOD destroyed, so the biological solids produced must be $Y \times ((1 - i) \times h S_0) = 0.5 \times 3192 = 1596$ kg/d.

Plugging the known and unknown information into the mass balance in kg per day yields

$$0 = [(1 - k)X_0] - [X_e + X_w] + [Y(1 - i)(h S_0)] - 0$$
$$0 = 1728 - [480 + X_w] + 1596 - 0$$

or

$$X_w = 2844 \text{ kg/d}$$

A great deal of money could be saved, and troubles averted, if sludge could be disposed of without further treatment, just as it is drawn off the main process train. Unfortunately, the sludges produced in wastewater treatment have three characteristics that make such

FOCUS ON

Biogas to Energy Using a Fuel Cell

Sewage treatment facilities process human waste 24 hours a day 365 days a year. The volume varies per community, but the supply is constant. Managing this waste is an engineering marvel. Current municipal wastewater treatment results in large amounts of solids. Many wastewater treatment plants use anaerobic digestion to reduce the volume of these solids, resulting in the production of biogas.

Biogas (also known as digester gas) is actually a mixture of gases. Depending on the process, it is made up of methane (60–70%), carbon dioxide (20–30%), and small amounts of hydrogen sulfide, nitrogen, hydrogen, methylmercaptans, and oxygen. In many plants, the methane generated from anaerobic digestion was treated as a pollutant and was burned in flares to convert it to CO_2. It was not unusual to see a large gas flare looming high as if the plant were the next site for the Olympics.

Flaring reduces the negative climate impact of releasing methane because methane has approximately 21 times the global warming potential (i.e., potency) of CO_2 in the environment. But flaring also wastes the energy value inherent in the methane that gives the biogas an energy content of about 600 BTU/ft^3. A more efficient alternative is to capture the energy of biogas using an engine or a fuel cell. For a fuel cell, the biogas, once treated, is converted to pure methane. Molten carbonate fuel cells operating at a high temperature (650 °C) allow the methane to be converted to hydrogen within the fuel cell, eliminating the typical external reforming process and its associated costs and emissions (Figure 11.30).

Like all fuel cells, electricity is generated without combustion by using an electrolyte sandwiched between an anode that receives the fuel and a cathode over which oxygen passes, typically as plain

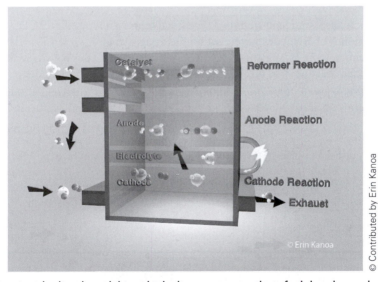

© Contributed by Erin Kanoa

Figure 11.30 An idealized model in which digester gas and air fuel the electrochemical reaction in the fuel cell to produce electricity and heat.

air. The fuel is oxidized at the anode, releasing electrons that move to the cathode via an external circuit. These electrons meet the hydrogen and push charged ions across the electrolyte. The charged ions then move across the ion-conducting electrolyte membrane, completing the electric circuit. The waste products from this process include pure H_2O and nominal amounts of CO_2, because the CO_2 is cycled back into the process.

Reforming reaction to : H_2 create the	$CH_4 + 2H_2O \rightarrow CO_2 + 4H_2$
Reaction at the anode:	$H_2 + CO_3^{2-} \rightarrow H_2O + CO_2 + 2e^-$
Reaction at the cathode:	$0.5O_2 + CO_2 + 2e^- \rightarrow CO_3^{2-}$
The overall cell reaction:	$H_2 + 0.5O_2 \rightarrow H_2O$

The heat generated is recovered to preheat waste sludge, optimizing the anaerobic digestion process. Direct fuel cell (DFC) power plants extract the hydrogen necessary for the electrochemical power generation process from a variety of fuels, including biogas generated in the wastewater treatment process. Because this process does not involve combustion, DFC power plants produce negligible amounts of harmful emissions, such as nitrogen oxides (NO_x) and sulfur oxides (SO_x), as well as significantly reduced CO_2, compared to traditional fossil fuel power plants of equivalent size. A DFC plant sized for a municipal sewage treatment plant could generate 1 MW of power, enough for about 1000 households.

Source: The term Direct Fuel Cell® (DFC®) is patented name and technology of FuelCell Energ. Contributed by Erin Kanoa, Hydrogen & Fuel Cell Specialist, Digital Artist & Designer, UrbanMarmot.

simple disposal unlikely: they are aesthetically displeasing, they are potentially harmful, and they contain too much water. The first two problems are often solved by stabilization, and the third problem requires sludge dewatering. The next three sections cover the topics of stabilization, dewatering, and ultimate disposal.

11.5.1 Sludge Stabilization

The objective of sludge stabilization is to reduce the problems associated with two of the detrimental characteristics listed: sludge odor and putrescence and the presence of pathogenic organisms. Three primary means are used:

- lime
- aerobic digestion
- anaerobic digestion.

Lime stabilization is achieved by adding lime (either as hydrated lime $Ca(OH)_2$ or as quicklime CaO) to the sludge, which raises the pH to about 11 or above. This significantly reduces the odor and helps in the destruction of pathogens. The major disadvantage of lime stabilization is that it is temporary. With time (days), the pH drops and the sludge once again becomes putrescent.

Aerobic digestion is merely a logical extension of the activated sludge system. Waste activated sludge is placed in dedicated aeration tanks for a very long time, and the concentrated solids are allowed to progress well into the endogenous respiration phase, in which food is obtained only by the destruction of other viable organisms (Figure 11.12). This results in a net reduction in total and volatile solids. Aerobically digested sludges are, however, more difficult to dewater than anaerobic sludges.

Anaerobic digestion is the third commonly employed method of sludge stabilization. The biochemistry of anaerobic decomposition of organics is illustrated in Figure 11.31. Note that this is a staged process, with the dissolving of organics by extracellular enzymes being followed by the production of organic acids by a large and hearty group of anaerobic microorganisms known, appropriately enough, as the *acid formers*. The organic acids are, in turn, degraded further by a group of obligate anaerobes called *methane formers* (Chapter 8). These microorganisms are the prima donnas of wastewater treatment, getting upset at the least change in their environment. The success of anaerobic treatment boils down to the creation of a suitable condition for the methane formers. Because they are obligate anaerobes, they are unable to function in the presence of oxygen and are very sensitive to environmental conditions, such as temperature, pH, and the presence of toxins. If a digester goes sour, the methane formers have been inhibited in some way. The acid formers, however, keep chugging away, making more organic acids. This has the effect of further lowering the pH and making conditions even worse for the methane formers. A sick digester is therefore difficult to cure without massive doses of lime or other antacids.

Often the reason for such problems is the difficulty of mixing the sludge in the digester. No good mixing techniques have been developed for the digesters usually used in American wastewater treatment plants. Most American treatment plants have two kinds of anaerobic digesters: primary and secondary (Figure 11.32 on the next page). The primary digester is covered, heated, and mixed to increase the reaction rate. The temperature of the sludge is usually about 35°C. Secondary digesters are not mixed or heated and are used for storage of gas and for concentrating the sludge by settling. As the solids settle, the liquid supernatant is pumped back to the main plant for further treatment. The cover of the secondary digester often floats up and down, depending on the amount of gas stored. The gas is high enough in methane to be used as a fuel and is, in fact, usually used to heat the primary digester.

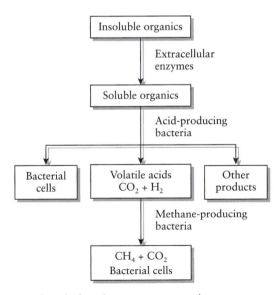

Figure 11.31   Anaerobic sludge digestion process dynamics.

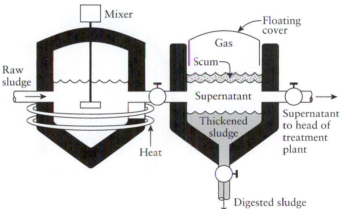

Figure 11.32   Two-stage anaerobic digestion: primary and secondary anaerobic digesters.

In Europe egg-shaped digesters (Figure 11.33 on the next page) have found favor mainly because of the ease of mixing. The digester gas is pumped into the bottom, and an effective circulation pattern is set up.

Anaerobic digesters should achieve substantial pathogen reduction because they are run at elevated temperatures, but the process is not perfect, and many pathogenic organisms survive. An anaerobic digester cannot therefore be considered a method of sterilization.

11.5.2   Sludge Dewatering

In most wastewater plants dewatering is the final method of volume reduction prior to ultimate disposal. In the United States, three dewatering techniques are presently widely used: sand beds, belt filters, and centrifuges. Each of these is now discussed.

Sand beds have been in use for a great many years and are still the most cost-effective means of dewatering when land is available and labor costs are not exorbitant. As shown

Figure 11.33   Egg-shaped anaerobic digesters.

in Figure 11.34, sand beds consist of tile drains in gravel covered by about 0.25 m of sand. The sludge to be dewatered is poured on the beds to about 15 cm deep. Two mechanisms combine to separate the water from the solids: seepage and evaporation. Seepage into the sand and through the tile drains, although important in the total volume of water extracted, lasts for only a few days. As drainage into the sand ceases, evaporation takes over, and this process is actually responsible for the conversion of liquid sludge to solid. In some northern areas sand beds are enclosed under greenhouses to promote evaporation as well as prevent rain from falling onto the beds.

For mixed digested sludge, the usual design is to allow for 3 months' drying time. Some engineers suggest that this period be extended to allow a sand bed to rest for a month after the sludge has been removed, which seems to be an effective means of increasing the drainage efficiency once the sand beds are again flooded. Raw primary sludge will not drain well on sand beds and will usually have an obnoxious odor. It is, therefore, seldom dried on sand beds. Raw secondary sludges have a habit of either seeping through the sand or

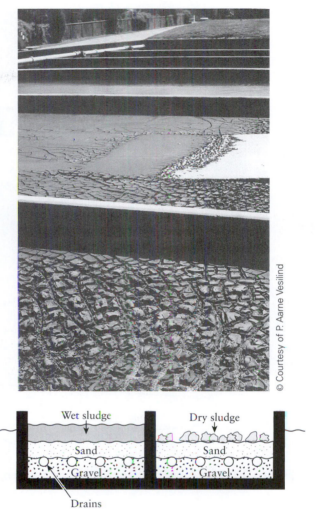

Figure 11.34 Sand drying bed for sludge dewatering.

clogging the pores so quickly that no effective drainage takes place. Aerobically digested sludge can be dried on sand but usually with difficulty.

If dewatering by sand beds is considered impractical, mechanical dewatering techniques must be employed. One mechanical method of dewatering is the *belt filter*, which operates as both a pressure filter and a gravity filter (Figure 11.35 on the next page). As the sludge is introduced onto a moving belt, the free water surrounding the sludge solid particles drips through the belt. The solids are retained on the surface of the belt. The belt then moves into a dewatering zone where the sludge is squeezed between two belts, forcing the *filtrate* from the sludge solids. The dewatered solids, called the *cake*, are then discharged when the belts separate.

Another widely used mechanical method of dewatering is the *centrifuge*. The most common centrifuge used is the solid bowl decanter, which consists of a bullet-shaped body

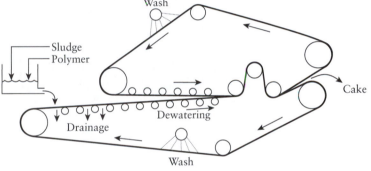

Figure 11.35 A belt filter used for sludge dewatering.

rotating around its long axis. When the sludge is placed into the bowl, the solids settle out under a centrifugal force that is about 500 to 1000 times gravity. They are then scraped out of the bowl by a screw conveyor (Figure 11.36). The solids coming out of a centrifuge are known as *cake*, as in filtration, but the decanted liquid is known as *centrate*.

The objective of a dewatering process is two-fold: to produce a solids cake of high solids concentration and to make sure all the solids, and only the solids, end up in the cake. Unfortunately, the centrifuge is not an ideal device, and some of the solids still end up in the centrate and some of the liquid in the cake. The performance of centrifuges is measured by sampling the feed, centrate, and cake coming out of an operating machine. The hardware is such that it is very difficult to measure the flow rates of the centrate and cake, and only the solids concentrations can be sampled for these flows. When solids recovery and cake solids (purity) are to be calculated, it is necessary to use mass balances.

The centrifuge is a two-material black box, as introduced in Chapter 3. Assuming steady-state operation and recognizing that no liquid or solids are produced or consumed in the centrifuge, the mass balance equation is

$$[\text{Rate of material IN}] = [\text{Rate of material OUT}]$$

Figure 11.36   A solid bowl decanter centrifuge used for sludge dewatering.

The volume balance in terms of the sludge flowing in and out is

$$Q_0 = Q_k + Q_c$$

and the mass balance in terms of the sludge solids is

$$Q_0 C_0 = Q_k C_k + Q_c C_c$$

where $Q_0 =$ flow of sludge as the feed, volume/unit time
$\qquad Q_k =$ flow of sludge as the cake, volume/unit time
$\qquad Q_c =$ flow of sludge as the centrate, volume/unit time

and subscripts 0, k, and c refer to feed, cake, and centrate solids concentrations, respectively.

The recovery of solids, as defined previously, is

$$\text{Solids recovery} = \frac{\text{Mass of dry solids as cake}}{\text{Mass of dry feed solids}} \times 100$$

$$= \frac{C_k Q_k}{C_0 Q_0} \times 100$$

Solving the first material balance for Q_c and substituting into the second material balance yields

$$Q_k = \frac{Q_0(C_0 - C_c)}{C_k - C_c}$$

Substituting this expression into the equation for solids recovery yields

$$\text{Solids recovery} = \frac{C_k(C_0 - C_c)}{C_0(C_k - C_c)} \times 100$$

This expression allows for the calculation of solids recovery by using only the concentration terms.

EXAMPLE
11.10

Problem   A wastewater sludge centrifuge operates with a feed of 40 L/min and a feed solids concentration of 1.2%. The cake solids concentration is 22% solids, and the centrate solids concentration is 500 mg/L. What is the recovery of solids?

Solution   The solids concentrations must first be converted to similar units. Recall that if the density of the solids is almost one (a good assumption for wastewater solids), 1% solids = 10,000 mg/L.

Solids Concentration (mg/L)

Feed solids	12,000
Cake solids	220,000
Centrate solids	500

$$\text{Solids recovery} = \frac{220,000(12,000 - 500)}{12,000(220,000 - 500)} \times 100 = 96\%$$

The centrifuge is an interesting solids separation device because it can operate at almost any cake solids and solids recovery, depending on the condition of the sludge and the flow rate. For any given sludge, a centrifuge will have an *operating curve* as shown in Figure 11.37. As the flow rate is increased, the solids recovery begins to deteriorate because the residence time in the machine is shorter, but at the same time the cake solids improve because the machine discriminates among the solids and spews out only those solids that are easy to remove, producing a high solids concentration in the cake. At lower feed flows, high residence times allow for complete settling and solids removal, but the cake becomes

Figure 11.37   Centrifuge operating curve.

wetter because all the soft and small particles that carry a lot of water with them are also removed.

This seems like a difficult problem. It appears that one can only move up and down on the operating curve, trading off solids recovery for cake solids. There are, however, two ways of moving the entire curve to the right, that is, obtaining a better operating curve so that both solids recovery and cake solids are improved. The first method is to increase the centrifugal force imposed on the solids.

The creation of the centrifugal force is best explained by first recalling Newton's law:

$$F = ma$$

where F = force, N
   m = mass, kg
   a = acceleration, m/s^2

If the acceleration is due to gravity,

$$F = mg$$

where g = acceleration due to gravity, m/s^2.

When a mass is spun around, it requires a force toward the center of rotation to keep it from flying off into space. This force is calculated as

$$F_c = m(w^2 r)$$

where F_c = centrifugal force, N
   m = mass, kg
   w = rotational speed, rad/s
   r = radius of rotation

The term $w^2 r$ is called centrifugal acceleration and has units of meters per seconds squared (m/s^2). The number of *gravities*, G, produced by a centrifuge is

$$G = \frac{w^2 r}{g}$$

where w = rotational speed, radians/s (recall that each rotation equals 2π radians)
   r = inside radius of the centrifuge bowl

Increasing the speed of the bowl therefore increases the centrifugal force and moves the operating curve to the right.

The second method of simultaneously improving cake solids and solids recovery is to condition the sludge with chemicals prior to dewatering. This is commonly done by using organic macromolecules called *polyelectrolytes*. These large molecules, with molecular weights of several million, have charged sites on them that appear to attach to sludge particles and bridge the smaller particles so that larger flocs are formed. These larger flocs are able to expel water and compact to small solids, yielding both a cleaner centrate and a more compact cake. This, in effect, moves the operating curve to the right.

11.5.3 Ultimate Disposal

The options for ultimate disposal of sludge are limited to air, water, and land. Strict controls on air pollution usually make incineration expensive, although this may be the only option for some communities. Disposal of sludges in deep water (such as oceans) is banned in the United States due to adverse detrimental effects on aquatic ecology. Land disposal can be achieved by either dumping into a landfill or spreading the sludge over land and allowing natural biodegradation to assimilate the sludge back into the environment. Finally, sludge can be disposed of by either giving it away or, better yet, selling it as a valuable fertilizer and soil conditioner.

Strictly speaking, incineration is *not* a method of disposal but rather a sludge treatment step in which the organics are converted to H_2O, CO_2, and many other partially oxidized compounds, and the inorganics drop out as a nonputrescent residue. Two types of incinerators have found use in sludge treatment: multiple-hearth and fluid-bed. The *multiple-hearth incinerator*, as the name implies, has several hearths stacked vertically, with rabble arms pushing the sludge progressively downward through the hottest layers and finally into the ash pit (Figure 11.38). The *fluid-bed incinerator* has a boiling cauldron of hot sand into which the sludge is pumped. The sand provides mixing and acts as a thermal flywheel for process stability.

The second method of disposal—land spreading—depends on the ability of land to absorb sludge and to assimilate it into the soil matrix. This assimilative capacity depends on such variables as soil type, vegetation, rainfall, slope, and the sludge composition (in particular, the amount of nitrogen, phosphorus, and heavy metals). Generally, sandy soils with lush vegetation, low rainfall, and gentle slopes have proven most successful. Mixed digested sludges have been spread from tank trucks, and activated sludges have been sprayed from both fixed and moving nozzles. The application rate has been variable, but 0.025 dry tonnes/m^2/yr is not an unreasonable estimate. Most unsuccessful land application systems can be traced to overloading the soil. Given enough time (and absence of toxic materials), most soils will assimilate sprayed liquid sludge.

Transporting liquid sludge is often expensive, however, and volume reduction by dewatering is necessary. The solid sludge can then either be deposited on land and worked into the soil or be deposited in trenches and covered. The sludge seems to assimilate rapidly, with no undue leaching of nitrates or toxins.

The toxicity of sludge can be interpreted in several ways: toxicity to vegetation, toxicity to animals who eat the vegetation (including people), and contamination of groundwater. Most domestic sludges do not contain sufficient toxins, such as heavy metals, to

Figure 11.38 A multiple-hearth incinerator.

cause harm to vegetation. The total body burden of toxins is of some concern, however, and has resulted in regulations (the part 503 Rule) limiting the types of sludges that can be applied to land (Chapter 1). The most effective means of controlling such toxicity seems to be to prevent these materials from entering the sewerage system in the first place by reducing the contribution from industrial waste. Strongly enforced sewer ordinances are, thus, necessary and can be cost-effective.

11.6 SELECTION OF TREATMENT STRATEGIES

Who is to choose which of these treatment strategies will be used for given wastewater and aquatic conditions? The simple (but wrong) answer is decision makers, those ambiguous, anonymous bureaucrats who decide how to spend our money. In fact, decisions such as which treatment strategy to adopt are made by engineers who define what is to be achieved and then propose the treatment to accomplish the objective.

This process is not unlike the way architects work. A building is commissioned, and the architect first spends considerable time with the client trying to understand just what the client intends to do with the building. But then the architect designs the building based on the architect's own tastes and ideas. In a similar manner, the design engineers establish what the objectives of treatment are and then design a facility to meet these objectives.

In this role engineers have considerable latitude and hence responsibility. They are asked by society to design something that not only works but something that works at the least possible cost, does not prove to be a nuisance to its neighbors, and looks nice as well. In this role engineers become the repositories of the public trust.

Because of this public and environmental responsibility, engineering is a profession, and as such, all engineers are expected to adhere to high professional standards. Historically, and especially overseas, engineering has been an honored and highly respected profession. Aleksandr I. Solzhenitsyn describes the Russian engineers of his day in this manner:

> An engineer? I had grown up among engineers, and I could remember the engineers . . . their open, shining intellects, their free and gentle humor, their agility and breadth of thought, the ease with which they shifted from one engineering field to another, and for that matter, from technology to social concerns and art.[9]

A lot to live up to, isn't it?

If engineers are indeed a special breed, do they also have social responsibilities that exceed their roles as private citizens? Is an engineer simply a robot that does certain chores and after 5 o'clock becomes a person with no more social responsibilities than any other person? Or is the engineer's special training, experience, and status in society such that he/she cannot help being a special person whom others will look up to and whose opinions will be respected?

If it is the latter, the role of the environmental engineer takes on special significance. Not only are environmental engineers responsible for performing a job, they also have another "client": the environment. Explaining the conflicts that arise in working with environmental concerns would be a major role of the environmental engineer, and his/her responsibilities would far exceed those of an ordinary citizen.

FOCUS ON

Getting Sanitation to Bolivia

In the developed world the availability of toilets is taken for granted. We notice them only when they are either unavailable or malfunctioning. But in much of the world proper toilets are a luxury, and most people live with no way to properly manage human waste. Because many diseases, such as cholera, typhoid, and dysentery, can be transmitted by improper sanitation, millions of people, particularly children, die each year due to the lack of proper sanitary facilities.[10]

For example, in rural Bolivia, many people do not have access to improved sanitation. At first, you might think that the solution is just to go to Bolivia and start setting up plumbing and sanitation systems, much like the ones we use in our society. It is not that easy. Besides the basic technological issues that come with implementing sanitation systems (such as the need for *consistent* water service for plumbed systems), many societal and cultural issues complicate the situation. For instance, many Bolivians take offense to digging and defecating into the earth; the planet is seen as a respectable entity, not as a vessel for human wastes. And toilets are viewed as a luxury item. Because many rural people are content with their current situations, the demand for improved sanitation is rather low. Furthermore, in the Bolivian political landscape, local governments decide how to allocate funds for communal use. Because toilets are viewed as a personal luxury, very few funds are available for people to install sanitation systems in their homes.

These challenges do not mean that it is impossible for Bolivia to build sanitation facilities. Groups such as Water for People and UNICEF are making strides to cut the number of people lacking sanitation in half by the year 2015.[11] Water for People is working extensively in Bolivia to help improve its sanitation conditions. The organization has made a variety of sanitation options available, including improved pit latrines, a water-based option, and a composting option, allowing communities to choose which is right for them. They are also working directly with communities to start dialogues on new technologies that may be best suited for particular areas and are promoting the view of human waste as a resource through ecological sanitation. This area-specific technique of working with people and providing options should make it easier to improve sanitation for all the people of Bolivia and elsewhere. We have to remember that understanding a different worldview is just as important, if not more so, than understanding the technical issues surrounding sanitation services. Constructing something doesn't mean it will be used.

Sometimes engineers are placed in what might appear to be anti-environmental positions. A classical case would be engineers involved in the construction of a massive secondary treatment plant for a large coastal city where the discharge is to the deep waters of the bay. Even though the efficiency of treatment provided by a primary treatment facility is quite adequate to meet all environmental concerns about public health and water quality, the political situation might demand the expenditure of public funds for the construction of secondary treatment facilities. These billions (yes, billions) of dollars could be well spent elsewhere with no detrimental effect on the aquatic environment. The new secondary plant, however, is a political necessity. Should the engineer speak out in favor of *less* environmental control?

PROBLEMS

11.1 The following data were reported on the operation of a wastewater treatment plant:

Constituent	Influent (mg/L)	Effluent (mg/L)
BOD_5	200	20
SS	220	15
P	10	0.5

a. What percent removal was experienced for each of these constituents?
b. What kind of treatment plant would produce such an effluent? Draw a block diagram showing one configuration of the treatment steps that would result in such a plant performance.

11.2 Describe the condition of a primary clarifier one day after the raw sludge pumps broke down. What do you think would happen?

11.3 One operational problem with trickling filters is *ponding*, the excessive growth of slime on the rocks and subsequent clogging of the spaces so that the water no longer flows through the filter. Suggest some cures for the ponding problem.

11.4 One problem with sanitary sewers is illegal connections from roof drains. Suppose a family of four, living in a home with a roof area of $21 \text{ m} \times 10 \text{ m}$ ft, connects the roof drain to the sewer. Assume that 3 cm of rain falls over 1 day.

a. What percentage increase will there be in the flow from their house over the dry weather flow (assumed at 200 L/capita-day)?
b. What is so wrong with connecting the roof drains to the sanitary sewers? How would you explain this to someone who has just been fined for such illegal connections? (Don't just say, "It's the law." Explain *why* there is such a law.) Use ethical reasoning to fashion your argument.

11.5 What secondary and, if necessary, tertiary unit operations are necessary to treat the following wastes to effluent levels of $BOD_5 = 20$ mg/L, SS = 20 mg/L, P = 1 mg/L.

Waste	BOD_5 (mg/L)	SS (mg/L)	P (mg/L)
a. Domestic	200	200	10
b. Chemical industry	40,000	0	0
c. Pickle cannery	0	300	1
d. Fertilizer mfg.	300	300	200

11.6 The success of an activated sludge system depends on the settling of the solids in the final settling tank. Suppose the sludge in a system starts to bulk (not settle very well), and the suspended solids concentration of the return activated sludge drops from 10,000 mg/L to 4000 mg/L. (Do not answer quantitatively.)

a. What will this do to the mixed-liquor suspended solids?
b. What will this, in turn, do to the BOD removal? Why?

11.7 The MLSS in an aeration tank is 4000 mg/L. The flow from the primary settling tank is 0.2 m³/s with a SS of 50 mg/L, and the return sludge flow is 0.1 m³/s with a SS of 6000 mg/L. Do these two sources of solids make up the 4000 mg/L SS in the aeration tank? If not, how is the 4000-mg/L level attained? Where do the solids come from?

11.8 A 1-L cylinder is used to measure the settleability of 0.5% suspended solids sludge. After 30 min the settled sludge solids occupy 600 mL. Calculate the SVI.

11.9 What measures of stability would you need if a sludge from a wastewater treatment plant was to be

a. placed on the White House lawn?
b. dumped into a trout stream?

c. sprayed on the playground?

d. spread on a vegetable garden?

11.10 A sludge is thickened from 2000 mg/L to 17,000 mg/L. What is the reduction in volume, in percent? (Use a black box and material balance!)

11.11 Sludge age (sometimes called mean cell residence time) is defined as the mass of sludge in the aeration tank divided by the mass of sludge wasted per day. Calculate the sludge age if the aeration tank has a hydraulic retention time of 2 hr, a suspended solids concentration of 2000 mg/L, the flow rate of wastewater to the aeration basin is 1.5 m³/min, the return sludge solids concentration is 12,000 mg/L, and the flow rate of waste activated sludge is 0.02 m³/min.

11.12 The block diagram in Figure 11.39 shows a secondary wastewater treatment plant.

a. Identify the various unit operations and flows, and state their purpose or function. Why are they there, or what do they do?

b. Suppose you are a senior engineer in charge of wastewater treatment for a metropolitan region. You have retained a consulting engineering firm to design the plant shown. Four firms were considered for the job, and this firm was selected, over your objection, by the metropolitan authority board. There are indications of significant political contributions from the firm to the members of the board. While this plant will probably work, it clearly will not be the caliber of plant you desired,

and you will probably spend a lot of (public) money upgrading it. What responsibilities do you, a professional engineer, have in this matter? Analyze the problem, identify the people involved, state the options you have, and come to a conclusion. All things considered, what ought you to do?

11.13 Suppose a law requires that all wastewater discharges to a watercourse meet the following standards:

BOD	Less than 20 mg/L
Suspended solids	Less than 20 mg/L
Phosphorus (total)	Less than 0.5 mg/L

Design treatment plants (block diagrams) for the following wastes:

Waste 1	BOD =	250	mg/L
	SS =	250	mg/L
	P =	10	mg/L
Waste 2	BOD =	750	mg/L
	SS =	30	mg/L
	P =	0.6	mg/L
Waste 3	BOD =	30	mg/L
	SS =	450	mg/L
	P =	20	mg/L

11.14 Draw a block diagram for a treatment plant, showing the necessary treatment steps for achieving the desired effluent.

	Influent (mg/L)	Effluent (mg/L)
BOD	1000	10
Suspended Solids	10	10
Phosphorus	50	5

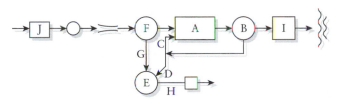

Figure 11.39   Schematic of a secondary wastewater treatment plant. See Problem 11.12.

Be sure to use only the treatment steps necessary to achieve the desired effluent. Extraneous steps will be considered wrong.

11.15 A 90 cm-diameter centrifuge for dewatering sludge is rotated at 1000 rotations per minute. How many gravities does this machine produce? (Remember that each rotation represents 2π radians.)

11.16 A gas transfer experiment results in the following data:

Time (min)	Dissolved Oxygen (mg/L)
0	2.2
5	4.2
10	5.0
15	5.5

The water temperature is 15°C. What is the gas transfer coefficient $K_L a$?

11.17 Derive the equation for calculating centrifuge solids recovery from concentration terms. That is, recovery is to be calculated only as a function of the solids concentrations of the three streams: feed, centrate, and cake. Show each step of the derivation, including the material balances.

11.18 Counterattacks by people who do not agree with the environmental movement have been an interesting phenomenon in the environmental ethics field. These people do not understand nor appreciate the values that most people hold about environmental concerns, and try to discredit the environmentalists. Often this is too much like shooting fish in a barrel. It's too easy. The environmental movement is full of dire warning, disaster predictions, and apocalypses that never happened. As an example of this genre of writing, read the first chapter in one of the classics in the field, *The Population Bomb* by Paul Erlich. Note the date it was written, and be prepared to discuss the effect this book probably had on public opinion.

11.19 A community with a wastewater flow of 10 mgd is required to meet effluent standards of 30 mg/L for both BOD_5 and SS. Pilot plant results with an influent of $BOD_5 = 250$ mg/L estimate the kinetic constants at $K_S = 100$ mg/L, $\hat{\mu} = 0.25$ d^{-1} and $Y = 0.5$. It is decided to maintain the MLSS at 2000 mg/L. What are the hydraulic retention time, the sludge age, and the required tank volume?

11.20 Wastewater from a peach packaging plant was tested in a pilot activated sludge plant, and the kinetic constants were found to be $\hat{\mu} = 3$ day^{-1}, $Y = 0.6$, $K_S = 450$ mg/L. The influent BOD is 1200 mg/L, and a flow rate of 19,000 m^3/day is expected. The aerators to be used will limit the suspended solids in the aeration tank to 4500 mg/L. The available aeration volume is 5100 m^3.

a. What efficiency of BOD removal can be expected?

b. Suppose we found that the flow rate is actually much higher, say 35,000 m^3/d, and the flow is more dilute, $S_0 = 600$ mg/L. What removal efficiency might we expect now?

c. At this flow rate and S_0, suppose we could not maintain 4500 mg/L solids in the aeration tank (why?). If the solids were only 2000 mg/L, and if 90% BOD removal was required, how much extra aeration tank volume would be needed?

11.21 An activated sludge system has a flow of 4000 m^3/day with $X = 4000$ mg/L and $S_0 = 300$ mg/L. From pilot plant work the kinetic constants are $Y = 0.5$, $\hat{\mu} = 3$ d^{-1}, $K_S = 200$ mg/L. We need to design an aeration system that will remove 90% of the BOD_5. Specifically, we need to know

a. the volume of the aeration tank
b. the sludge age
c. the amount of waste activated sludge

11.22 A centrifuge manufacturer is trying to sell your city a new centrifuge that is supposed to dewater your sludge to 35% solids. You know, however, that their machines will most likely achieve only about 25% solids in the cake.

a. How much extra volume of sludge cake will you have to be able to handle and dispose of (35% versus 25%)?

b. You know that none of the competitors can provide a centrifuge that works any better, and you like this machine and convince the city to buy it. A few weeks before the purchase is final, you receive a set of drinking glasses with the centrifuge company logo on them, along with a thank you note from the salesman. What do you do? Why? Use ethical reasoning to develop your answer.

11.23 In the Westminster Abbey in London are honored many of the immortal characters of English history, each of whom made a significant contribution to Western civilization. Numbered among these, perhaps inadvertently, is one Thomas Crapper, the inventor of the pull-and-let-go flush toilet. A manhole cover, shown in Figure 11.40, is Crapper's memorial. The theory is that the American soldiers during the First World War were so enamored and impressed with Crapper's invention that they brought both the idea and the name back to the United States. In England, if you ask someone for directions to the crapper they will not understand you. You have to ask for the loo. So in his own country, poor Thomas is without honor. But in the colonies his fame and glory live on.

For this problem, investigate the flushing mechanism of an ordinary toilet and explain its operation, using pictures and a verbal description.

11.24 Your city is contemplating placing its wastewater sludge on an abandoned strip mine to try to reclaim the land for productive use, and you, as the city consulting engineer, are asked for an opinion. You know that the city has been having difficulties with its anaerobic digesters, and although it can probably meet the Class B standards on pathogens most of the time, it probably will not be able to do so consistently. Reclaiming the strip mine is a positive step, as is the low cost of disposal. The city engineer tells you that if they cannot use the strip mine disposal method, they must buy a very expensive incinerator to meet the new EPA regulation. The upgrade of the

© Courtesy of P. Aarne Vesilind

Figure 11.40   Thomas Crapper's manhole cover from the Westminster Abbey.

anaerobic digesters will also be prohibitively expensive for this small town.

What do you recommend? Analyze this problem in terms of the affected parties, possible options, and final recommendations for action.

11.25 The state had promised to come down hard on Domestic Imports Inc. if it violates its NPDES permit one more time, and Sue is on the spot. It is her treatment works, and she is expected to operate it. Her requests for improvements and expansion had been turned down, and she had been told that the treatment works had sufficient capacity to treat the waste. This is true, if the average flow is used in the calculation. Unfortunately, the manufacturing operation is such that slug flows could come unexpectedly, and the biological system simply cannot adapt fast enough. The effluent BOD can shoot sky high for a few days, and then settle back again to below NPDES standards. If the state shows up on one of the days during the upset, she is in big trouble.

One day she is having lunch with a friend of hers, Emmett, who works in the quality control lab. Sue complains to Emmett about her problem. If only she could figure out how to reduce the upsets when the slug loads come in. She has already talked to the plant manager to get him to assure her that the slug loads would not occur, or to build an equalization basin, but both requests were denied.

"The slug load is not your problem, Sue," Emmett suggested. "It's the high BOD that results from the slug load upsetting your plant."

"OK, wise guy. You are right. But that doesn't help me."

"Well, maybe I can. Your problem is that you are running high BOD, and you need to reduce it. Suppose I told you that you could reduce the BOD by simply bleeding a chemical additive into the line, and that you can add as much of this chemical as you need to reduce your BOD. Would you buy it?" asked Emmett.

"Just for fun, suppose I would. What is this chemical?"

"I have been playing around with a family of chemicals that slows down microbial metabolism, but does not kill the microbes. If you add this stuff to your line following the final clarifier, your effluent BOD will be reduced because the metabolic activity of the microbes will be reduced; hence, they will use less oxygen. You can set up a small can of it and whenever you see one of those slugs coming down from manufacturing you start to bleed in this chemical. Your BOD will stay within the effluent limits, and in a few days when things have calmed down, you turn it off. Even if the state comes to visit, there is no way they can detect it. You are doing nothing illegal. You are simply slowing down the metabolic activity."

"But is this stuff toxic?" asks Sue.

"No, not at all. It shows no detrimental affect in bioassay tests. You want to try it out?"

"Wait a minute. This is complicated. If we use this magic bullet of yours, our BOD will be depressed, we pass the state inspection, but we haven't treated the wastewater. The oxygen demand will still occur in the stream."

"Yes, but many miles and many days downstream. They will never be able to associate your discharge with a fish kill—if, in fact, this would occur. What do you say? Want to give it a try?"

Assuming that it is highly unlikely that Sue will ever get caught adding this chemical to the effluent, why should she not do it? What values is she struggling with? Who are the parties involved, and what stakes do they have in her decision? What are all her options, and which one do you recommend?

11.26 What are the principal mechanisms to remove and transform pollutants in surface water flow wetlands?

11.27 What are the principal mechanisms to remove and transform pollutants in subsurface water flow wetlands?

11.28 How many gallons per day of sludge will be generated if a 110 L/s water treatment plant thickens 7 tonnes/d of sludge to 9% solids?

11.29 A municipal water treatment plant uses two sedimentation tanks operating in parallel as part of a lime-soda softening process. Each clarifier has a diameter of 12 m, a sidewater depth of 5 m, and an outlet weir length of 80 m. The design flow rate is 66 L/s. Do these two clarifiers meet the requirements of at least 4 hr detention time and surface overflow rate no more than 30 $m^3/d/m^2$?

11.30 A conventional wastewater treatment plant receives 90 L/s with an average BOD of 250 mg/L. The aeration basin is 3000 m^3. The MLSS is 2,800 mg/L and effluent SS is 25 mg/L. The waste activated sludge is 2 m^3/s from the recycle line. The SS of the recycle flow is 9,000 mg/L.

    a. What is the F/M ratio?
    b. What is the MCRT?

11.31 Estimate the sludge wasting rate and the required reactor volume of an activated sludge plant being designed. The flow to the plant is 1100 L/s. The influent BOD_5 to the aeration basin is 200 mg/L, and the effluent BOD_5 from the plant is 8 mg/L. The design is for a MCRT of 10 days. The maximum yield coefficient is estimated to be 0.6 mg cells/mg substrate, and the endogenous decay coefficient is estimated to be 0.09/d. The aeration basin is expected to operate at an MLSS of 2,500 mg/L, and the return sludge and wasting concentrations are expected to be 9,000 mg/L.

11.32 You are designing an additional sand filter for a local wastewater treatment plant to remove a bottleneck at the plant. However, first you have to convince the state environmental protection agency that the plant does not need an additional secondary clarifier. There is no room on the site to build the clarifier and the plant's effluent has had only minor violations the entire time it has been operating. You have the original design specifications, including the estimated flowrates and wastewater composition, from when the plant was built in the early 1970s. You must send the letter to the agency today. Should you use the original design specifications? Why or why not?

11.33 What changes can occur over time in a community that will affect wastewater composition and/or flowrate?

END NOTES

1. Crites, Ron, and George Tchobanoglous. 1998. *Small and Decentralized Wastewater Management Systems*. Boston: WCB McGraw-Hill.
2. Klankrong, Thongchai, and Thomas S. Worthley. 2001. Rethinking Bangkok's wastewater strategy. *Civil engineering* 71, No. 6: 72–7.
3. Brock, Thomas D., Michael T. Madigan, John M. Martinko, and Jack Parker. 1994. *Biology of Microorganisms*. Englewood Cliffs, NJ: Prentice Hall.
4. Recycling nutrients into environmentally safe commercial fertilizer. 2008. Water Environment Federation. 20, No. 5. Accessed at http://www.wef.org/ScienceTechnologyResources/Publications/WET/08/08May/08MayProblemSolvers.htm in September 2008.

5. Smith, Gar, 2003. Why fluoride is an environmental issue. Earth Island Institute. *Earth Island Journal*.

6. Centre Europeen d'Etudes des Polyphosphates. Accessed at http://www.ceep-phosphates.org/Documents/shwList.asp?NID=4&HID=33 in September 2008.

7. Rittmann, Bruce E., and Perry L. McCarty. 2001. *Environmental Biotechnology: Principles and Applications*. Boston: McGraw-Hill.

8. Reed, Sherwood C., E. Joe Middlebrooks, and Ronald W. Crites. 1988. *Natural Systems for Waste Management and Treatment*. New York: McGraw-Hill.

9. Solzhenitzyn, Aleksandr I. 1974. *The Gulag Archipelago*. New York: Harper-Row.

10. Sanitation. 2008. World Health Organization. Accessed at http://www.who.int/topics/sanitation/en/ on 13 September, 2008.

11. Aruquipa, Abraham. 2008. Targeting sanitation in Bolivia. EWRI Currents. Accessed at http://email.asce.org/ewri/SanitationBolivia.html on 13 September, 2008.

Air Quality

© JL images/Alamy

Tiananmen Square (Beijing, China) on a good air quality day

© Martyn Vickery/Alamy

Tiananmen Square (Beijing, China) on a poor air quality day

Environmental engineers consider the potential local and global impacts of designs on air quality.

409

The air we breathe, like the water we drink, is necessary to life. And as with water, we want to be assured that the air will not cause us harm. We expect to breathe clean air.

But what exactly is clean air? This question is just as difficult to answer as defining clean water. Recall from Chapter 9 that many parameters are needed to describe the quality of water and that only with the selective and judicious use of these parameters is it possible to describe what is meant by water quality. Recall also that water quality is a relative term and that it is unrealistic to ask for all water to be pure H_2O. In many cases, such as in streams and lakes, pure water would actually be unacceptable.

An analogous situation exists with air quality. Pure air is a mixture of gases, containing, on a dry volume (or molar) basis,

- 78.1% nitrogen
- 20.9% oxygen
- 0.9% argon
- 0.04% carbon dioxide
- 0.002% neon
- 0.0005% helium

and so on. But such air is not found in nature and is of interest only as a reference, like pure H_2O.

If this is pure air, then it may be useful to define as pollutants those materials (gases, liquids, or solids) that, when added to pure air at sufficiently high concentrations, will cause adverse effects. For example, sulfur compounds emitted into the atmosphere reduce the pH of rain and result in acidic rivers and lakes, causing widespread damage. This is clearly unacceptable, and the sulfur compounds can be (without much argument) classified as air pollutants. Yet the problem is not so easily settled because some sulfur may be emitted from natural sources, such as volcanos and hot springs. It is therefore not valid simply to classify sulfur as a pollutant without specifying its sources.

The pollutants emitted into the atmosphere must travel through the atmosphere to reach people, animals, plants, or things to have an effect. Whereas in water pollution this carriage of pollutants is by water currents, in air pollution wind is the means for transport of pollutants.

This chapter first discusses some basic meteorology to illustrate how the transport and dispersion of pollutants takes place. Then the methods of air quality measurement are introduced, followed by a discussion of the sources and effects of some major air pollutants. Finally, some air pollution law is introduced to show how the government can influence the attainment of quality air.

12.1   METEOROLOGY AND AIR MOVEMENT

Earth's atmosphere can be divided into easily recognizable strata, depending on the temperature profile (similar to lake stratification, as discussed in Chapter 8). Figure 12.1 shows a typical temperature profile for four major layers. The troposphere, where most of our weather occurs, ranges from about 5 km at the poles to about 18 km at the equator. The temperature here decreases with altitude. Over 80% of the air is within this well-mixed layer. On top of the troposphere is the stratosphere, a layer of air where the temperature profile is inverted and in which little mixing takes place. Pollutants that migrate up

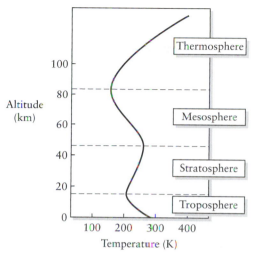

Figure 12.1   Earth's atmosphere.

to the stratosphere can stay there for many years. The stratosphere has a high ozone concentration, and the ozone adsorbs the sun's short-wave ultraviolet radiation. Above the stratosphere are two more layers, the mesosphere and the thermosphere, which contain only about 0.1% of the air.

Other than the problems of global warming and stratospheric ozone depletion, air pollution problems occur in the troposphere. Pollutants in the troposphere, whether produced naturally (such as terpenes in pine forests) or emitted from human activities (such as smoke from power plants), are moved by air currents that we commonly call wind. Meteorologists identify many different kinds of winds, ranging from global wind patterns caused by the differential warming and cooling of Earth as it rotates under the sun to local winds caused by differential temperatures between land and water masses. A sea breeze, for example, is a wind caused by the progressive warming of the land during a sunny day. The temperature of a large water body, such as an ocean or large lake, does not change as rapidly during the day, and the air over the warm land mass rises, creating a low-pressure area toward which air coming horizontally over the cooler large water body flows.

Wind not only moves the pollutants horizontally, but it causes the pollutants to disperse, reducing the concentration of the pollutant with distance away from the source. The amount of dispersion is directly related to the *stability* of the air, or how much vertical air movement is taking place. The stability of the atmosphere is best explained by using an ideal parcel of air.

As an imaginary parcel of air rises in earth's atmosphere, it experiences lower and lower pressure from surrounding air molecules and, thus, it expands. This expansion lowers the temperature of the air parcel. Ideally, a rising parcel of air cools at about 1°C/100 m (or warms at 1°C/100 m if it is coming down). This warming or cooling is termed the *dry adiabatic lapse rate** and is independent of prevailing atmospheric temperatures.

*Recall from Chapter 7 that "adiabatic" is a term denoting no heat transfer (e.g., between the air parcel and the surrounding air).

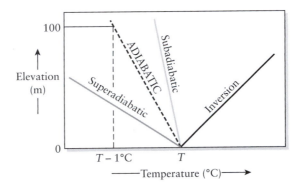

Figure 12.2 Prevailing and dry adiabatic lapse rates.

The 1°C/100 m *always* holds (for dry air), regardless of what the actual temperature at various elevations might be. When there is moisture in the air, the lapse rate becomes the *wet adiabatic lapse rate* because the evaporation and condensation of water influences the temperature of the air parcel. This is an unnecessary complication for the purpose of the following argument, so a moisture-free atmosphere is assumed.

The actual temperature-elevation measurements are called *prevailing lapse rates* and can be classified as shown in Figure 12.2. A *superadiabatic lapse rate*, also called a *strong lapse rate*, occurs when the atmospheric temperature drops more than 1°C/100 m. A *subadiabatic lapse rate*, also called a *weak lapse rate*, is characterized by a drop of less than 1°C/100 m. A special case of the weak lapse rate is the *inversion*, a condition that has warmer air above colder air.

During a superadiabatic lapse rate, the atmospheric conditions are unstable; a subadiabatic lapse rate, especially an inversion, characterizes a stable atmosphere. This can be demonstrated by depicting a parcel of air at 500 m (Figure 12.3A). In this case the air temperature at 500 m is 20°C. During a superadiabatic condition, the air temperature at ground level might be 30°C, and at 1000 m it might be 10°C. Note that this represents a change of more than 1°C/100 m.

If a parcel of air at 500 m and 20°C is moved upward to 1000 m, what would be its temperature? Remember that, assuming an adiabatic condition, the parcel would cool 1°C/100 m. The temperature of the parcel at 1000 m is, thus, 5°C less than 20°C, or 15°C. The *prevailing* temperature (the air surrounding the parcel), however, is 10°C, and the air parcel finds itself surrounded by cooler air. Will it rise or fall? Obviously, it will rise because warm air rises. We then conclude that once a parcel of air under superadiabatic conditions is displaced upward, it keeps right on going, an unstable condition.

Similarly, if a parcel of air under superadiabatic conditions is displaced downward, say to ground level, the air parcel is 20°C + (500 m × [1°C/100 m]) = 25°C. It finds the air around it a warm 30°C, and thus, the cooler air parcel would just as soon keep going down if it could. Because any upward or downward movement tends to continue and not be dampened out under superadiabatic conditions, superadiabatic atmospheres are characterized by a great deal of vertical air movement and turbulence. In other words, they are unstable.

The subadiabatic prevailing lapse rate is, by contrast, a very stable system. Consider again (Figure 12.3B) a parcel of air at 500 m and at 20°C. A typical subadiabatic system has a ground-level temperature of 21°C, and 19°C at 1000 m. If the parcel is displaced

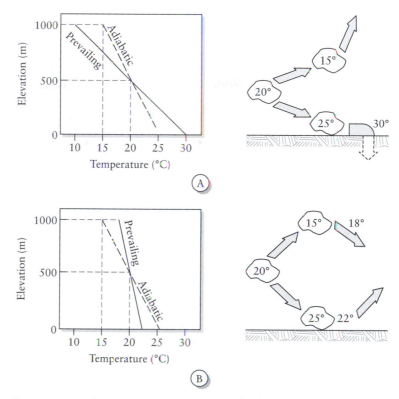

Figure 12.3   Prevailing lapse rates for a parcel of air moving in the atmosphere: (A) superadiabatic conditions (unstable) and (B) subadiabatic conditions (stable).

to 1000 m, it will cool by 5°C to 15°C. But, finding the air around it a warmer 19°C, it will fall right back to its point of origin. Similarly, if the air parcel is brought to ground level, it would be at 25°C, and finding itself surrounded by 21°C air, it would rise back to 500 m. Thus, the subadiabatic system would tend to dampen out vertical movement and is characterized by limited vertical mixing.

An inversion is an extreme subadiabatic condition, and the vertical air movement within an inversion is almost nil. Some inversions, called *subsidence inversions*, are due to the movement of a large warm air mass over cooler air. Such inversions, typical in Los Angeles, last for several days and are responsible for serious air pollution episodes. A more common type of inversion is the *radiation inversion*, caused by the radiation of heat to the atmosphere from earth. During the night, as Earth cools, the air close to the ground loses heat, thus causing an inversion (Figure 12.4 on the next page). The pollution emitted during the night is caught under this lid and does not escape until earth warms sufficiently to break the inversion. Such a radiation inversion occurred in Donora during the 1948 episode (Chapter 1). A layer of fog over the valley prevented the sunlight from breaking the inversion.

In addition to inversions, serious air pollution episodes are almost always accompanied by fogs. These tiny droplets of water are detrimental in two ways. In the first place, fog makes it possible to convert SO_3^- to H_2SO_4. Secondly, fog sits in valleys and prevents the

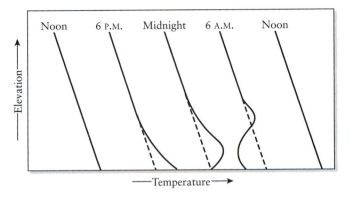

Figure 12.4 Atmospheric inversion caused by thermal radiation.

sun from warming the valley floor and breaking inversions, often prolonging air pollution episodes.

The movement of plumes from smokestacks is governed by the lapse rate into which they are emitted, as illustrated by the example below.

E X A M P L E

12.1   **Problem**   A stack 100 m tall emits a plume at 20°C. The prevailing lapse rates are shown in Figure 12.5. How high will the plume rise (assuming perfect adiabatic conditions)?

Solution   Note that the prevailing lapse rate is subadiabatic to 200 m and an inversion exists above 200 m. The plume at 20°C finds itself surrounded by colder (18.5°C) air, so it rises. As it rises, it cools at the dry adiabatic lapse rate, so at 200 m, it is 19°C. At about 220 m the surrounding air is at the same temperature as the plume (about 18.7°C), and the plume ceases to rise.

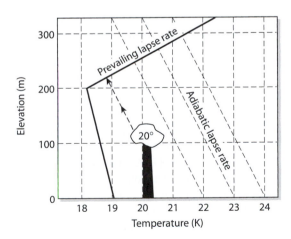

Figure 12.5 A typical prevailing lapse rate.

12.2   MAJOR AIR POLLUTANTS

12.2.1   Particulates

An air pollutant can be a gas or a particulate. Particulate pollutants can be further classified as dusts, fumes, mists, smoke, or spray. Approximate size ranges of the various types of particulate pollutants are shown in Figure 12.6. The EPA defines particulate matter as shown in Table 12.1.

Dust is defined as solid particles that are

Figure 12.6   (A) Definition of particulate pollutants; (B) illustration of sizes. (From Particulate Matter Research Program: Five Years of Progress, Paul Gilman, Ph.D., EPA Research and Development. Presentation to Stakeholders. September 9 2004.)

Table 12.1 Particulate Matter Sizes

Terminology	Abbreviation	Diameter Range (μm)
Ultrafine	–	<0.1
Fine	$PM_{2.5}$	<2.5
Coarse	PM_{10}	2.5–10

FOCUS ON

Abraded Tires as a Source of Air Pollution

Air quality is affected by both natural and anthropogenic (human-caused) sources. One source of air pollution currently gaining attention is particulate matter generated by the wear of automotive vehicle tires. As cars travel down the road, tires rub against the pavement. This contact shears away very small particles of tire, which go into the air. Some of the fragments stay suspended in the air, while others fall to the road, where they collect until another vehicle stirs them up into the air again. Additionally, the particulate matter on the road may also be washed into waterways by rain.

Passenger tires are typically made from synthetic rubber (~14%), natural latex rubber (~27%), carbon black (~28%), and steel wire (~14–15%) along with smaller amounts of sulfur and sulfur compounds from vulcanization, silica, phenolic resin, aromatic, naphthenic, and/or paraffinic oils, polyester, nylon, and other fabrics, petroleum waxes, pigments including zinc oxide and titanium dioxide, fatty acids and other inert materials added to increase the strength and durability of an average tire.[1] These ingredients, while helpful to tire performance, may be harmful to humans when suspended in air as tiny particulates. Studies show that there is a correlation between high-traffic areas and the number of people in the area who are asthmatic. For years, this link was attributed to the emissions that come from a vehicle's exhaust system because it was thought that abraded tire particles were too large to enter the

lung. Now it appears that is not the case. Up to 60% of the particulates from abraded tires can be very small, allowing them to penetrate the deepest part of the human lung.[2] Studies suggest that tire dust is equally or more hazardous than the exhaust from car tailpipes. One reason is due to the latex in the rubber. Latex is a natural substance known to cause allergic reactions ranging from a runny nose to asthma to anaphylactic shock. Latex is a sensitizing agent, that is, it can induce allergic responses in people. Current estimates are that almost 10% of health care professionals have latex allergies that were initiated by exposure to latex gloves. Latex in tire dust has been shown to be allergenic in laboratory tests.[3]

Because this is a fairly new area of research, data on the amount of tire dust in the air and the extent of exposure to it are limited. Studies are being conducted linking heavy traffic areas to latex allergies, but more studies are needed. A major challenge is determining how to cut down on the amount of tire dust generated. Cars will continue to fill the roadways into the foreseeable future, and people will continue to live close to high-traffic areas. How can tires, roads, or even vehicles be designed to prevent tire dust from forming? Can we find ways to design tires using materials that will not be harmful to human health when they abrade? These are questions that the next generation of engineers will likely face.

a. entrained by process gases directly from the material being handled or processed (e.g., coal, ash and cement)
b. direct offspring of a parent material undergoing a mechanical operation (e.g., sawdust from woodworking)
c. entrained materials used in a mechanical operation (e.g., sand from sandblasting).

Dust consists of relatively large particles. Cement dust, for example, is about 100 μm in diameter.*

A *fume* is also a solid particle, frequently a metallic oxide, formed by the condensation of vapors by sublimation, distillation, calcination, or chemical reaction processes. Examples of fumes are zinc and lead oxide resulting from the condensation and oxidation of metal volatilized in a high-temperature process. The particles in fumes are quite small, with diameters from 0.03 to 0.3 μm.

A *mist* is an entrained liquid particle formed by the condensation of a vapor and perhaps by chemical reaction. Mists typically range from 0.5 to 3.0 μm in diameter.

Smoke is made up of entrained solid particles formed as a result of incomplete combustion of carbonaceous materials. Although hydrocarbons, organic acids, sulfur oxides, and nitrogen oxides are also produced in combustion processes, only the solid particles resulting from the incomplete combustion of carbonaceous materials are called smoke. Smoke particles have diameters from 0.05 to approximately 1 μm.

Finally, a *spray* is a liquid particle formed by the atomization of a parent liquid. Sprays settle under gravity.

12.2.2   Measurement of Particulates

The measurement of PM_{10} has been historically done by using the *high-volume sampler* (or hi-vol). The high-volume sampler (Figure 12.7 on the next page) operates much like a vacuum cleaner, forcing upto 2,434,000 L of air through a filter in 24 h. The analysis is gravimetric; the filter is weighed before and after, and the difference is the particulates collected.

The air flow is measured by a small flow meter, usually calibrated in cubic feet of air per minute. Because the filter gets dirty during the 24 h of operation, less air goes through the filter during the latter part of the test than in the beginning, and the air flow must therefore be averaged over the 24-h sampling period.

EXAMPLE
12.2

Problem   A clean filter is found to weigh 10.00 g. After 24 h in a hi-vol, the filter plus dust weighs 10.10 g. The air flows at the start and end of the test are 1700 L/min and 1100 L/min, respectively. What is the particulate concentration?

Solution

$$\text{Weight of the particulates (dust)} = (10.10 - 10.00) \text{ g } (10^6 \ \mu\text{g/g})$$
$$= 0.1 \times 10^6 \ \mu\text{g}$$

---

*In air pollution control parlance, a micrometer (μm) is often referred to as a micron (μ). This usage is adopted here.

$$\text{Average air flow} = \frac{(1700 + 1100)}{2} = 1400 \text{ L/min}$$

$$
\begin{aligned}
\text{Total air through the filter} &= (1400 \text{ L/min})(60 \text{ min/hr})(24 \text{ hr/d}) \,(1 \text{ d}) \\
&= 2{,}016{,}000 \text{ L} \\
&= 2016 \text{ m}^3
\end{aligned}
$$

$$\text{Total suspended particulates} = \frac{(0.1 \times 10^6 \text{ µg})}{2016 \text{ m}^3} = 50 \text{ µg/m}^3$$

Woodburning stoves are apparently particularly effective emitters of small particulates of the PM_{10} variety. In one Oregon city, the concentration of PM_{10} reached above 700 µg/m^3, whereas the national ambient air quality standard (discussed in Section 12.4) is only 150 µg/m^3. Cities where the atmospheric conditions prevent the dispersal of wood smoke have been forced to pass woodburning bans during nights when the conditions could lead to high particulate levels in the atmosphere.

The particulate concentration measured by a high-volume sampler is often referred to as *total suspended particulates* (TSP) to differentiate it from other measurements of particulates. Another widely used measure of particulates in the environmental health

© Courtesy P. Aarne Vesilind

Figure 12.7 Hi-vol sampler.

area is *respirable particulates*, or particulates that could be respired into lungs. The respirable particle fraction has 50% of the particle mass associated with 4-μ particles. The respirable particles are separated using centrifugal forces. The measurement of PM$_{2.5}$ is done similarly to PM$_{10}$ but with the addition of a specially shaped inlet and an impactor.

12.2.3   Gaseous Pollutants

In the context of air pollution control, gaseous pollutants include substances that are gases at normal temperature and pressure as well as vapors of substances that are liquid or solid at normal temperature and pressure. Among the gaseous pollutants of greatest importance in terms of present knowledge are carbon monoxide, hydrocarbons, hydrogen sulfide, nitrogen oxides, ozone and other oxidants, and sulfur oxides. Carbon dioxide and methane have been added to this list because of their effect on climate. These and other gaseous air pollutants are listed in Table 12.2.

12.2.4   Measurement of Gases

While the units of particulate measurement are consistently in terms of micrograms per cubic meter, the concentration of gases can be either parts per million (ppm) on a volume-to-volume basis or micrograms per cubic meter. As explained in Chapter 3, the conversion from one to the other is:

$$\mu g/m^3 = \frac{MW \times 1000}{24.5} \times ppm$$

where MW = molecular weight of the gas. This equation is applicable for conditions of 1 atmosphere and 25°C. For 1 atmosphere and 0°C, the constant becomes 22.4.

EXAMPLE

12.3

Problem   A stack gas contains carbon monoxide (CO) at a concentration of 10% by volume. What is the concentration of CO in $\mu g/m^3$? (Assume 25°C and 1 atmosphere pressure.)

Solution   The concentration in ppm makes use of the fact that 1% by volume is 10,000 ppm (Chapter 3). Therefore, 10% by volume is 100,000 ppm. The molecular weight of CO is 28 g/mol, so the concentration in micrograms per cubic meter is

$$\frac{28 \times 1000}{24.5} \times 100,000 = 114 \times 10^6 \ \mu g/m^3$$

The earliest gas-measurement techniques almost all involved the use of a *bubbler*, shown in Figure 12.8 on page 421. The gas is literally bubbled through the liquid, which either reacts chemically with the gas of interest or into which the gas is dissolved. Wet chemical techniques are then used to measure the concentration of the gas.

A simple (but now seldom used) bubbler technique for measuring SO$_2$ is to bubble air through hydrogen peroxide, causing the following reaction to occur:

$$SO_2 + H_2O_2 \rightarrow H_2SO_4$$

Table 12.2 Some Gaseous Air Pollutants

Name	Formula	Properties of Importance	Significance as Air Pollutant
Sulfur dioxide	SO_2	Colorless gas, intense choking, odor, highly soluble in water—forming sulfurous acid, H_2SO_3	Damage to property, health, and vegetation
Sulfur trioxide	SO_3	Soluble in water—forming sulfuric acid H_2SO_4	Highly corrosive
Hydrogen sulfide	H_2S	Rotten egg odor at low concentrations, odorless at high concentrations	Highly poisonous
Nitrous oxide	N_2O	Colorless gas, used as carrier gas in aerosol bottles	Relatively inert; not produced in combustion
Nitric oxide	NO	Colorless gas	Produced during high-temperature, high-pressure combustion; oxidizes to NO_2
Nitrogen dioxide	NO_2	Brown to orange gas	Major component in the formation of photochemical smog
Carbon monoxide	CO	Colorless and odorless	Product of incomplete combustion; poisonous
Carbon dioxide	CO_2	Colorless and odorless	Formed during complete combustion; greenhouse gas
Ozone	O_3	Highly reactive	Damage to vegetation and property; produced mainly during the formation of photochemical smog
Hydrocarbons	C_xH_y or HC	Many	Emitted from automobiles and industries; formed in the atmosphere
Methane	CH_4	Combustible, odorless	Greenhouse gas
Chlorofluorocarbons	CFC	Nonreactive, excellent thermal properties	Deplete ozone in upper atmosphere

The amount of sulfuric acid formed can be determined by titrating the solution with a base of known strength.

One of the better third-generation methods of measuring SO_2 is the colorimetric pararosaniline method, in which SO_2 is bubbled into a liquid containing tetrachloromercurate (TCM). The SO_2 and TCM combine to form a stable complex. Pararosaniline is then added to this complex, with which it forms a colored solution. The amount of color is proportional to the SO_2 in the solution, and the color is measured with

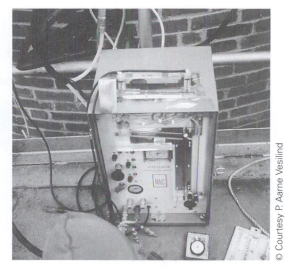

© Courtesy P. Aarne Vesilind

Diffuser

Figure 12.8 A typical bubbler for measuring gaseous air pollutants.

a spectrophotometer. (See Chapter 9 for ammonia measurement—another example of a colorimetric technique.)

12.2.5 Measurement of Smoke

Air pollution has historically been associated with smoke—the darker the smoke, the more pollution. We now know, of course, that this isn't necessarily true, but many regulations (e.g., for municipal incinerators) were written on the basis of smoke density.

The density of smoke has for many years been measured on the *Ringlemann scale*, devised in the late 1800s by Maxmilian Ringlemann, a French professor of engineering. The scale ranges from 0 for white or transparent smoke to 5 for totally black, opaque smoke. The test is conducted by holding a card, such as shown in Figure 12.9, and comparing the blackness of the card to the smoke. A fairly dark smoke is, thus, said to be Ringlemann 4, for example.

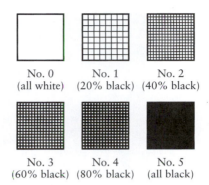

No. 0   No. 1   No. 2
(all white)   (20% black) (40% black)

No. 3   No. 4   No. 5
(60% black) (80% black) (all black)

Figure 12.9 The Ringlemann scale.

12.2.6 Visibility

One of the obvious effects of air pollutants is a reduction in visibility. Loss of visibility is often defined as the condition when it is just possible to identify a large object, such as a building, in bright daylight or it is just possible to see a moderately bright light at night. This is, of course, a vague definition of visibility, but it is especially useful for defining the limits of visibility.

Reductions in visibility can occur due to natural air pollutants, such as terpenes from pine trees (which is why the Smoky Mountains are smoky), or from human-produced emissions. Many constituents can cause attenuated visibility, such as water droplets (fog) and gases (NO_2), but the most effective reduction in visibility is by small particulates. Particulates reduce visibility both by adsorbing the light as well as scattering the light. In the first case the light does not enter the eye of the viewer, and in the second case the scattering reduces the contrast between light and dark objects. Mathematically, the intensity of an object illuminated by a beam of intensity I at a distance x from the observer is attenuated as

$$\frac{dI}{dx} = -\sigma I$$

where I = intensity of the light beam as viewed by the observer
       x = distance from the observer
       σ = constant that takes into account the atmospheric condition, often called the extinction coefficient

Integrated, this equation reads

$$\ln\left(\frac{I}{I_0}\right) = -\sigma x$$

where I_0 is the intensity of the beam when the distance x approaches zero. It has been found that the lower limit of visibility for most people occurs when the light intensity has been reduced to about 2% of the unattenuated light. If this value is substituted in the above equation, the distance x is then the limit of visibility, L_V, or

$$\ln(0.02) = -\sigma L_V$$

$$L_V = \frac{3.9}{\sigma}$$

As a rough approximation, the extinction coefficient can be thought of as being directly and linearly proportional to the concentration of particulates (for atmospheres containing less than 70% moisture). An approximate expression for visibility can then be written as

$$L_V = \frac{1.2 \times 10^3}{C}$$

where the constant 1.2×10^3 takes into account the conversion factor if L_V is in kilometers and C is the particulate concentration in micrograms per cubic meter. It should be emphasized that this is an approximate relationship and should be used only for atmospheres with less than 70% moisture.*

12.3   SOURCES AND EFFECTS OF AIR POLLUTION

Unwanted constituents in air, or air pollution, can have a detrimental effect on human health, on the health of other creatures, on the value of property, and on the quality of life. Some of the pollutants of human health concern are formed and emitted through natural processes. For example, naturally occurring particulates include pollen grains, fungus spores, salt spray, smoke particles from forest fires, and dust from volcanic eruptions. Gaseous pollutants from natural sources include carbon monoxide as a breakdown product in the degradation of hemoglobin, hydrocarbons in the form of terpenes from pine trees, hydrogen sulfide resulting from the breakdown of cysteine and other sulfur-containing amino acids by bacterial action, nitrogen oxides, and methane (natural gas).

People-made sources of pollutants can be conveniently classified as stationary combustion, transportation, industrial process, and solid waste disposal sources. The principal pollutant emissions from stationary combustion processes are particulate pollutants (such as fly ash and smoke) and sulfur and nitrogen oxides. Sulfur oxide emissions are, of course, a function of the amount of sulfur present in the fuel. Thus, combustion of coal and oil, both of which contain appreciable amounts of sulfur, yields significant quantities of sulfur oxide.

Much of the knowledge of the effects of air pollution on people comes from the study of acute air pollution episodes. The two most famous episodes occurred in Donora, Pennsylvania (described in Chapter 1), and in London, England. In both episodes the pollutants

---

*This discussion is based in part on Ross, R. D., ed. 1972. *Air Pollution and Industry*. New York: VanNostrand Reinhold, and Wark, K., and C. F. Warner. 1981. *Air Pollution*. New York: Harper & Row.

affected a specific segment of the public—those individuals already suffering from dis-eases of the cardiorespiratory system. Another observation of great importance is that it was not possible to blame the adverse effects on any one pollutant. This observation puzzled the investigators (industrial hygiene experts), who were accustomed to studying industrial problems in which one could usually relate health effects to a specific pollu-tant. Today, after many years of study, it is thought that the health problems during the episodes were attributable to the combined action of a particulate matter (solid or liquid particles) and sulfur dioxide, a gas. No one pollutant by itself, however, could have been responsible.

Except for these episodes, scientists have little information from which to evaluate the health effects of air pollution. Laboratory studies with animals are of some help, but the step from a rat to a person (anatomically speaking) is quite large.

Four of the most difficult problems in relating air pollution to health are unanswered questions concerning (1) the existence of thresholds, (2) the total body burden of pollutants, (3) the time-versus-dosage problem, and (4) synergistic effects of various combinations of pollutants.

1. *Threshold.* The existence of a threshold in health effects of pollutants has been debated for many years. As discussed further in Chapter 15, several dose-response curves are possible for a dose of a specific pollutant (e.g., carbon monoxide) and the response (e.g., reduction in the blood's oxygen-carrying capacity). One possibility is that there will be no effect on human metabolism until a critical concentration (the threshold) is reached. On the other hand, some pollutants can produce a detectable response for any finite concentration. Nor do these curves have to be linear. In air pollution the most likely dose-response relationship for many pollutants is nonlinear without an identifiable threshold but with a minimal response up to a higher concen-tration, at which point the response becomes severe. The problem is that, for most pollutants, the shapes of these curves are unknown.

2. *Total Body Burden.* Not all of the dose of pollutants comes from air. For exam-ple, although a person breathes in about 50 μg/d of lead, the intake of lead from water and food is about 300 μg/d. In the setting of air quality standards for lead, it must, therefore, be recognized that most of the lead intake is from food and water.

3. *Time versus Dosage.* Most pollutants require time to react, and the time of con-tact is as important as the level. The best example of this is the effect of carbon monoxide, as illustrated in Figure 12.10. Human hemoglobin (Hb) has a carbon monoxide affinity 210 times greater than its affinity for oxygen, and thus, CO com-bines readily with hemoglobin to form carboxyhemoglobin (COHb). The formation of COHb reduces the hemoglobin available to carry oxygen. At about 60% COHb concentration, death results from lack of oxygen. The effects of CO at sublethal con-centrations are usually reversible. Because of this type of time–response problem, ambient air quality standards are set at maximum allowable concentrations for a given time.

4. *Synergism.* Synergism is defined as an effect that is greater than the sum of the parts. For example, black lung disease in coal miners occurs only when the miner is also a

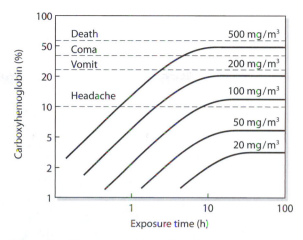

Figure 12.10   Effect of carbon monoxide on health. (After W. Agnew, 1968. *Proceedings of the Royal Society* A307: 153.)

cigarette smoker. Coal mining by itself and cigarette smoking by itself will not cause black lung, but the synergistic action of the two puts miners who smoke at high risk. [While there is overwhelming evidence that air pollution can increase the risk of diseases (such as lung cancer, emphysema and asthma) especially when tied synergistically to cigarette smoking, the actual cause-and-effect is not medically proven. Therefore, it is incorrect to assert that air pollution, no matter how bad (or cigarettes, for that matter) *cause* lung cancer or other respiratory diseases.]

Of course, the major target of air pollutants in humans (and other animals) is the respiratory system, pictured in Figure 12.11 on the next page. Air (and entrained pollutants) enter the body through the throat and nasal cavities and pass to the lungs through the trachea. In the lungs the air moves through bronchial tubes to the alveoli, small air sacs in which the gas transfer takes place. Pollutants are either absorbed into the bloodstream or moved out of the lungs by tiny hair cells called cilia, which are continually sweeping mucus up into the throat. The respiratory system can be damaged by both particulate and gaseous pollutants.

Particles greater than 0.1 µm will usually be caught in the upper respiratory system and swept out by the cilia. However, particles less than 0.1 µm in diameter can move into the alveoli, where there are no cilia, and they can stay there for a long time and damage the lung. Droplets of H_2SO_4 and lead particles represent the most serious forms of particulate matter in air.

Perhaps the most important gas with reference to health is sulfur dioxide. It acts as an irritant, restricts air flow, and slows the action of the cilia. Because it is highly soluble, it should be readily removed by the mucus membrane, but it has been shown that it can get to the deep reaches of the lung by first adsorbing onto tiny particles and then using this transport to reach the deep lung—a classic case of synergistic action.

The amount of sulfur oxide produced can be calculated if the sulfur content of the fuel is known, as the example below illustrates.

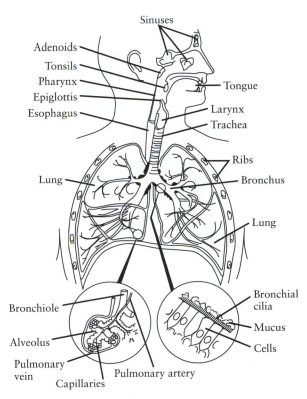

Sinuses

Adenoids

Tonsils

Pharynx

Epiglottis

Esophagus

Tongue

Larynx

Trachea

Ribs

Lung

Bronchus

Lung

Bronchiole

Bronchial cilia

Mucus

Alveolus

Cells

Pulmonary vein

Pulmonary artery

Capillaries

Figure 12.11 The respiratory system. (Courtesy of the American Lung Association.)

EXAMPLE

12.4

Problem During a 2-wk-long major air pollution episode in London in 1952, it is estimated that 25,000 tonnes of coal that had an average sulfur content of about 4% were burned per week. The mixing depth (height of the inversion layer or cap over the city that prevented the pollutants from escaping) was about 150 m over an area of 1200 km². If initially there was no SO_2 in the atmosphere (a conservative assumption), what is the expected SO_2 concentration at the end of the 2 wk?

Solution Use the box model for calculating the concentration. That is, consider the volume over London a black box, and use the well worn material balance equation.

$$
\begin{bmatrix} \text{Rate of } SO_2 \\ \text{ACCUMULATED} \end{bmatrix} = \begin{bmatrix} \text{Rate of } SO_2 \\ \text{IN} \end{bmatrix} - \begin{bmatrix} \text{Rate of } SO_2 \\ \text{OUT} \end{bmatrix}
$$
$$
+ \begin{bmatrix} \text{Rate of } SO_2 \\ \text{PRODUCED} \end{bmatrix} - \begin{bmatrix} \text{Rate of } SO_2 \\ \text{CONSUMED} \end{bmatrix}
$$

The rate of SO_2 out was zero because nothing escaped from under this *haze hood*. The rate produced and consumed is also zero if it is assumed that sulfur oxide was not created or destroyed in the atmosphere. The rate of SO_2 concentration increasing in the box is constant, so the equation is zero-order (Chapter 5)

$$\begin{bmatrix} \text{Rate of SO}_2 \\ \text{ACCUMULATED} \end{bmatrix} = \begin{bmatrix} \text{Rate of SO}_2 \\ \text{IN} \end{bmatrix} - 0 + 0 - 0$$

$$\frac{dA}{dt} = k \quad \text{or} \quad A = A_0 + kt$$

where A = the mass of SO_2.

The SO_2 emitted per week is

$$\left(\frac{25{,}000 \text{ metric tons coal}}{\text{wk}} \right) \left(\frac{4 \text{ parts S}}{100 \text{ parts coal}} \right) \left(\frac{\text{mol S}}{32 \text{ g S}} \right)$$

$$\times \left(\frac{1 \text{ mol SO}_2}{1 \text{ mol S}} \right) \left(\frac{64 \text{ g SO}_2}{\text{mol SO}_2} \right) = 2000 \frac{\text{tonnes}}{\text{week}}$$

If the initial concentration, A_0, is assumed to be zero,

$$A = 0 + (2000 \text{ metric tons/wk})(2 \text{ wk}) = 4000 \text{ tonnes of SO}_2$$

accumulated in the atmosphere over London.

The volume into which this is mixed is

$$150 \text{ m} \times 1200 \text{ km}^2 \times 10^6 \text{ m}^2/\text{km}^2 = 180{,}000 \times 10^6 \text{ m}^3$$

The concentration of SO_2 at the end of 2 wk should have been

$$\frac{4000 \text{ metric tons} \times 10^6 \text{ g/metric ton} \times 10^6 \text{ μg/g}}{180{,}000 \times 10^6 \text{ m}^3} = 22{,}000 \text{ μg/m}^3$$

The actual peak concentration of SO_2 during the London episode was less than 2000 μg/m^3. Yet the calculation in Example 12.4 shows more than 10 times that. Where did all of the SO_2 go? The answer lies in the assumption used in the calculation that no SO_2 was consumed. In fact, there is a continuous scavenging of SO_2 from the atmosphere by the SO_2 coming into contact with buildings, vegetation, wildlife, and human beings. Most importantly, sulfur oxides seem to be a major constituent of poorly buffered precipitation, more commonly known as acid rain.

12.3.1 Sulfur and Nitrogen Oxides and Acid Rain

One way in which SO_2 is removed from the atmosphere is the formation of acid rain. Normal, uncontaminated rain has a pH of about 5.6 (due to CO_2), but acid rain can be pH 2 or even lower. Acid rain formation is a complex process, and the dynamics are not fully understood. In its simplest terms SO_2 is emitted from the combustion of fuels containing sulfur and then reacts with atmospheric components

$$S + O_2 \xrightarrow{\text{heat}} SO_2$$

$$SO_2 + O \xrightarrow{\text{sunlight}} SO_3$$

$$SO_3 + H_2O \longrightarrow H_2SO_4$$

H_2SO_4 is sulfuric acid. Sulfur oxides do not literally produce sulfuric acid in the clouds, but the idea is the same. The precipitation from air containing high concentrations of sulfur oxides is poorly buffered, and its pH readily drops.

Nitrogen oxides, emitted mostly from automobile exhaust but also from any other high-temperature combustion, contribute to the acid mix in the atmosphere. The chemical reactions that apparently occur with nitrogen are

$$N_2 + O_2 \longrightarrow 2NO$$

$$NO + O_3 \longrightarrow NO_2 + O_2$$

$$NO_2 + O_3 + H_2O \longrightarrow 2HNO_3 + O_2$$

where HNO_3 is nitric acid.

The effect of acid rain has been devastating. Hundreds of lakes in North America and Scandinavia became so acidic that they no longer supported fish life. In a study of Norwegian lakes, more than 70% of the lakes having a pH less than 4.5 contained no fish, and nearly all lakes with a pH of 5.5 and above contained fish. The low pH not only affects fish directly but contributes to the release of potentially toxic metals, such as aluminum, thus magnifying the problem.

In North America acid rain wiped out all fish and many plants in 50% of the high mountain lakes in the Adirondacks. The pH in many of these lakes has reached such levels of acidity as to replace the trout and native plants with acid-tolerant mats of algae.

The deposition of atmospheric acid on freshwater aquatic systems prompted EPA to suggest a limit of 10 to 20 kg SO_4^{2-} per hectare per year. If Newton's law of air pollution is used (what goes up must come down), it is easy to see that the amount of sulfuric and nitric oxides emitted is vastly greater than this limit. For example, for the state of Ohio alone, the total annual emissions were 2.4×10^6 metric tons of SO_2 per year. If all this were converted to SO_4^{2-} and deposited on the state of Ohio, the total would be 360 kg per hectare per year.[4]

But not all of this sulfur falls on the folks in Ohio. In fact, much of it is exported by the atmosphere to places far away. Similar calculations for the sulfur emissions for the northeastern United States indicate that the rate of sulfur emission is 4 to 5 times greater than the rate of deposition. Where does it all go?

The Canadians have a ready and compelling answer. They have for many years blamed the United States for the formation of most of the acid rain that falls within their borders. Similarly, much of the problem in Scandinavia can be traced to the use of tall stacks in Great Britain and the lowland countries of continental Europe. For years British industry simply built taller and taller stacks as a method of air pollution control, reducing the immediate ground level concentration but emitting the same pollutants into the higher atmosphere. The air quality in the United Kingdom improved but at the expense of acid rain in other parts of Europe.

Pollution across political boundaries is a particularly difficult regulatory problem. The big stick of police power is no longer available. Why *should* the United Kingdom worry about acid rain in Scandinavia? Why *should* the Germans clean up the Rhine before it flows through The Netherlands? Why *should* Israel stop taking water out of the Dead Sea, which it shares with Jordan? Laws are no longer useful, and threatened retaliation is unlikely.

What forces are there to encourage these countries to do the right thing? Is there such a thing as international ethics?

12.3.2 Photochemical Smog

An important approach to classification of air pollutants is that of *primary* and *secondary pollutants*. A primary pollutant is defined as one that is emitted as such to the atmosphere, whereas secondary pollutants are actually produced in the atmosphere by chemical reactions. The components of automobile exhaust are particularly important in the formation of secondary pollutants. The well-known and much discussed Los Angeles smog is a case of secondary pollutant formation (Figure 12.12). Table 12.3 lists in simplified form some of the key reactions in the formation of photochemical smog.

The reaction sequence illustrates how nitrogen oxides formed in the combustion of gasoline and other fuels and emitted to the atmosphere are acted upon by sunlight to yield ozone (O_3), a compound not emitted as such from sources and, hence, considered a secondary pollutant. Ozone, in turn, reacts with hydrocarbons to form a series of compounds that includes aldehydes, organic acids, and epoxy compounds. The atmosphere can be

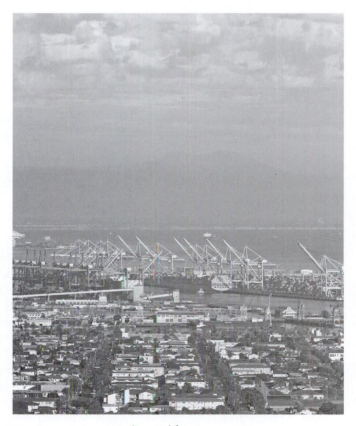

Figure 12.12 Smog in Los Angeles, California (Jens Peermann/Shutterstock)

Table 12.3   Simplified Reaction Scheme for Photochemical Smog*

$NO_2 + Light$	$\longrightarrow$	$NO + O$
$O + O_2$	$\longrightarrow$	O_3
$O_3 + NO$	$\longrightarrow$	$NO_2 + O_2$
$O + HC$	$\longrightarrow$	HCO°
$HCO^{\circ} + O_2$	$\longrightarrow$	HCO_3°
$HCO_3^{\circ} + HC$	$\longrightarrow$	Aldehydes, ketones, etc.
$HCO_3^{\circ} + NO$	$\longrightarrow$	$HCO_2^{\circ} + NO_2$
$HCO_3^{\circ} + O_2$	$\longrightarrow$	$O_3 + HCO_2^{\circ}$
$HCO_x^{\circ} + NO_2$	$\longrightarrow$	Peroxyacetyl nitrates

*NO_2 = nitrogen dioxide, NO = nitric oxide, O = atomic oxygen, O_2 = molecular oxygen, O_3 = ozone, HC = hydrocarbon, $^{\circ}$ = radical.

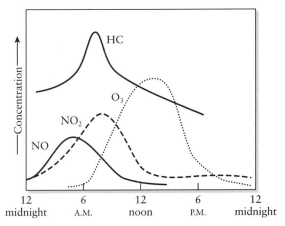

Figure 12.13   Formation of photochemical smog during a sunny 24-h period.

viewed as a huge reaction vessel wherein new compounds are being formed while others are being destroyed.

The formation of photochemical smog is a dynamic process. Figure 12.13 is an illustration of how the concentrations of some of the components vary during the day. Note that, as the morning rush hour begins, the NO levels increase, followed quickly by NO_2. As the latter reacts with sunlight, O_3 and other oxidants are produced. The hydrocarbon level similarly increases at the beginning of the day and then drops off in the evening.

The reactions involved in photochemical smog remained a mystery for many years. Particularly baffling was the formation of high ozone levels. As seen from the first three reactions in Table 12.3, for every mole of NO_2 reacting to make atomic oxygen and, hence, ozone, one mole of NO_2 was created from reaction with the ozone. All these reactions are fast. How, then, could the ozone concentrations build to such high levels?

FOCUS ON

Discovering the Cause of Photochemical Smog

In the 1940s, as the number of clear days in Southern California became fewer and fewer, concern was being expressed as to what the cause of this bad visibility might be. But during World War II, little was done and the situation became progressively worse. When a butadiene plant was built in downtown Los Angeles, it experienced severe upsets and produced noxious fumes that caused office buildings to be evacuated. Although we now know that the butadiene plant had little to do with the poor visibility, the event caused a public outcry and a demand to do something about the air pollution. Experts were sent for, and they came with their instruments for measuring SO_2 and smoke, and these soon proved worthless because the levels of sulfur oxides were very low. In 1947 the County of Los Angeles got the police power to do something about the air pollution and the LA County Air Pollution District was formed. Early efforts to reduce air pollution in LA all centered still on the reduction of SO_2 from such sources as backyard burning, but all to no avail. Finally it was decided that some research was necessary.

Arie Haagen-Smit (Figure 12.14), a Dutch-born biologist working at CalTech on the fumes emitted by pineapples, decided to distill the contents of the Los Angeles air. He discovered peroxy-organic substances, which were no doubt the source of the eye irritation. The source, if this was true, had to have been the gasoline-powered automobile, and the publication of this research resulted in a firestorm of protest by the auto industry. Scientists at the Stanford Research Institute, which had been doing smog research on behalf of the transportation industry, presented a paper at CalTech accusing Haagen-Smit of bad science. This made him so angry that he abandoned his pineapple work and began to work full-time on the smog problem.

Some of his previous research had been on the damage caused by ozone on plants, and he discovered that the effect of automobile exhaust on plants produced a similar injury, suggesting that the

Figure 12.14   Arie Haagen-Smit. (Courtesy of the Archives, California Institute of Technology)

exhaust contained ozone. But ozone was not emitted by automobiles, so where did it come from? He finally hit on the idea of mixing automobile exhaust with hydrocarbons in a large air chamber and subjecting the mixture to strong light, in effect modeling the atmosphere over the city. With this experiment, he was able to demonstrate that ozone is not emitted directly from any source but is formed by reactions in the atmosphere, and the ozone in the LA smog is created by reactions that begin with the oxides of nitrogen in automobile exhaust.

The powerful automobile and gasoline industries denied that this could be the cause of the smog and suggested instead that the ozone had to be in the smog because it somehow descended from the stratosphere, which of course was nonsense. Haagen-Smit's courageous work, taking on the most powerful political forces in California, paved the way for an eventual (but not yet present!) solution to the smog problem in Los Angeles.

Source: Biographical Memoir of ARIE JAN HAAGEN-SMIT by James Bonner. Biographical Memoirs of the National Academy of Sciences Vol. 58. Washington DC; National Academies Press 1989.

One answer is that NO enters into other reactions, especially with various hydrocarbon radicals and, thus, allows excess ozone to accumulate in the atmosphere (the seventh reaction in Table 12.3). In addition, some hydrocarbon radicals react with molecular oxygen and also produce ozone.

The chemistry of photochemical smog is complex. This was witnessed by the attempt to reduce hydrocarbon emissions to control ozone (O_3) levels. The thinking was that, if HC is not available, the O_3 will be used to oxidize NO to NO_2, thus using the available ozone. Unfortunately, this control strategy was a failure, and the answer seems to be that all primary pollutants involved in photochemical smog formation must be controlled.

12.3.3   Ozone Depletion

Ozone (O_3) is an eye irritant at usual urban levels, but urban O_3 should not be confused with stratospheric O_3, 11 to 16 km above the earth's surface. The latter acts as an ultraviolet radiation shield, and its alteration can increase the risk of skin cancer as well as change the ecology in unpredictable ways.

The problem with the depletion of upper atmospheric ozone is due to the manufacture and discharge of a class of chemicals called chlorofluorocarbons (CFCs). These chemicals that found wide use in aerosols and refrigeration systems and are responsible for climate change as well as the depletion of the protective ozone layer in the stratosphere. Two of the most important CFCs are trichlorofluoromethane, $CFCl_3$, and dichlorodifluoromethane, CF_2Cl_2, both of which are inert and not water soluble and, therefore, do not wash out of the atmosphere. They drift into the upper atmosphere and are eventually destroyed by short-wave solar radiation, releasing chlorine, which can react with ozone. The depletion of ozone allows the ultraviolet radiation to pass through unimpeded, and this can have a serious potential effect on the incidence of skin cancer. While basking in the sun is not a good idea in the first place, doing so when the ozone layer is not effective in screening out much of the ultraviolet rays is even more risky.

Because the problem with ozone is not at ground level and because the atmospheric pressure varies, it is difficult to measure ozone concentration. *Dobson units* (*DU*) were developed to solve this problem, where

$$1 \text{ DU} = 0.01 \text{ mm of } O_3 \text{ at 1 atm and } 0°C$$

At midlatitudes the ozone concentration is typically about 350 DU, at the equator it is 250 DU, and in the Antarctic region it is only 100 DU.

Ozone in the upper atmosphere is created when oxygen reacts with light energy (hv):

$$O_2 + hv \longrightarrow O + O$$
$$O_2 + O \longrightarrow O_3$$

Light energy also destroys ozone:

$$O_3 + hv \longrightarrow O_2 + O$$

This is the mechanism by which ozone prevents ultraviolet radiation from reaching earth's surface. These reactions also heat the atmosphere, causing inversion conditions that result in very stable conditions.

When CFCs are introduced into the upper atmosphere, ozone is destroyed. First, the CFCs react with light energy and release chlorine. One form of a CFC reacts as follows:

$$CF_2Cl_2 + hv \longrightarrow CF_2Cl + Cl$$

The atomic chlorine acts as a catalyst speeding the destruction of ozone:

$$Cl + O_3 \longrightarrow ClO + O_2$$
$$ClO + O \longrightarrow Cl + O_2$$

A single Cl atom can make the loop thousands of times before it finally reacts with some other chemical compound, such as methane.

The destruction of ozone was first noted in Antarctica. The reason it appeared was at first a mystery. The answer appears to be that a polar vortex forms over the South Pole during the winter. This whirling mass of very cold air isolates the air above the pole from the rest of the atmosphere. The vortex disappears during spring, releasing the trapped air. The air in the vortex is extremely cold, about $-90°C$, and atmospheric clouds are formed composed of ice crystals. On the surface of these crystals the following reactions appear to occur:

$$ClONO_2 + H_2O \longrightarrow HOCl + HNO_3$$
$$HOCl + HCl \longrightarrow Cl_2 + H_2O$$
$$ClONO_2 + HCl \longrightarrow Cl_2 + HNO_3$$

With the coming of spring, the chlorine is released and is split by light as before:

$$Cl_2 + hv \longrightarrow 2Cl$$

and the process of destroying ozone begins anew but now with a huge influx of stored chlorine. This causes the formation of the annual ozone hole. In late spring ozone-rich air flows in, and the hole is again plugged.

The concern with the reduction of stratospheric ozone worldwide is that the ultraviolet radiation can cause skin damage, particularly the formation of skin cancers. Estimates generally project that a 1% decrease in ozone will result in a 2% increase in non-melanoma skin cancers. The relationship between O_3 levels and melanoma, a more lethal form of skin cancer, is uncertain. The most at-risk people are those with fair skin who stay out in the sun too much.

In addition to skin cancer, increased ultraviolet radiation can cause eye damage, can suppress the immune system, and can even reduce plant photosynthesis. Changes in natural ecosystems, such as the mysterious malady suffered by freshwater frogs, may be attributed to the increased incidence of ultraviolet radiation.

The reduction of CFCs in the stratosphere obviously takes a multinational effort, and this is one rare instance when many nations united to reduce the effect of global pollution. This process took some time, of course. CFCs were first invented in the 1930s, and it wasn't until 1970 that the USEPA banned CFCs in nonessential aerosols. In 1974 scientists predicted that CFCs would reduce ozone, but 15 years had to pass before British

FOCUS ON

Believe Thine Own Data

Ozone levels in the stratosphere had been measured since the 1970s using satellites (Figure 12.15). In addition to satellite data, a British team of scientists in the Antarctic began to measure the concentration of ozone in the stratosphere using sophisticated instruments at the base near the pole. In 1985 they discovered that there had been a dramatic drop in the ozone levels in the stratosphere, creating a large hole over the South Pole. The satellite data, however, did not show such a hole, and the scientists concluded that their instruments must be faulty. They had a new set of instruments flown in, and even with careful calibration, they had to conclude that this huge hole had indeed developed. They studied the satellite data and found that the software was designed to ignore very low levels of ozone and report it simply as no data. The satellite had seen this hole for many years, but the data reduction did not recognize the presence of the hole. When the scientists adjusted the program to report low levels of ozone, the satellite data agreed with the data obtained on the ground. There indeed was a huge monster hole right in the middle of the South Pole.

Source: Reprinted from *Controlling Environmental Pollution*, Vesilind, P. A. and DiStefano, T. D., 2006. Lancaster, PA: DEStech Publications, Inc. Reproduced with permission of DEStech Publications, Inc.

Figure 12.15   The ozone hole.

meteorologists at the South Pole discovered the ozone hole and the public became concerned. This concern was translated into the Montreal Accord in 1987, which sought to voluntarily cut CFCs by 50% by 1999 and then at an accelerated rate phase out CFCs completely. DuPont, the largest manufacturer of CFCs, voluntarily stopped their manufacture in 1988. Many refrigeration systems, such as car air conditioners, however, required CFCs, so the remaining stocks of CFCs became very dear. A considerable amount of CFCs was smuggled into the United States before the refrigeration systems were able to be adapted to other refrigerants.

12.3.4   Global Warming (Climate Change)

Earth acts as a reflector to the sun's rays, receiving the radiation from the sun, reflecting some of it into space, and absorbing the rest, only to radiate this into space as heat. In effect, earth acts as a wave converter, receiving the high-energy, high-frequency radiation from the sun and converting most of it into low-energy, low-frequency heat to be radiated back into space. In this manner earth maintains a balance of temperature, so that

$$\begin{bmatrix} \text{Energy from the sun} \\ \text{IN} \end{bmatrix} = \begin{bmatrix} \text{Energy radiated back to space} \\ \text{OUT} \end{bmatrix}$$

This is a very simple equation, so global temperature change ought to be able to be modeled because we know the primary variables, such as the size of earth, the energy coming from the sun, and so on. Such a simple model is developed by Masters, and much of the following discussion is credited to him.[5] The model depends on an energy balance as noted.

Consider earth as receiving light from the sun, as shown in Figure 12.16. The amount of radiation arriving from the sun is calculated on the basis of intensity, called a *solar constant*, S. This constant has been found to be about 1370 W/m^2. The area receiving this radiation is measured as the area of earth's shadow, which is calculated as πR^2, where R is the radius of earth. Another way of picturing this is as a circle in front of earth through which all of the sun's radiation must travel.

The rate at which solar radiation hits earth can then be calculated as

$$B = S\pi R^2$$

where B = rate at which solar light energy strikes earth, W
   S = the solar constant, estimated as 1370 W/m^2

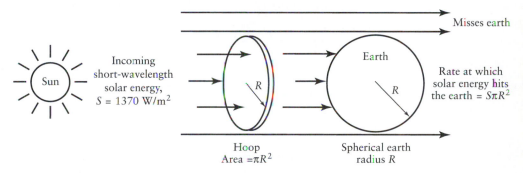

Figure 12.16   Solar energy from the sun hitting earth. (After Masters, G. 1998. *Introduction to Environmental Engineering and Science.* Upper Saddle River, NJ: Prentice-Hall. Used with permission.)

Because earth is a reflector (that's why we can see it from space!), some of the light energy is reflected back. The fraction of energy reflected back is called the *albedo*, and this has been estimated as being about 31%. The albedo obviously varies with groundcover, ice and snow, and other variables. If earth's average albedo is α, then the amount of light energy reflected back into space is

$$F = S\pi R^2 \alpha$$

where F = light energy reflected into space, W
    α = albedo, or the fraction of light energy reflected, estimated as 31%

Using the energy balance, the amount of energy absorbed by earth is the difference between B and F, or

$$A = S\pi R^2 (1 - \alpha)$$

where A = light energy absorbed by earth, W.

Because earth is a wave converter, the light energy absorbed is converted to heat. Every object radiates energy at a rate proportional to its surface area and its absolute temperature to the fourth power, or

$$E = \sigma A T^4$$

where E = energy radiated back to space
    σ = the Stefan–Boltzmann constant = 5.67×10^{-8} W/m^{2}K^4
    T = absolute temperature, K

Making some fairly outrageous assumptions, such as assuming that the temperature of earth is the same everywhere on the surface and that radiation is perfect, we can estimate that the heat energy radiated back into space is

$$E = \sigma 4\pi R^2 T_e^4$$

where T_e = earth's average temperature, K

We now have an estimate of the rate of (light) energy coming to earth and the rate of (heat) energy emitted back into space, and this has to be in balance, so that

$$A = E$$
$$S\pi R^2 (1 - \alpha) = \sigma 4\pi R^2 T_e^4$$

Solving for the temperature:

$$T_e = \left[\frac{S(1 - \alpha)}{4\sigma} \right]^{1/4}$$

Substituting some typical values, the temperature of earth's surface can be calculated as

$$T_e = \left[\frac{(1370 \text{ W/m}^2)(1 - 0.31)}{(4)(5.67 \times 10^{-8} \text{ W/m}^2\text{K}^4)} \right]^{1/4} = 254 \text{ K} = -19°\text{C}$$

But we know that, on average, earth's temperature is about 15°C. Why the huge discrepancy in the model? If earth's temperature was, indeed, −19°C, then there would be little life here, so there must be some error in the model. The error, it turns out, is that the reflection of some heat off earth's atmosphere was ignored. The atmosphere is like a greenhouse in that it allows light through but prevents some of the heat from escaping back into space, and this is appropriately called the *greenhouse effect*.

It is important to realize that this greenhouse effect occurs in an unpolluted (by people) atmosphere and has been happening for thousands of years. It is the greenhouse effect that keeps earth 34°C warmer than it would otherwise be and is responsible for the temperate climates that allowed life to develop.

The light energy from the sun has a spectrum somewhat like that shown in Figure 12.17A. Almost all the energy is at a wavelength less than 3 μm. The outgoing heat energy, however, has a very different spectrum, as shown in Figure 12.17B. Almost all this energy has a wavelength greater than 3 μm. For this reason we often refer to the light energy coming from the sun as *high-frequency*, *short-wavelength*, and the heat energy emitted to space as *low-frequency*, *long-wavelength*.

Earth's atmosphere is made up of various gases, and each of these absorbs heat energy at specific wavelengths. Figure 12.18 on the next page shows the absorptivity of various gases as a function of wavelength. Also shown in the figure are the spectra for sunlight and heat radiation (from Figure 12.17A and B). In Figure 12.18 note, for example, the effect of carbon dioxide, CO_2. Carbon dioxide absorbs almost none of the sunlight coming to earth because its absorptive effect is most pronounced at wavelengths greater than about 1.5 μm, missing most of the sunlight spectrum. Looking to the right side of Figure 12.18, however, it is clear that carbon dioxide can be an effective energy absorber at the frequencies normal to heat radiation from earth. The same is true for water vapor, methane, and nitrous oxide. Methane especially has a very large "bump" at about 8 μm, which is very close to

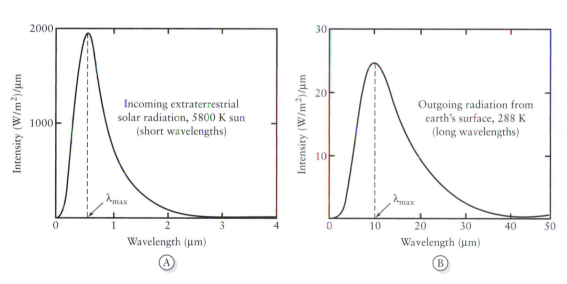

Figure 12.17 Incoming solar radiation (light energy) from the sun and outgoing heat energy from earth's surface into space. (After Masters, G. 1998. *Introduction to Environmental Engineering and Science*. Upper Saddle River, NJ: Prentice-Hall. Used with permission.)

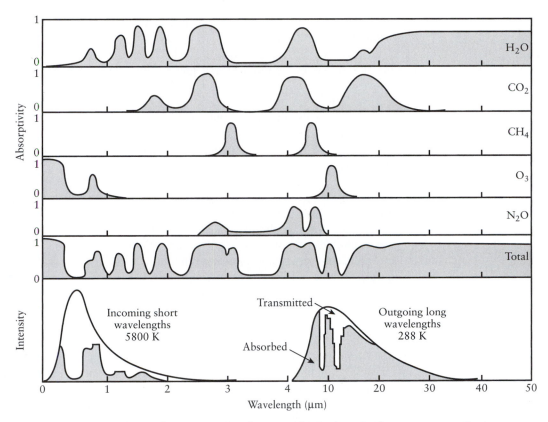

Figure 12.18 Absorptivity as a function of wavelength of various atmospheric gases. At the bottom are the spectra for incoming light energy and outgoing heat energy. The gases are far more efficient at absorbing the outgoing heat energy than in attenuating the incoming light energy. (After Masters, G. 1998. *Introduction to Environmental Engineering and Science.* Upper Saddle River, NJ: Prentice-Hall. Used with permission.)

the peak of the heat energy spectrum, and, thus, small increases in CH_4 have significant effects on the ability of earth to retain heat energy while doing nothing to shield Earth from light energy. The increase in the concentration of the so-called greenhouse gases is what constitutes the problem with global temperature change, much like it is the increase in the phosphorus level in lakes that causes eutrophication.

While it is fairly easy to measure the concentrations of greenhouse gases, such as CO_2, CH_4, and N_2O, and to show that the levels of these gases have been increasing during the past 50 years, it is another matter to argue from empirical evidence that the increase in these gases is causing the increases in earth's temperature. The temperature of earth undergoes continual change, with fluctuation frequencies ranging from a few years to thousands of years.

However, there is near consensus among scientists that humans are causing the current temperature increases, resulting in the phenomenon known as global warming. (Note that the phrase "climate change" is increasingly preferred to "global warming" because it conveys that there are changes in addition to rising temperatures.) The primary controversy that continues is about the likely impacts and what to do, if anything. Peoples' opinions lie all along the spectrum. Some still believe global warming is a natural phenomenon that is not an issue. Others believe that humans are a causative factor but that global warming won't really matter. Others believe it is a looming catastrophe that requires urgent action, and still others believe it is an inevitable catastrophe and it is too late to do anything about it. But there is growing momentum by government and industry to take action–to find ways to measure and significantly reduce emissions in order to lessen the impacts.

FOCUS ON

Carbon Footprinting

Figure 12.19   Jet Trails, 2007—Depicts 11,000 jet trails, equal to the number of commercial flights in the U.S. every eight hours.

Businesses are subject to increasing environmental legislation and growing expectations for corporate social responsibility and stakeholder engagement. These drivers are leading to the development of sustainable and "environmentally friendly" products and to a growing number of organizations now looking to assess their impact on the environment. The onset of climate change and the increasing cost of fuel means that energy use is now a key consideration when conducting these impact assessments or similar studies, such as life cycle analysis (LCA) (Figure 12.19). A carbon footprint is the standard format for expressing global warming impact, taking into account not only emissions from energy use but also other gases said to play a major role in climate change [i.e., CO_2, CH_4, N_2O, PFCs (perfluorocarbons), HFCs (hydrofluorocarbons), and SF_6].

Although carbon footprints usually show the greenhouse gas (GHG) emissions for which a product or organization is responsible, they can also be calculated for events, specific sites, or individuals. For clarity, and because carbon dioxide is the most important greenhouse gas, other GHGs are reported relative to carbon dioxide (CO_2), to provide a total footprint in terms of CO_2 equivalent (CO_2e). The impact from 1 tonne of methane, for example, would be equivalent to 21 tonnes of CO_2.

When calculating a footprint, GHG emissions will fall into one of three areas, or scopes. These are based on definitions from the Greenhouse Gas Protocol with the methodology for corporate emissions reporting now covered by international standard ISO 14064. Other documents, such as PAS2050, are looking to provide further clarity on how a carbon footprint should be calculated.

- *Scope 1, direct emissions.* These are emissions that the company has direct control of, such as gas used for heating or emissions from company-owned vehicles.
- *Scope 2, imported power/heat/steam.* Most organizations use electricity, and, though some of it may be produced from renewable sources, fossil fuels still supply the major portion. Although the company cannot control the emissions produced offsite, it is still indirectly responsible for the CO_2 produced by power stations.
- *Scope 3, indirect emissions.* The most challenging of the three areas includes emissions upstream and downstream of a product or service. Producing a carton of milk, for example, has indirect GHG emissions from a number of sources, including transport, packaging, processing, refrigeration, waste disposal, and farm animals.

Calculating a footprint can be a complex task. The level of detail needed will ultimately depend on its use. A company looking to calculate its own footprint as part of an internal energy management policy may decide on a more basic approach than that required for a product or requested by a third party. It is important that a company clearly identify boundaries, dictating the terms of the footprint and the overall approach, including any source of GHGs that were not included in calculations. Factors such as shared office space, rented buildings, and ownership or control of the company may also impact who is responsible for emissions and should be incorporated into the boundary conditions. Not only will this provide transparency, it will also allow the figure to be updated over time because a carbon footprint is a dynamic measurement.

In Europe, carbon footprints have already started appearing on major brands, such as

- Tesco, which has calculated footprints on 20 of its grocery products
- Walkers Crisps, which has calculated footprints across their entire range of potato chips
- Boots Pharmacy, which has calculated the footprint of their Botanics Shampoo.

North American companies looking for a competitive edge and a way to become energy conscious are also calculating their carbon footprints, so expect to see more of these measurements in the media and on product labels.

Source: Provided by Phil Kingston and Margaret H. Whittaker of Linden Consulting Partnership (Essex, England).

The most remarkable thing about global heat might be that it has been so constant over millions of years. Somehow earth has been able to develop the thermal conditions suitable for the creation and support of life. Speculating on this idea, James Lovelock suggested the Gaia hypothesis, that earth is actually a single entity (Mother Nature) that lives much like any other creature and has to adapt to changing conditions as well as to fight off diseases.[6] The name "Gaia" comes from the Greek name for the nurturing earth goddess. Some Gaians have interpreted this notion in its broadest and most spiritual sense, taking the view that earth really is one organism, albeit an unusual one, and that it has many of the characteristics of other organisms. Others (including James Lovelock, who has been a little embarrassed about this Gaia business) see the hypothesis as nothing but a feedback response model and do not ascribe mysticism to it.

But suppose humans are simply one part of a whole living organism, earth, much like brain cells are a part of an animal. If this is true, then it makes no sense to destroy one's own body, and therefore, it makes no sense for humans to destroy the rest of the creatures that coinhabit earth. To think of earth as a living organism suggests a spiritual rather than a scientific approach, especially if it is taken literally.

If we accept this notion, some interesting extensions can result. One could speculate (and this is pure whimsy!) that earth (Gaia) is still developing and is going through adolescent stages. Most notably, it has not settled on the carbon balance. Millions of years ago, most of the carbon on earth was in the atmosphere as carbon dioxide, and the preponderance of CO_2 promoted the development of plants. But as the plants grew, they soon started to rob the atmosphere of carbon dioxide, replacing it with oxygen. This change in atmospheric gases promoted the growth of animals, which converted the oxygen back to carbon dioxide, once again seeking a balance. Unfortunately, Gaia made a mistake and did not count on the animals dying in such huge numbers, trapping the carbon in deep geologic deposits that eventually became coal, oil, and natural gas.

How to get this carbon out? What is needed is some semi-intelligent being with opposable digits. So Gaia invented humans! It can therefore be argued that our sole purpose on earth is to dig up the carbon deposits as rapidly as possible and to liberate the carbon dioxide. This is the long-sought-for meaning of life.

Unfortunately, Gaia again messed up and did not count on humans becoming so prolific and destructive and especially did not count on humans inventing nuclear power, which eliminates their sole purpose for existing. The sheer numbers of humans (especially if they are not going to dig up the carbon) become a problem, much as a pathogenic bacterium causes an infection in an organism. It is not a coincidence that the growth of the human population on earth is not at all unlike the growth of cancer cells in a malignancy.

So next Gaia has to limit the number of humans, and she will do so by developing antihumatics (as in antibiotics) that will kill off a sufficient number of humans and once again bring Earth to a proper balance. The increase in highly resistant strains of bacteria and viruses may be the first indications of a general culling of the human population.

The application of the Gaia hypothesis to environmental ethics leads to existentialism—the notion that life is meaningless (except as a burner of fossil fuel) and that one should not concern oneself with the environment. Grim stuff.

12.3.5   Other Sources of Air Pollutants

Combustion processes create a range of air pollutants. Nitrogen oxides are formed by the thermal fixation of atmospheric nitrogen in high-temperature processes. Accordingly, almost any combustion operation will produce nitric oxide (NO). Other pollutants of interest from combustion processes are organic acids, aldehydes, ammonia, and carbon monoxide. The amount of carbon monoxide emitted relates to the efficiency of the combustion operation; that is, a more efficient combustion operation will oxidize more of the carbon present to carbon dioxide, reducing the amount of carbon monoxide emitted. Of the fuels used in stationary combustion, natural gas contains practically no sulfur, and particulate emissions are almost nil.

Transportation sources, particularly automobiles using the internal combustion engine, constitute a major source of air pollution. Particulate emissions from the automobile include smoke and used to include lead particles, the latter usually as halogenated compounds. Smoke emissions, as in any other combustion operation, are due to the incomplete combustion of carbonaceous material. On the other hand, lead emissions related directly to the addition of tetraethyl lead to the fuel as an antiknock compound, which was successfully phased out. Gaseous pollutants from transportation sources include carbon monoxide, nitrogen oxides, and hydrocarbons. Hydrocarbon emissions result from incomplete combustion and evaporation from the crankcase, carburetor, and gasoline tank.

Pollutant emissions from industrial processes reflect the ingenuity of modern industrial technology. Thus, nearly every imaginable form of pollutant is emitted in some quantity by some industrial operation.

Although solid waste disposal operations need not be a major source of air pollutants, some communities still permit backyard burning of solid waste. Other communities use incinerators for solid waste management. Burning, whether in the backyard or in incinerators, attempts to reduce the volume of waste but may produce instead a variety of difficult-to-control air pollutants. Inefficient systems contribute to many pollutants, some with undesirable odors; among these are carbon monoxide, small amounts of nitrogen oxides, organic acids, hydrocarbons, aldehydes, and smoke. Landfills, which are widely used to dispose of municipal solid waste, generate large quantities of methane in addition to other air pollutants.

12.3.6 Indoor Air

A sticky problem for the USEPA has been control of indoor air quality. Does USEPA's mandate include the indoor environment, and should they, in effect, become a health agency, or should they limit their concern to outdoor air? Pragmatically, if they do not address the problems of indoor air, no other federal agency seems ready to do so.

Indoor air quality is of importance to health simply because we spend so much time indoors, and the quality of the air we breathe is seldom monitored. Contaminated indoor air can cause any number of health problems, including eye irritation, headache, nausea, sneezing, dermatitis, heartburn, drowsiness and many other symptoms. These problems can arise as a result of breathing harmful pollutants, such as

- asbestos—from fireproofing and vinyl floors
- carbon monoxide—from smoking, space heaters, and stoves
- formaldehyde—from carpets, ceiling tile, and paneling
- particulates—from smoking, fireplaces, dusting
- nitrogen oxides—from kerosene stoves and gas stoves
- ozone—from photocopying machines
- radon—from diffusion out of the soil

- sulfur dioxide—from kerosene heaters
- volatile organics—from smoking, paints, solvents, and cooking.

While the air exchange and cleaning in office buildings and many apartments are controlled, private residences usually depend on natural ventilation to provide air exchange.

Most houses are, in fact, poorly sealed, and air leakage takes place at many locations, including through doors and windows, exhaust vents, and chimneys. In warmer weather some homes also have forced ventilation, with whole-house fans or individual window fans.

In calculating the ventilation of any enclosure, such as a home or office, engineers use the concept of *air change*, defined as

$$a = \frac{Q}{V}$$

where a = number of air changes per hour, 1/h
Q = flow rate, m³/hr
V = volume of enclosure, m³

**EXAMPLE
12.5**

Problem   A small room is to be used for a copy machine, and there is concern that the ozone level may be too high unless the room is ventilated. The volume of the room is 20,000 L, and it is recommended that the number of air changes per hour be 30. What flow of air must the fan deliver?

Solution   From the preceding equation:

$$Q = (a)(V)$$

Fan air-handling capacity $= (30 \text{ hr}^{-1})(20,000 \text{ L}) = 600,000 \text{ L/h}$

The emission of indoor air pollutants can be handily analyzed by using the black box technique, mainly because an enclosure literally is a box. Pollutants are emitted within the box and are completely and ideally mixed while clean air is brought in and contaminated air is flushed out. The material balance in terms of the pollutants is

$$\begin{bmatrix} \text{Rate of} \\ \text{pollutants} \\ \text{ACCUMULATED} \end{bmatrix} = \begin{bmatrix} \text{Rate of} \\ \text{pollutants} \\ \text{IN} \end{bmatrix} - \begin{bmatrix} \text{Rate of} \\ \text{pollutants} \\ \text{OUT} \end{bmatrix}$$

$$+ \begin{bmatrix} \text{Rate of} \\ \text{pollutants} \\ \text{PRODUCED} \end{bmatrix} - \begin{bmatrix} \text{Rate of} \\ \text{pollutants} \\ \text{CONSUMED} \end{bmatrix}$$

FOCUS ON

The Mystery of Legionnaire's Disease

The bicentennial convention of the American Legion was held in Philadelphia in 1976 at the elegant but aging Bellevue Stratford Hotel (Figure 12.20). Toward the end of the conference, a number of the attendees were suffering from flu-like symptoms, and 4 days after leaving Philadelphia, an Air Force veteran died from pneumonia. During the next week, 30 additional Legionnaires who had been at the conference died from what became known as Legionnaire's Disease.

The Center for Disease Control in Atlanta was called in to find out what was going on. Theories ranged from viral pneumonia to terrorism (before this became a popular alternative) to a communist plot (after all, these were mostly veterans who died). The first step was to find out if there were some common organisms that might explain the disease. Using tissues from lung biopsies and samples of sputum, microbiologists discovered a gram-negative rod-shaped bacterium that was present in all of the samples. The next step was to try to grow this organism, now called *Legionella pneumophila*, in the laboratory and then to use it to infect an animal that would exhibit symptoms similar to pneumonia. Through trial and error, microbiologists were able to grow *L. pneumophila* and use it to infect guinea pigs. There was now little doubt that this organism was the cause of the Legionnaire's disease, but where did it come from, and why and how did it infect the Legionnaires in Philadelphia?

Microbiological detective work suggested that the organism is thermophilic, and thus if it is to be found in nature, it should exist in warm places. Sure enough, the *L. pneumophila* turned out to be similar to organisms cultured from warm springs and geysers. Further work showed that the organism liked warm, dark places and grew in a biofilm associated with specific species of algae. The next step was to find how the Legionnaires came into contact with the pathogen. Recognizing that this must be an airborne infection, the epidemiologists focused on the air quality in the hotel and discovered that biofilms were growing in the air ducts of the hotel heating and ventilating system. Within these biofilms they isolated *L. pneumophila*, and the mystery was solved.

Source: Reprinted from *Controlling Environmental Pollution*, Vesilind, P. A. and DiStefano, T. D., 2006. Lancaster, PA: DEStech Publications, Inc. Reproduced with permission of DEStech Publications, Inc.

© AP/Wide World Photos

Figure 12.20   The Bellevue Stratford Hotel.

If steady state is assumed, then the first term is zero. The second term can be assumed zero if the incoming air is clean, and if the pollutants are not consumed, the last term drops out. Hence:

$$0 = 0 - [\text{Rate OUT}] + [\text{Rate PRODUCED}] - 0$$

The rate of pollutants leaving is equal to

$$M = CQ = CaV$$

where M = rate of pollutant leaving the room, mg/h
$\quad\quad\,\,\, C$ = concentration of pollutants in the room, mg/m^3
$\quad\quad\,\,\, a$ = air change rate, 1/h
$\quad\quad\,\,\, V$ = volume of the enclosure, m^3

The rate of pollutants being produced is called the *source strength*, S, expressed as mg of pollutants emitted per hour. From the material balance, the concentration of the pollutant in the room and in the exiting air is then

$$C = \frac{S}{aV}$$

EXAMPLE

12.6

Problem   Smokers in a room (5 m × 5 m × 2.5 m) are smoking so that there is on average one cigarette burning at all times. If the emission rate of a cigarette is 86 mg/hr of CO and if the ventilation is 0.5 change per hour, what is the level of CO in the room?

Solution

$$C = \frac{S}{aV} = \frac{86 \text{ mg/hr}}{(0.5/\text{hr})(5 \text{ m})(5 \text{ m})(2.5 \text{ m})} = 2.752 \text{ mg/m}^3$$

This converts to 2752 μg/m^3. Is this a concentration that would be of concern? (Refer to Figure 12.10.)

If the pollutant is decaying or being removed by adhering to surfaces, then the "CONSUMED" term cannot be assumed to be zero. The decay or removal of the pollutant can be assumed to be first order, or proportional to the concentration in the enclosure (Chapter 5). Similarly, if the ventilation air is not perfectly clean and contains some of the pollutant, the "IN" term cannot be zero. The steady state equation then must read

$$0 = [\text{Rate IN}] - [\text{Rate OUT}] + [\text{Rate PRODUCED}] - [\text{Rate CONSUMED}]$$

$$0 = C_0 aV - CaV + S - KCV$$

or

$$C = \frac{S/V + C_0 a}{a + K}$$

where C_0 = ambient air quality, air used for ventilation
   a = air change rate
   V = volume of enclosure
   C = enclosure or exhaust concentration
   S = source strength
   K = rate of decay constant

One of the most insidious indoor air pollutants is secondary smoke from cigarettes and pipes. Most of the particulates in a cigarette are emitted into the room without being inhaled by the smoker, and these are then inhaled by everyone else in the room. Cigarette smoke also contains high amounts of CO, and when several people are smoking in a room, the CO level can become high enough to affect performance. One of the smallest rooms we commonly live in is our car, and cigarette smokers can significantly affect the CO level in a car. Smokers exhale high levels of CO even when they are not smoking, and this has been blamed for the sleepy driver syndrome in commuter cars.

Another important and troublesome indoor air pollutant is *radon*, a naturally emitted gas entering homes through basements, well water, and even building materials. Radon and its radioactive daughters are part of the natural decay chain that begins with uranium and ends with lead. The decay products of radon—polonium, lead, and bismuth—are easily inhaled and can readily find their way into lungs. The most important health effect of radon therefore is lung cancer. Radon control is discussed further in Chapter 15.

There are no effective means of correlating the inhalation of radon gas and incidence of lung cancer, and all studies depend on very high exposure rates, such as to uranium miners. Relatively speaking, however, the risk of radon in the home is considerable when compared to other potential sources of cancer. Because radon is a carcinogen, it is considered to have no threshold, and therefore, there is no safe level of radon in the home. Nevertheless, the USEPA recommends that residents living in homes that contain 4 pCi/L (pico Curies of radon per liter of air) take corrective action. Mitigating action usually involves the sealing of the basement and other soil contact areas and the ventilation of basements to prevent the radon gas from seeping upstairs into the living quarters.

It is estimated that the risk of contracting lung cancer due to living for 70 years in a house containing 1.5 pCi/L is about 0.3%.[7] This is a high risk when compared to other regulated carcinogens, which are in the range of 1×10^{-4}%. Radon exposure, in fact, ranks high in USEPA's list of problem areas, responsible for 20,000 lung cancer deaths annually.

While this may seem to represent a major problem, the risk of harm by radon still pales when compared to the voluntary risks we routinely impose on ourselves, such as driving cars, drinking alcohol, and smoking. In fact, living all of one's life in a home with between 10 and 20 pCi/L represents an equivalent risk of smoking one pack of cigarettes per day.

12.4   AIR QUALITY STANDARDS

Historically, local communities have been afraid to enact strict air pollution ordinances for fear of driving away industry. While there are cheaper and less restrictive locations, industrial plants could threaten to leave town, taking much needed jobs with them. In the United States only federal legislation has been able to prevent this type of blackmail.

But even with the present federal legislation, industry has the option to move off shore, to China or to Latin America, where pollution control regulations are not as stringent. This seems to be a clearly unethical trend, transporting pollution to other countries less able to resist such contamination.

On the other hand, the United States grew rich, in part, because we were able to produce better products at cheaper prices, and one reason the cost was low was because we did not bother with pollution. The water and air were free, and there was so much of it that it made little sense to clean up. Nature seemed to do a pretty good job by itself.

Now countries considerably poorer than the United States are saying that they would like to become just as wealthy and would like to use their water and air the same way—that is, pollute it. They welcome American industry that promises to build plants in their countries and decline to implement strict pollution regulations.

What should be the ethical stance of American industry in this case? If industry declines to invest in the poorer countries, these countries are deprived of income and benefits, and the gap between the poor and rich nations will increase. If, on the other hand, industry invests in these countries and builds manufacturing facilities, industry must have some advantage, such as minimal pollution control regulations, to make it worth their while, and hence, the air and water will be polluted to what we would consider unacceptable levels. What ethical responsibility, if any, does industry have to implement American pollution control standards for facilities located in countries that do not require such strict standards?

12.4.1 Air Quality Legislation in the United States

The history of air quality standards has been long and often tumultuous. At first all air pollution was thought of as smoke pollution, and the first legislation governing smoke was passed in Los Angeles in 1905. Los Angeles sits in a basin that is frequently covered by a thermal inversion, and even from the days of the conquistadors it was evident that smoke from campfires rose only to a set level and remained there.

By 1943 the number of automobiles in Los Angeles had increased to where they were experiencing the world's first photochemical smog. This phenomenon was thought at first to be due to industries and unique to Los Angeles. It took many years for both of these myths to be exposed. Angelenos simply did not want to believe that their much-beloved cars were the cause of such putrid air. It became obvious that something had to be done about automobile emissions. This was a classical uphill battle for the scientists and regulators, however. The automobile lobby, which includes the automobile manufacturers, the oil industry, and the people who wanted cheap personal transportation, engaged in a now notorious campaign of footdragging, denial, and even obfuscation. They launched public relations and lobbying campaigns to prevent or slow down any tampering with cars or the fuels. When it finally became evident to all that smog in Los Angeles was indeed due mostly to internal combustion engines, the industry insisted that this was a special case for Los Angeles and that national legislation was not necessary. Photochemical smog could not happen anywhere else, they claimed.

Finally, as it became evident that both California and the United States were about to pass technology driving legislation restricting auto emissions, the American auto industry bemoaned about their inability to meet these goals and that cars would become undrivable.

Much of their bombast was blunted when Honda came out with a vehicle that not only met all of the stringent exhaust requirements but also got 16 km to a liter of gas!

The 1963 Clean Air Act, a major piece of legislation that for the first time anywhere set both emission and ambient air quality limits, was passed—in great part due to the efforts of Senator Edmund Muskie of Maine. The act required the federal government to set ambient air quality standards for seven major air pollutants. These were not implemented until 1970, and the states were allowed until 1975 to meet them. In order to not drag out this gruesome tale, suffice it to say that many states have yet to meet these standards, and the city of Los Angeles may have to go to extreme measures to reduce its photochemical smog.

The 1990 amendments to the Clean Air Act added over 180 hazardous materials (now known as *air toxics*) to the ambient air quality standards and required significant cuts in the emissions of sulfur and nitrogen oxides, precursors of acid rain. The act also extended the deadline for meeting the ambient ozone standard until the year 2007—44 years since the original 1963 Clean Air Act!

12.4.2 Emission and Ambient Air Quality Standards

The emission standards are regulated (and for the most part set) by the individual state air quality offices. As an example of an emission standard, an incinerator might not be allowed to exceed emissions of X $\mu g/m^3$ of particulates at a specified CO_2 minimum level. The latter is necessary because particulate concentrations can be reduced by simply diluting with excess air. But because air contains negligible CO_2, the minimum CO_2 level prevents emission standard attainment by simple dilution.

The USEPA also has the authority to set national emission performance standards for hazardous air pollutants. The list of pollutants is continually growing, and once a pollutant is *listed*, the USEPA can enforce the emission standards for larger facilities. Presently listed are hundreds of chemicals, such as asbestos, benzene, ethylene glycol, methanol, phenol, styrene, and vinyl chloride.

The second type of air quality standards parallel the stream standards in water quality and specify the minimum quality of ambient air. Included in these standards are particulates, sulfur dioxide, carbon monoxide, photochemical oxidants, hydrocarbons, nitrogen oxides, and lead. As data become available, other standards will be developed.

Similar to drinking water standards, the ambient air quality standards are of two kinds: primary and secondary. Primary standards are intended to relate to human health while secondary standards address other problems, such as corrosion, animal health, and visibility. Table 12.4 is a listing of the present national ambient air quality standards (NAAQS) for six principal pollutants, known as *criteria pollutants*.

Areas in the United States where the NAAQSs are exceeded on average more than twice a year for any of the pollutants are known as *nonattainment areas* for those pollutants. In such areas air pollution control programs must be initiated to bring the area back into compliance, and industries contemplating expansion must show how they can improve the air quality by reducing other emissions. In some areas automobiles are required to change fuels to reduce emissions. Because the internal combustion engine is the largest contributor of air pollution, a replacement that uses hydrogen or electricity would be a significant contribution to cleaner air.

Table 12.4 National Ambient Air Quality Standards *

Pollutant and Averaging Time	Primary Standard	Secondary Standard
CO		
1 hr	35 ppm (40 mg/m²)	None
8 hr	9 ppm (10 mg/m³)	None
Pb		
Rolling 3-mg average	0.15 µg/m³	0.15 µg/m³
Quaterly	1.5 µg/m³	1.5 µg/m³
NO₂		
Annual (arithmetic mean)	100 µg/m³	100 µg/m³
*PM₁₀***		
24 hr	150 µg/m³	150 µg/m³
*PM₂.₅***		
24 hr	35 µg/m³	35 µg/m³
Annual (arithmetic mean)	15.0 µg/m³	15.0 µg/m³
O₃		
1 hr (for 8-hr nonattainment areas)	0.12 ppm	0.12 ppm
8 hr (2008 standard)	0.075 ppm	0.075 ppm
8 hr (1997 standard)	0.08 ppm	0.08 ppm
SO₂		
3 hr	None	0.5 ppm (1300 µg/m³)
24 hr	0.14 ppm	
Annual (arithmetic mean)	0.03 ppm	

*If interested, see the USEPA's website for important footnotes regarding calculating pollutant concentrations to evaluate against the standards.
**"PM_{10}" and "$PM_{2.5}$" stand for particulate matter less than 10 µm and 2.5 µm in diameter, respectively.

PROBLEMS

12.1 A 1974 car is driven an average of 1500 km/month. The EPA 1974 emission standards were 3.4 g/km for HC and 30 g/km of CO.

a. How much CO and HC would be emitted during the year?

b. How long would it take to exceed lethal concentration of CO in a common double-car garage, $6 \times 7.5 \times 2$ m?

12.2 A 2.5% level of CO in hemoglobin (COHb) has been shown to cause impairment in time-interval discrimination. The level of CO on crowded city streets sometimes hits 10 µg/m³. An approximate relationship between CO and COHb (after prolonged exposure) is

$$COHb \ (\%) = 0.5 + (0.16)(10)(CO \ µg/m^3)$$

What level of COHb would a traffic cop be subjected to during a working day of directing traffic on a city street?

12.3 Photochemical smog is a serious problem in many large cities.

a. Draw a graph showing the concentration of NO, NO_2, HC, and O_3 in the Los Angeles area during a sunny, smoggy day.

b. Draw another graph showing how the same curves appear on a cloudy day. Explain the difference.

c. The only feasible way of reducing the formation of photochemical smog in Los Angeles seems to be to prevent automobiles from entering the city. Draw the same curves as they might appear if *all* cars were banned from LA streets.

d. Would this ever happen? Why or why not?

12.4 Give three examples of synergism in air pollution.

12.5 If SO_2 is so soluble in water, how can it get to the deeper reaches of the lung without first dissolving in the mucus?

12.6 A hi-vol clean filter weighs 18.0 g, and the dirty filter weighs 18.6 g. The initial and final air flows are 2000 L/min and 1100 L/min. What volume of air went through the filter in 24 h? What was the particulate concentration?

12.7 A high-volume sampler draws air in at an average rate of 2000 L/min. If the particulate reading is 200 $\mu g/m^3$, what was the weight of the dust on the filter?

12.8 Research and report on one of the classical air pollution episodes such as Meuse Valley, London, or Donora.

12.9 What does Ringlemann 5 tell you about smoke being emitted from a chimney?

12.10 The data for a hi-vol are as follows:
Clean filter: 20.0 g
Dirty filter: 20.5 g
Initial air flow: 2000 L/min
Final air flow: 1400 L/min
Time: 24 h

a. What volume of air was put through the filter? (Answer in liter.)

b. How is the air flow in a hi-vol measured?

c. What is the weight of the particulates collected?

d. Was the NAAQS exceeded?

12.11 The primary ambient air quality standard for NO_2 is 100 $\mu g/m^3$. What is this in ppm? (Assume 298 K and 1 atm pressure.)

12.12 The concentration of carbon monoxide in a smoke-filled room can reach as high as 500 ppm.

a. What is this in $\mu g/m^3$? (Assume 1 atm and 298 K.)

b. What effect would this have on people who are sitting around having a political discussion for 4 h?

12.13 Figure 12.21 shows schematically the global average energy flow among space, the atmosphere, and earth's surface. The units in this figure are in watts per square meter of surface area.

a. Using space, the atmosphere, and earth as black boxes, check if the numbers represent valid balances.

b. Suppose earth experienced the eruption of several large volcanos, spewing dust into the atmosphere. How might these numbers change, and what effect would such an event have on earth's temperature?

c. Since most of the greenhouse gases are emitted from industrialized nations in the northern hemisphere, what arguments might the director of an environmental protection agency in an equatorial country present to argue for the curtailment of greenhouse gas emissions? Write a short radio commentary for the environmental protection agency director, one that he could read for National Public Radio in the United States. Remember that emotionalism will not win nearly the public support that a solid ethical argument will.

12.14 A woodburning stove operates for 4 hr in a room measuring $5 \times 5 \times 2.5$ m, with a ventilation of 0.5 air change per hour. After 1 h the CO level reaches 5 mg/m^3 and remains there for 3 h. Assume that the ventilation

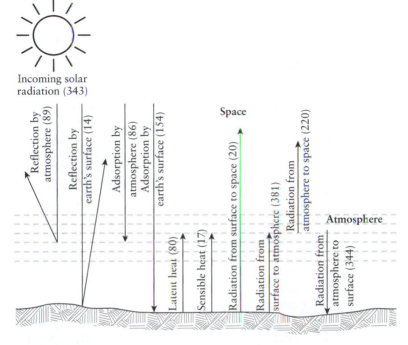

Figure 12.21 Global average energy flows. Units are watts per square meter of surface area. (After Harte, J. 1985. *Consider a Spherical Cow*. Menlo Park, CA: Wm. Kaufman.)

air has negligible CO and that CO does not decay.

a. At what rate does the stove emit CO?
b. If this stove is used in a small hut measuring $2 \times 3 \times 3$ m, what would be the CO concentration in the hut?
c. Is burning wood any better environmentally than using natural gas or oil? Analyze the use of wood as alternative energy for heating based on presumed environmental impact and cost.

12.15 The CEO of a prominent environmental consulting firm, Steven Fisher of Brown & Caldwell, was quoted as saying, "We have to realize that all of the work in the environmental area is really being done because of fear. That in turn leads to public pressure, legislation, and then enforcement of that legislation."[8]

Fear of what? Do you think he is right? If so, is this fear well founded? If not, what is driving environmental legislation and enforcement? Consider these questions in light of the history of air pollution control legislation in the United States, and write a one-page discussion of why this legislation was passed, and why it is taking EPA so long to enforce the regulations.

12.16 If the rate of coal excavation is decreasing by about 1.2% per year while the use of oil and gas is increasing by about 3% per year, estimate how long it would take to double the rate of carbon emissions.

12.17 The particulate concentration in an urban atmosphere is $160 \ \mu g/m^3$.

a. What would be the visibility at the airport?
b. Suppose the measurement was made early in the morning when the atmosphere

was 90% saturated with moisture. Would the visibility have been lower or higher then?

12.18 A pilot reports that the visibility at the airport is 5 km.

   a. If it is a dry day, what might be the concentration of the particulates in this city?
   b. Does this concentration exceed the national ambient air quality standards for particulates?

12.19 Metals are frequently coated with thin films of light hydrocarbon oil to prevent oxidation of the raw metal feedstock during shipping and storage. The oil is removed with a solvent. Unfortunately, about 90% of the solvent eventually escapes into the atmosphere.

   Trichlorethylene (TCE) used to be the most widely used solvent for degreasing and metal cleaning. TCE, however, is a major contributor to photochemical smog, acting as one of the hydrocarbons that reacts with nitric oxides, thereby allowing high concentrations of ozone to build up. 1,1,1-trichloroethane (TCA) has been substituted for TCE. The substitution is not without problems, however, since TCA has potential for both stratospheric ozone depletion and global warming. The following table illustrates a comparison of TCE and TCA relative to their environmental effects.

	Smog Formation Potential per Tonne	Global Warming Potential per Tonne	Ozone Depletion Potential per Tonne
TCE	350	6.9	negligible
TCA	3.0	390	1000

The increasingly stringent regulation of TCE, a photochemically reactive chlorinated solvent, encouraged industry to convert to TCA. The results indicate that, whereas TCA plays a relatively minor role in photochemical smog formation, its contribution to global warming and to the depletion of stratospheric ozone is relatively significant. A dilemma arises from the conflicting criteria used in choosing a solvent. For instance, should we replace a photochemically reactive solvent that breaks down easily in the troposphere with a solvent that is persistent and makes its way up into the stratosphere? Once there, the solvent will absorb infrared radiation and contribute to global warming or release a chlorine atom, which then participates in the catalytic destruction of the ozone layer.

Write a one-page paper outlining your criteria for making the decision between TCE and TCA.[9]

12.20 Insulation is often rated on the basis of the R-value, defined by the equation

$$q = \frac{A(T_i - T_o)}{R}$$

where q = heat transfer rate through a surface, W/hr
A = area of surface, m^2
T_i = temperature on the surface, °C
T_o = temperature of the air, °C
R = thermal resistance, m^2–°C/W

The thermal resistance, R, is a useful number because it expresses the resistance of a surface to conduct heat. A typical fiberglass insulation, for example, has an R-value of 1 h-m^2 − K/kj. The higher the R-value, the greater is the insulating ability.

The mass balance of [heat IN] = [heat OUT] must then govern the amount of energy needed to heat a house to a given temperature.

   a. Assume a house with a surface area of 100 m^2 is poorly insulated, with an average R-value of 0.5 h-m^2−K/kj. How much energy, in Btu/hr, is needed to heat the house to 21°C when the temperature outside is 12°C?
   b. If the house was subsequently insulated to $R = 20$, how much energy would the owners save?

12.21 A family member is considering buying an ozone generator to improve indoor air quality. Why is this purchase a bad idea?

12.22 How have engineers inadvertently decreased indoor air quality?

12.23 Vehicles were the largest air pollution source in 2003 and 2004 in Illinois according to the *Illinois EPA 2003–2004 Biennial Report* (June 2005). How can the profession of civil engineering reduce the impact of vehicles on air quality?

END NOTES

1. Rubber Manufacturers of America. Accessed at http://www.rma.org/scrap_tires/scrap_tire_markets/scrap_tire_characteristics/#anchor135840 in September 2008.

2. Thomas, P. 2005. Tyre dust: Where rubber meets road. *The Ecologist*. Accessed at http://www.theecologist.org/pages/archive_detail.asp?content_id=1002 on 27 September, 2008.

3. Miguel, A. G., G. G. Cass, J. Weiss, and M. M. Glovsky, 1996. Latex allergens in tire dust and airborne particles. *Environmental Health Perspectives* 1180–6. Accessed at http://www.pubmedcentral.nih.gov/picrender.fcgi?artid=1469526&blobtype=pdf on 27 September, 2008.

4. Schwartz, Stephen. 1989. Acid deposition: Unraveling a regional phenomenon. *Science* (February) 243.

5. Masters, Gilbert M. 1998. *Introduction to Environmental Engineering and Science*. Englewood Cliffs, NJ: Prentice-Hall.

6. Lovelock, James. 1979. *Gaia: A New Look at Life on Earth*. New York: Oxford University Press.

7. Nero, A. 1986. The indoor radon story. *Technology Review*. January.

8. Quoted in The Environmental Age. 1992. *Engineering News Record* 228: 25.

9. This problem is based on a similar problem in Allen, David, N. Bakshani, and Kirsten Sinclair Rosselot. 1991. *Pollution Prevention: Homework and Design Problems for Engineering Curricula*. New York: American Institute of Chemical Engineers and other societies.

Air Quality Control

Seal Bay sea lion, Australia

Coober Pedy mine blower, Australia

Environmental engineers keep the air fit for ourselves as well as animals and plants.

The easiest way to control air pollution is to eliminate the source of the pollution. Surprisingly, this is often also the most economical solution to an air pollution problem. Decommissioning a sludge incinerator and placing the sludge on dedicated land, for example, may be a great deal more economical than installing air cleaning equipment for the incinerator. In other cases a modification of the process, such as switching to natural gas instead of coal in an electrical power plant, will eliminate the immediate air pollution problem. Most often, however, control is achieved by some form of air treatment similar in concept to water treatment. In this chapter some of the alternatives available for treating emissions are discussed, followed by a review of the last control strategy—dispersion.

13.1   TREATMENT OF EMISSIONS

Selection of the correct treatment device requires matching characteristics of the pollutant with features of the control device. It is important to remember that the sizes of air pollutants range many orders of magnitude, and it is, therefore, not reasonable to expect one device to be effective and efficient for all pollutants. In addition, the types of chemicals in emissions often will dictate the use of some devices. For example, a gas containing a high concentration of SO_2 could be cleaned by water sprays, but the resulting sulfuric acid might present serious corrosion problems.

The various air pollution control devices are conveniently divided into those applicable for controlling particulates and those used for controlling gaseous pollutants. The reason, of course, is the difference in sizes. Gas molecules have diameters of about $0.0001\ \mu m$ particulates range up from $0.1\ \mu m$.

Any air pollution control device can be treated as a black box separation device. The removal of the pollutant is calculated by using the principles of separation covered in Chapter 4. If a gas (such as air) contains an unwanted constituent (such as dust), the pollution control device is designed to remove the dust. The recovery efficiency of any separating device is expressed by Equation 4.2:

$$R_1 = \frac{x_1}{x_0} \times 100$$

where R_1 = recovery, %
   x_1 = amount of pollutant collected by the treatment device
         per unit time, kg/s
   x_0 = amount of pollutant entering the device, per unit time, kg/s

Difficulties may arise in defining the x_0 term. Some dust, for example, may be composed of particles so small that the treatment device cannot be expected to remove this fraction. These very small particles should not, in all fairness, be included in x_0. Yet, it is common practice to simply let x_0 equal the contaminants to be removed, such as total particulates.

FOCUS ON

West Texas Lead Smelter

In 1954 the Dallas, Texas, Housing Authority built a large public housing project on land immediately adjacent to a lead smelter. The project had 3500 living units and became a low income, predominantly African American community. During the 1960s the smelter recovered lead from as many as 10,000 car batteries per day and emitted over 200 tons of lead annually into the air. The lead emissions were associated with blood lead levels in the housing project's children that were 35% higher than in children from comparable areas.

Lead is a particularly insidious pollutant because it can result in developmental damage. Study after study showed that the children at this project were in danger of higher lead levels, but nothing was done for over 20 years. Finally, in the early 1980s, the city brought suit against the lead smelter, and the smelter immediately initiated control measures that reduced its emissions to allowable standards. The smelter also agreed to clean up the contaminated soil around its location and to pay compensation to people who had been harmed.

This case illustrates two issues of environmental injustice. First, the housing units should never have been built next to a lead smelter. The reason for locating the units there would have been justified on the basis of economics. The land was inexpensive, thus saving the government money. The second issue was the foot dragging by the city in insisting that the smelter clean up the emissions. When the city finally acted, the plant was in compliance within two years. By 2003, blood lead levels in the area were below the national average. Why did it take so long for the city to do the right thing?

Source: Adapted from *Socially Responsible Engineering* by Daniel A. Vallero, P. Aarne Vesilind, 2007, John Wiley & , Inc. Used by permission.

EXAMPLE

13.1

Problem   An air pollution control device is to remove a particulate that is being emitted at a concentration of 125,000 $\mu g/m^3$ at an air flow rate of 180 m^3/s. The device removes 0.48 metric ton per day. What are the emission concentration and the collection recovery?

Solution   A black box and a material balance with regard to particulates is first set up and the old material balance equation used once again:

$$
\begin{bmatrix} \text{Rate of} \\ \text{particulates} \\ \text{ACCUMULATED} \end{bmatrix} = \begin{bmatrix} \text{Rate of} \\ \text{particulates} \\ \text{IN} \end{bmatrix} - \begin{bmatrix} \text{Rate of} \\ \text{particulates} \\ \text{OUT} \end{bmatrix}
$$
$$
+ \begin{bmatrix} \text{Rate of} \\ \text{particulates} \\ \text{PRODUCED} \end{bmatrix} - \begin{bmatrix} \text{Rate of} \\ \text{particulates} \\ \text{CONSUMED} \end{bmatrix}
$$

Since this is in steady state and there are no particulates produced or consumed, the equation reduces to

$$[\text{Rate of particulates IN}] = [\text{Rate of particulates OUT}]$$

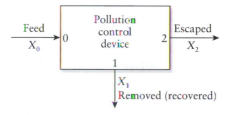

Figure 13.1   An air pollution control device as a black box. See Example 13.1.

Figure 13.1 shows the black box. The particulates IN is the feed and is calculated as the flow rate times the concentration, which, of course, yields the mass flow rate:

$$(180 \text{ m}^3/\text{s})(125{,}000 \text{ } \mu\text{g/m}^3)(10^{-6} \text{ g/}\mu\text{g}) = 22.5 \text{ g/s}$$

The particulates OUT consists of the particles that escape and those that are collected. The latter is calculated as

$$\frac{(0.48 \text{ metric ton/day})(10^6 \text{ g/metric ton})}{(86{,}400 \text{ s/d})} = 5.6 \text{ g/s}$$

The balance is

$$22.5 = 5.6 + [\text{Particulates that escape}]$$

$$\text{Escaped particulates} = 16.9 \text{ g/s} \cong 17 \text{ g/s}$$

$$\text{The emission concentration} = \frac{(16.9 \text{ g/s})(10^6 \text{ } \mu\text{g/g})}{(180 \text{ m}^3/\text{s})}$$

$$\cong 94{,}000 \text{ } \mu\text{g/m}^3$$

The recovery is

$$R_1 = \frac{5.6 \text{ g/s}}{22.5 \text{ g/s}}(100) = 25\%$$

13.1.1   Control of Particulates

The simplest devices for controlling particulates are *settling chambers* consisting of nothing more than wide places in the exhaust flue where larger particles can settle, usually with a baffle to slow the emission stream. Obviously, only very large particulates (>100 μm) can be efficiently removed in settling chambers.

Possibly the most popular, economical, and effective means of controlling particulates is the *cyclone*. Figure 13.2 on the next page shows a simple schematic. The dirty air is blasted into a conical cylinder, but off centerline. This creates a violent swirl within the cone, much like a centrifuge. The heavy solids migrate to the wall of the cylinder, where

© Courtesy of P. Aarne Vesilind

Figure 13.2 Cyclone used for dust collection.

they slow down due to friction, slide down the cone, and finally exit at the bottom. The clean air is in the middle of the cylinder and exits out the top.

Bag (or *fabric*) *filters* used for controlling particulates (Figure 13.3) operate like the common vacuum cleaner. Fabric bags are used to collect the dust, which must be periodically shaken out of the bags. The fabric will remove nearly all particulates, including submicron sizes. Bag filters are widely used in many industrial applications but are sensitive to high temperatures and humidity.

The basic mechanism of dust removal in fabric filters is thought to be similar to the action of sand filters in water treatment. The dust particles adhere to the fabric due to entrapment and surface forces. They are brought into contact by impingement and/or Brownian diffusion. Since fabric filters commonly have an air space-to-fiber ratio of 1:1, the removal mechanisms cannot be simple sieving.

The *spray tower* or *scrubber*, pictured in Figure 13.4 on page 460, is an effective method for removing large particulates. More efficient scrubbers promote the contact

© Courtesy of P. Aarne Vesilind

Figure 13.3   Bag filter used for control of particulate air pollutants.

between air and water by violent action in a narrow throat section into which the water is introduced. Generally, the more violent the encounter, the smaller the gas bubbles or water droplets and the more effective the scrubbing. Wet scrubbers are efficient devices but have two major drawbacks.

- They produce a visible plume, albeit only water vapor. The lay public seldom differentiates between a water vapor plume and any other visible plume, and hence, public relations often dictate no visible plume.
- The waste is now in liquid form, and some manner of water treatment is necessary.

Electrostatic precipitators are widely used in power plants, mainly because power is readily available. The particulate matter is removed by being charged by electrons jumping

© Courtesy of McGill AirClean LLC

Figure 13.4   Scrubber. In the photograph, the scrubber is the high round tower.

© Courtesy of P. Aarne Vesilind

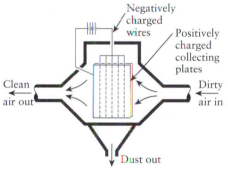

Figure 13.5   Electrostatic precipitator used for control of particulate air pollutants.

from one high-voltage electrode to the other and then migrating to the positively charged collecting electrode. The type of electrostatic precipitator shown in Figure 13.5 consists of a pipe with a wire hanging down the middle. The particulates collect on the pipe and must be removed by banging the pipe with a hammer. Electrostatic precipitators have no moving parts, require only electricity to operate, and are extremely effective in removing submicron particulates.

The efficiencies of the various control devices obviously vary widely with the particle size of the pollutants. Figure 13.6 on the next page shows approximate collection efficiency curves, as a function of particle size, for the various devices discussed.

13.1.2   Control of Gaseous Pollutants

The control of gases involves the removal of the pollutant from the gaseous emissions, a chemical change in the pollutant, or a change in the process producing the pollutant.

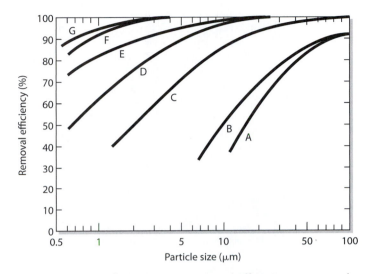

Figure 13.6 Comparison of approximate removal efficiencies. A = settling chamber, B = simple cyclone, C = high-efficiency cyclone, D = electrostatic precipitator, E = spray tower wet scrubber, F = venturi scrubber, G = bag filter. (Based on C. E., Lappe, 1951. Processes use many collection types. *Chemical Engineering* 58: 145.)

Wet scrubbers, as already discussed, can remove gaseous pollutants by simply dissolving them in the water. Alternatively, a chemical (such as lime) may be injected into the scrubber water that then reacts with the pollutants. This is the basis for most SO_2 removal techniques, as discussed below.

Adsorption is a useful method when it is possible to bring the pollutant into contact with an efficient adsorber, such as activated carbon, as shown in Figure 13.7.

Incineration or *flaring* is used when an organic pollutant can be oxidized to CO_2 and water (Figure 13.8). A variation of incineration is *catalytic combustion*, in which the temperature of the reaction is lowered by the use of a catalyst that mediates the reaction.

Figure 13.7 Adsorber for removing gaseous air pollutants.

Figure 13.8   Incinerator used for burning gaseous pollutants.

EXAMPLE

13.2

Problem   An industrial plant generates approximately 1000 kg/month of waste pentachlorophenol (C_6Cl_5OH). The waste is incinerated. How much oxygen is theoretically required to completely combust the waste? How much sodium hydroxide is theoretically required to neutralize the acid created? Assume a purity of 98%.

Solution   Remember how to calculate theoretical oxygen demand (Chapter 9)? It's the same process. First, determine the chemical reaction.

$$C_6Cl_5OH + a\ O_2 \longrightarrow b\ CO_2 + c\ H_2O + d\ HCl$$

Then balance the reaction. In this case, $c = -2$, so it's actually a reactant rather than a product.

$$C_6Cl_5OH + 4.5\ O_2 + 2\ H_2O \rightarrow 6\ CO_2 + 5\ HCl$$

Now we can use the stoichiometric conversion factors to calculate the amount of oxygen and hydrochloric acid generated.

$$M_{O_2} = \left[\frac{1000\ \text{kg/mo penta}}{266.5\ \text{g/mol penta}} \right] \left[\frac{4.5\ \text{mol } O_2}{\text{mol penta}} \right] (32\ \text{g/mol } O_2) \cong 540\ \text{kg/month}$$

$$M_{HCl} = \left[\frac{1000\ \text{kg/mo penta}}{266.5\ \text{g/mol penta}} \right] \left[\frac{5\ \text{mol HCl}}{\text{mol penta}} \right] (36.5\ \text{g/mol HCL}) \cong 685\ \text{kg/month}$$

Repeat the process to determine the amount of NaOH.

$$HCl + NaOH \rightarrow NaCl + H_2O$$

$$M_{\text{NaOH}} = \left[\frac{685 \text{ kg/mo HCl}}{36.5 \text{ g/mol HCl}} \right] \left[\frac{1 \text{ mol NaOH}}{\text{mol HCl}} \right] (40 \text{ g/mol NaOH}) \left[\frac{1}{0.98} \right] \cong 770 \text{ kg/month}$$

During operation, the quantities would be different due to a variety of reasons, including incomplete combustion, nonideal mixing, and nonideal reaction kinetics.

13.1.3 Control of Sulfur Oxides

As noted earlier, sulfur oxides (SO_2 and SO_3) are serious and yet ubiquitous air pollutants. The major source of sulfur oxides (or SO_x as they are often referred to in shorthand) is coal-fired power plants. The increasingly strict standards for SO_x control have prompted the development of a number of options and techniques for reducing their emission. Among these options are the following.

- *Change to low-sulfur fuel.* Natural gas and oil are considerably lower in sulfur than coal. However, uncertain and expensive supplies make this option risky.
- *Desulfurize the coal.* Sulfur in coal can be organic or inorganic. The inorganic form is iron pyrite (FeS_2), which can be removed by washing because it occurs in discrete particles. The removal of the organic sulfur (generally about 60% of the total) requires chemical reactions and is most economically accomplished if the coal is gasified (changed into a gas resembling natural gas).
- *Build tall stacks.* A short-sighted, albeit locally economical, method of SO_x control is to build incredibly tall smokestacks and disperse the SO_x. This option was employed in Great Britain and was in part responsible for the acid rain problem in Scandinavia.
- *Desulfurize the flue gas.* The last option is to reduce the SO_x emitted by cleaning the gases coming from the combustion, the so-called flue gases. Many systems have been employed, and a great deal of research has been undertaken to make these processes more efficient. The most widely used method of SO_x removal in flue gas is to contact the sulfur with lime. The reaction for SO_2 is

$$SO_2 + CaO \longrightarrow CaSO_3$$

or, if limestone, is used:

$$SO_2 + CaCO_3 \longrightarrow CaSO_4 + CO_2$$

Both the calcium sulfite and sulfate (gypsum) are solids that have low solubilities and can be separated in gravity settling tanks. The calcium salts thus formed represent a staggering disposal problem. It is also possible to convert the sulfur to H_2S, H_2SO_4, or elemental sulfur and market these raw materials. Unfortunately, the total possible markets for these chemicals are far less than the potential production from desulfurization.

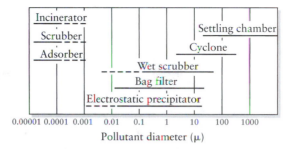

Figure 13.9   The effectiveness of various air pollution control devices depends on particle size.

Before moving on to the next topic, it might again be useful to reiterate the importance of matching the type of pollutant to be removed with the proper process. Figure 13.9 reemphasizes the importance of particle size in the application of control equipment. Other properties, however, may be equally important. A scrubber, for example, removes not only particulates but gases that can be dissolved in the water. Thus, SO_2, which is readily soluble, is removed in a scrubber while NO, which is poorly soluble, is not. The selection of the proper control technology is an important component of the environmental engineering profession.

13.2   DISPERSION OF AIR POLLUTANTS

Recall from the discussion of meteorology in Chapter 12 that atmospheric conditions primarily determine the dispersion of air pollutants. If the conditions are superadiabatic, a great deal of vertical air movement and turbulence is produced, so dispersion is enhanced. The subadiabatic prevailing lapse rate is by contrast a very stable system. An inversion is an extreme subadiabatic condition, and the vertical air movement within an inversion is almost nil.

The effect of atmospheric stability on a plume can be illustrated as in Figure 13.10 on the next page. A superadiabatic lapse rate produces atmospheric instability and a *looping* plume while a neutral lapse rate produces a *coning* plume. If the plume is emitted into an inversion layer, a *fanning* plume will result—a highly descriptive name because, from above, it can be seen that the plume fans out without any vertical dispersion. A particularly nasty situation is the fumigation condition—when an inversion cap is placed on the plume but a superadiabatic lapse rate under the inversion causes mixing and high ground level concentrations.

The distance a plume rises is also of importance if dispersion is to be attained. While there have been no theoretical models that consistently predict plume rise, a number of empirical models have been suggested. Briggs developed a model that seems to effectively predict plume rise from power stations.[1]

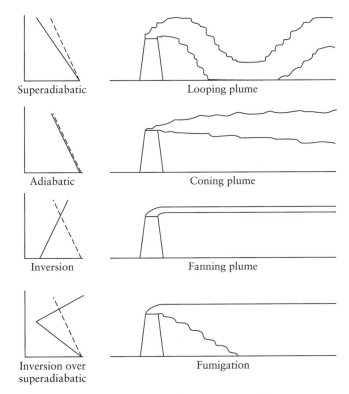

Figure 13.10 Prevailing lapse rates produce signature plumes (temperature versus elevation).

$$\Delta h = 2.6 \left(\frac{F}{\bar{u}S} \right)^{1/3} \tag{13.1}$$

$$F = \frac{g V_s d^2 (T_s - T_a)}{4(T_a + 273)}$$

$$S = \frac{g}{(T_a + 273)} \left(\frac{\Delta T}{\Delta z} + 0.01 \right)$$

where

Δh = plume rise above top of stack, m

$\bar{u}$ = average wind speed, m/s

$\Delta T / \Delta z$ = prevailing lapse rate, the change in temperature with elevation, °C/m

V_s = stack gas exit velocity, m/s

d = stack exit diameter, m

g = gravitational acceleration, 9.8 m/s²

T_a = temperature of the atmosphere, °C

T_s = temperature of the stack gas, °C

F = buoyancy flux, m⁴/s³

EXAMPLE

13.3

Problem   A stack at a power station has an emission exiting at 3 m/s through a stack diameter of 2 m. The average wind speed is 6 m/s. The temperature at the top of the stack is 28°C, and the temperature of the emission is 167°C. The atmosphere is at neutral stability. What is the expected rise of the plume?

Solution   For neutral stability, $\Delta T / \Delta z = 1°C/100$ m $= 0.01°C/m$ (Chapter 12). The expected plume rise is Δh:

$$\Delta h = 2.6 \left(\frac{F}{6S} \right)^{1/3}$$

First, F and S must be calculated:

$$F = \frac{9.8(3)(2)^2(167 - 28)}{4(28 + 273)} = 13.6$$

$$S = \frac{9.8}{28 + 273} (0.01 + 0.01) = 6.5 \times 10^{-4}$$

$$\Delta h = 2.6 \left(\frac{13.6}{6(6.5 \times 10^{-4})} \right)^{1/3} \cong 40 \text{ m}$$

Dispersion is the process of spreading the emission over a large area, thereby reducing the concentration of the specific pollutants. The plume spread or dispersion is in two dimensions: horizontal and vertical. It is assumed that the greatest concentration of the pollutants is in the plume centerline, that is, in the direction of the prevailing wind. The farther away from the centerline, the lower the concentration. If the spread of a plume in both directions is approximated by a Gaussian probability curve, as introduced in Chapter 3, the concentration of a pollutant at any distance x downwind from the source can be calculated as

$$C_{(x,y,z)} = \frac{Q}{2\pi\bar{u}\sigma_y\sigma_z} \exp\left(-\frac{1}{2} \left[(y/\sigma_y)^2 + (z/\sigma_z)^2 \right] \right)$$

where $C_{(x,y,z)}$ = concentration at some point in the coordinate space, kg/m³
Q = source strength, or the emission rate, kg/s
$\bar{u}$ = average wind speed, m/s
σ_z and σ_y = standard deviation of the dispersion in the z and y directions
y = distance crosswind horizontally, m
z = distance vertically, m

FOCUS ON

Seveso, Italy

The accident was blamed on a broken valve. On a Saturday afternoon on 10 July, 1976, the plant, owned by the Industrie Chimiche Meda Societa Azionaria, started releasing a toxic cloud into the atmosphere. At first there was no alarm and little concern as the 3000 kg of various chemicals including 2,4,5-trichlorophenol used in the manufacture of herbicides wafted on their way toward the municipality of Seveso. The emission, however, also carried with it between 10 and 20 kg of dioxin, which had been found to be an incredibly potent toxin, acting in the parts per trillion range.

After the toxic cloud passed through the city, the medical personnel were on the scene estimating the effect on the people of Seveso. The first sign of human health problems was a high incidence of chloracne, a burn-like skin lesion, that appeared on children a few hours after the accident. The problems were thus minimized and the residents assured that there was no risk, but then the presence of dioxin was identified, two weeks after the accident. This caused the evacuation of nearly 400 people from the most severely affected areas, which was, of course, much too late to prevent exposure to the dioxin. Eventually, about 37,000 people are believed to have been exposed to the chemical. The hardest hit were the animals, which either died from the exposure or were killed to prevent the transmission of dioxin up the food chain.

The exposure to dioxin suffered by these people ought to have killed them outright or, at the very least, caused a very high rate of cancer. Nothing like this happened, however, proving that the extrapolation of toxins from mouse to human is not simply a case of multiplying by the ratio of the body weights. What happened is that, during the first seven years after the accident, a statistically unlikely number of female babies were born to the exposed population. Other effects reported by the people who were exposed to the toxins were immune system and neurological disorders as well as spontaneous abortions, but these problems do not seem to be linked to dioxin.

The highly contaminated area was dug up and the soil stored in two massive concrete tanks, along with the slaughtered animals. The area was covered with clean soil and is now a park for the people of Seveso.

Source: Reprinted from *Controlling Environmental Pollution*, Vesilind, P. A. and DiStefano, T. D., 2006. Lancaster, PA: DEStech Publications, Inc. Reproduced with permission of DEStech Publications, Inc.

The coordinates are shown in Figure 13.11. Note that z is in the vertical direction, y is horizontal crosswind, and x is downwind.

The standard deviations are measures of how much the plume spreads. If σ_y and σ_z are large, the spread is great, and the concentration is, of course, low. The opposite is also true; if the deviations are small, the spread is small, and the concentration is higher. The dispersion is dependent on both atmospheric stability (as noted earlier) and the distance from the source. Figure 13.12 is one approximation for the dispersion coefficients. Atmospheric stability is denoted in Figure 13.12 by

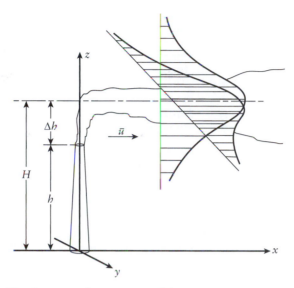

Figure 13.11   The Gaussian dispersion model.

Figure 13.12   Dispersion coefficients. (After Turner, D.B. *Workbook of Atmospheric Dispersion Estimates*. U.S. Department of Health, Education and Welfare, Public Health Service, National Center for Air Pollution Control, Publication No. 999-Ap-28.)

Table 13.1 Atmospheric Stability Key for Figure 13.12

| Surface Wind Speed (at 10 m) (m/s) | Day Incoming Solar Radiation (Sunshine) | | | Night Mostly Overcast or ≥4/8 Cloud Cover | Night Mostly Clear or ≤3/8 Cloud Cover |
	Strong	Moderate	Slight		
<2	A	A–B	B	—	—
2–3	A–B	B	C	E	F
3–5	B	B–C	C	D*	E
5–6	C	C–D	D	D	D
>6	C	D	D	D	D

*The neutral category, D, should be assumed for overcast conditions day or night.
Source: Pasquill F. The estimation of the dispersion of windborne material, *The Meteorological Magazine*, Vol. 90, No. 1063, pp. 33–49. © British Crown Copyright 1961, the Met Office.

letters ranging from A to F. Table 13.1 is a key for selecting the proper stability condition.*

A plume emitted from a stack has an effective height H, which is calculated as the stack height plus the plume rise Δh as calculated from Equation 13.1 and as shown in Figure 13.11. The elevation of the plume centerline is then $z = H$, and the diffusion equation is

$$C_{(x,y,z)} = \frac{Q}{2\pi\bar{u}\sigma_y\sigma_z}\exp\left(-\frac{1}{2}\left[\left(\frac{y}{\sigma_y}\right)^2 + \left(\frac{z-H}{\sigma_z}\right)^2\right]\right)$$

These two equations hold as long as the ground does not influence the diffusion. This is usually not a good assumption because the ground is not an efficient sink for the pollutants, and the levels must be higher at ground level due to the inability of the plume to disperse into the ground. The ground effect can be taken into account by assuming an imaginary mirror image source at elevation $z - H$, as shown in Figure 13.13. Taking the reflection of the ground into account, the dispersion equation reads

$$C_{(x,y,z)} = \frac{Q}{2\pi\bar{u}\sigma_z\sigma_y} \times$$

$$\exp\left[-\frac{1}{2}\left(\frac{y}{\sigma_y}\right)^2\right]\left\{\exp\left[-\frac{1}{2}\left(\frac{z+H}{\sigma_z}\right)^2\right] + \exp\left[-\frac{1}{2}\left(\frac{z-H}{\sigma_z}\right)^2\right]\right\} \tag{13.2}$$

*These curves and the table are based on some data, but a lot of it is extrapolation. This is especially evident when one notes that under stability category A, it is possible to have a vertical standard deviation of 10,000 km. It gets pretty silly.

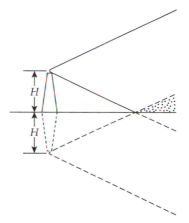

Figure 13.13   Source and image source. In the shaded area the concentration is doubled due to the image source.

Equation 13.2 is the most general dispersion equation, taking into account the reflection of the ground and the emission of the pollutant at some effective stack height H. We can simplify this equation by making various assumptions. For example, if the measurement is taken at ground level and the plume is also emitted at ground level, both the z and H terms are zero, yielding the equation

$$C_{(x,y,0)} = \frac{Q}{2\pi\bar{u}\sigma_z\sigma_y}\exp\left[-\frac{1}{2}\left(\frac{y}{\sigma_y}\right)^2\right] \quad \text{at } H = 0 \text{ and } z = 0$$

If the emission is at ground level ($H = 0$), the measurement is also at ground level ($z = 0$), and the pollutant is measured on its centerline ($y = 0$), then:

$$C_{(x,0,0)} = \frac{Q}{2\pi\bar{u}\sigma_z\sigma_y} \quad \text{at } y = 0, \ H = 0, \ \text{and } z = 0$$

E X A M P L E

13.4   **Problem**   Given a sunny summer afternoon with average wind, $\bar{u} = 4$ m/s, emission $Q = 0.01$ kg/s, and the effective stack height $H = 20$ m, find the ground level concentration at 200 m from the stack.

Solution   From Table 13.1 the atmospheric stability is Type B for a wind speed of 4 m/s and strong sunshine. From Figure 13.12 at 200 m and stability B, $\sigma_y \cong 30$ m and $\sigma_z \cong$

22 m. Note that maximum concentrations occur on the plume centerline, at $y = 0$. Using Equation 13.2:

$$C_{(200,0,0)} = \frac{0.01 \text{ kg/s}}{2(\pi)(4 \text{ m/s})(30 \text{ m})(22 \text{ m})}$$

$$\times \left\{ \exp\left(-\frac{1}{2}(0/30 \text{ m})^2\right) \times \left[\exp\left(-\frac{1}{2}[(0 - 20 \text{ m})/22 \text{ m}]^2\right)\right.\right.$$

$$\left.\left. + \exp\left(-\frac{1}{2}[(0 + 20 \text{ m})/22 \text{ m}]^2\right)\right]\right\}$$

$$C_{(200,0,0)} = 8 \times 10^{-7} \text{ kg/m}^3$$

$$\text{or } C_{(200,0,0)} = 800 \text{ } \mu\text{g/m}^3$$

Finally, it should be pointed out that the accuracy of this plume dispersion analysis is very poor. Air pollution modelers are usually pleased to find their models predicting concentrations to within an order of magnitude!

13.3 CONTROL OF MOVING SOURCES

Although many of the above control techniques can apply to moving sources as well as stationary ones, one very special moving source—the automobile—deserves special attention. Although the automobile has many potential sources of pollution, only a few important points require control (Figure 13.14):

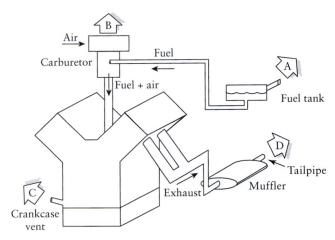

Figure 13.14 Internal combustion engine, showing sources of pollution.

- evaporation of hydrocarbons (HC) from the fuel tank
- evaporation of HC from the carburetor
- emissions of unburned gasoline and partially oxidized HC from the crankcase
- the NO_x, HC, and CO from the exhaust.

The evaporative losses from the gas tanks have been reduced by the use of gas tank caps, which prevent the vapor from escaping. Losses from carburetors have been reduced by activated carbon canisters, which store the vapors emitted when the engine is turned off and the hot gasoline in the carburetor vaporizes. When the car is restarted, the vapors can be purged by air and burned in the engine (Figure 13.15).

The third source of pollution, the crankcase vent, has been eliminated by closing off the vent to the atmosphere and recycling the blowby gases into the intake manifold. The positive crankcase ventilation (PCV) valve is a small check valve that prevents the buildup of pressure in the crankcase.

The most difficult control problem is the exhaust, which accounts for about 60% of the HC and almost all the NO_x and CO. One immediate problem is how to measure these emissions. It is not as simple as sticking a sampler up the tailpipe because the quantity of pollutants emitted changes with the mode of operation. The effect of operation on emissions is illustrated in Table 13.2. Note that, when the car is accelerating, the combustion is efficient (low CO and HC) and the high compression produces a lot of NO_x. On the other hand, decelerating results in low NO_x and very high HC due to partially burned fuel. Because of these difficulties, the EPA has instituted a standard test for measuring emissions. This test procedure includes a cold start, acceleration and cruising on a dynamometer to simulate a load on the wheels, and a hot start.

Emission control techniques for the internal-combustion automobile engine include:

- tuning the engine to burn fuel efficiently
- installation of catalytic reactors
- engine modifications.

Figure 13.15   Internal-combustion engine with pollution control devices.

Table 13.2 Effect of Engine Operation on Automotive Exhaust Characteristics, Shown as Fraction of Idling Emissions

	Component		
	CO	**HC**	**NO$_x$**
Idling	1.0	1.0	1.0
Accelerating	0.6	0.4	100
Cruising	0.6	0.3	66
Decelerating	0.6	11.4	1.0

A tune-up can have a significant effect on emission components. For example, a high air-to-fuel ratio (a lean mixture) will reduce both CO and HC but will increase NO$_x$. Typical emissions resulting from changing the air-to-fuel ratio are shown in Figure 13.16. A well-tuned car is the first line of defense for controlling automobile emissions, regardless of what other devices and/or processes are used.

The second control strategy, now used in all gasoline-powered cars sold in the U.S., is the catalytic converter, which oxidizes the CO and HC to CO_2 and H_2O. The most popular catalyst is platinum, which can be fouled by some gasoline additives, such as lead (furthering the cause of eliminating leaded gasoline). The second problem with catalytic converters is that the sulfur compounds in gasoline are oxidized to particulate SO_3 and, thus, increase the sulfur levels in urban environments and the global acid rain problem.

The greatest advance in engine development, however, has been the complete redesign of engines to produce fewer emissions. For example, the geometric configuration of the cylinder in which the combustion occurs is important because complete combustion

Figure 13.16 Variability in auto emissions with changes in air-to-fuel ratio.

requires that all the gasoline ignite together and burn as a steady flame for the very short time required. Cylinders that have nooks and crannies in which the air/gasoline mixture can hide will produce partially combusted emissions, such as hydrocarbons and carbon monoxide. A second advance has been fuel injection, which accurately measures the exact amount of gasoline needed by the engine and pumps this in, avoiding the problem of the engine sucking in too much fuel from the carburetor and then emitting it in the exhaust.

Even with extensive investment in engineering, however, it will be difficult to manufacture a totally clean internal-combustion engine. Electric cars are clean but can store only limited power; thus, their range is limited. In addition, the electricity used to power such vehicles must be generated in power plants, thereby creating more pollution. Fuel cell technology is also being explored to replace the internal-combustion engine.

The diesel engines used in trucks and buses also are important sources of pollution. In the United States, there are, currently fewer of these vehicles in operation than gasoline-powered cars. Diesel engines typically produce a visible smoke plume and odors, two characteristics that have led to considerable public irritation with diesel-powered vehicles and concerns for public health. However, diesel engines are typically more fuel efficient than gasoline-powered engines and therefore emit less CO_2 per Km. Design improvements have resulted in cleaner emissions, and USEPA restrictions on the sulfur content of fuel will further improve emissions. Therefore, in the future diesel-powered vehicles may become more desirable from both the fuel efficiency and environmental perspectives.

PROBLEMS

13.1 Taking into account cost, ease of operation, and ultimate disposal of residuals, what type of control device would you suggest for the following emissions?

a. A dust with particle range of 5–10 μ?
b. A gas containing 20% SO_2 and 80% N_2?
c. A gas containing 90% HC and 10% O_2?

13.2 A stack emission has the following characteristics: 90% SO_2, 10% N_2, no particulates. What treatment device would you suggest and why?

13.3 An industrial emission has the following characteristics: 80% N_2, 15% O_2, 5% CO_2, and no particulates. You are called in as a consultant to advise on the type of air pollution control equipment required. What would be your recommendation? How much would you charge for your time?

13.4 A whiskey distillery has hired you as a consultant to design the air pollution equipment for their new plant, to be built in a residential area. A major environmental effect of a whiskey distillery is the odor produced by the process, but many people think that this odor is pleasant (it's a sweet, musty aroma). How would you handle this assignment?

13.5 Given the following temperature readings:

Elevation (m)	Temperature (°C)
0	20
50	15
100	10
150	15
200	20
250	15
300	20

what type of plume would you expect if the exit temperature of the plume is 10°C and the smoke stack is

a. 50 m tall?
b. 150 m tall?
c. 250 m tall?

13.6 Consider a prevailing lapse rate that has these temperatures: ground $= 21°C$, 500 m $= 20°C$, 600 m $= 19°C$, 1000 m $= 20°C$. If a parcel of air is released at 500 m and at 20°C, would it tend to sink, rise, or remain where it is? If a stack is 500 m tall, what type of plume would you expect to see?

13.7 A power plant burns 1000 tonnes of coal per day, 2% of which is sulfur, and all of this is emitted from the 100-m stack. For a wind speed of 10 m/s, calculate:

a. the maximum ground level concentration of SO_2, 10 km downwind from the plant
b. the maximum ground level concentration and the point at which this occurs for stability categories A, C, and F.
(Part b should be done with a computer.)

13.8 The odor threshold of H_2S is about 0.7 $\mu g/m^3$. If an industry emits 0.08 g/s of H_2S out of a 40-m stack during an overcast night with a wind speed of 3 m/s, estimate the area (in terms of x and y coordinates) where H_2S would be detected. (This is much too tedious to do by hand. Write a computer program or use a spreadsheet.)

13.9 A power plant emits 300 kg/h of SO_2 into a neutral atmosphere with a wind speed of 2 m/s. The σ_y and σ_z values are assumed to be 100 m and 30 m, respectively.

a. What will be the ground level concentration of SO_2, expressed in micrograms per cubic meter, if the town is 2 km off the plume centerline and the plant has a stack that produces an effective stack height of 20 m?
b. What will be the ground level concentration if the power plant installs a 150-m stack?

13.10 A temperature-sounding balloon feeds back the following data:

Elevation (m above ground level)	Temperature (°C)
0	20
20	20
40	20
60	21
80	21
100	20
120	17
140	16
160	14
200	12

What type of plume would you expect to see out of a stack 70 m high? Why?

13.11 A coal-fired power plant burns 1000 metric tons (1 metric ton $= 1000$ kg) of coal per day, at 2% sulfur.

a. What is the SO_2 emission rate?
b. What is the SO_2 concentration 1.0 km downwind ($x = 1000$ m), 0.1 km off centerline in y direction ($y = 100$ m), and at ground level ($z = 0$) if the wind speed is 4 m/s and it's a cloudy day? The effective stack height is 100 m. Ignore reflection.
c. What is the ground level concentration if the reflection of the ground is included?
d. If the source is at ground level ($H = 0$), and it is necessary to measure the concentration at ground level ($z = 0$), what is the concentration? (Include reflection.)

13.12 A stack is 300 m tall, and emits smoke at 305 K. The ground level temperature is 300 K, and the temperature at 600 m is 310 K (assume a straight-line lapse rate between these two temperatures).

a. Draw a picture of the expected plume and name the type of plume.
b. If in the same situation the plume temperature is 306 K, how high would the plume rise? (Assume zero stack velocity and perfect adiabatic conditions.)

13.13 Suppose you were asked to design equipment for controlling emissions from an industry. For the three emissions below, draw block diagrams showing which equipment you might choose in order to obtain about 90% efficiency.

Emission 1: Particle size range—70 to 200 μm
No gaseous emissions
Temperature—366 K
No space limitations

Emission 2: Particle size range—0.1 to 200 μm
No gaseous emissions
Temperature—366 K
No space limitations
Visible plume not acceptable

Emission 3: Particle size range—5 to 40 μm
Gaseous emission—SO_2
Temperature—922 K
Contaminated water not acceptable (no treatment facilities)
No space limitations

13.14 A furniture manufacturer emits air pollution that consists of mostly particulate material of diameter 10 μm and bigger.

a. As the consulting engineer what type of treatment system would you suggest for this factory?

b. The collected pollutant is to be disposed of in the local landfill. After the facility is constructed, and the system is placed into operation, you notice that the material headed for the landfill has a strange organic odor to it. You ask the plant manager, and she tells you it's probably xylene (a highly toxic chemical). You tell the plant manager that this material is potentially dangerous to the workers and it's probably a hazardous waste and should not be placed in the landfill. She tells you to mind your own business. You have been fully paid for the job and are no longer

officially retained as their engineer. What do you do? Be realistic with your answer, not idealistic.

13.15 Assume that a power plant is emitting 400 tons per day of particulates and that a town is 10 km north of the plant. The wind is 25% of the time from the south, 25% of the time from the north, and 50% of the time from the west. There is no east wind. The wind speed is always 5 km/h. (How's that for simplification?) Assume neutral stability (O) condition for all west winds, unstable conditions for north winds (A), and inversion conditions for south winds (F). Calculate the average ambient particulate air quality in the town. Does this air quality meet the EPA National Ambient Air Quality Standards?

13.16 A sewage sludge incinerator has a scrubber for removing the particulate material from its emissions, but the scrubber has been acting up, and a small pilot bag filter is being tested for applicability for treating the particulate emissions. The experimental setup is to split the emissions from the incinerator so that the 200 m³/s total flow is split with 97% going to the scrubber and only 3% going to the pilot bag filter. The particulate concentration of the untreated emission is 125 mg/m³. The solids collected at the baghouse are collected hourly and averaged out to 2.6 kg/h. The water does not have any particulates in it when it enters the scrubber at a flow rate of 2000 L/min. There is negligible evaporation in the scrubber, and the scrubber water is found to carry solids at a rate of 52 kg/h.

a. What is the efficiency of each method of air pollution control?

b. Regardless of the outcome in the efficiency calculations, is there any reason why the scrubber would still be superior as a pollution control device?

13.17 What is the particulate removal efficiency of a cyclone if it has to remove particles of diameter

a. 100 µm?

b. 10 µm?

c. 1 µm?

d. 0.1 µm?

13.18 A new coal-fired power plant is to burn coal of 3% sulfur, and a heating value of 25,000 KJ/kg. What is the minimum efficiency that a SO_2 scrubbing device will need to have in order to meet standard of 0.5 kg $SO_2/10^6$ BTU?

13.19 Why can't the gasoline engine be tuned so that it results in the minimum production of all three pollutants, CO, HC, and NO_x simultaneously?

13.20 A paper mill has a plant in a valley, and it wants to build a stack to push the plume centerline over the mountain so it can reduce the sulfur dioxide concentration in its own valley. The mountain is 4 km away, and its elevation is 1000 m. The valley floor, where the plant is, is at 1400 ft. If the emission temperature is 200°C, the prevailing temperature in the valley is 20°C, the wind velocity can be assumed at 2 m/s, and a prevailing lapse rate of 0.006°C/m can be assumed, what stack height must they build to achieve their objective? (*Note*: You will have to assume some numbers, and each solution will be unique.)

13.21 If a car is designed to burn ethanol (C_2H_5OH), what would be the stoichiometric air-to-fuel ratio?

13.22 A fertilizer manufacturer emits HF at a rate of 0.9 kg/s from a stack with an effective height of 200 m, the average wind speed is 4.4 m/s, and the stability is category B.

a. What is the concentration of HF, in micrograms per cubic meter ($\mu g/m^3$), at ground level, on plume centerline, at 0.5 km from the stack?

b. Calculate also the ground level concentration at 1.0, 1.5, 2.0, 2.5, 3.0, 3.5, and 4.0 km from the stack. Plot the results as concentration versus distance.

13.23 For the conditions in Problem 13.22, draw a three-dimensional picture of the HF concentration at a distance 2 km from the stack.

13.24 The concentration of SO_2 at ground level is not allowed to exceed 80 $\mu g/m^3$. If the wind speed is 4 m/s on a clear day, and if the source emits 0.05 kg SO_2/s, what must the effective stack height be to meet this requirement?

13.25 A chemical plant expects to emit 200 kg/d of a highly odoriferous but nondangerous gas. Downwind from the plant at a distance of 2000 m is a town. The plant decides to build a stack that will be tall enough so that they would not attain concentration levels of more than 10 $\mu g/m^3$ in town. How tall will the stack have to be? Assume the gas exit velocity of 5 m/s, the gas exit temperature of 150°C, the stack diameter of 1.5 m, the ambient temperature of 20°C, a wind velocity of 4 m/s, and a prevailing lapse rate of 0.004°C/m.

13.26 A major catalog auto parts company advertised a book with the title of *How to Bypass Emission Controls*. For merely $7.95 (plus shipping and handling) you could find how to get from 14% to 140% better gas mileage, increase acceleration and performance, run cooler and smoother, and have longer engine life. The book included easy-to-follow instructions for both amateurs and professionals.

It is not illegal to tamper with your own car. You may not pass the emission inspections in most states, but modifying your car is not against the law.

When permission was asked from the mail order company to reproduce the ad in this book, they refused permission. If it is perfectly legal, and they are providing a public service, why would they not allow their ad to be reproduced?

Reconstruct the management board meeting where this request was discussed.

Make up characters such as the company CEO, the VP for marketing, the legal counsel, etc. Create a dialogue that may have occurred when this request was discussed, ending with the decision not to allow the ad to be reproduced in this book. How did ethics come into the discussion (if it did at all)?

13.27 How could planting trees and using lighter construction materials (e.g., light rather than dark roofs) reduce smog formation?

13.28 A 400 m, one-way, two-lane rectangular tunnel (9 m × 4 m) filled with cars is blocked by an accident at the exit. The ventilation system involves large exhaust fans blowing ambient air out of the exit end of the tunnel. Assume that the tunnel is full of cars, all with engines idling. One car occupies 6 linear meter in one lane and emits 2.0 g/min CO. The fans pull 0.42×10^6 m^3/min of fresh air containing 0.5 mg/m^3 CO into the tunnel entrance. What is the concentration of CO in the exit air? Assume complete mixing.

END NOTES

1. Briggs, G.A. 1969. *Plume rise*. AEC Critical Review Series, TID–25075. (As modified and discussed in Wark, K., and C.F. Warner. 1981. *Air pollution*. New York: Harper & Row.)
2. Golding, M.P., and M.H. Golding. 1979. Why preserve landmarks? A preliminary inquiry. In Goodpaster and Jayne, eds. *Ethics and the problems of the 21st century*. South Bend, IN: University of Notre Dame Press.
3. Stone, C.D. 1975. *Should trees have standing*? New York: Avon Books.

Solid Waste

Cars abandoned in conservation park, Australia

Dumpster and recycling containers at park, Illinois

Environmental engineers design systems to manage solid waste properly and consider the impacts or designs on the end-of-life management options.

It doesn't sound like much—2 kg/d—until it is multiplied by the total number of people in the United States.[1] Suddenly 0.63 trillion kg of garbage *per day* sounds like what it is, a humongous lot of trash.

What is to be done with these solid residues from our effluent society? The search for an answer represents a monstrous challenge to the engineering profession.

Common ordinary household and commercial waste, called *refuse* or *municipal solid waste (MSW)*, is the subject of this chapter. Technically, refuse is made up of *garbage*, which is food waste, and *rubbish*, almost everything else in your garbage can. *Trash* is larger items, such as old refrigerators, tree limbs, mattresses, and other bulky items, that are not commonly collected with the household refuse. A very important subcategory of solid waste, called *hazardous waste*, is covered in the next chapter.

The municipal solid waste problem can be separated into three steps:

1. collection and transportation of household, commercial, and industrial solid waste
2. recovery of useful fractions from this material
3. disposal of the residues into the environment.

14.1   COLLECTION OF REFUSE

In many locations in the United States and other countries, solid waste from households and commercial establishments is collected by trucks. Sometimes these are open-bed trucks that carry trash or bagged refuse, but more often these vehicles are *packers*, trucks that use hydraulic rams to compact the refuse to reduce its volume and make it possible for the truck to carry larger loads (Figure 14.1). Commercial and industrial collections are facilitated by the use of containers (dumpsters or roll-offs) that are either emptied into the truck by using a hydraulic mechanism or carried by the truck to the disposal site (Figure 14.2 on the next page). Vehicles for collecting separated materials, such as newspaper, aluminum cans, and glass bottles, are also used (Figure 14.3 on the next page).

Household collection of mixed refuse is usually by a packer truck with one or two workers: one driver and one loader. These workers dump the refuse from the garbage cans into the truck and then drive the full truck to the disposal area. The entire operation is a study in inefficiency and hazardous work conditions. The safety record of solid waste

Figure 14.1   Domestic solid waste collection vehicle.

Figure 14.2 Commercial solid waste collection vehicle.

collection personnel is by far the worst of any group of workers (three times as bad as coal miners, for example).

Various modifications to this collection method have been implemented to cut collection costs and reduce accidents, including the use of compactors and garbage grinders in the kitchen and semiautomated and fully automated collection. The householder is given a large plastic can, often 240–360 L, and is asked to roll the can to the curb every week for collection. The truck is equipped with a hydraulic lift or mechanical arm that empties the can into the truck (Figure 14.4 and Table 14.1).

Other alternate systems have been developed for collecting refuse, one especially interesting one being a system of underground pneumatic pipes. The pneumatic collection system at Disney World in Florida has collection stations scattered throughout the park that receive the refuse, and the pneumatic pipes deliver the waste to a central processing

Figure 14.3 Collection vehicle for separated materials.

Figure 14.4 (A) Semiautomated and (B) fully automated collection vehicles.

plant. In areas with low population density, drop-off, or convenience, centers are common (Figure 14.5 on the next page). They are also common for collecting recyclables.

The selection of a proper route for collection vehicles, known as *route optimization*, can result in significant savings to the hauler. The problem of route optimization was first addressed in 1736 by the famous Swiss mathematician Leonard Euler (pronounced "oiler") (1707–1783). He was asked to design a parade route for Köningsberg such that the parade would not cross any bridge over the River Pregel more than once and would return to its starting place. Euler's problem is pictured in Figure 14.6 on the next page.

Table 14.1 Typical Vehicle and Container Sizes

| Type of Collection | Typical Size | |
	Vehicle (yd³)	Container (gal)
Manual	6–39	N/A
Semiautomated	6–39	30–110
Fully automated	15–31	32–440

Source: Hickman, H. Lanier, Jr. 1999. *Principles of Integrated Solid Waste Management.* American Academy of Environmental Engineers; Annapolis, MD. 1999. Reprinted by permission of Packard & Anderson Companies, Inc.

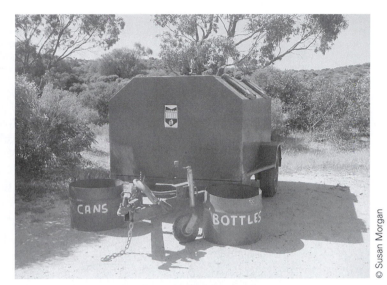

Figure 14.5   A drop-off, or convenience, center.

Not only did Euler show that such a route was impossible for the king's parade, but he generalized the problem by specifying what conditions are necessary to establish such a route, now known as an *Euler's tour*. The objective of truck routing is to create an Euler's tour, wherein a street is traversed only once and *deadheading*—traveling twice down the same street—is eliminated. An Euler's tour is also known as a *unicoursal route* because the traveler courses each street only once.

Travel takes place along specific *links* (streets and bridges in Euler's problem) that connect *nodes* (intersections). An Euler's tour is possible only if the number of links entering all the nodes is even numbered. The nodes in Figure 14.6B are A through D, and *all* of

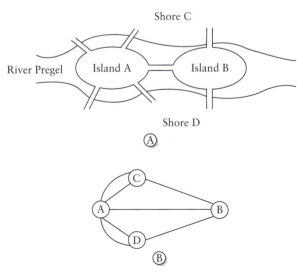

Figure 14.6   Euler's problem in Köningsberg.

them have an odd number of links, so the parade the king wanted was not possible. Euler's principle is illustrated by the example below.

E X A M P L E

14.1   **Problem**   A street network is shown in Figure 14.7A. If refuse from homes along these streets is to be collected by a vehicle that intends to travel down each street only once (solid waste being collected along both sides of the street, a situation that would typically occur in a residential neighborhood), is an Euler's tour possible? If refuse is to be collected only on the block face (refuse is collected on only one side of a street, a situation that would typically occur in heavy traffic with large streets), is an Euler's tour possible?

Solution   The street network is reduced to a series of links and nodes in Figure 14.7B for the case wherein the truck travels down a street only once. Note that eight of the nodes have an odd number of links, so an Euler's tour is impossible.

In the second case the links are shown in Figure 14.7C, and all of the nodes have an even number of links, indicating that an Euler's tour is possible. But what is it?

Figure 14.7   Routing of trucks.

Computer programs are available for developing the most efficient route possible, but often these are not used because:

- it takes too much time to write and debug the computer program, or the available software is not appropriate for the specific situation
- it's possible to develop a very good solution (maybe not the absolute optimal solution) by commonsense means
- the collection crews will change the routes around to suit themselves anyway!

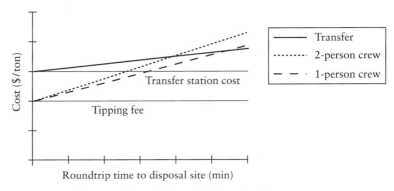

Figure 14.8 Comparison of transfer versus direct haul costs.

Common sense routing is sometimes called *heuristic routing*, which means the same thing. Some sensible rules of thumb, when followed, will go a long way toward producing the best collection solution. For example:

- try to avoid heavy traffic
- try to always make right-hand turns
- try to travel in long, straight lines
- try not to leave a one-way street as an exit from a node.

Most of these rules are, like the name suggests, common sense. Tremendous savings can often be realized by seemingly minor modifications to the collection system because this is usually the greatest cost center for refuse management.

There is a point at which the cost for individual collection vehicles to travel directly to the disposal site (known as direct haul) costs more than to use a transfer system (Figure 14.8). A transfer system consists of a transfer station, where waste is consolidated, and transfer vehicles, which are tractor-trailers or rail cars. The transfer vehicles can carry a much larger load than the packer trucks.

14.2 GENERATION OF REFUSE

The science of ecology (Chapter 8) teaches us that, if dynamic ecosystems are to remain healthy, they must reuse and recycle materials. In simple ecosystems, such as ponds and lakes, for example, phosphorus is used during photosynthesis by the aquatic plants to build high-energy molecules, which are then used by the aquatic animals, and when both produce waste and die, phosphorus is released so it can be reused. The health of natural ecosystems can be measured by their diversity, resilience, and ability to maintain homeostasis (steady state).

The flow of materials through the human ecosystem is not unlike the flow of nutrients or energy through natural ecosystems and can be similarly analyzed. Figure 14.9 shows a black box that represents human society, just as we use a black box to represent an ecosystem. In an ecosystem, nutrients are extracted from earth, used by the living organisms, and then redeposited on earth. Similarly, human society uses raw materials extracted from earth

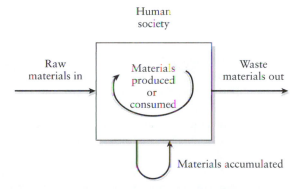

Figure 14.9   A black box showing the flow of materials through human society.

that are manufactured into products to be used by human beings and then discarded. The mass balance shows the flow of materials through this black box.

$$\begin{bmatrix} \text{Rate of} \\ \text{materials} \\ \text{ACCUMULATED} \end{bmatrix} = \begin{bmatrix} \text{Rate of} \\ \text{materials} \\ \text{IN} \end{bmatrix} - \begin{bmatrix} \text{Rate of} \\ \text{materials} \\ \text{OUT} \end{bmatrix}$$

$$+ \begin{bmatrix} \text{Rate of} \\ \text{materials} \\ \text{PRODUCED} \end{bmatrix} - \begin{bmatrix} \text{Rate of} \\ \text{materials} \\ \text{CONSUMED} \end{bmatrix}$$

If a raw material, such as iron, is considered at steady state, the amount of iron ore extracted from earth must equal the amount of iron discarded as ferrous materials (IN = OUT).

FOCUS ON

Bottled Water Versus Tap Water

Bottled water has exploded in popularity in the United States and around the world. In 2005, Americans spent $8.8 billion on 7.2 billion gal (27 billion L) of bottled water, up from 4 billion gal (15 billion L) in 1999.[2] This increase in sales has been driven in part by marketing that capitalizes on concerns about the safety of tap water and presents bottled water as the only viable alternative. Despite costing over a thousand times more than tap water, however, it is actually less rigorously tested for contaminants.

Almost 70% of bottled water in the United States is exempt from the Food and Drug Administration (FDA) regulations. Only bottles that are sold over state lines must meet even the most basic FDA standards. The National Resource Defense Council (NRDC) estimates that between 25% and 40% of bottles sold in the United States contain nothing but city tap water. Some are actually less pure. An NRDC study found unsafe levels of arsenic, synthetic organic carcinogens, and bacteria in nearly one in four of the 103 brands of water tested.

Convenience is another reason people buy bottled water. It is very easy simply to open a bottle and drink it. The reality of producing and transporting a bottle of water, however, is not easy at all and has negative impacts on the environment. Most water bottles are made from polyethylene terephthalate (PET), which is derived from crude oil. It takes over 1.5 million barrels of oil to make the bottles used by Americans each year, the same amount that would be needed to drive 100,000 cars for the same period of time. The carbon dioxide emissions from the production of these plastics can reach 1 billion lb per year. Compounding this problem is the issue of transportation. Some bottled waters travel hundreds or thousands of miles before being drunk, using even more fossil fuel.

Use of bottled water also aggravates problems of water and waste management. Each pound of plastic used to make water bottles requires over 24 gal (91 L) of water to produce, meaning that, when drinking one 16 oz (0.5 L) bottle of water, a person actually consumes up to 10 times that much water.

The inefficiency of this process is increased by the fact that in the United States, it is estimated that 86% of water bottles are thrown away instead of being recycled (Figure 14.10). In contrast, the recycling rate for all beverage containers in Germany and the Scandinavian countries of Norway, Sweden, Denmark, and Finland, where national container deposit legislation mandates bottle deposits for refunds, is much higher, ranging from 80% to nearly 100%.

While some countries do not have tap water that is safe to drink, the United States as a whole has remarkably safe tap water that is better on average than most bottled waters. Americans could save money and prevent great environmental damage by simply filling reusable water bottles at the tap. Although a few places in the United States do indeed have unsafe tap water, the NRDC estimates that the money spent annually on bottled water would be more than sufficient to fix any faults in the current water purification and distribution infrastructure.[3] People who are particularly sensitive to taste and enjoy drinking the purest water can simply purchase and install filters to polish tap water in their own home.

Source: H$_2$O Conserve. 2008 Bottled Water: Bottles of Trouble. July 12. Located at http://www.h2oconserve.org/index.php?page_id=15&pd=bottle and at Natural Resources Defense Council. 2008. Bottled Water: Pure Drink or Pure Hype? July 10. Located at http://www.nrdc.org/water/drinking/bw/bwinx.asp.

Figure 14.10 Every 5 minutes in the United States, 2 million plastic bottles are used. (© Chris Jordon/Chris Jordon Photographic Arts).

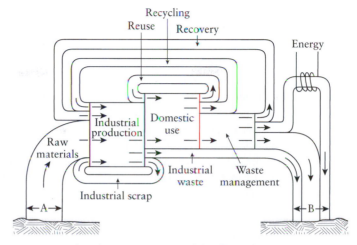

Figure 14.11   A graphical representation of the flow of materials through human society.

Figure 14.11 is a more detailed representation of materials flow through human society. The width of the bands is intended to indicate the mass rate of flow (e.g., millions of tons per year). The wider the band, the larger the flow. All the materials originate from earth, and the amount of material extracted is represented in the figure by the letter "A." These are the raw (primary) materials, such as iron ore and oil, that feed industry. These materials are extracted and fed into the manufacturing sector for the production of useful goods.

Not all the extracted materials can be used, however, and some become industrial waste that must be disposed of into the environment. In addition, some materials become industrial scrap and can be used by the same industry or shipped to some other industry through a waste exchange. The primary distinction between scrap and other types of secondary materials is that scrap never enters the public sector. It is never used by the public, and thus, the public does not have a decision to make about its eventual destination.

The materials in the form of manufactured goods flow from industry to the public. The public uses these materials and then discards some of them after their usefulness is finished (e.g., batteries) or when the public is sufficiently wealthy to choose to use other products (e.g., CDs replacing LP records). The material the public chooses to classify as waste becomes domestic waste as shown in the figure, and this material is given to the waste management sector.

14.3   REUSE AND RECYCLING OF MATERIALS FROM REFUSE

The public can exercise three alternate means for getting rid of its unwanted material once it is generated—reuse, recycling, and disposal. In *reuse* an individual either uses products again for the same purpose or puts products to secondary, often imaginative, use. An example of the former is buying milk in glass containers and returning the containers to the store for the milk distributor to clean them and put fresh product into them. In the latter case the use of paper and plastic shopping bags for the disposal of trash is a secondary use for the bags. Creating a bird feeder from a plastic beverage container is a secondary use. Storing car cleaning supplies in a laundry detergent bucket is a secondary use. Making a cat toy from an old sock is a secondary use. And so on. Reuse extends the life of the original product, thus reducing waste.

Recycling, or *material recovery*, on the other hand, involves the collection of waste and subsequent processing of that waste into new products—for example, turning plastic food containers into park benches or clothing or turning used aluminum cans back into aluminum cans. (The former is actually termed *downcycling* because the waste material, in this case plastic, cannot be turned into as "high" a product as the original product, in this case food containers.) As shown in Figure 14.11, material recovery uses domestic waste rather than raw materials as the feedstock. Promoted in the 1970s as resource recovery, this process includes such diverse operations as the recovery of steel from old automobiles and the production of compost for nurseries.

FOCUS ON

Biological Nutrients and Their Engineered Metabolisms

In Chapter 1 we describe cradle-to-cradle design and the idea of developing products that flow in appropriate metabolisms (i.e., cycles). Products can be designed to provide service as technical nutrients flowing in technical metabolisms or as biological nutrients to cycle within biological metabolisms. Some materials can flow in both.

One such product is polylactic acid or polylactide (PLA), a biodegradable, thermoplastic, aliphatic polyester derived from renewable resources (such as corn starch) that is both biodegradable and recyclable. Although PLA has been known for more than a century, it has gained commercial interest in recent years in part due to innovations in its processing. Its benefits include reduced greenhouse gas emissions as determined by comparative life cycle assessment.

One company, Natureworks LLC, manufactures PLA for use primarily in two forms: as fiber for use in the Ingeo™ brand and as plastic for use in a broad range of packaging applications, such as high-value films, rigid thermoformed food and beverage containers, and coated papers and boards. When used in food packaging, composting is practical because the packaging can be composted along with waste food. However, according to the Natureworks' website, "while this (composting) is superior to landfilling—the fate of most food waste and packaging—it is still not a high form of recycling for the carbon and energy inputs in a PLA package. The real cradle-to-cradle potential for NatureWorks biopolymer is in higher-value applications. ... In these applications, NatureWorks biopolymer can be recycled back into virgin resin through chemical hydrolysis . . . ."

One of the challenges with PLA is its newness to the waste management infrastructure. In the case of beverage containers, PLA bottles can pose a threat to PET recycling. PLA containers can look similar to PET bottles, so they may become mixed during recycling. However, PLA is chemically incompatible with PET. If it contaminates the PET recycling feedstock, it disrupts recycling and is detrimental to the economics.

It is important to understand what is meant by "compostable" or "biodegradable." Most materials based on organic chemicals are biodegradable under the right conditions, given enough time (perhaps hundreds of years). Standards have been developed to certify materials as biodegradable, meaning that they will decompose within a specific time period and under specific treatment conditions into carbon dioxide, methane, water, inorganic compounds, or biomass via microbial assimilation. The American Society of Testing and Materials (ASTM) created definitions for what constitutes biodegradability in various disposal environments. Natureworks biopolymer was certified to ASTM D6400-99, verifying that it is compostable in municipal and industrial aerobic composting facilities. In Europe, Natureworks biopolymer passed the equivalent standardized test method EN 13432, which applies to composting in industrial and semi-industrial facilities. In Japan, PLA was certified to meet the requirements of the GreenPla standard.

Note that a material certified as industrially compostable will not necessarily degrade in non-industrial or under nonideal conditions, e.g., as litter, in soil, in water, or even in home composting units. Belgium is unique in offering The OK Compost mark, which guarantees that products will biodegrade in home composting systems.

The importance of biodegradation in the marine environment has increased with growing plastic pollution in the oceans. An estimated 3 million tonnes are swirling around in the Pacific Ocean. Recent speculation that the value of waste plastic will continue to rise with increasing petroleum prices makes harvesting them economically feasible within a decade! But much damage would be avoided if plastic trash in the marine environment was biodegradable.

Environmental engineers increasingly find themselves working at all stages of product or material lifecycles—from product design to developing end-of-life infrastructures or strategies for recovering value from discarded products. Whatever the point of influence, environmental engineers need to consider how best to optimize material flows to protect human health and the environment and to recover matter or energy at the highest possible value.

Source: http://www.natureworksllc.com/;   http://www.timesonline.co.uk/tol/news/environment/article45 10201.ece

Manufacturers can enhance the feasibility of recycling and recovery of material by consciously producing products that are simple and inexpensive to recover or recycle or that can be reused. At some point society may choose to mandate such rules, but voluntary efforts might be far more effective than governmental dictates.

A central processing facility is known as a *material recovery facility*, or MRF (pronounced "murf"). The main feature of material recovery in a MRF is that the recovered materials are produced from mixed or source-separated domestic waste and are then reintroduced into industrial use. Note that the recycling process must include collection, processing, transport, and sale of the material to an industry that then uses the material. This last step is very important; the recycling process depends on the willingness of industry to purchase the recycled material. Without the development of dependable markets, the process cannot be viable.

14.3.1   Processing of Refuse

In both methods, reuse and recycling, the primary goal is purity. For example, the daily refuse from a city of 100,000 would contain perhaps 200 tonnes per day of paper. Secondary paper has sold for about $20/tonne (but it fluctuates greatly), so the income to the community could be about $4,000 per day, or about $1.5 million per year! So why isn't every community recovering the paper from its refuse and selling it? The answer is elementary—because processing the refuse to recover the dirty paper costs more than producing paper from trees.

The obvious solution is to never dirty the paper (and other materials that might have market value) in the first place. This requires the public to separate their waste, a practice known as *source separation*. But recycling programs rely on voluntary cooperation, and such cooperation can be fickle.

One way to increase cooperation might be to adopt laws that *require* people to separate their refuse. But this is not a successful approach in a democracy because public officials advocating unpopular regulation can be removed from office. Source separation in totalitarian regimes, on the other hand, is easy to implement, but this is hardly a compelling argument to abolish democracy. A further complication is that, in some places, it is difficult to convince people to put refuse in trash cans, much less convince them to separate the refuse into the various components.

Another option is to get rid of source separation and let the MRF handle all the separation. This reduces the public's role and reduces collection costs (because only one type of vehicle is required and only one trip is required to each stop) but increases the complexity of and the processing costs at the MRF.

Theoretically, however, vast amounts of materials can be reclaimed from refuse, but this is not an easy task regardless of how it is approached. Using source separation, a person about to discard an item must first identify it by some characteristics and then manually separate the various items into separate containers. In the MRF the search for purity requires either a manual or mechanical identification and separation of materials. The separation in both cases relies on some readily identifiable characteristic or properties of the specific material that distinguish it from all others. This characteristic is known as a *code*, and this code is used to separate the material from the rest of the mixed refuse by using a *switch*, as originally introduced in Chapter 4.

In recycling, the code is usually simple and visual. Anyone can identify newspapers from aluminum cans. But sometimes confusion can occur, such as identifying aluminum cans from steel cans or newsprint from glossy inserts, especially if the glossy inserts are wrapped in the Sunday paper.

The most difficult operation in recycling is the identification and separation of plastics. Because mixed plastic has few economical uses, plastic recycling is economical only if the different types of plastic are separated from each other. However, most people cannot distinguish one type of plastic from another. The plastics industry has responded by marking most consumer products with a code that identifies the type of plastic, as shown in Figure 14.12. Examples of the uses for the different types of plastic are listed in Table 14.2.

Theoretically, all a person about to discard an unwanted plastic item has to do is to look at the code and separate the various types of plastic accordingly. Realistically, sometimes finding and reading the code is a challenge, and there is almost no chance that a domestic

Figure 14.12   Identification markings on plastics.

Table 14.2 Examples of Plastics

Code Number	Chemical Name	Nickname	Typical Uses
1	Polyethylene terephthalate	PETE	Soft drink bottles
2	High-density polyethylene	HDPE	Milk containers
3	Polyvinyl chloride	PVC	Food packaging, wire insulation, and pipe
4	Low-density polyethylene	LDPE	Plastic film used for food wrapping, trash bags, grocery bags, and baby diapers
5	Polypropylene	PP	Automobile battery casings and bottle caps
6	Polystyrene	PS	Food packaging, foam cups and plates, and eating utensils
7	Mixed plastic, polylactic acid	PLA	Fence posts, benches, pallets, and compostable packaging

household will have seven different waste receptacles for plastics, nor are there enough of these to be economically collected and used. Typically, only the most common types of plastic are recycled, including PETE (polyethylene terephthalate), the material out of which the 2-L soft drink bottles are made, and HDPE (high-density polyethylene), the white plastic used for milk bottles.

Mechanical recovery operations have a chance of succeeding if the material presented for separation is clearly identified by a code and if the switch is then sensitive to that code. Currently, no such technology exists. It is impossible, for example, mechanically to identify and separate all the PETE soft drink bottles from refuse. In fact, most recovery operations employ *pickers*, human beings who identify the most readily separable materials, such as corrugated cardboard and HPDE milk bottles, before the refuse is mechanically processed.

Many items in refuse are not made of a single material, and to be able to use mechanical separation, these items must be separated into discrete pieces consisting of a single material. A common tin can, for example, contains steel in its body, zinc on the seam, a paper wrapper on the outside, and perhaps an aluminum top. Other common items in refuse provide equally challenging problems in separation.

One means of producing single-material pieces and, thereby, assisting in the separation process is to decrease the particle size of refuse by grinding up the larger pieces. This will increase the number of particles and achieve many so-called clean, or single-material, particles. The size-reduction step, although not strictly materials separation, is commonly a second step in a MRF, after the picking process.

Size reduction is followed by various other processes, such as air classification (which separates the light paper and plastics) and magnetic separation (for the iron and steel). Various unit operations used in a typical MRF are shown in Figure 14.13 on the next page.

14.3.2 Markets for Processed Refuse

The recovery of materials, although it sounds terribly attractive, is still a marginal option. The most difficult problem faced by engineers designing such facilities is the availability of firm markets for the recovered product. Occasionally, the markets are quite volatile, and secondary material prices can fluctuate wildly. One example is the secondary paper market.

© Courtesy of P. Aarne Vesilind

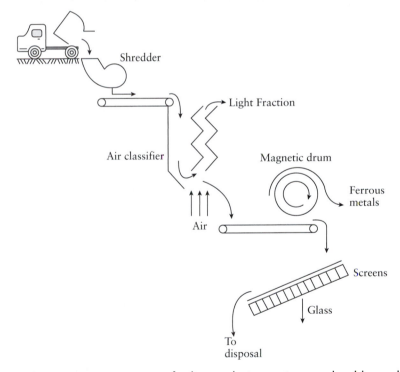

Figure 14.13 A resource recovery facility producing various marketable products from municipal solid waste.

Figure 14.14   A paper company is vertically integrated, producing from trees the paper sold to consumers. Short-term fluctuations in demand are met by the purchase of secondary paper.

Paper industry companies are what is known as vertically integrated, meaning that the company owns and operates all the steps in the papermaking process. They own the lands on which the forests are grown; they do their own logging and take the logs to their own papermill. Finally, the company markets the finished paper to the public. This is schematically shown in Figure 14.14.

Suppose a paper company finds that it has a base demand of 100 million tonnes of paper. It then adjusts the logging and pulp and paper operations to meet this demand. Now suppose there is a short-term fluctuation of 5 million tonnes that has to be met. There is no way the paper company can plant the trees necessary to meet this immediate demand, nor are they able to increase the capacity of the pulp and paper mills on such short notice. What they do then is to go to the secondary paper market and purchase secondary fiber to meet the incremental demand. If several large paper companies find that they have an increased demand, they will all try to purchase the secondary paper, and suddenly the demand will greatly increase the price of waste paper. When either the demand decreases or the paper company has been able to expand its capacity, it no longer needs the secondary paper, and the price of waste paper plummets. Because paper companies purchase waste paper on the margin, secondary paper dealers are always in either a boom or a bust situation. When paper companies that use only secondary paper to produce consumer products increase their production (due to the demand for recycled or recovered paper), these extreme fluctuations are dampened.

14.4 COMBUSTION OF REFUSE

One product that always has a market is energy. Because refuse is about 80% combustible material, it can be burned as is, or it can be processed to produce a *refuse-derived fuel* (RDF). A cross section of a typical *waste-to-energy* (WTE) facility is shown in Figure 14.15. The refuse is dumped from the collection trucks into a pit to mix and equalize the flow over the 24-hr period since such facilities must operate around the clock. A crane

© Courtesy of P. Aarne Vesilind

Figure 14.15 A waste-to-energy facility.

lifts the refuse from the pit and places it in a chute that feeds the furnace. The grate mechanism moves the refuse, tumbling it and forcing in air from the bottom and the top as the combustion takes place. The hot gases produced from the burning refuse are cooled with a bank of tubes filled with water. As the gases are cooled, the water is heated, producing low-pressure steam. The steam can be used for heating and cooling or for producing electricity in a turbine. The cooled gases are then cleaned by pollution control devices, such as electrostatic precipitators (as described in Chapter 13), and discharged through a stack.

FOCUS ON

Cement Manufacturing and the Use of Alternative Fuels and Resources

Cement kilns are among the largest contributors of greenhouse gases, producing up to 5% of the total CO_2 emissions worldwide. The CO_2 comes from both the calcining of limestone, a major feedstock for the cement, and from the tonnes of fuel (primarily coal) burned to provide energy.[4,5]

The cement-making process begins with the crushing of raw materials that contain appropriate amounts of calcium compounds, silica, alumina, and iron oxide. The materials are then screened and placed, either wet or dry, in a rotary kiln that is 3–8 m in diameter and 45–225 m long (Figure 14.16A). Flame jets at the exit end further heat the materials to 2700–3000°F (1480–1650°C). This high heat drives off, or calcines, water and CO_2 from the raw materials, which fuse into marble-sized pellets called "clinker." After several hours, the clinker emerges from the kiln and falls into the cooling unit. For each ton of material that goes into the feed end of the kiln, two-thirds of a ton of clinker is produced. Portland cement is then produced using finely ground clinker.

Air quality impacts of cement manufacturing primarily stem from the high need for energy and the combustion, particularly of coal, to meet that need. Emissions include dust, CO_2, NO_x, SO_2, and toxics. Traditionally the emission of dust, particularly from kiln stacks, has been the main environmental concern. The primary sources of dust are the operations in which large volumes of gases flow through dusty materials: the kilns, raw mills, clinker coolers and cement mills.[6] Approximately 60% of CO_2 emissions originate in the calcining process, and the remaining 40% is related to fuel combustion. The intense heat required by the process also results in significant emissions of NO and NO_2 (NO > 90% of the nitrogen oxides). The sources of the nitrogen are the air and the fuel. SO_2 emissions are primarily determined by the content of the volatile sulphur in the raw materials. Toxic emissions include dioxins, furans, and various hydrocarbon compounds referred to as products of incomplete combustion (PICs); acid gases generated by the combustion of elements such as chlorine and bromine; and heavy metals. These pollutants are found in both stack and fugitive emissions (i.e., dust and gas stirred up from plant operations and emitted from locations other than stacks).[7]

To address its sustainability, the cement industry has engaged in initiatives to improve energy efficiency, environmental performance, and transparency, all of which are considered indicators of sustainable performance. Under the umbrella of the Working Group Cement of the World Business Council for Sustainable Development (WGC-WBCSD), major world cement makers—including companies like CEMEX, Holcim and Lafarge—developed a set of sustainable strategies, including the establishment of a common framework for reporting greenhouse emissions, a pledge to publish individual performance data and targets for CO_2 emission reductions, stakeholder dialogues to develop guidelines on fuel use and other measures to make their plants greener.[8]

Other initiatives include programs to recycle concrete (the end product, e.g., as aggregate in roads), to improve worldwide monitoring and reporting of emissions, and to conduct activities related to the responsible use of fuels and raw materials. One controversial program is the use of alternative fuels and resources (AFR) in the cement manufacturing process. AFR are the waste and byproducts of other industries and can be solid or liquid, hazardous or nonhazardous. The volume of AFR (tires, solid hazardous waste, and liquid hazardous and nonhazardous wastes) used by the United States cement sector was 15% in 2000.[9] In Switzerland, the share of alternative fuels was 47.8% of the total thermal energy consumption of the cement industry in 2002.[10] On average in European cement plants, alternative fuels provide about 12% of thermal energy consumption, but in some regions AFR provide up to 72%.[11]

Concerns related to AFR focus primarily on the use of hazardous wastes. Hazardous wastes often contain toxic organic chemicals, heavy metals, and chlorine, trace amounts of which may be released into the atmosphere. Kilns burning hazardous wastes (especially halogenated wastes) emit more particles than kilns burning conventional fuel, providing a pathway for metals in a form that is particularly dangerous to humans (attached to fine particles). The fly ash from kilns burning hazardous wastes is loaded with heavy metals. Another source of problems may be chemical releases resulting from transporting hazardous wastes.

On the other hand, the use of alternative fuels is compatible with the principles of sustainable development. In terms of the cement manufacturing process, AFR can provide a source of energy with cleaner emissions than the coal it displaces. Certain plastics, polymers, and agricultural products are nonhazardous, primarily made from carbon and hydrogen, and they do not contain metals such as mercury typically found in coal. What is important is to ensure that wastes are screened not only for their energy value and for the fees that the waste generator will pay for their disposal but also for the inherent chemistry. For example, halogenated materials, such as polyvinyl chloride (PVC) and halogenated solvents, are undesirable for a number of reasons. They form acids; precipitate out as salts that build up in the kiln, shutting down production; and under nonideal conditions (i.e., starting up and shutting down) can facilitate dioxin formation and release.

In terms of a waste management system, cement kilns offer an alternative to conventional disposal of waste in dedicated incinerators and results in an overall reduction of CO_2 and other emissions compared to incinerating the waste in dedicated incinerators (Figure 14.16B). The use of waste in European cement plants saves the equivalent of about 3 million tonnes of coal per year. A ruling delivered by the European Court of Justice in February 2003 held that using waste as a fuel in cement kilns is classified as recovery, whereas burning municipal waste in dedicated incinerators, even with energy recovery, is classified as disposal.[12] This ruling is in contrast to the USEPA determination that burning hazardous waste in a cement kiln is considered disposal.

The use of AFR raises some thorny questions. Properly handled and applied to appropriate types of wastes, the combustion can be carried out fully and safely. And the energy extracted from the wastes can save on fuel that would otherwise be necessary. It would seem to be a classic synergy: solve a waste disposal problem and save resources. But the public has concerns and expectations. Do cement companies, whose expertise lies in high-volume production of a low-profit margin commodity, have comparable expertise in waste management? Would alternative ways of dealing with the waste be better or worse? Would it be better to force those who generate the waste in the first place to work harder at source reduction by denying them the option of relatively easy disposal afforded by cement kilns?[12] Use of carefully screened AFR in cement kilns can be a good interim strategy, especially if it offsets the need for the waste incinerators and the use of dirty fuel. The kilns are flexible with respect to fuels, unlike municipal solid waste incinerators which become institutionalized in their demand for waste to operate.

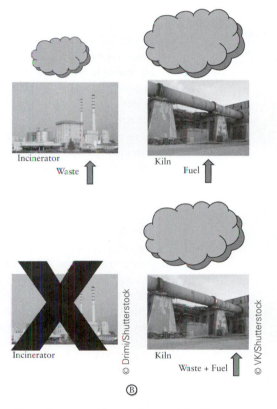

Figure 14.16 (A) Model of a dry process kiln with preheater and precalciner; (B) Efficiencies gained by combusting waste in cement kilns.

Because solid waste can be combusted as is and because it can also be processed in many ways before combustion, there might be confusion as to what exactly is being burned. The American Society for Testing and Materials (ASTM) developed a scheme for classifying solid waste destined for combustion:

RDF-1   unprocessed MSW
RDF-2   shredded MSW (but no separation of materials)
RDF-3   organic fraction of shredded MSW (usually produced in a MRF or from source-separated organics, such as newsprint)

RDF-4   organic waste produced by a MRF that has been further shredded into a fine, almost powder, form, sometimes called "fluff"

RDF-5   organic waste produced by a MRF that has been densified by a pelletizer or a similar device and that can often be fired with coal in existing furnaces

RDF-6   organic fraction of the waste that has been further processed into a liquid fuel, such as oil

RDF-7   organic waste processed into a gaseous fuel.

RDF-4 has been seldom used because of the extra cost of processing. RDF-6 was attempted some years ago, but the full-scale facility failed to operate properly. Similarly, RDF-7 has been unsuccessful so far. Organic MSW can be processed into a gas by anaerobic digestion, but while the process is deceptively (and seductively) simple and appears to use common components, all experiences to this date have been failures (including Harrelson's infamous attempt as described in Chapter 1). The most difficult problems are the preprocessing of the waste to achieve a high-quality organic fraction, the mixing of the high solids digesters, and the presence of toxic material that can severely hinder the operation of the anaerobic digester.

One reason WTE facilities have not found greater favor is concern with the air emissions, but this seems to be a misplaced concern. Studies have shown that the risk of MSW combustion facilities on human health is negligible, far below the risks resulting from the combustion of gasoline in automobiles, for example.

Of particular concern to many people is the production of dioxin in waste combustion. Dioxin is actually a family of organic compounds called polychlorinated dibenzodioxins (PCDD). Members of this family are characterized by a triple-ring structure of two benzene rings connected by a pair of oxygen atoms (Figure 14.17). A related family of organic chemicals are the polychlorinated dibenzofurans (PCDF), which have a similar structure except that the two benzene rings are connected by only one oxygen. Because any of the carbon sites are able to attach either a hydrogen or a chlorine atom, the number of possibilities is great. The sites that are used for the attachment of chlorine atoms are identified by number, and this signature identifies the specific form of PCDD or PCDF. For example, 2,3,7,8-tetrachloro-dibenzo-*p*-dioxin (or 2,3,7,8-TCDD in shorthand) has four chlorine atoms at the four outside corners, as shown in Figure 14.17. This form of dioxin is especially toxic to laboratory animals and is often identified as a primary constituent of contaminated pesticides and emissions from WTE plants.

All the PCDD and PCDF compounds (referred to here as dioxins) have been found to be extremely toxic to animals. Neither PCDD nor PCDF compounds have found any commercial use and are not manufactured. They do occur, however, as contaminants in other organic chemicals. Various forms of dioxins have been found in pesticides (such as

Figure 14.17   2, 3, 7, 8-tetrachloro-dibenzo-*p*-dioxin.

Agent Orange, widely used during the Vietnam War) and in various chlorinated organic chemicals (such as chlorophenols).

Curiously, evidence has not borne out the same level of toxicity to humans, and it seems less likely that dioxins are actually as harmful as they might seem from laboratory studies. A large chemical spill in Italy was expected to result in a public health disaster based on the extrapolations from animal experiments, but thus far, this does not seem to have materialized. Nevertheless, dioxins are able, at very low concentrations, to disrupt normal metabolic processes, and this has caused the USEPA to continue to place severe limitations on the emission of dioxins from incinerators.

There is little doubt that WTE plants emit trace amounts of dioxins, but nobody knows for sure how the dioxins originate. It is possible that some of them are in the waste and simply not combusted, being emitted with the off-gases. On the other hand, it is also likely that *any* combustion process that has even trace amounts of chlorine produces dioxins and that these are simply an end product of the combustion process. The presence of trace quantities of dioxins in emissions from wood stoves and fireplaces seems to confirm this view.

It might be well to remember here that the two sources of risk, incinerators versus fireplaces, are clearly unequal. The effect of the latter on human health is greater than the effect of incinerator emissions. But the fireplace is a *voluntary* risk, whereas the incinerator is an *involuntary* risk. People are willing to accept voluntary risks 1000 times higher than involuntary risks, and they are, therefore, able vehemently to oppose incinerators while enjoying a romantic fire in the fireplace.

14.5   ULTIMATE DISPOSAL OF REFUSE: SANITARY LANDFILLS

The disposal of solid wastes is a misnomer. Our present practices amount to nothing more than hiding the waste well enough so it cannot be readily found.

The only two realistic options for disposal are in the oceans (or other large bodies of water) and on land. The former is presently forbidden by federal law in the United States and is similarly illegal in most other developed nations. Little else needs to be said of ocean disposal, except perhaps that its use was a less-than-glorious chapter in the annals of public health and environmental engineering.

Although the volume of the refuse is reduced by over 90% in WTE facilities, the remaining 10% still has to be disposed of somehow, along with the materials that cannot be incinerated, such as old refrigerators. A landfill is, therefore, necessary even if the refuse is combusted, and a WTE plant is, therefore, not an ultimate disposal facility. A landfill for ash is a great deal simpler and smaller than a landfill for refuse, and the problem with siting landfills has resulted in more WTE facilities because the volume reduction significantly extends the lives of existing landfills.

The placement of solid waste on land is called a *dump* in the United States and a *tip* in Great Britain (as in "tipping"). The dump is by far the least expensive means of solid waste disposal and thus was the original method of choice for almost all inland communities. The operation of a dump is simple and involves nothing more than making sure that the trucks empty at the proper spot. Volume is often reduced by setting the dumps on fire, thus prolonging dump life.

Rodents, odor, air pollution, and insects at the dump, however, can result in serious public health and aesthetic problems, and alternate methods for disposal are necessary. Larger communities can afford to use an incinerator for volume reduction, but smaller towns cannot afford such capital investment, so this has led to the development of the *sanitary landfill.*

The sanitary landfill differs markedly from open dumps in that the latter are simply places to dump wastes while sanitary landfills are engineered operations, designed and operated according to accepted standards. The basic principle of a landfill operation is to prepare a site with liners to deter pollution of groundwater, deposit the refuse in the pit, compact it with specially built heavy machinery with huge steel wheels, and cover the material at the conclusion of each day's operation (Figure 14.18). Siting and developing a proper landfill requires planning and engineering design skills.

© Courtesy of P. Aarne Vesilind

Figure 14.18 A sanitary landfill.

Table 14.3   Approximate Density of Municipal Solid Waste

Location/Condition	Density (kg/m^3)
As generated, in the kitchen trash can	60
In the garbage can	120–240
In the packer truck	240–480
In the landfill, after compaction by overlaying refuse	480–960

Even though the tipping fees paid to use landfills are charged on the basis of weight of refuse accepted, landfill capacity is measured in terms of volume, not weight. Engineers designing the landfills first estimate the total volume available to them and then estimate the density of the refuse as it is deposited and compacted in the landfill. The density of refuse increases markedly from when it is first generated in the kitchen to when it is finally placed into the landfill. Table 14.3 is a crude estimate of the density of household refuse.

EXAMPLE
14.2

Problem   Imagine a town where 10,000 households each fill up one 320 L container of refuse per week. To what density would a 15 m^3 packer truck have to compact the refuse to be able to collect all the households during one trip?

Solution   There is, of course, (what else) a mass balance involved here. Imagine the packer truck as a black box; the refuse goes in at the households and out at the landfill.

$$[\text{Mass IN}] = [\text{Mass OUT}]$$
$$V_L C_L = V_P C_P$$

where V and C are the volume and density of the refuse and the subscripts L and P denote loose and packed refuse. Assume that the density in the cans is 120 kg/m^3 (Table 14.3).

$$[(10{,}000 \text{ households})(320 \text{ L/household})(0.001 \text{ } m^3/L)]$$
$$\times [120 \text{ kg/m}^3] = (15 \text{ } m^3)C_P$$
$$C_P = 25{,}600 \text{ kg/m}^3 (!)$$

Clearly impossible. Obviously, more than one truck and/or more than one trip is required.

An added complication in the calculation of landfill volume is the need for the daily cover, which may be removable (such as a plastic 'tarp') and not use any of the volume or may not be removable (such as dirt). The more permanent cover material (e.g., dirt) that is placed on the refuse, the less volume there is available for the refuse itself, so the shorter the life of the landfill. Commonly, engineers estimate that the volume occupied by cover dirt is one-fourth the total landfill volume.

However, sanitary landfills are not inert. The buried organic material decomposes, first aerobically and then anaerobically. The anaerobic degradation produces various gases, such as methane and carbon dioxide, and liquids (known as *leachate*) that have extremely high pollutional capacity when they enter the groundwater. Liners made of either impervious clay or synthetic materials, such as plastic, are used to try to prevent the movement of leachate into the groundwater. Figure 14.19 shows how a synthetic landfill liner is installed in a prepared pit. The seams have to be carefully sealed and a layer of soil placed on the liner to prevent landfill vehicles and waste from puncturing it.

Synthetic landfill liners are useful in capturing most of the leachate, but they cannot be perfect. No landfill is sufficiently tight that groundwater contamination by leachate is totally avoided. Wells have to be drilled around the landfill to check for groundwater

© Courtesy of P. Aarne Vesilind

Figure 14.19 Synthetic liner being installed in a landfill.

contamination from leaking liners, and if such contamination is found, remedial action is necessary.

The use of plastic liners substantially increased the cost of landfills to the point where a modern landfill costs nearly as much per ton of refuse as a WTE plant. And, of course, the landfill never disappears—it will be there for many years to come, limiting the use of the land for other purposes.

Modern landfills also require the gases to be collected and either burned or vented to the atmosphere. The gases are about 50% carbon dioxide and 50% methane, both of which are greenhouse gases (Chapter 12). In the past, when gas control in landfills was not practiced, the gases were known to cause problems with odor, soil productivity, and even explosions. Now the larger landfills use the gases for running turbines to produce electricity to sell to the power company. Smaller landfills simply burn the gases at flares.

No matter what the propaganda, sanitary landfills are at best an interim solution to our solid waste problem. Environmental concerns will soon dictate material and/or energy recovery as the disposal method of choice for solid waste management.

Obviously, we must attack the solid waste problem from both ends—reducing the total quantity of wastes (e.g., by making materials more recyclable) and developing more environmentally acceptable disposal methods. We are still many years away from the development and use of fully recyclable or biodegradable materials. The only truly disposable package available today is the ice cream cone.

14.6   REDUCING THE GENERATION OF REFUSE: SOURCE REDUCTION

There is a third way we can affect the quantity and content of the solid waste stream: carefully select the materials and products we use and, hence, have to throw away. There is a lot to be said for being selective in the type of packaging accepted for various products. For example, foam plastic wrappers for fast foods have little utility. A piece of paper works just as well, as many fast food chains discovered once their customers started complaining (and in some cases passing local ordinances prohibiting such packaging). Rejecting useless bags at stores when a bag is not needed is not bad manners. We can all do a lot of little things to make a big impact on the quantity and composition of the solid waste stream.

14.6.1   Why?

The question, of course, is, why should we? Consider the utility of a person rejecting an unnecessary bag at a store. The effort may be significant (you sometimes have to argue) and the convenience is less because now the purchase is not carried as conveniently. The bag would not have cost anything personally to dispose of if the collection is a municipal service. What, then, is the benefit to you to reject the bag? Essentially, none at all. The quantity of landfill space you personally save by your actions is negligible, and your choice of purchases that reduce the quantity of waste will not benefit you at all (unless, of course, the product with less packaging costs less). So the question remains, why ought you to do something that is not to your measurable benefit?

The answer might be that "it is the right thing to do," and this may be perfectly satisfying to you. It makes you feel a part of the community, all pulling together to achieve some good end, with everyone's individual actions totaling to a significant effect. You could argue that, if you did not do this, you would have no basis for expecting others to do the same, and then there would be no way to achieve anything by community action. But if someone says, "Prove to me that I ought to participate in such altruistic activity," you have to admit that there is no such proof. There are arguments, but all of them depend on the basic goodness of humans. When that is no longer the norm, then there is no hope for human civilization.

14.6.2 Life Cycle Analysis

One means of getting a handle on questions of material and product use is to conduct what is known as a *life cycle analysis* (LCA). Such an analysis is a holistic approach to pollution prevention by analyzing the entire life cycle of a product, process, or activity, encompassing raw materials, manufacturing, transportation, distribution, use, maintenance, recycling, and final disposal. In other words, LCA should yield a complete picture of the environmental impact of a product.

LCAs are performed for several reasons, including the comparison of products for purchasing and the comparison of products by industry. In the former case the total environmental effect of, say, glass returnable bottles could be compared to the environmental effect of nonrecyclable plastic bottles. If all the factors going into the manufacture, distribution, and disposal of both types of bottles are considered, one container might be shown to be clearly superior. In the case of comparing the products of an industry, we might determine if the use of phosphate builders in detergents is more detrimental than the use of substitutes that have their own problems in treatment and disposal.

One problem with such studies is that they are often conducted by industry groups or individual corporations, and (surprise!) the results often promote their own product. For example, Proctor & Gamble, the manufacturer of a popular brand of disposable baby diapers, found in a study done for them that cloth diapers consume three times more energy than disposable diapers. But a study by the National Association of Diaper Services found disposable diapers consume 70% more energy than cloth diapers. The difference was in the accounting procedure. If one uses the energy contained in the disposable diaper as recoverable in a WTE facility, then the disposable diaper is more energy efficient.[16]

LCAs also suffer from a dearth of data. Some of the information critical to the calculations is virtually impossible to obtain. For example, something as simple as the tonnage of solid waste collected in the United States is not readily calculable or measurable. And even if the data were there, the procedure suffers from the unavailability of a single accounting system. Is there an optimal level of pollution, or must all pollutants be removed 100% (a virtual impossibility)? If there is both air pollution and water pollution, how must these be compared?

A simple example of the difficulties in LCA is in finding the solution to the great coffee cup debate: whether to use paper coffee cups or polystyrene coffee cups. The answer most people would give is not to use either but instead to rely on the permanent mug. But there nevertheless are times when disposable cups are necessary (e.g., in hospitals), and a decision must be made as to which type to choose.[17] So let's use LCA to make a decision.

FOCUS ON

The Social Side of Trash

The Town of Wellesley is a predominantly residential community located outside of Boston, Massachusetts. The town is also a college community, having within its boundaries Wellesley College, Babson College, and Massachusetts Bay Community College. Surprisingly, one area of notoriety for the town is the Wellesley Recycling and Disposal Facility (RDF), aka the dump. The RDF is an 88-acre (0.36 km^2) facility for use by Town of Wellesley residents only.[14] About 85% of Wellesley residents regularly frequent the dump. There is no town pickup service, although private hauling services are available. Nearly 40 categories of products and materials are collected for recycling (Figure 14.20 and Table 14.4). The Book Exchange and Take-It-or-Leave-It swap areas are, in particular, very popular. And recycling in Wellesley is targeting 35% through 2009, which is above the national average for recycling (estimated at between 26.5 percent and 27.2 percent[15]).

What accounts for the success and popularity of the dump? Convenience and the wide scope of recycling options are cited as major reasons for its success. Another reason is that the dump has become one of the hottest spots in town to socialize and to shop for goods. Residents joke that, upon visiting the dump, they typically come back with as much stuff as they took because junkyard treasures are plentiful. Affluence may play a role also because Wellesley is a relatively affluent community. Many people can't be bothered to have yard sales, and others are quick to toss items that are gently used, if used at all. Notable finds include a Weber grill, a Bianchi racing bike, and a rare book worth $12,000. One man claims he scored a baseball signed by Ted Williams. The swaps have become so popular in Wellesley and in other towns as well that in some cases officials have taken measures to keep away out-of-town scavengers and imposed new rules to discourage aggressive tactics.

But there is more. While dump fever has clearly spread, the Wellesley dump has always been a social place. Prior to elections, politicians

Figure 14.20   Example of reuse.

frequent the dump to meet voters, and residents mingle with their neighbors. The dump's 88 acres are so nicely landscaped that rumor has it that even weddings have been held there, perhaps symbolizing the everlasting cycles of love and material flows. Maybe it is time for an enterprising barrista to open a coffee shop, with reusable or swappable mugs, of course.

Costs associated with the RDF (such as labor, equipment, hardware, transportation, and education/promotion) are appropriated from tax dollars. All revenues earned are then returned to the Town General Fund. In fiscal year 2007 the recycling program returned approximately $644,000. A crew of more than 100 volunteers rescued the town swap from budget cuts and now runs it as a nonprofit outfit, sorting and categorizing items into stacks ranging from skis to suitcases. At a time when middle-class Americans are cutting back on their discretionary spending, when environmentally conscious consumers are questioning whether their children really need more cheap plastic toys, when craigslist.com has made used shopping fun and fashionable, dump site swap shops are thriving, as exemplified by Wellesley residents, who with the spirit of Yankee thrift are supporting their dump.

Table 14.4 Materials Accepted at the Wellesley RDF

Household Recycling	Composting	Metals	Reuse	Special Hazards
Aluminum cans	Compost	Aluminum	Book exchange	Appliances
Aluminum foil and trays	Wood waste	Copper	Clothing/textiles	Automotive batteries
Boxboard and chipboard		Metal/light iron	Health care equipment	Computers and monitors
Brown paper bags				Fluorescent bulbs
Cardboard and corrugated cartons				Paint
Cell phones				Propane tanks
Electronic media				Tires
Eye glasses				Waste oil
Glass (brown/green/clear)				
Household batteries				
Inkjet cartridges				
Mixed paper				
Mixed office paper				
Newspapers				
Plastic bags				
Plastic bottles (clear/colored/natural)				
Plastic nonbottles				
Refundable containers				
Steel/tinned cans				

Source: Based on Town of Wellesley Department of Public Works. RDF Recycling Guide. Accessed at http://wellesleyma.virtualtounhall.net/pages/WellesleyMA_DPW/rdf/recyclingposter.pdf on 9 January, 2009.

The paper cup comes from trees, but the act of cutting trees results in environmental degradation. The foam cup comes from hydrocarbons, such as oil and gas, and this also results in adverse environmental impacts, including the use of nonrenewable resources. The production of the paper cup results in significant water pollution, with 30–50 kg of BOD per cup produced, while the production of the foam cup contributes essentially no BOD. The production of the paper cup results in the emission of chlorine, chlorine dioxide, reduced sulfides, and particulates while the production of the foam cup results in none of these. The paper cup does not require chlorofluorocarbons (CFCs), but neither do the foam cups, ever since the CFCs in polystyrene were phased out. The foam cups, however, contribute from 35–50 kg per cup of pentane emissions while the paper cup contributes none. The recyclability of the foam cup is much higher than that of the paper cup because the latter is made from several materials, including the plastic coating on the paper. They both burn well, although the foam cup produces 40,000 kJ/kg while the paper cup produces only 20,000 kJ/kg. In the landfill the paper cup degrades into CO_2 and CH_4, both greenhouse gases, while the foam cup is inert. Since it is inert, it will remain in the landfill for a very long time while the paper cup will eventually (but very slowly!) decompose. If the landfill is considered a waste storage receptacle, then the foam cup is superior because it does not participate in the reaction while the paper cup produces gases and probably leachate. If, on the other hand, the landfill is thought of as a treatment facility, then the foam cup is highly detrimental.

So which cup is better for the environment? If you wanted to do the right thing, which cup should you use? This question, like so many others in this book, is not an easy one to answer.

14.7   INTEGRATED SOLID WASTE MANAGEMENT

The EPA developed a national strategy for the management of solid waste called "integrated solid waste management" (ISWM). The intent of this plan is to assist local communities in their decision making by encouraging strategies that are the most environmentally acceptable but providing flexibility to manage wastes efficiently. It is based on the solid waste management hierarchy, with the most-to-least-desirable solid waste management strategies being

- source reduction
- recycling
- combustion
- landfilling.

That is, when an ISWM plan is implemented for a community, the first means of attacking the problem should be by source reduction. This is an unfortunate term because it is both incorrect and misleading. One is not reducing sources but rather reducing the amount of waste coming from a source and, thus, the term really should be waste reduction.

On a household level it is possible through judicious buying to significantly reduce the amount of waste generated in each household. For example, buying produce unwrapped instead of on plastic trays with shrink wrap is one simple way of reducing waste. Purchasing products that have a longer life is a second way of achieving waste reduction. For example, buying automobile tires that last 130,000 km instead of those that wear out at 65,000 reduces the production of waste tires by 50%. In addition, getting your name off mailing lists is a big help because it reduces junk mail. This option of solid waste management is even more effective on the industrial and commercial level. Unfortunately, it's difficult to measure residential source reduction, and you don't need any new equipment to implement it, so it is often ignored.

Instead, the focus has been on recycling, the second option, even though in ISWM recycling (and reuse) should be undertaken after most of the waste reduction options have been implemented because there are costs associated with collecting and processing the recyclable material. Recycling and reuse do, however, reduce some waste. Figure 14.11 is useful in showing this effect. An increase in material flows through the reuse and recycling loops will reduce waste. For example, the use of paper shopping bags for trash disposal presupposes that the public chooses not to purchase plastic bags for trash disposal. If this occurs, fewer plastic bags will be produced and sold, and the total amount of waste destined for disposal will likewise be reduced. Glass bottles can be separated from refuse, collected, and remanufactured into new glass bottles, thereby increasing the recycling loop. Finally, mixed refuse can be hand sorted, with corrugated cardboard removed and recovered, increasing materials flow in the recovery loop.

FOCUS ON

Vehicles

Examples of environmental concerns being neglected for the sake of short-term economic gain abound. The use and manufacture of nonrecyclable beverage containers is perhaps the most ubiquitous example, but the most environmentally unfriendly product ever manufactured has to be the Trabi.

The Trablant, affectionately known as the Trabi, was manufactured in East Germany during the 1970s and 1980s (Figure 14.21A). This homely looking car was designed to be the East German version of the Volkswagen (the People's Car), and its design objectives were to make it as cheap as possible. So, to power the Trabi, the engineers used a two-stroke engine that is notoriously polluting. All of the components were designed at the least cost, and few survived normal use. Worst of all, the body was made of a fiberglass composite that was impossible to fix (except with duct tape!) and, as far as solid waste management was concerned, had absolutely no recycling value because it could not be melted down or reprocessed in any other way; nor could it be burned in incinerators.

After the reunification of Germany, thousands of Trabis were abandoned on the streets and had to be disposed of in landfills. The Trabi is the best example of engineering design when the sole objective is production cost and when environmental concerns are nonexistent, and the Trabi deserves to be immortalized as the worst example of environmentally destructive design.

Contrast the Trabi with the intent of relatively recent European legislation. In part because there is greater pressure on landfill space than in the United States, more is being done to maximize the amount of material recycled from the estimated 14 million motor vehicles that reach the end of their roadworthy lives and end up as 8–9 million tonnes of waste in the member states of the European Union (EU) annually. EU Directive 2000/53/EC, the End of Life Vehicle (ELV) Directive, was adopted in May 2000. The legislation aims to protect the environment and prevent waste by promoting the collection, reuse, and recycling of vehicle components. By 2015, the legislation will ensure EU member states reuse or recycle 85% of a scrapped vehicle by weight.

While the directive's main objective is to manage the flow of ELVs into the recycling stream,

© wim wiskerke/Alamy

(A)

© AP/Wide World Photos

(B)

Figure 14.21   (A) One method of disposing of a Trabi hulk; (B) Ford Model U concept car.

it also includes stipulations about designing more environmentally friendly vehicles. For example, Although ELVs are full of easily recyclable material, they have their share of troublesome components. Composite plastics can be a challenge because they are not designed to be disassembled, and parts like windshields with glass and plastic laminated covers are hard to separate. The ELV Directive is causing automakers to reduce the amount of mixed plastics in vehicles. The directive is also causing automakers to balance the ease of dismantling and recycling with basic vehicle functionality and driver and passenger safety. A car that is too easily dismantled is potentially problematic!

While some view the regulations as restrictive, effective regulations can be drivers for innovation. However, an innovative concept vehicle not driven by legislation but by an environmental (and business) ethic is the Ford Model U (Figure 14.21B). In contrast to the Trabi, the Model U includes:[1]

- a modular interior and exterior that allows for different components to be mounted, moved around, or added later
- a "2.3-liter, four-cylinder supercharged, inter-cooled hydrogen internal combustion engine coupled with a hybrid electric transmission. . . [that provides] the equivalent of 19 km/L and about 450 km of range plus near-zero regulated emissions and a 99-percent reduction in carbon dioxide"
- a system that allows voice operation of on-board systems
- advanced safety features, such as precrash sensing, adaptive front headlights, and a night vision system
- materials designed to meet cradle-to-cradle specifications.

And for those who long nostalgically for a simpler mode of transportation, there is always the bicycle. . .

The third level of the ISWM plan is solid waste combustion, which really should mean all methods of treatment. The idea is to take the solid waste stream and to transform it into a nonpolluting product. This may be by combustion, but other thermal and chemical treatment methods may eventually prove just as effective. Of course, it's preferable to recover some product (e.g., energy) from the treatment.

Finally, if all of the above techniques have been implemented and/or considered and there is still waste left over (which there will be), the final solution is landfilling. There really is no alternative to landfilling (except disposal in deep water, which is now illegal), and therefore, every community must develop some landfilling alternative.

While this ISWM strategy is useful, it can lead to problems if taken literally. Some communities cannot, as hard as they might try, implement recycling. In other communities the only option right from the start is landfilling. In others the intelligent thing to do would be to provide one type of treatment for one kind of waste (e.g., composting for yard waste) and a second type for another (landfilling for refuse). This is where engineering judgment comes into play and where the solid waste engineer really earns his or her salary. We have to juggle all the options and integrate these with the special features (economics, history, politics) of the community. The engineer is, after all, there to serve the needs of the people.

PROBLEMS

14.1 The street plan for Example 14.1 in Figure 14.7A is analyzed as Figure 14.7C if the collection has to be by *block faces*—that is, the truck must travel along each block to collect and the crew cannot cross the street. Since it is possible to construct an Euler's tour, find one solution.

14.2 Using the street networks in Figure 14.22, design the most efficient route possible if the truck is

a. collecting only one side of a street at a time
b. collecting both sides of a street at the same time

14.3 The objective of this assignment is to evaluate and report on the solid waste you personally generate. For one day, selected at random, collect *all* of the waste you would have normally discarded. This includes food, newspapers, beverage containers, etc. Using a scale, weigh your solid waste and report it as follows:

Component	Weight	Percentage of Total Weight
Paper		
Plastics		
Aluminum		
Steel		
Glass		
Garbage		
Other	_____	_____
	100	

× Start and end here

Figure 14.22 See Problem 14.2.

Your report should include a data sheet and a discussion. Answer the following questions.

a. How do your percentage and total generation compare to national averages?
b. What in your refuse might have been reusable (as distinct from recoverable), and if you had reused it, how much would this have reduced the refuse?
c. What in your refuse is *recoverable*? How might this be done?
d. How typical was this day in terms of your trash's composition and quantity?

14.4 One of the costliest aspects of municipal refuse collection is the movement of the refuse from the household to the truck. Suggest a method by which this can be improved. Originality counts heavily, practicality far less.

14.5 Suggest how you would approach the Trabi problem discussed in the chapter if suddenly you are placed in charge of this project. List all the feasible alternatives, and discuss each, finally choosing the one you favor.

14.6 Paper or plastic? They both have their advocates and detractors (Figure 14.23). For example, recycling advocates often point to the plastic bags as the prototype of wastage and pollution, as stuff that clogs up our landfills. In retaliation plastic bag manufacturers developed a public relations campaign to promote their product. On one of the flyers (printed on paper) they said:

> The (plastic) bag does not emit toxic fumes when properly incinerated. When burned in waste-to-energy plants, the resulting by-products from combustion are carbon dioxide and water vapor, the very same by-products that you and I produce when we breathe. The bag is inert in landfills, where it does not contribute to leaching, bacterial or explosive gas problems. The bag photodegrades in sunlight

Figure 14.23 (A) Paper Bags, 2007—Depicts 1.14 million brown paper supermarket bags, the number used in the United States every hour; (B) Plastic Bags, 2007—Depicts 60,000 plastic bags, the number used in the United States every five seconds.

to the point that normal environmental factors of wind and rain will cause it to break into very small pieces, thereby addressing the unsightly litter problem.

Critique this statement. Is all of it true? Is anything misleading? Do you agree with their evaluation? Write a one-page response.

14.7 The siting of landfills is a major problem for many communities. This is often an exasperating job for engineers because the public is so intimately involved. A prominent environmental engineer, Glenn Daigger of CH2M Hill, is quoted as follows:

> Environmental matters are on the front page today because we, the environmental industry, are not meeting people's expectations. They're telling us that accountability and quality are not open questions that ought to be considered. It's sometimes difficult to

grasp in the face of all the misinformation out there: Ultimately, we're responsible for accommodating the public's point of view, not the other way around.[19]

Do you agree with him? Should the engineer "accommodate the public's point of view," or should the engineer impose his/her own point of view on the public since the engineer has a much better understanding of the problem? Write a one-page summary of what you believe should be the engineer's role in the siting of a landfill for a community.

14.8 The recycling symbols shown in Figure 14.24 are taken from various forms of packaging.

   a. What purpose do you suppose the container manufacturers had in mind in placing the symbols on their packages?
   b. Was this use ethical in all cases? Argue your conclusion.
   c. Cut out another recycling symbol from a package and discuss its merits.

14.9 A high school has a student body of 660 students. Studies have shown that, on average, each student is expected to contribute 0.1 kg of solid waste each day from all sources except the cafeteria, which contributes 150 kg daily. If the density of refuse in a dumpster is 120 kg/m^3, and if trash pickup occurs only once a week, how many dumpsters does the school need if each dumpster can hold 15 m^3?

14.10 Figure 14.25 shows an isolated development that is to be serviced by refuse pickup. Numbers on the blocks indicate the number of households that need to be picked up on that block. Assume that each household is 4 people, and that the waste generation is 2 kg/capita/d. The density of refuse in an old-style packer truck is about 240 kg/m^3.

   a. If the pickup is to occur once a week, how many trucks would be needed?
   b. If the trucks entered and exited the development at the point shown by E, what would be one efficient routing for these trucks? (Count the number of deadheads as an estimate of efficiency.)
   c. If the old trucks were replaced with trucks achieving 600 kg/m^3 how would the collection system change?

14.11 How do you think the composition of municipal solid waste will change in 20 yr. Why?

14.12 A commercial establishment has a private hauler who weighs the solid waste transported to the landfill. The data for 10 consecutive weeks are as follows:

Week Number	Refuse (kg)
1	270
2	310
3	460
4	205
5	156
6	410
7	270
8	210
9	140
10	390

Figure 14.24 Recycling symbols off consumer packaging. See Problem 14.8.

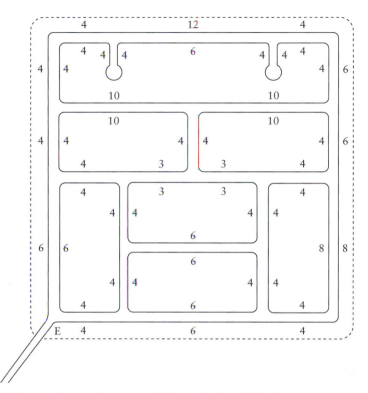

Figure 14.25 Trucking and routing problem. See Problem 14.10.

a. What is the average waste generation for this establishment?
b. What might be the *lowest* week in the entire year, based on these data?
c. If the dumpster is able to handle 1000 kg, will it ever be full? (This is a curve ball. Careful with your answer.)

14.13 A solid waste incinerator has a maximum operating capacity of 100 tonnes/day. The facility receives MSW during the work week, but not during the weekends or on holidays. Typically, it receives the following daily loads:

	Received (tonnes/d)
Mondays	180
Tuesdays	160
Wednesdays	150
Thursdays	120
Fridays	80
Saturdays	0
Sundays	0

How large should the receiving pit be to be able to hold enough waste to operate through a three-day weekend? (There is no one right answer to this problem. Think about what you would do if you were the engineer having to make this decision.) Justify your answer.

14.14 Suppose you are the city engineer of a small community, and the town council tells you to design a recycling program that will achieve at least 50% diversion from the landfill. What would be your response to the town council? If you agree to try to achieve such a diversion, how would you do it? For this problem you are to prepare a formal response to the town council, including a plan of action and what would be required for its success. This response would include a cover letter addressed to the council and a report of several pages, all bound in a report format with title page and cover.

14.15 Design, using flow diagrams, how you would produce RDF-1, RDF-2, RDF-3, and RDF-4 as defined by the ASTM standards for refuse-derived fuel.

14.16 Briefly answer the questions posed in the following quote from *Rubbish! The Archaeology of Garbage* by William Rathje and Cullen Murphy (2001).

> The advantages brought to us by the automobile, we now know, come at a cost of 46,000 lost lives every year, and many times that number of devastating injuries; the price we pay for the automobile in terms of pollution has, of course, been staggering. Had we known in advance that this might be the case, would production of this machine ever have been allowed to proceed? Should it have been?

14.17 You are the city engineer for Kattown, which manages the solid waste program for its residents.

a. You determine that 2500 residences per week need to be served by each waste collection vehicle. Now you must determine the size of vehicle in cubic yards to buy. Each vehicle will make 3 trips per day, 4 days per week and will compact the refuse to 300 kg/m^3. Each residence contains approximately 2.5 people. Clearly state all assumptions you make.

b. Why do you need local solid waste generation and composition data rather than relying on state or national averages to develop a solid waste management plan?

END NOTES

1. The 4.6 lb/c/d generation rate includes residential and commercial waste, USEPA estimates that 55–65% of the waste generated is from residences and 35–45% is from businesses and institutions.

2. H2O Conserve. 2008. *Bottled Water: Bottles of Trouble*. Accessed at http://www.h2oconserve. org/index.php?page_id=15& pd=bottle on 12 July, 2008.

3. Natural Resources Defense Council. 2008. *Bottled Water: Pure Drink or Pure Hype?* July 10. Accessed at http://www.nrdc.org/water/drinking/bw/bwinx.asp on 10 July, 2008.

4. Portland Cement Association. Accessed at http://www.cement.org/basics/concretebasics_faqs.asp on 8 December, 2008.

5. European Commission. December 2001. Integrated Pollution Prevention and Control (IPPC): Reference Document on Best Available Technique in the Cement and Line Manufacturing Industries. European IPPC Bureau at the Institute for Prospective Technological studies: Seville, Spain.

6. Reference Document on Best Available Techniques in the Cement and Lime Manufacturing Industries (BREF Report). March 2000. Accessed at http://eippcb.jrc.es/pages/FActivities.htm on 8 December, 2008.

7. Recycling or Disposal? Hazardous Waste Combustion in Cement Kilns. Accessed at http://www.mindfully.org/ Air/Cement-Kilns-Burning-Waste5.htm on 8 December, 2008.

8. Energy Use in the Cement Industry in North America: Emissions, Waste Generation and Pollution Control, 1990–2001, pp. 33–4. Accessed at http://www.texascenter.org/publications/ cement.pdf on 8 December, 2008.

9. Ibid.

10. Sevler, Christina, Stefanie Hellweg, and Michel Monteil. March 2005. Life Cycle inventory for use of waste solvent as fuel substitute in the cement industry. A Multi-Input Allocation Model *The International Journal of Life Cycle Assessment*. vol 10, no. 2, pp. 120–130.

11. Cement Bureau. Accessed at http://www. cembureau.be/on 8 December, 2008.

12. Ibid.

13. Impacts, Risks and Recommendations. Accessed at http://ccm.ncms/ERI/new/IRR cement.htm on 8 December, 2008.

14. Wellesley, MA Recycling and Disposal Facility. Accessed at http://wellesleyma.virtualtown hall. net/Pages/WellesleyMA_DPW/rdf/index

15. Simmons, Phil, Scott M. Kaufman, and Nickolas J. Themelis. 2006. State of garbage in American recycling data analysis. *BioCycle* 47, No. 10 on 9 January, 2009. (October): 21.

16. Life cycle analysis measures greenness, but results may not be black and white. *Wall Street Journal*, 28 February, 1991.

17. Hocking, M.B. 1991. Paper versus polystyrene: a complex choice. *Science* 251 (February): 1.

18. Model U concept: A model for change. 2008. Ford Motor Company. Accessed at http://media. ford.com/article_display.cfm? article_id=14047 on 8 December, 2008.

19. Glenn Daigger, quoted in The new environmental age. *Engineering news record* 228: 25, 22 June, 1992.

CHAPTER FIFTEEN

Hazardous Waste

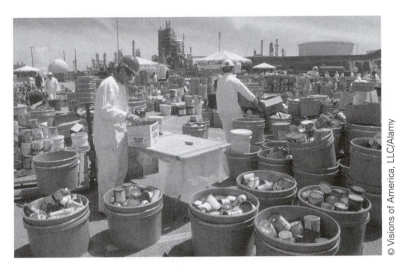

Household hazardous waste collection event

Desert scene

Environmental engineers design systems to manage hazardous waste properly and consider the impacts of designs on the end-of-life management.

What is so amazing, in retrospect, is the total nonchalance with which the American people tolerated the disposal of highly toxic wastes ever since industry first started producing such materials. Open dumps, huge waste lagoons, blatant dumping into waterways—all this existed for decades. Only since the early 1980s, however, has the public become conscious of what such indiscriminate dumping can do. Since then, we have made heroic efforts to clean up the most odious examples of environmental insults and health dangers and to regulate industry so as to prevent future problems.

15.1 DEFINING HAZARDOUS WASTE

A *hazardous substance* is defined in the United States by the USEPA as any substance that

> because of its quantity, concentration, or physical, chemical, or infectious characteristics may cause, or significantly contribute to, an increase in mortality; or cause an increase in serious irreversible or incapacitating reversible illness; or pose a substantial present or potential hazard to human health and the environment when improperly treated, stored, transported, or disposed of, or otherwise managed.

Hazardous waste is a name given to material that, when intended for disposal, meets one of two criteria (Table 15.1).

Table 15.1 Examples of RCRA Hazardous Waste Codes (40 CFR Part 261)

Contaminant	RCRA Hazardous Waste Code
Characteristic Waste	
Ignitable waste	D001
Corrosive waste	D002
Reactive waste	D003
Toxic waste	
Arsenic	D005
Mercury	D009
Benzene	D018
Waste from nonspecific sources	
Wastewater treatment sludges from electroplating	F006
Quenching bath sludge from metal heat treating operations	F010
Waste from specific sources	
Wastewater treatment sludges from the production of chlordane	K032
Ammonia still lime sludge from coking production	K060
Off-specification and discarded chemicals, spill and container residues	
Arsenic trioxide	P012
Tetraethyl lead	P110
Creosote	U051
Mercury	U151

FOCUS ON

Global System to Classify and Label Chemicals

The Globally Harmonized System (GHS) of Classification and Labeling of Chemicals is a worldwide initiative to promote standard criteria for classifying chemicals according to their health, physical, and environmental hazards. It uses pictograms, hazard statements, and signal words such as "Danger" and "Warning" to communicate hazard information on product labels and material safety data sheets (MSDSs) (Table 15.2). The GHS system defines how to classify hazards based on scientific information and standardized tests. It addresses hazards associated with individual chemicals as well as with chemical mixtures.

Table 15.2 Examples of Pictograms

Hazard	Symbol	
The most severely toxic chemicals that may be toxic or fatal through ingestion, inhalation or dermal absorption	Skull and crossbones	
Chemicals that cause (or in some cases are suspected of causing) allergy or asthma symptoms or breathing difficulties if inhaled, genetic defects, cancer, damage to fertility or to the unborn child, organ damage; or chemicals that are fatal or harmful if swallowed and enter airways	Health hazard	
Chemicals with less severe toxicity, such as those that cause skin or eye irritation, allergic skin reactions or those that are harmful via contact with the skin or if swallowed or inhaled	Exclamation mark	
Chemicals that are corrosive to metals, that cause severe skin burns and eye damage, or that cause serious eye damage	Corrosion	
Chemicals that are acutely hazardous to fish, crustacea, or aquatic plants	Environment	
Chemicals that are flammable	Flame	

Source: Pictograms extracted from the Globally Harmonized System of Classification and Labelling of Chemicals (GHS), second revised edition. Copyright © United Nations, 2009. Reproduced with the kind permission of the United Nations Economic Commission for Europe.

The primary goal of GHS is to better protect human health and the environment worldwide by providing chemical users and handlers with enhanced and consistent information on chemical hazards. Sound management of chemicals requires that chemical manufacturers identify the relevant hazards and then communicate them. GHS will help ensure more consistency in this process, thereby improving and simplifying hazard communication. This improved communication system will inform users of the presence of a hazard and the need to minimize risk whether by minimizing exposure or by finding safer alternatives. The end result will be safer transportation, handling, and use of chemicals.

In addition GHS will reduce the inefficiencies caused by the need to comply with multiple classification and labeling systems. Companies that do business in multiple countries have typically been tasked with complying with the hazard label requirements specific to each country. Sometimes product labels cannot contain all of the information needed to properly label a product for sale in multiple countries. When each country requires different systems of hazard communication and labeling, the product may not be big enough to contain the label needed to sell it! With GHS, there will now be a harmonized basis for the communication of hazards. For countries without well developed regulatory systems, the GHS will be particularly useful because countries can simply adopt it and automatically have a system equivalent to the rest of the world.

GHS will also provide a standardized approach to Material Safety Data Sheets [called MSDSs in the United States and Safety Data Sheets (SDSs) in Europe]. (M)SDSs accompany products when sold. (*Note*: In the United States, MSDSs are mandatory only for industrial products. Products purchased for the home do not require them, making it difficult for consumers to know more about the chemical composition and hazards associated with the products they buy. GHS will not change that.)

Source: United States Environmental Protection Agency, http://www.epa.gov/oppfead1/international/globalharmon.htm.

---

1. It contains one or more of the *criteria pollutants* or those chemicals that have been explicitly identified (*listed*) as hazardous. Over 50,000 chemicals are thus identified.
2. The waste can be defined (by laboratory tests) to have at least one of the following characteristics:

   - flammability
   - reactivity
   - corrosivity
   - toxicity.

Flammable materials are defined as those liquids with flash points below 60°C or those materials that are "easily ignited and burn vigorously and persistently." *Corrosive* materials are those which, in an aqueous solution, have pH values outside the range 2.0 to 12.5 or any liquid that exhibits corrosivity to steel at a rate greater than 6.35 mm per year. *Reactive* wastes are classified as unstable and either can form toxic fumes or can explode. The greatest difficulty in defining hazardous waste comes from establishing what is and what is not *toxic*. Toxicity is almost impossible to measure. Toxic to which animals (or plants?), at what concentrations, over what time periods? The USEPA defines toxicity in terms of four criteria:

- bioconcentration
- LD_{50}
- LC_{50}
- phytotoxicity.

Bioconcentration is the ability of a material to be retained in animal tissue to the extent that organisms higher up the trophic level will have increasingly higher concentrations of this chemical. Many pesticides, for example, will reside in the fatty tissues of animals and will not break down very quickly. As the smaller creatures are eaten by the larger ones, the concentration in the fatty tissues of the larger organisms can reach toxic levels for them. Of most concern are aquatic animals and birds that feed on fish, such as seals and pelicans, as well as other carnivorous birds such as eagles, falcons, and condors.

LD_{50} is a measure of the amount of a chemical that is needed to kill half of a group of test specimens, such as mice. The animals in a toxicity study are fed progressively higher doses of the chemical until half of them die, and this dose is then known as the median lethal dose (50%), or LD_{50}. The lower the amount of the toxin used to kill 50% of the specimens, the higher the toxic value of the chemical. Some chemicals, such as dioxin and PCBs, show incredibly low LD_{50}s, suggesting that these chemicals are extremely dangerous to small animals and other test species.

There are, however, some serious problems with such a test. First, the tests are conducted on laboratory specimens, such as mice, and the amount of the toxin that would produce death in humans is extrapolated based on body weight, as illustrated in the example below.

EXAMPLE

15.1   **Problem**   A toxicity study on the resistance of mice to a new pesticide has been conducted, with the following results:

Amount Ingested (mg)	Fraction That Died after 4 h
0 (control)	0
0.01	0
0.02	0.1
0.03	0.1
0.04	0.3
0.05	0.7
0.06	1.0
0.07	1.0

What is the LD_{50} of this pesticide for a mouse? What is the LD_{50} for a typical human being weighing 70 kg? Assume a mouse weighs 20 g.

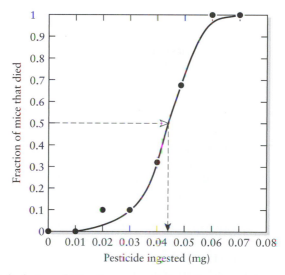

Figure 15.1   Calculation of LD_{50} from mouse data. See Example 15.1.

Solution   The data are plotted as in Figure 15.1 (known as a dose-response relationship), and the point at which 50% of the mice die is identified. The mouse LD_{50} is, therefore, about 0.043 mg. The human lethal dose is estimated as

$$\frac{70,000 \text{ g}}{20 \text{ g}}(0.043 \text{ mg}) = 150 \text{ mg}$$

This is a shaky conclusion, however. First, the physiology of a person is quite different from that of a mouse, so a person may be able to ingest either more or less of a toxin before showing adverse effect. Second, the effect measured is an *acute* effect, not a long-term (*chronic*) effect. Thus, toxicity of chemicals that affect the body slowly, over years, is not measured, since the mouse experiments are done in hours. Finally, the chemical being investigated may act synergistically with other toxins, and this technique assumes that there is only one adverse effect at a time. As an example of such problems, consider the case of dioxin (discussed in Chapter 14), which has been shown to be extremely toxic to small laboratory animals. All available epidemiological data, however, show that humans appear to be considerably more resilient to dioxins than the data would suggest.

LC_{50} is the concentration at which some chemical is toxic, and this is used where the amount ingested cannot be measured, such as in the aquatic environment or in evaluating the quality of air. Specimens, such as goldfish, are placed in a series of aquariums, and increasingly higher concentrations of a toxin are administered. The fraction of fish dying within a given time is then recorded, and the LC_{50} is calculated. As a rough guideline, a waste is considered toxic if it is found to have a LD_{50} of less than 50 mg/kg body weight or if the LC_{50} is less than 2 mg/kg.

Finally, a chemical is considered toxic if it exhibits *phytotoxicity*, or toxicity to plants. Thus, all herbicides are, by this definition, toxic materials, and when they must be disposed of, they must be treated as hazardous wastes.

A final criterion for being hazardous is if the material is *radioactive*. Radioactive wastes are, however, handled separately and are governed by separate rules and regulations. Radioactive waste management is covered in Section 15.3.

One concern in hazardous waste disposal is the speed with which the chemical can be set free to produce toxic effects in plants or animals. For example, one oft-used method of hazardous waste disposal is to mix the waste with a slurry consisting of cement, lime, and other materials (a process known as stabilization/solidification). When the mixture is allowed to harden, the toxic material is safely buried inside the block of concrete, from which it cannot escape and cause trouble.

Or can it? This question is also relevant in cases such as the disposal of incinerator ash. Many of the toxic materials, such as heavy metals, are not destroyed during incineration and escape with the ash. If these metals are safely tied to the ash and cannot leach into the groundwater, then there seems to be no problem. If, however, they leach into the water when the ash is placed in a landfill, then the ash would have to be treated as a hazardous waste and disposed of accordingly.

How then is it possible to measure the rate at which such potential toxins can escape the material in which they are presently embedded? As a crude approximation of such potential leaching, the EPA uses an extraction procedure, where the solidified waste is crushed, mixed with weak acetic acid, and shaken for a number of hours. This process is known as the Toxicity Characteristic Leaching Procedure (TCLP).

Once the leaching has taken place, the leachate is analyzed for possible hazardous materials. The USEPA has determined that there is a list of contaminants that constitute possible acute harm and that the level of these contaminants must not be exceeded in the leachate. Table 15.3 includes a few such chemicals. Many of the numbers on this list are USEPA Drinking Water Standards multiplied by 100 to obtain the leaching standard.

Table 15.3   A Few Examples of EPA's Maximum Allowable Concentrations in Leachates from the TCLP Test

Contaminant	Allowable Level (mg/L)
Arsenic	5.0
Benzene	0.5
Cadmium	1.0
Chromium	5.0
Chloroform	6.0
2,4-D	10.0
Heptachlor	0.008
Lead	5.0
Pentachlorophenol	100.0
Trichloroethylene	0.5
Vinyl chloride	0.2

The REACH Directive

The landmark European Union (EU) REACH (Registration, Evaluation & Authorization of Chemicals) Directive came into force on 1 June, 2007. The REACH Directive focuses on registering chemical substances to enable fast and accurate recognition and classification. Under REACH, substances are registered and their intrinsic physical and associated human health and environmental properties are made accessible to the public on the REACH Register, which is controlled by the European Chemicals Agency (ECHA). Registration with the ECHA is required under the directive for

- substances greater than 1 tonne per year imported or placed on the EU market
- articles containing substances that are intentionally released
- substances of very high concern (SVHC).

SVHCs are chemicals that can be attributed to the development of cancer, infertility, birth defects, or gene mutations and chemicals that are persistent and can accumulate in the human body. Under the legislation, companies will have to substitute less dangerous chemicals for SVHCs or obtain authorization to continue using SVHCs. Over the next 10 years, at least 30,000 chemicals in commerce are expected to be placed on the REACH Register.

REACH also requires the traceability and identification of chemical substances within articles, which is a regulatory term for products. In effect, products will have to be labeled. For example, a computer will be labeled indicating which heavy metals are included. If a product contains an SVHC and the SVHC has not been authorized for that product, then the product can't be sold in the EU.

The Directive is significant because it represents a comprehensive overhaul of existing chemicals legislation. REACH will require public disclosure of health and environmental effects data on chemicals, even those that were previously grandfathered. (Grandfathered refers to chemicals that were in commerce prior to enactment of the 1967 Directive on Dangerous Substances and were permitted to continue to be used, even if little or no data on them were available.) It will also track chemicals in products and start to eliminate SVHCs. It will have a major impact on the import and manufacture of chemicals and their use in products.

REACH is beneficial to both consumers and producers because providing information on a chemical's intrinsic properties, including its human health and environmental properties, can enhance understanding and ensure safe handling. It also makes markets more competitive by leveling the playing field with respect to the availability of data and provides a service to consumers by shifting the onus of responsibility for chemical substance property information and safety toward producers.

REACH affects pure chemical substances, preparations (i.e., mixtures), and products in consumer goods from numerous sectors, including chemicals, textiles, electronics, and food contact materials. However, certain classes of substances are exempt, such as radioactive substances, wastes, foods, medicinal substances, and polymers. REACH encourages in vitro testing of chemicals to characterize health and environmental risks, gaining support from animal rights groups.

In time, REACH will enable the quick and efficient establishment of bans or restrictions on the manufacture, distribution, and sale of certain substances and further contribute to the correct assessment and detection of risks by ensuring that needed chemical data and associated product information are available to support decision making.

Source: Contributed by Imogen Sambrook and Margaret H. Whittaker of Linden Consulting Partnership, Essex, England.

Critics of the test point out that calculating toxicity on the basis of drinking water standards is compounding a series of potential errors. The drinking water standards are, after all, based on scarce data and often set on the basis of expediency (discussed in Chapter 1). How is it possible then to multiply these spurious numbers by 100 and with a straight face say that something is or is not toxic?

The TCLP also is a test with a lot of loose ends. The conditions, such as the water pH and temperature, are, of course, important as is the turbulence of the mixing and the condition of the solid placed into the mixer. The financial implications to an industry can be staggering if one of their leaching tests shows contaminant levels exceeding the allowable concentrations. The test results that mean so much to industrial firms often have a feeble epidemiological base. Unfortunately, until we can come up with something better, this is all we've got. Yes, it is conservative, but this technique gives us a crude handle on the possible detrimental effects of hazardous wastes.

In summary, a waste is hazardous if it flunks any of the tests that would keep it from being so classified (and is not specifically exempted in the regulations). Using this criterion, the EPA has developed a list of chemicals it considers hazardous. The list is long and growing as new chemicals and waste streams are identified. Once a chemical or process stream is listed, it needs to be treated as a hazardous waste and is subject to all the applicable regulations. Getting delisted is a difficult and expensive process, and the burden of proof is on the petitioner. This is, in a way, a situation wherein the chemical is considered guilty until proven innocent.

If a waste is listed, its disposal becomes a true headache because only a few hazardous waste disposal facilities are available. Usually, long and costly transportation is necessary, forcing the industry either to dispose of the waste surreptitiously (such as the 300 km of roadway in North Carolina contaminated by PCBs when the drivers just opened the drain valves and drained the trucks instead of making the long trips to the disposal facilities) or to redesign their plant so as not to create waste. The latter requires skill and capital, and more than one industry has had to close when one or both were unavailable.

15.2 HAZARDOUS WASTE MANAGEMENT

Hazardous wastes are controlled in the United States by the USEPA. This mandate is under one of several pieces of legislation, most notably the Toxic Substances Control Act (TSCA); the Comprehensive Environmental Response, Compensation, and Liability Act (CERCLA); its amendment, the Superfund Amendments and Reauthorization Act (SARA); the Resource Conservation and Recovery Act (RCRA); and its amendment, the Hazardous and Solid Waste Amendments (HSWA). TSCA is aimed at preventing the creation of materials that may eventually prove damaging or difficult to dispose of safely, whereas RCRA addresses the disposal of hazardous wastes by establishing standards for secure landfills and treatment processes. CERCLA is directed at correcting the mistakes of the past by cleaning up old hazardous waste sites and is usually referred to as Superfund because it set up a large trust fund paid for by chemical and extractive industries for the purpose of providing resources to clean up abandoned sites. The USEPA uses these funds to address acute problems due to improper hazardous waste disposal or accidental discharge and the cleanup of old sites.

15.2.1 Cleanup of Old Sites

As part of the Superfund program, the USEPA established a National Priorities List (NPL) of hazardous waste sites that are in need of immediate cleanup and are eligible for funding under CERCLA. Estimates of the number of these sites vary, but there are tens—if not hundreds—of thousands of sites in need of some degree of cleanup, at a cost of millions to billions of dollars (!). The NPL still includes about 1100 nonfederal sites that are in immediate need of attention. As of this writing, about 980 total sites have been cleaned up.

Because of the large number of potential Superfund sites, the USEPA has developed a ranking system so that the worst situations can be cleaned up first and the less critical can wait until funds, time, and personnel become available. This Hazard Ranking System (HRS) tries to incorporate the more sensitive effects of a hazardous waste area, including potential adverse health effects, the potential for flammability or explosions, and the potential for direct contact with the substance. These three modes are calculated separately and then used to obtain the final score. Being conscious of public opinion, the scores tend to be overwhelmed by situations wherein human health is in danger, especially for explosive materials stored in densely populated areas. The potential for groundwater contamination in areas where water supplies are from groundwater sources also receives priority status. This technique minimizes the criticism often faced by USEPA that they have not done enough to clean up the Superfund sites. They have, in fact, tackled the most difficult and most sensitive sites first and should be commended for the rapid response to a problem that has been festering for generations.

In addition to the USEPA list, the Department of Energy has about 160 sites that require cleanup, and the Department of Defense has thousands more. The Department of Defense itself has spent *billions* of dollars on cleanups. Much of this is the legacy of nuclear weapons testing and manufacturing because many of these facilities were strictly secret, and the contracting operators could do almost anything they wished with the waste. The states also have lists of sites to be remediated. (New Jersey's Jersey City chromium site was discussed in Chapter 1.)

The type of work conducted at these sites depends on the severity and extent of the problem. In some cases where there is an imminent threat to human health, the USEPA can authorize a *removal action*, which results in the hazardous material being removed to safe disposal or treatment. In less acute instances the USEPA authorizes *remedial action*, which may consist of the removal of the material or more often the stabilization of the site so that it is less likely to cause health problems. For example, in the case of the notorious Love Canal in Niagara Falls, the remedial action of choice was to essentially encapsulate the site and monitor all groundwater and air emissions. The cost of excavating the canal would have been outrageous, and there was no guarantee that the eventual disposition of the hazardous materials would have been any safer than leaving it there. In addition, the actual act of removal and transport might have resulted in significant human health problems. Remedial action, therefore, simply implies that the site has been identified and action taken to minimize the risk of having the hazardous material cause human health problems.

In the most common situation some chemical has already contaminated the groundwater, so remediation is necessary. For example, a dry cleaner may accidentally (or purposefully) discharge its waste cleaning fluid, and this may find its way into a drinking water well for a nearby residence. Once the problem is detected, the first question is whether or not the contamination is life-threatening or if it poses a significant threat to the environment. Then a series of tests are run using monitoring wells or soil samples to determine the geology of the area and the size and shape of the plume or range of the contaminated area.

FOCUS ON

Love Canal

It was a grand dream. William T. Love wanted to build a canal between the two great lakes, Erie and Ontario, to allow ships to pass around Niagara Falls. The project started in the 1890s but soon floundered due to inadequate financing. Love walked away from a big hole in the ground, and in 1942 Hooker Chemical found this to be the perfect place to dump its industrial waste. This was wartime in the United States, and there was little concern for possible environmental consequences. Hooker Chemical (which became Occidental Chemical Corporation) disposed of over 21,000 tonnes of chemical wastes, including halogenated pesticide and chlorobenzines, into the old Love Canal. The disposal continued until 1952, at which time the company covered the site with soil and deeded it to the City of Niagara Falls, which wanted to use it for a public park. In the transfer of the deed, Hooker specifically stated that the site was used for the burial of hazardous materials and warned the city that this fact should govern future decisions on the use of the land. Everything Hooker Chemical did during those years was legal and aboveboard.

About this time the Niagara Falls Board of Education was looking around for a place to construct a new elementary school, and the old Love Canal seemed like a perfect spot. This area was a growing suburb, with densely packed single-family residences on streets paralleling the old canal. A school on this site seemed like a perfect solution.

In the 1960s the first complaints began, and they intensified during the early 1970s. The groundwater table rose during those years and brought to the surface some of the buried chemicals. Children in the school playground were seen playing with (55 gallon) drums that popped out of the ground. The contaminated liquids started to ooze into the basements of the nearby residents, causing odor and health problems. More importantly, perhaps, the contaminated liquid was found to have entered the storm sewers and was being discharged upstream of the water intake for the Niagara Falls water treatment plant. The situation reached a crisis point and President Jimmy Carter declared an environmental emergency in 1978, resulting in the evacuation of 950 families in an area of 10 square blocks around the canal.

But the solution presented a difficult engineering problem. Excavating the waste would have been dangerous work and would probably have caused the death of some of the workers. Digging up the waste would also have exposed it to the atmosphere, resulting in uncontrolled toxic air emissions. Finally, there was the question as to what would be done with the waste. Because it was all mixed up, no single solution such as incineration would have been appropriate. The USEPA finally decided that the only thing to do with this dump was to isolate it and continue to monitor and treat the groundwater. The contaminated soil on the school site was excavated, detoxified, and stabilized, and the building itself was razed (Figure 15.2). All of the sewers were cleaned, removing 62,000 tonnes of sediment that had to be treated and transported to a remote site. The groundwater was pumped and treated to prevent further contamination. The cost, of course, was staggering, paid for by Occidental Chemical, the Federal Emergency Management

© Joe Traver/Contributor/Getty Image News/Getty Images

(A)

© MIKE SEGAR/Reuters Landov

(B)

Figure 15.2   Love Canal (A) before and (B) after.

Agency, and the U.S. Army (which was found to have contributed waste to the canal).

The Love Canal story had the effect of galvanizing the American public into understanding the problems of hazardous waste and was the impetus for the passage of several significant pieces of legislation, such as the Resource Conservation and Recovery Act, the Comprehensive Emergency Response, Control, and Liability Act, and the Toxic Substances Control Act.

Source: Reprinted from *Controlling Environmental Pollution*, Vesilind, P. A. and DiStefano, T. D., 2006. Lancaster, PA: DEStech Publications, Inc. Reproduced with permission of DEStech Publications, Inc.

Depending on the seriousness of the situation, several remedial action options are available. If there is no threat to life and if it can be expected that the chemical will eventually metabolize into harmless end products, then one solution is to do nothing except continued monitoring—a method known as *natural attenuation*. In most cases this is not so, and direct intervention is necessary.

Containment is used where there is no need to remove the offending material and/or if the cost of removal is prohibitive, as was the case at Love Canal. Containment is usually the installation of slurry walls, which are deep trenches filled with bentonite clay or some other highly nonpermeable material, and continuous monitoring for leakage out of the containment. With time the offending material might slowly biodegrade or chemically change to a nontoxic form, or new treatment methods may become available for detoxifying this waste.

Extraction and treatment is the pumping of contaminated groundwater to the surface for either disposal or treatment or the excavation of contaminated soil for disposal or treatment. Sometimes air is blown into the ground and the contaminated air collected.

The characteristics of a hazardous chemical often determine its location underground and, therefore, dictate the remediation process. If the chemical is immiscible with water and is lighter than water, it most likely will be floating on top of the aquifer, as shown in Figure 15.3A. Pumping this chemical out of the ground is relatively simple. If the chemical readily dissolves in water, however, then it can be expected to be mixed in a contaminated plume, as shown in Figure 15.3B. In such situations it will be necessary to contain the plume either by installing barrier walls and pumping the waste to the surface for treatment or by sinking a discharge well to reverse the flow of water. A third option is that the chemical is denser than water and is immiscible, in which case it would be expected to lie on an impervious layer somewhere under the aquifer, as shown in Figure 15.3C.

Commonly, wells are drilled around the contaminated site so that the groundwater flow can be reversed or the groundwater can be contained in the area, as discussed in Chapter 10. Once the contaminated water is extracted, it must be treated, and the choice of treatment obviously depends on the nature of the problem. If the contamination is from hydrocarbons, such as trichloroethylene, it is possible to remove this with activated carbon. Some type of biological treatment system or distillation process may also be used. If the contamination is from metals, then a precipitation or redox process may be used.

Some soils may be so badly contaminated that the only option is to excavate the site and treat the soil *ex-situ*. This is usually the case with PCB contamination because no other method seems to work well. The soil is dug out and usually incinerated to remove the PCBs, and then returned to the site or landfilled. Depending on the contaminant, biodegradation in reactors or piles may be used. (Excavation and landfilling are also commonly used to meet construction and real estate transaction schedules.)

In-situ treatment of the contaminated soil involves the injection of either bacteria or chemicals that will destroy the offending material. If heavy metals are of concern, these can be tied up chemically, or fixed, so that they will not leach into the groundwater. Organic solvents and other chemicals can be degraded by injecting freeze-dried bacteria or by making conditions suitable for indigenous bacteria to degrade the waste (e.g., by injecting air and nutrients). Microorganisms that are able to decompose materials that were previously thought to be refractory or even toxic have been found.

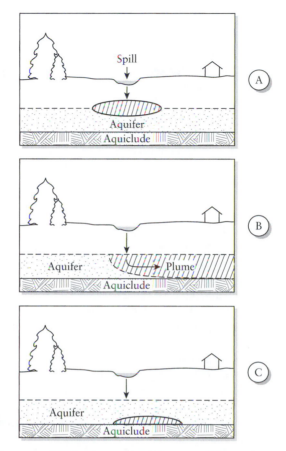

Figure 15.3 Three possible locations for underground hazardous waste.

15.2.2 Treatment of Hazardous Wastes

The treatment of materials deemed hazardous is obviously specific to the material and the situation. Therefore, there are a number of alternatives that engineers may consider in such treatment operations.

Chemical treatment is commonly used, especially for inorganic wastes. In some cases a simple neutralization of the hazardous material will render the chemical harmless. In other cases oxidation is used, such as for the destruction of cyanide. Ozone is often used as the oxidizing agent. In a case wherein heavy metals must be removed, precipitation is the method of choice. Most metals become extremely insoluble at high pH ranges, so the treatment consists of the addition of a base, such as lime or caustic, and the settling of the precipitate (similar to the lime-soda softening process described in Chapter 10). Other physical–chemical methods employed in industry include reverse osmosis, electrodialysis, solvent extraction, and ion exchange.

If the hazardous material is organic and is readily biodegradable, most often the least expensive and most dependable treatment is biological, using techniques described in

Chapter 11. The situation becomes interesting, however, when the hazardous material is an anthropogenic compound (created by people). Because these combinations of carbon, hydrogen, and oxygen are new to nature, there may not be any microorganisms that can use them as an energy source. In some cases it is still possible to find a microorganism that will use this chemical as a food source, and treatment would then consist of a biological contact tank in which the pure culture is maintained. Alternatively, it is now increasingly likely that specific microorganisms can be designed by gene manipulation to attack certain especially difficult-to-treat organic wastes.

Finding the specific organism can be a difficult and arduous task. There are millions to choose from, so how would we know that *Corynebacterium pyrogens* just happens to like toxaphene, a particularly refractory organic pesticide? Secondly, often the pathway is convoluted, and a single microorganism can only break down the chemical to another refractory compound, which would then be attacked by a different organism. For example, DDT is metabolized by *Hydrogenomonas* to *p*-chlorophenylacetic acid, which is then attacked by various *Arthrobacter* species. Tests in which only a single culture is used to study metabolism would, therefore, fail to note the need for a sequence of species.

One interesting development has been the use of cometabolism to treat organic chemicals that were once considered biologically nonbiodegradable. With this technique the hazardous material is mixed with a nonhazardous and at least partially biodegradable material, and the mixture is treated in a suspended- or fixed-film bioreactor. The microorganisms apparently are so busy making the necessary enzymes for the degradation of the biodegradable material (the metabolite) that they forget that they cannot treat the toxic material (the cometabolite). The enzymes produced, however, will biodegrade both the chemicals. The addition of phenol to the soil, for example, will trick some types of microorganisms, such as fungus, into decomposing chlorophenol, even though chlorophenol is usually toxic to the fungus.

One of the most widely used treatment techniques for organic wastes, however, is incineration. Ideally, hazardous waste incinerators produce carbon dioxide, water vapor, and an inert ash. In actuality, no incinerator will achieve complete combustion of the organics. It will discharge some chemicals in the emissions, concentrate others in the bottom ash, and produce various compounds called products of incomplete combustion (PIC). For example, polychlorinated biphenyls (PCBs) are thought to decompose within the incinerator to highly toxic chlorinated dibenzo furans (CDBF), which, although organic, do not oxidize at normal incinerator temperatures. Despite these problems, hazardous waste incinerators must achieve high levels of removal efficiencies, often 99.99% or higher, which is commonly referred to as "four nines." In some cases removal efficiencies require five or even six nines.

EXAMPLE
15.2

Problem   A hazardous waste incinerator is to burn 100 kg/h of a PCB waste that is 22% PCB and 78% organic solvents. In a trial run the concentration of PCB in the emission is measured as 0.02 g/m^3, and the stack gas flow rate is 40 m^3/min. No PCB is detected in the ash. What removal efficiency (destruction of PCB) does the incinerator achieve?

Solution   A black box is once again useful. What comes in must go out. The rate in is

$$(100 \text{ kg/h})(0.22)(1000 \text{ g/kg}) = 22,000 \text{ g/h}$$

The rate out of the stack is

$$QC = (40 \text{ m}^3/\text{min})(0.02 \text{ g/m}^3)(60 \text{ min/h}) = 48 \text{ g/h}$$

As there are no PCBs in the ash, the amount of PCBs combusted must be the difference between what enters and what leaves:

$$22,000 - 48 = 21,952 \text{ g/h}$$

so the combustion efficiency is

$$\text{Efficiency} = \frac{21,952 \text{ g/hr}}{22,000 \text{ g/hr}} \times 100 = 99.78\%$$

There are only two nines, so this test does not meet the four nines criterion.

15.2.3   Disposal of Hazardous Waste

The disposal of hazardous waste is similar in many ways to the disposal of nonhazardous solid waste. Because disposal in the oceans is prohibited and outerspace disposal is still far too expensive, the final resting place has to be on land.

Deep well injection has been used in the past and is still the method of choice in the petrochemical industry. The idea is to inject the waste so deep into earth that it could not in any conceivable time reappear and cause damage. This is, of course, the problem. Once deep in the ground it is impossible to tell where its final destination will be and what groundwater it will eventually contaminate.

A second method of land disposal is to spread the waste on land and allow the soil microorganisms the opportunity to metabolize the organics. This technique was widely used in oil refineries and seemed to work exceptionally well. The USEPA, however, has restricted the practice because there is little control over the chemical once it is on the ground.

The method most widely used for the disposal of hazardous waste is the secure landfill. Such landfills essentially look like the RCRA Subtitle D landfills discussed in the previous chapter, only more so. Instead of one impervious liner, secure landfills require multiple liners. Liquid waste is banned; all waste must be stabilized or in containers. Similar to sanitary landfills, leachate is collected, and a cap is placed on the landfill once it is complete. Continued care is also required, although the USEPA presently requires only 30 years of monitoring.

Most hazardous waste engineers agree that there is no such thing as a secure landfill and that eventually all the material will once again find its way into the water or air. What we are betting on, therefore, is that the wastes in these storage basins will degrade on their own or that the storage basins will eventually be dug up and the material treated. We find that it is inexpedient to do so today and leave this as a legacy to future generations.

15.3 RADIOACTIVE WASTE MANAGEMENT

A special type of hazardous material emits ionizing radiation, and in high doses this radiation can be highly detrimental to human health. Environmental engineers do not usually get involved in radiation safety, which is a specialized field, but they nevertheless ought to be knowledgeable about both the risk and the disposal technology of radioactive materials. In addition to the USEPA, the Nuclear Regulatory Commission and Department of Energy have authority over management of radioactive waste. The USEPA is particularly involved when radioactive waste is mixed with RCRA hazardous waste (substances known as *mixed wastes*).

15.3.1   Ionizing Radiation

Radiation is a form of energy due to the decay of isotopes. An *isotope* of an element has the same atomic number (number of protons) but a different mass number (number of neutrons and protons) than the standard element. (In other words, it has a different number of neutrons.) Remember that the atomic number defines an element; for example, the atomic number of uranium is 92. Uranium 235 (U-235), therefore, is an isotope of uranium (U-238).

To regain equilibrium, the isotopes decay by emitting protons, neutrons, or electromagnetic radiation to carry off energy. This natural spontaneous process is *radioactivity*. The isotopes that decay in this manner are called *radioisotopes*. The energy emitted by this decay that is strong enough to strip electrons and sever chemical bonds is called *ionizing radiation*.

There are four kinds of ionizing radiation: alpha particles, beta particles, gamma (or photon) rays, and X-rays. Alpha particles consist of two protons and two neutrons, so they are the equivalent of the nucleus of a helium atom being ejected. Alpha particles are quite large and do not penetrate material readily; therefore, they can be readily stopped by skin and paper and are not of major health concern, unless they are taken internally, in which case they can cause a great deal of damage. The decay changes the parent element into a different element, known as a daughter product.

Beta radiation results from an instability in the nucleus between the protons and neutrons. Having too many neutrons, some of the neutrons decay into a proton and an electron. The proton stays in the nucleus to reestablish the neutron/proton balance while the electron is ejected as the beta particle. The beta particle is much smaller than the alpha particle and can penetrate living tissue. However, about 1 cm of material will shield against beta particles. Beta decay, as with alpha decay, changes the parent into a new element.

Both alpha and beta decay are accompanied by gamma radiation, which is a release of energy from the change of a nucleus in an excited state to a more stable state. The element thus remains the same, and energy is emitted as gamma radiation. Related to gamma radiation are X-rays, which result from an energy release when electrons transfer from a higher to a lower energy state. Both types of radiation have more energy and penetrating power than the alpha and beta particles, so a dense material, such as lead or concrete, is required to stop them.

All radioactive isotopes decay and will eventually reach stable energy levels. As introduced in Chapter 5, the decay of radioactive material is first order, in that the change in the activity during the decay process is directly proportional to the original activity present, or

$$\frac{dA}{dt} = -kA$$

where A = activity
t = time
k = radioactive decay constant, inverse time units

As before, integration yields

$$A = A_0 e^{(-kt)}$$

where A_0 = the activity at time zero.

Of particular interest is the half-life of the isotope, meaning that half of the nuclei have decayed in this time. Inserting $A = A_0/2$ into the preceding equation and solving, the half-life is calculated as

$$t_{1/2} = \frac{\ln 2}{k} = \frac{0.693}{k}$$

Half-lives of radioisotopes can range from the almost instantaneous (e.g., polonium-212 with a half-life of 3.03×10^{-7} s!) to very long (e.g., carbon-14 with a half-life of 5,570 yr) (Table 15.4). The half-life is characteristic of an isotope. Therefore, if you know the isotope, you know the half-life and vice versa.

EXAMPLE

15.3   **Problem**   What will be the activity after 5 days of a 1.0-curie (Ci) Rn-222 source? (A Ci, and a becquerel, Bq, are measures of the emission, or decay, rate.)

Solution   To determine the activity, we need to use

$$A = A_0 e^{(-kt)}$$

We are given A_0 and t, and we can calculate the constant k because we know the element, Rn-222. Using Table 15.4, we see that Rn-222 has a half-life of 3.824 days. Using

$$t_{1/2} = \frac{\ln 2}{k}$$

gives $k = 0.181$/d. So, the activity will be 0.40 Ci in 5 days.

Table 15.4 Examples of Half-Lives

Radioisotope	Half-Life
Carbon C-14	5,570 yr
Gold Au-195 Au-198	183 d 2.696 d
Hydrogen H-3	12.28 yr
Iron Fe-55 Fe-59	2.7 d 44.63 d
Lead Pb-210 Pb-212	22.26 y 10.643 h
Mercury Hg-203	46.6 d
Polonium Po-210 Po-212 Po-214 Po-218	138.38 d 3.03×10^{-7} s 1.64×10^{-4} s 3.05 min
Radon Rn-220 Rn-222	55.61 s 3.824 d
Uranium U-234 U-235 U-238	2.45×10^5 yr 7.04×10^5 yr 4.47×10^9 yr

15.3.2 Risks Associated with Ionizing Radiation

Soon after Wilhelm Röntgen (1845–1923) discovered X-rays, the detrimental effect of ionizing radiation on humans became known, and the emergent field of health physics set about creating safety standards, with the assumption of identifying the level of exposure that would be safe. They figured that there had to be a threshold, an exposure below which no effects would be found. Over about 30 years of studying both acute and chronic cases, they found that their threshold kept getting lower and lower. While small amounts of radiation above background levels have been found to stimulate biological systems (a process known as radiation hormesis), they finally concluded that assuming there is no threshold in radiation is the safest course of action.

The exposure of human tissue to ionizing radiation is complicated by the fact that different tissues absorb radiation differently. To get a grip on this, however, we must first define some unit of energy that expresses ionizing radiation. Historically, this was established as the *roentgen*, defined as the exposure from gamma or X-ray radiation equal to a unit quantity of electrical charge produced in air. The roentgen is a purely physical quantity measuring the rate of ionization; it has nothing to do with the absorption or effect of the radiation.

Different types of radioactivity can create different effects, and not all tissues react the same way to radiation. Thus, the *rem*, or roentgen equivalent man, was invented. The rem takes into account the biological effect of absorbed nuclear radiation, so it measures the extent of biological injury. The rem, in contrast to the roentgen, is a biological dose. When different sources of radiation are compared for possible damage to human health, rems are used as the units of measurement.

Modern radiological hygiene has replaced the roentgen with a new unit, the *gray* (Gy), defined as the quantity of ionizing radiation that results in absorption of one joule of energy per kg of absorbing material. But the same problem exists, in that the absorption may be the same, but the damage might be different. Hence, the *sievert* (Sv) was invented. A Sv is an absorbed radiation dose that does the same amount of biological damage to tissue as one Gy of gamma radiation or X-ray. One Sv is numerically equal to 100 rem.

The damage from radiation is chronic as well as acute. Radiation poisoning, as witnessed in the Hiroshima and Nagasaki bombings at the end of World War II, showed that radiation can kill within a few hours or days. Over time, lower levels of radiation exposure can lead to cancer and mutagenic effects, and these are the exposures that are of most concern to the public. The sources of exposure of humans to radioactivity can be classified as involuntary background radiation, voluntary radiation, involuntary incidental radiation, and involuntary radiation exposure due to accidents.

Background radiation is due mostly to cosmic radiation from space, the natural decay of radioactive materials in rocks (terrestrial), and radiation from living inside buildings (internal). A very special kind of background radiation is radon. Radon 222 is a natural isotope with a half-life of about 3.8 days and is the product of uranium decay in earth's surface. Radon is a gas, and because uranium is so ubiquitous in soil and rock, there is a lot of radon present. With its relatively long half-life, it stays around long enough to build up high concentrations in buildings. Its decay products are known to be dangerous isotopes that, if breathed into lungs, can produce cancer. The best technique for reducing the risk from radon is to first monitor to see if a problem exists and, if necessary, ventilate the radon to the outside (Figure 15.4 on the next page).

Voluntary radiation can occur from such sources as diagnostic X-rays. One dental X-ray can produce hundreds of times the background radiation, and these should be avoided unless critically necessary. Following the discovery of X-rays (called Roentgen rays everywhere in the world except the United States), the damaging effects of X-rays began to be discovered due to the early deaths of many of Roentgen's friends. Another source of voluntary exposure is from high-altitude flights in commercial airlines. Earth's atmosphere is a good filter for cosmic radiation, but there is little filtering at high altitudes.

Involuntary incidental radiation would be from such sources as nuclear power plants, weapons facilities, and industrial sources. While the Nuclear Regulatory Commission allows fairly high exposure limits for workers within these facilities, it is fastidious about allowing radiation off premises. In all probability the amount of radioactivity produced in such facilities would be within the background.

The fourth type of radiation exposure is from accidents, and this is a different matter. The most publicized accidental exposure to radiation has been from accidents or

(A)

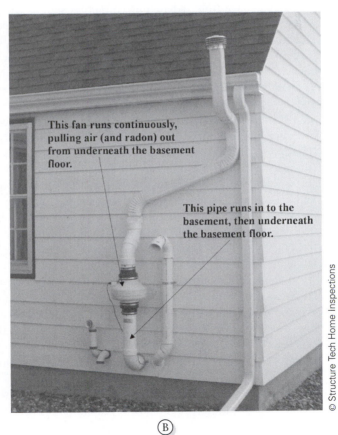

This fan runs continuously, pulling air (and radon) out from underneath the basement floor.

This pipe runs in to the basement, then underneath the basement floor.

(B)

Figure 15.4 (A) Radon test kit and (B) removal system.

near-accidents at nuclear power plants. The 1979 accident near Harrisburg, Pennsylvania at the Three Mile Island nuclear power facility was serious enough to bring to a halt the already staggering nuclear power industry in the United States. Following a series of operating errors, a large amount of fission products were released into the containment structure. The amount of fission products released to the atmosphere was minor, so at no time were the acceptable radiation levels around the plant exceeded. The release of radiation during this accident was so small, in fact, that it is estimated that it would result in one additional cancer death over the 80-km radius surrounding the Three Mile Island facility. This would be impossible to detect because within that radius there can be expected to develop over 500,000 cancers during the lifetime of the people living in the area.

FOCUS ON

Radiation Poisoning in Goiânia, Brazil

In the early 1980s a small cancer clinic was opened in Goiânia (Figure 15.5), but business was not good, and the clinic closed five years later. Left in the abandoned building was a radiation therapy machine and some canisters containing waste radioactive material: 1400 Ci of cesium 137, which has a half-life of 30 years. In 1987 the container of cesium 137 was discovered by local residents and was opened, revealing a luminous blue powder. The material was a local curiosity, and children even used it to paint their bodies, which caused them to sparkle—something that children adore. One of the little girls went home for lunch and ate a sandwich without first washing her hands. Six days later she was diagnosed with radiation illness, having received a radiation dose estimated to be five to six times the lethal exposure for adults. The ensuing investigation identified the true content of the curious barrel. In all, over 200 people had been contaminated, and 54 developed illnesses serious enough to be hospitalized, with four people dying from the exposure (including the little girl who ate the sandwich). By the time the government mobilized to respond to the disaster, the damage was done. A large fraction of the population had received excessive radiation, and the export of produce from Goiânia dropped to zero, creating a severe economic crisis. The disaster is now recognized as the second worst radiation accident in the world, second only

Figure 15.5   Location of Goiânia, Brazil.

to the explosion of the nuclear power plant in Chernobyl.

Source: Adapted from *Socially Responsible Engineering* by Daniel A. Vallero, P. Aarne Vesilind, 2007, John Wiley & Sons, Inc. Used by permission.

© Charobnica/Shutterstock

A much more far-reaching accident occurred in Chernobyl in the Ukraine. This actually wasn't an accident in the strict sense because the disaster was caused by several engineers who wanted to conduct some unauthorized tests on the reactor. The warning signals went off as they proceeded with their experiment, and they systematically turned off all the alarms until the reactor experienced a meltdown. The core of the reactor was destroyed; the reactor caught on fire, and the flimsy containment structure was blown off. Massive amounts of radiation escaped into the atmosphere, and while much of it landed within the first few kilometers of the facility, some reached as far as Estonia, Sweden, and Finland. In fact, the accident was first discovered when radiation safety personnel in Sweden detected unusually high levels of radioactivity. Only then did the officials of the former Soviet Union acknowledge that, yes, there had been an accident. In all, 31 people died of acute radiation poisoning, many by heroically going into the extremely highly radioactive facility to try to extinguish the fires. The best estimates are that the accident will produce up to 9000 excess cancer deaths among the population in the Ukraine, Byelorussia, Lithuania, and the Scandinavian countries. But once again, this number is dwarfed by the expected total cancer deaths of 10 million during the lifetime of the people in these areas.

The total radiation we receive is, of course, related to how we live our lives, where we choose to work, if we choose to smoke cigarettes, whether we have radon in our basements, and so on. It is possible, nevertheless, to come up with some estimates of annual radiation dose for the average citizen of the United States. These figures, compiled by the National Academy of Sciences, are shown in Table 15.5. It is worth highlighting that this tabulation

Table 15.5 Average Annual Dose of Ionizing Radiation to an Average Person in the United States

Source of Radiation	Dose mSv
Involuntary background	
Radon	24
Cosmic radiation	0.27
Terrestrial	0.28
Internal	0.39
Voluntary	
Medical: diagnostic X-ray	0.39
Medical: nuclear medicine	0.14
Medical: consumer products	0.10
Occupational	0.009
Involuntary incidental	
Nuclear power production	<0.01
Fallout from weapons testing	<0.01
Miscellaneous	<0.01
Total	25.64

Source: Upton, A., ed. 1990. *National Academy of Sciences Health Effects of Exposure to Low Levels of Ionizing Radiation (BEIR V)*. Washington, DC: National Academy Press. Reprinted with permission from the National Academies Press, Copyright 1990, National Academy of Sciences.

shows the *average* exposure to ionizing radiation for a person living in the United States. Obviously, the total radiation is much higher for some people. The other point worth noting is the overwhelming importance of radon. This becomes especially true when it is understood that only a fraction of people in the United States live in homes with basements, where radon problems generally occur.

15.3.3   Treatment and Disposal of Radioactive Waste

The most important distinction to be made in radioactive waste disposal is the level of radioactivity emitted. While there appears to be an increasingly complex system of characterization for radioactive wastes, the broad classification is as high-level, intermediate-level, and low-level waste. High-level wastes occur mostly from the production of electric power, and these are identified by activities in the range of curies per liter. Intermediate-level wastes are produced by weapons manufacture, and although their activities are in the range of millicuries, the particular isotopes are long-lived, so these wastes require long-term storage. Low-level wastes, characterized as those with activities in the range of microcuries per liter, are produced in hospitals and research laboratories.

In nuclear power plants, nuclear fission occurs when fissionable material, such as uranium 237, is bombarded with neutrons and a chain reaction is set up. The fissionable material then splits (hence the term "fission") to release huge quantities of heat, used in turn, to produce steam and then electricity. As the U-237 decays, it produces a series of daughter isotopes that themselves decay until the rate and hence heat output are reduced. What is left over, commonly known as the fission fragments, represent the high-level radioactive material that requires cooling down, long-term storage, and eventual disposal. The long-term storage of high-level radioactive waste has been debated for decades. Over protests, a site in Nevada at Yucca Flats has been selected and is being prepared.

Low-level radioactive waste should not represent a disposal problem. Because the activity levels of these wastes are so low that they can be handled by direct contact, it would seem that, with judicious volume reduction such as incineration, any secure landfill would be adequate. The U.S. Congress passed the Low-Level Waste Policy Act in 1980 that stipulated that states form compacts so that each state will take its turn in providing a disposal facility for a certain time. Unfortunately, the very mention of the word "radioactivity" is enough to arouse public opinion and prevent the siting of these disposal facilities.

15.4   SUSTAINABLE MATERIALS MANAGEMENT

In the book, *Materials Matter*, Kenneth Geiser identifies two strategies to support sustainable materials management.[1] One involves developing ways to use less material to provide the same service to satisfy human needs, a process known as "dematerialization." The second strategy, called "detoxification", is used to lower the inherent hazard associated with chemicals used in products and industrial processes.

Dematerialization means reducing the amount of material in a product without decreasing the quality of the service it provides. It reduces the flow of virgin materials into and through the industrial/economic system. Strategies for dematerialization include reducing the weight or thickness of packaging materials, developing organic chemicals from biomass wastes rather than fossil fuels, and promoting recycling and secondary materials industries to keep materials cycling in industrial loops rather than relying on virgin resources. A variation on dematerialization is rematerialization – switching to the use of materials that are inherently more sustainable rather than using less of the initial material. Designing products for repair, upgrade, and/or recycling supports the need for fewer resources and helps keep materials within the industrial/economic system. Dematerialization as a strategy for sustainable materials management is challenged by the growing demand for materials, even if at a decreased per capita rate, through growth. An example of dematerialization is the Soda Club system.

EXAMPLE
15.4

Problem Many people enjoy drinking club soda (also called sparkling water, carbonated water, or seltzer water). The Soda Club system requires the purchase of one 1-L bottle, a refillable CO_2 cartridge that will produce 110 L of soda and the housing that holds it. The bottle is refillable until it becomes cracked or otherwise fails. To make club soda, you fill the bottle with your own tap or filtered water, screw the bottle into the CO_2 cartridge housing, and then press a button to release CO_2 into the bottle. After a few moments, a loud noise generated from pressure in the bottle indicates that the liquid is adequately carbonated and ready to drink. Assume that total cost to get started is $79; it includes a CO_2 cartridge, the bottle, and the housing. When the CO_2 cartridge is empty, you can return it to the point of purchase to refill it for $15. Assume that your family currently drinks 6 L of club soda each week at $1.50 per bottle and that you purchase them by the case at 12 bottles per case and one trip to the grocery store per case. Compare the cost, transportation footprint, and waste generated from purchasing individual bottles of club soda to use of the Soda Club system.

Solution First determine the situation for purchasing individual bottles.

$$\text{Number of individual bottles} = \frac{(6 \text{ L/wk})(52 \text{ wk/yr})}{1 \text{ L/bottle}} = 312$$

$$\text{Cost} = \frac{312 \text{ bottles/yr}}{\$1.50/\text{bottle}} = \$468/\text{yr}$$

$$\text{Trips} = \frac{(312 \text{ bottles/yr})(1 \text{ trip/case})}{12 \text{ bottles/case}} = 26 \text{ trips}$$

Next determine the situation for purchasing the system.

$$\text{Number of cartridges} = \frac{(6 \text{ L/wk})(52 \text{ wk/yr})}{110 \text{ L/cartridge}} = 2.8 = 3$$

$$\text{First year's cost} = \$79 + \$15(2) = \$109$$

$$\text{Subsequent years' costs} = \$15(3) = \$45$$

The data are summarized in the following table.

Option	Cost ($ per Year)	Number of Bottles	Number of Trips to Store
Purchase bottles by the case	468	312	26
Soda Club system (first year)	109	1	3
Soda Club system (following year)	45	0	3

While not a complete life cycle assessment, the data make it evident that the Soda Club system would dematerialize your family's consumption of club soda. That is, you would need far fewer bottles to provide the same amount of beverage. Nice side benefits include reducing your direct costs and transportation footprint. If adopted broadly, the impact could affect the market and the amount of waste generated by soda beverages, reducing the need for solid waste management.

The strategy of *detoxification* includes:

- substituting benign alternatives for problematic chemicals (e.g., substituting rapidly biodegradable surfactants for surfactants that degrade into toxic byproducts)
- phasing out the use of chemicals of concern such as persistent, bioaccumulative, and toxic substances (PBTs)
- generating toxics on demand or just-in-time to avoid the need for storage and transportation
- designing new chemicals and materials that are benign with respect to human health and the environment.

The design of such chemicals is the domain of the growing field of green chemistry. Green chemistry has emerged as an effective strategy for detoxification and as a means to design chemical products and processes to create sustainable materials and products.

15.4.1   Green Chemistry

Green chemistry is the *design* of chemical products and processes that reduce or eliminate the use and generation of *hazardous* substances.[2] The focus of green chemistry is on design because decisions are made at the level of design that impact the performance, toxicity, and fate of chemicals. The focus on hazard rather than risk is also significant. Risk is a function of hazard and exposure. The term "hazard" is used broadly to include any of the attributes inherent in a chemical if you are exposed to it. Hazard may include acute human toxicity, chronic human toxicity (i.e., carcinogenicity, reproductive or developmental toxicity, skin or eye irritation, sensitization of the skin or respiratory system), ecological toxicity, safety issues (e.g., explosiveness or flammability), potential for damage to the ozone layer, destruction of biological habitat, and more.

Green chemistry addresses risk but from a different angle than conventional practices. Traditional risk management focuses on controlling risk by controlling the potential for

FOCUS ON

Bhopal, India

Union Carbide, with the cooperation of the government of India, built a large fertilizer manufacturing plant in Bhopal, India, providing jobs for the people and inexpensive fertilizers for the local farmers. During the early hours of December 13, 1984, an accident occurred in the plant and a toxic cloud of methyl isocyanate swept down into a sleeping city, killing thousands of people (Figure 15.6). When the extent of the disaster was finally tallied, about 3800 people had died, with thousands more permanently disabled.

The plant originally imported the methyl isocyanate used in the manufacture of the pesticide with a trade name of Sevin. With the objective of being more independent of imports, the Indian government encouraged the plant to start the manufacture of its own methyl isocyanate. Because of the involvement of the Indian government, a number of restrictions were placed on the company, including a requirement that only Indians be employed at the plant. As American engineers and other workers left the plant, they were replaced by Indians, many of whom built homes all around the plant. These homes were unfortunately merely shanties, and there was no governmental infrastructure to support the influx of people to the area. The fact that the workers and their families were packed close to the plant in largely open structures contributed to the severity of the accident.

In addition, when the gas escaped, there was no way to call for help. The city of nearly 750,000 people had only two phone trucks, which could handle only a few calls at a time. The news of the disaster was slow in reaching the government and Union Carbide. As a result, the initial response was slow and uncertain. The Indian govenment was not fully cooperative at first and made it difficult to send relief, thus exacerbating the situation.

The cause of the accident was uncertain until Union Carbide did a thorough study and discovered that large volumes of water had been put into the methyl isocyanate holding tanks, which created

Figure 15.6 Union Carbide plant, Bhopal, India.

© Sandro Tucci/Contributor/Getty Images News/Getty Images

an uncontrolled chemical reaction. The question still existed as to why this was done. Union Carbide's answer was sabotage. Some "disgruntled employee" intentionally filled the tank with water. This is possible but somewhat improbable because the employee would have known that the resulting reaction would be deadly to thousands of people, including most likely his own friends and family. A more likely scenario is that it was poor plant management and/or incompetence, for which both the Indian govenment and Unior Carbide would be responsible.

Could a disaster like Bhopal happen in the United States? Within a year of the tragedy in India, another Union Carbide plant in Industry, West Virginia, had an accidental release of methyl isocyanate into the Kanawha Valley, eerily mirroring the events in Bhopal. Fortunately, nobody died in this accident, but the episode brought home to Americans the risks and dangers of living close to plants that produce or use toxic chemicals, and it contributed to the passing of right-to-know legislation.

Source: Reprinted from *Controlling Environmental Pollution*, Vesilind, P. A. and DiStefano, T. D., 2006. Lancaster, PA: DEStech Publications, Inc. Reproduced with permission of DEStech Publications, Inc.

---

exposure. However, exposure controls can and do fail. Failures to control exposure have led to deadly incidents, such as in Bhopal, India when a release of methyl isocyanate killed thousands of people in a nearby town. Because methyl isocyanate is highly hazardous, when the exposure controls failed and released the chemical, it created a high-risk situation. By addressing and reducing the inherent hazard of a chemical, it is possible to reduce risk and even to eliminate the need for exposure controls, which can be very expensive.

Green chemistry is a tool for chemists, chemical engineers, and others who design chemicals and materials to help move society toward the goal of sustainability. It is defined by a set of 12 principles articulated by Anastas and Warner in 1998 (Table 15.6.). The wisdom behind the green chemistry approach has been gaining popularity in recent years, and we are now seeing the emergence of policy and legislation to support green chemistry. In the United States, Michigan's governor issued a Green Chemistry Executive Order; California is creating a policy platform called the Green Chemistry Initiative to support green chemistry and cradle-to-cradle design; and in the U.S. Congress, Senate Bill 2669, the Green Chemistry Research and Development Act of 2008, was introduced.

Several examples of green chemistry follow, to illustrate how materials can be designed to be environmentally compatible and to fit within a sustainable materials system. These examples are derived from entries to and winners of the Presidential Green Chemistry Challenge Awards program, the only presidential level chemistry award program in the United States.

In 1998, Rohm and Haas Company won a Presidential Green Chemistry Challenge Award for designing the marine antifoulant called Sea-Nine™. Fouling, the unwanted growth of plants and animals on a ship's surface, costs the shipping industry approximately $3 billion a year. A significant portion of this cost is the increased fuel consumption needed to overcome hydrodynamic drag. The main compounds historically used worldwide to control fouling are based on organotins, such as tributyltin oxide (TBTO). While they are effective at preventing fouling, they are toxic and persistent in the environment. The

Table 15.6 The 12 Principles of Green Chemistry

Prevention	It is better to prevent waste than to treat or clean up waste after it has been created.
Atom economy	Synthetic methods should be designed to maximize the incorporation of all materials used in the process into the final product.
Less hazardous chemical syntheses	Wherever practicable, synthetic methods should be designed to use and generate substances that possess little or no toxicity to human health and the environment.
Designing safer chemicals	Chemical products should be designed to effect their desired function while minimizing their toxicity.
Safer solvents and auxiliaries	The use of auxiliary substances (e.g., solvents, separation agents, etc.) should be made unnecessary wherever possible and innocuous when used.
Design for energy efficiency	The energy requirements of chemical processes should be recognized for their environmental and economic impacts and should be minimized. If possible, synthetic methods should be conducted at ambient temperature and pressure.
Use of renewable feedstocks	A raw material or feedstock should be renewable rather than depleting, whenever technically and economically practicable.
Reduce derivatives	Unnecessary derivatization (use of blocking groups, protection/deprotection, temporary modification of physical/chemical processes) should be minimized or avoided if possible, because such steps require additional reagents and can generate waste.
Catalysis	Catalytic reagents (as selective as possible) are superior to stoichiometric reagents.
Design for degradation	Chemical products should be designed so that at the end of their function they break down into innocuous degradation products and do not persist in the environment.
Real-time analysis for pollution prevention	Analytical methodologies need to be further developed to allow for real-time, in-process monitoring and control prior to the formation of hazardous substances.
Inherently safer chemistry for accident prevention	Substances and the forms of substances used in a chemical process should be chosen to minimize the potential for chemical accidents, including releases, explosions, and fires.

Source: Anastas, P. T., and J. Warner. 1998. *Green Chemistry: Theory and Practice*. New York: By permission of Oxford University Press.

negative environmental effects associated with TBTO include acute toxicity, bioaccumulation, decreased reproductive viability, and increased shell thickness in shellfish. Rohm and Haas Company sought to develop an antifoulant that would protect ships from fouling by a wide variety of marine organisms without causing harm to nontarget organisms.

Company scientists selected compounds from the 3-isothiazolone class. Specifically, 4,5-dichloro-2-*n*-octyl-4-isothiazolin-3-one was selected because it was found to be effective and to degrade very rapidly in seawater and even faster (1 hr) in sediment. While TBTO is considered highly bioaccumulative, Sea-Nine™ does not bioaccumulate at all. Both TBTO and Sea-Nine™ are acutely toxic to marine organisms, which is a desirable characteristic of a marine antifoulant. But Sea-Nine™ antifoulant degrades rapidly and is not chronically toxic. Thus, the chemical does its job but does not cause harm beyond its immediate function. TBTO persists and bioaccumulates in the environment so that it continues to have toxic effects.

EDEN Bioscience Corporation received a Presidential Green Chemistry Challenge Award in 2001 for the product Messenger®. Messenger® is based on a class of nontoxic, naturally occurring proteins called "harpins." Harpin proteins trigger the natural defense systems of plants to protect themselves against diseases and pests. It also activates certain plant growth systems without altering the plant's DNA. When applied to crops, harpins support plant health, resulting in increased plant biomass, photosynthesis, nutrient uptake, and root development. Ultimately such increases lead to greater crop yield and quality of product.

Estimated crop losses from pests are estimated at $300 billion worldwide. Growers have typically pursued two strategies to limit economic losses and to increase yields: (1) using traditional chemical pesticides or (2) growing crops that are genetically engineered for pest resistance. Each of these approaches has negative consequences, particularly on nontarget organisms and in terms of public perception. In contrast, harpins have been shown to have virtually no adverse effect on any of the organisms tested, including mammals, birds, honeybees, plants, fish, aquatic invertebrates, and algae. Because it is a fragile molecule that is degraded rapidly by UV and natural microorganisms, it has no potential to bioaccumulate, and it will not contaminate surface or groundwater resources. Unlike most agricultural chemicals, harpin-based products are produced in a water-based fermentation system that uses no harsh solvents or reagents, requires only modest energy inputs, and generates no hazardous chemical wastes. From the life-cycle perspective, harpin is made via green chemistry processes, has low hazard characteristics, and supports sustainable agriculture.

In 2008, Battelle and its partners, Advanced Image Resources and the Ohio Soybean Council, received the Presidential Green Chemistry Challenge award for developing a soy-based toner. Laser printers and copiers use over 200 million kg of toner each year in the United States alone. Traditional toners fuse so tightly to paper that they are difficult to remove from waste paper for recycling. They are also made from petroleum-based starting materials, like styrene acrylates and styrene butadiene. Battelle and its partners developed a soy-based toner that performs as well as traditional ones but is much easier to remove. The new toner technology can save significant amounts of energy and allow more paper fiber to be recycled. The new toner will be sold under the trade names BioRez® and Rezilution®. Once commercial, it will provide users with seamless printing and copying using soy-based chemistry.

There are many more examples of green chemical innovation. Some of them reflect changes in chemical processes, such as eliminating hazardous solvents or using catalysts to eliminate the number of steps in a chemical synthesis. Other examples illustrate the generation of new chemicals to replace chemicals of concern. And others demonstrate innovation by generating desired chemicals from biobased feedstocks rather than from petroleum-based feedstocks, which have been the backbone of organic chemistry. Green chemistry is important because it demonstrates that it is possible deliberately to design new molecules and chemical processes with the intent to reduce overall hazard and to produce sustainable products and processes. Although technology is not a panacea for sustainability, it can provide innovative solutions.

According to Arnulf Grübler, in his book, *Technology and Global Change*[3], "Through technology, humanity has for 300 years increasingly liberated itself from the environment. The job is not yet complete as billions of people continue to be excluded from the benefits of technology. The next immediate task is to assure their inclusion. But what next? We need a utopian vision, a new technology manifesto, to set our goals for the 21st century." Grübler argues that the task ahead is to progressively liberate the environment from adverse human interference.

> For this century-long journey, we will need more technology, not less. As the world's population grows to 10 billion or more, we recognize that nature cannot be shielded perfectly from human intervention any more than we can be shielded perfectly from nature. But technological change can relax our grip and lighten our tread on the natural world. Our choices are constrained by what already exists and the environmental legacy of the past. But over the long term the capacity for social choices to shape technology is endless."

15.4.2 Pollution Prevention

A subset but more widely practiced aspect of sustainable materials management is pollution prevention. The EPA defines *pollution prevention* as the following:

> The use of materials, processes, or practices that reduce or eliminate the creation of pollutants or wastes at the source. It includes practices that reduce the use of hazardous materials, energy, water or other resources and practices that protect natural resources through conservation or more efficient use.

In the widest sense, pollution prevention is the idea of eliminating waste, regardless of how this might be done. It is the same concept here as in the management of municipal solid waste as described in Chapter 14, but the incentives for using it are much greater.

Originally, pollution prevention was applied to industrial operations with the idea of either reducing the amount of the wastes being produced or changing their characteristics to make them more readily disposable. Many industries changed to water-soluble paints, for example, thereby eliminating organic solvents, cleanup time, etc., and often ended up saving considerable money. In fact, the concept was first introduced as "pollution prevention pays," emphasizing that many of the changes would actually save the companies money. In addition, the elimination or reduction of hazardous and otherwise difficult wastes has a long-term effect. It reduces the liability the company carries as a consequence of its disposal operations.

With the passage of the Pollution Prevention Act of 1990, the EPA was directed to encourage pollution prevention by setting appropriate standards for pollution prevention activities, assist federal agencies in reducing wastes generated, work with industry to promote the elimination of wastes by creating waste exchanges and other programs, seek out and eliminate barriers to the efficient transfer of potential wastes, and to do this with the cooperation of the individual states.

In general, the procedure for the implementation of pollution prevention activities is to

- recognize a need
- assess the problem
- evaluate the alternatives
- implement the solutions.

Contrary to the attitude towards most pollution control activities, industries generally have welcomed this governmental action, recognizing that pollution prevention can, and often does, result in the reduction of costs to the industry. Thus, recognition of the need quite often is internal, and the company seeks to initiate the pollution prevention procedure.

During the assessment phase, a common procedure is to perform a waste audit, which is nothing but the black box mass balance, using the company as the black box. Consider Example 15.5.

E X A M P L E

15.5   **Problem**   A manufacturing company is concerned about the air emissions of volatile organic carbons (VOCs). These chemicals can volatilize during the manufacturing process, but there is no way of estimating just how much or which chemicals. The company conducts an audit of three of their most widely used VOCs with the following results. (The average influent flow rate to the treatment plant is 0.076 m³/s.)

Purchasing Department Records

Material	Purchase Quantity (barrels)
Carbon tetrachloride (CCl_4)	48
Methylene chloride (CH_2Cl_3)	228
Trichloroethylene (C_2HCl_3)	505

Wastewater Treatment Plant Influent

Material	Average Concentration (mg/L)
Carbon tetrachloride	0.343
Methylene chloride	4.04
Trichloroethylene	3.23

Hazardous Waste Manifests (indicating what leaves the company for a hazardous waste treatment facility)

Material	Barrels	Concentration (%)
Carbon tetrachloride	48	80
Methylene chloride	228	25
Trichloroethylene	505	80

Unused Barrels at the End of the Year

Material	Barrels
Carbon tetrachloride	1
Methylene chloride	8
Trichloroethylene	13

How much VOC is escaping?

Solution Conduct a black box mass balance:

$$\begin{bmatrix} \text{Mass of} \\ A \text{ per} \\ \text{unit time} \\ \text{ACCUMULATED} \end{bmatrix} = \begin{bmatrix} \text{Mass of} \\ A \text{ per} \\ \text{unit time} \\ \text{IN} \end{bmatrix} + \begin{bmatrix} \text{Mass of} \\ A \text{ per} \\ \text{unit time} \\ \text{OUT} \end{bmatrix}$$

$$+ \begin{bmatrix} \text{Mass of} \\ A \text{ per} \\ \text{unit time} \\ \text{PRODUCED} \end{bmatrix} - \begin{bmatrix} \text{Mass of} \\ A \text{ per} \\ \text{unit time} \\ \text{CONSUMED} \end{bmatrix}$$

The material A is, of course, each of the three VOCs.

We need to know the conversion from barrels to cubic meters and the density of each chemical. Assume each barrel is 0.12 m^3, and the densities of the three chemicals are 1548, 1326, and 1476 kg/m^3, respectively.

The mass per year of carbon tetrachloride accumulated is

$$[\text{Mass ACCUMULATED}] = (1 \text{ barrel/yr})(0.12 \text{ m}^3/\text{barrel})(1548 \text{ kg/m}^3)$$

$$= 186 \text{ kg/yr}$$

Similarly,

$$[\text{Mass IN}] = (48 \text{ barrels/yr})(0.12 \text{ m}^3/\text{barrel})(1548 \text{ kg/m}^3) = 8916 \text{ kg/yr}$$

The mass out is in three parts: mass discharge to the wastewater treatment plant, mass leaving on the trucks to the hazardous waste disposal facility, and the mass volatilizing. So, assuming the plant operates 365 days/yr:

$$[\text{Mass OUT}] = [(0.343 \text{ g/m}^3)(0.076 \text{ m}^3/\text{s})(86{,}400 \text{ s/day})(365 \text{ days/yr})(10^{-3} \text{ kg/g})]$$

$$+ [(48 \text{ barrels/yr})(0.12 \text{ m}^3/\text{barrels})(1548 \text{ kg/m}^3)(0.80)] + A_{\text{air}}$$

$$= 822 + 7133 + A_{\text{air}}$$

There is no carbon tetrachloride consumed or produced, so in kg/yr

$$186 = 8916 - [822 + 7133 + A_{\text{air}}] + 0 - 0$$

and

$$A_{\text{air}} = 775 \text{ kg/yr}$$

If a similar balance is done on the other chemicals, it appears that the atmospheric losses are about 16,000 kg/yr of methylene chloride and about 7800 kg/yr of trichloroethylene. If the intent is to cut total VOC emissions, the first target may be the methylene chloride. However, it will depend on many other factors, such as the ease of reducing the losses, the relative hazards of the solvents, and the availability of alternative, less hazardous solvents for the particular applications involved.[4]

Once we know what and where the problems are, the next step is to figure out useful options. These options fall generally into three categories:

1. operational changes
2. material changes
3. process modifications.

Operational changes might consist simply of better housekeeping: plugging up leaks, eliminating spills, etc. A better schedule for cleaning as well as segregating the waste streams might similarly yield a large return on a minor investment.

Material changes often involve the substitution of a chemical that is less toxic or requires less hazardous materials for cleanup than another. The use of trivalent chrome for chrome plating instead of the much more toxic hexavalent chrome has found favor as has the use of water-soluble dyes and paints. In some instances ultraviolet radiation has been substituted for biocides in cooling water, resulting in better-quality water and no waste cooling water disposal problems. In one North Carolina textile plant, biocides were used in air washes to control algal growth. Periodic "blow down" and cleaning fluids were discharged to the stream, but this discharge proved toxic to the stream, so the State of North Carolina revoked the plant's discharge permit. The town would not accept the waste into its sewers, rightly arguing that this may have serious adverse effects on its biological wastewater treatment operations. The industry was about to shut down when it decided to try ultraviolet radiation as a disinfectant in its air wash system. Fortunately, they found that the ultraviolet radiation effectively disinfected the cooling water and that the biocide was no longer needed. This not only eliminated the discharge, but it eliminated the use of biocides all together, thus saving the company money. The payback was 1.77 years.[5]

Process modifications usually involve the greatest investments but can result in the greatest rewards. For example, countercurrent wash water use instead of a once-through batch operation can significantly reduce the amount of wash water needing treatment, but such a change requires pipes, valves, and a new process protocol. In industries wherein materials are dipped into solutions, such as in metal plating, the use of drag-out recovery tanks, an intermediate step, has resulted in saving plating solution and reducing waste generated.

In any case the most marvelous thing about pollution prevention is that most of the time the company not only eliminates or greatly reduces the discharge of hazardous wastes but it also saves money. Such savings are in several forms, including, of course, the direct savings in processing costs as with the ultraviolet disinfection example above. But there are other savings, including the savings in not having to spend time on submitting compliance permits and suffering potential fines for noncompliance. Future liabilities weigh heavily where hazardous wastes have to be buried or injected. Additionally, there are the intangible benefits of employee relations and safety. And, of course, finally, there is the benefit that comes from doing the right thing, something not to be sneezed at.

In hazardous waste management, doing the right thing often involves the elimination of a potential hazard for future generations. Much has been written about this commitment. Do we, indeed, owe anything to these yet-to-be-born people?

15.5 HAZARDOUS WASTE MANAGEMENT AND FUTURE GENERATIONS

Why should we even worry about all that? Is it not true that we will certainly personally never be affected adversely by hazardous wastes, and it is unlikely even that any of our children will suffer from inadequate disposal of these materials? Why are we so concerned with the future and with future generations?

RCRA requires waste facility operators to provide perpetual care for depositories; however, this is defined as a period of only 30 years. "Perpetual care" means only one half of a generation! The primary hazardous waste law in the United States apparently does not care about future generations. So why should we?

After all, these people do not exist, nor might they ever exist. They are not real people, and ethics can apply only to interactions between real people. Also, we don't know what the future will bring. We cannot assume that future people will have the same preferences or needs; therefore, we cannot predict the future. We may be well-intentioned but utterly wrong. Also, history teaches us that at least to this point in the existence of civilization, each successive generation has been better off than the previous one. We have better health, better communication, better food, more time, and greater opportunity for personal growth and the enjoyment of the quality of life than ever before. Because we can expect future generations to continue this trend and to have an ever improving life, we should do nothing for them. Besides, as the old saying goes, "What has posterity ever done for me?" So it may be very noble to take account of future generations, but there appears to be no basis for the claim that we ought to do so.

For all these reasons, then, it may seem that there is nothing wrong with our deciding to generate nuclear wastes in order to enjoy the benefits of electricity, or using up the nonrenewable resources such as oil, or interpreting our obligations of perpetual care as extending for only 30 years. Planning for hazardous waste management (or for anything else), therefore, needs to take into account only the interests of present generations.

But are there not counter-arguments? Surely we cannot just blunder along hedonistically without concern for future humanity. What reasons can we give for a concern for future generations?

Alastair Gunn[6] argues that there are indeed strong reasons for our obligations to future generations. First, it is true that (by definition) future generations do not exist, but that does not mean we cannot have obligations to them. Certainly, we do not have the kinds of one-on-one obligations that identifiable individuals have to each other. For instance, debts and promises can be owed only by one person to another. But we also have obligations to anyone who might turn out to be harmed by our actions even though we do not know their identity.

FOCUS ON

"If it's not in your product, you don't have to worry about it."

One of the mottos of the USEPA's Design for the Environment (DfE) Program is, "If it's not in your product, you don't have to worry about it." That means that there is wisdom in designing products that avoid hazardous chemicals, even if no one is expected to be exposed to them. Take for example a 2006 tragedy in which a 4-year-old boy from Minneapolis died after he swallowed a "silver" heart-shaped charm on a bracelet that was given away with the purchase of Reebok children's shoes. The charm was almost entirely made of lead. Of course, the charm was not intended for swallowing, but that did not prevent it from dissolving in his stomach acid and poisoning him. The toy was manufactured in China, where corporate brand owners can produce products at much lower costs than in the United States due to cheap labor. But they have much less control over product quality and production processes. Lead in the charm was just the tip of the iceberg. It has also been found in paints in what has seemed like every other doll and truck, and it has created a great deal of alarm (Figure 15.7).

Before the alarm about lead in toys, the thought of the U.S. Congress banning phthalates in children's playthings would have seemed impossible. But in what has been hailed as the first important reform in the 30-year history of the Consumer Product Safety Commission (CPSC), not only has such a bill passed, but with a veto-proof majority. The Consumer Product Safety Improvement Act ensures that toys are not only free of neurotoxic lead, but it also targets six types of phthalates (chemicals that may disrupt the hormonal development of children). Phthalates are used to make plastics soft and can be absorbed when children put toys in their mouths. Three phthalates have long been prohibited in the European Union—di-(2-ethylhexyl) phthalate (DEHP), dibutyl phthalate (DBP), and benzyl butyl phthalate (BBP)—and will be banned in toys outright. Three others—diisononyl phthalate (DINP), diisodecyl phthalate (DIDP), and di-n-octyl phthalate (DnOP)—are banned pending a study to assess their potential health impacts on children and pregnant women.

The new bill also greatly improves the monitoring of toy safety. Typically, untested toys enter the marketplace and are recalled only if they are found to contain lead once they hit store shelves. The bill turns this process around and requires that adequate samples of all toys be independently tested for lead and other hazards before they can be sold.

Avoiding hazardous chemicals in products, whether they are added intentionally or unintentionally, assumes a level of stewardship that is not always evident. In fact, many retailers and manufacturers have no knowledge of the chemical composition of the products they sell. Often, the information is considered proprietary to protect

Figure 15.7   (A) The lead charm from the Reebok promotion; (B) recalled toys.

trade secrets. The European REACH Legislation (discussed in Section 15.1) is the first comprehensive legislation that requires manufacturers to know what is in their products, to assess the hazards associated with the chemicals in their products, and to substitute safer alternatives for highly hazardous chemicals. In the United States, California is leading the charge with the California Green Chemistry Initiative, which is intended to support green chemistry and cradle-to-cradle products.

Awareness of the hazards associated with chemicals in products has increased in recent years, along with awareness of the need for sustainability and life-cycle thinking. Environmental engineers know that products are not made in magic black boxes, nor are they disposed of in a mythical place called "Away." Although some argue that tiny amounts of a hazardous chemical in a product creates low risk, others note that somewhere in the manufacturing supply chain the chemical was likely transported and used in larger volumes and concentrations at much higher levels of risk. In addition, chemicals found in diverse products can leach out of the products into dust and air over time and accumulate in people's bodies. Little is known about the cumulative effects of mixtures. At the end of a product's life, the chemical could be released, transformed, or concentrated again depending on how it is managed. It may even become hazardous waste. Some large retailers like Wal-Mart spend tens of millions of dollars annually disposing of products that were sold to the public as paints, cleaners, etc.—only to legally become hazardous waste when they are returned. Clearly, taking heed of the DfE motto may result not only in products that are better for public health and the environment throughout their life-cycle, but in reduced hazardous waste and increased financial savings.

Gunn uses an analogy of a terrorist who plants a bomb in a primary school.

Plainly the act is wrong, and in breach of a general obligation not to cause (or recklessly risk) harm to our fellow citizens. This is so even though the terrorist does not know who the children are, or even whether any children will be affected. (The bomb may not explode, or may be discovered in time, or the school may be closed down.) It is equally wrong to place a time bomb, set to go off in the school in 20 years time, even though any children that may be in the school when it goes off are not yet born. An unsafe hazardous waste dump may be seen as analogous to a time bomb.

The analogy is not perfect because the owners of the chemical company do not set out to kill children. But intent is not necessary if there is good reason to believe that an action would cause harm. A person throwing a cherry bomb into a crowd may not intend to harm anyone, but if harm does occur, he or she is as guilty as if the original intent was to hurt people. Whether the people harmed are here today or will be here in the future is irrelevant. It is the harm that counts, and causing the harm is unethical.

The counter-arguments for the second claim, that we will never know what future generations will value, state that, yes, we do, in fact, have some pretty good ideas of what they will *not* value. For example, it is highly unlikely that they will value cancer, or AIDS, or polluted groundwater. We can expect future generations, in fact, to be much more similar to ourselves than different from ourselves. To suggest that we should contaminate earth because we don't know what future people will value is not a valid argument.

The third objection to caring about future generations is that, because they will be better off than we are, there is no need to worry about their well-being. This makes no sense because the economic condition of a person is irrelevant to the harm that might be done. It is not legal or ethical to rob a person just because he or she might be wealthy. Besides, there is no guarantee that enhanced technology will bring continued greater well-being. There are more starving people in the world today than ever before, and the continued growth of human population and the destruction of natural habitats may well bring the world into an economic and ecological collapse. We have stretched the carrying capacity of Earth to its limits, and we have no guarantee that technology (or social and political progress) will enable us to even *have* a future generation, much less be able to care for it. As Gunn[6] concludes:

> We have an obligation to avoid creating detrimental environmental conditions for future generations that we do not want to create for ourselves. This means that in terms of waste management, it is necessary to develop methods for very long-term acceptably safe storage and, more importantly, steadily reducing the quantity, toxicity and persistence of the wastes we produce so that acceptably safe management becomes possible. Further, there seems little doubt that future generations will appreciate clean air and water, adequate wilderness areas, and the preservation of natural monuments, and that this is an obligation of our present generation.

PROBLEMS

15.1 The following paragraph appeared as part of a full-page ad in a professional journal:

Today's Laws
A reason to act now

The Resource Conservation and Recovery Act provides for corporate fines up to $1,000,000 and jail sentences up to five years for officers and managers, for the improper handling of hazardous wastes. That's why identifying waste problems, and developing economical solutions is an absolute necessity. O'Brien & Gere can help you meet today's strict regulations, and today's economic realities, with practical, cost-effective solutions.

What do you think of this ad? Do you find anything unprofessional about it? It is, after all, factual. Suppose you are the president of O'Brien & Gere (one of the most reputable environmental engineering firms in the business) and you saw this in the journal. Write a one-page memo to your marketing director commenting on it.

15.2 Some old electrical transformers were stored in the basement of a university maintenance building, and were "forgotten." One day a worker entered the basement and saw that some sticky, oily substance was oozing out of one of the transformers, and into a floor drain. He notified the director of grounds, who immediately realized the severity of the problem. They called in hazardous waste consulting engineers who first of all took out the transformers and eliminated the source of the polychlorinated biphenyls (PCBs) that were leaking into the storm drain. Then they traced the drain to a little stream, and started taking water samples and soil samples in the stream. They discovered that the water was at 0.12 mg/L PCB, and the soil ranged from 32 mg PCB/kg (dry soil) to 0.5 mg/kg. The state environmental management required that streams contaminated with PCBs be cleaned so that the stream is "free" of PCBs. Recall that PCBs are very toxic and extremely stable in the environment, and will biodegrade very slowly. If nothing was done, the contaminant in the soil would remain for perhaps hundreds of years.

a. If you were the engineer, how would you approach this problem? What would you do? Describe how you would clean this stream. Be as detailed as possible. Will your plan disturb the stream ecosystem? Is this of concern to you?

b. Will it ever be possible for the stream to be free of PCBs? If not, what do you think the state means by this? How would you know that you have done all you can?

c. There is no doubt that if we wait long enough, someone will eventually come up with a really slick way of cleaning PCB-contaminated streams, and that this method will be substantially cheaper than anything we have available today. Why not just wait, then, and let some future generation take care of the problem? The stream does not threaten us directly, and the worst part of it can be cordoned off so people would not be able to get close to the stream. Then someday, when technology has improved, we can clean it up. What do you think of this approach? (It is, of course, illegal, but this is not the question. Is it ethical? Would you be prepared to promote this solution?)

15.3 In the above example if the cleanup resulted in the PCB concentration in the water being at 0.000073 mg/L, what is the percentage reduction in the contaminant? How many nines have been achieved?

15.4 A dry cleaning establishment buys 2000 L of carbon tetrachloride every month. As a result of the dry cleaning operation, most of this is lost to the atmosphere, and only 200 L remain to be disposed of. What is the emission rate of carbon tetrachloride from this dry cleaner, in pounds per day? (The density of carbon tetrachloride is about 1.6 g/mL.)

15.5 A 60 m free aquifer is contaminated with a hazardous material. The aquifer is composed of sand with a permeability (K) of 6×10^{-4} m/s. The contaminated groundwater is being pumped up for treatment at a steady state pumping rate of 0.02 m^3/s, and a drawdown of 6 m occurs in a monitoring well 9 m away

from the extraction well. If there is a second monitoring well, how far away is it if its drawdown is 2 m?

15.6 A hazardous waste landfill is to be constructed, and some of the clay onsite is being considered for a clay liner. The requirement is that the permeability of the clay be less than 1×10^{-7} cm/s. A permeameter is set up and a test is conducted. If the depth of the sample in the permeameter is 10 cm, a head of 1 m water is placed on the sample, the diameter of the permeameter is 4 cm, and it takes 240 h to collect 100 mL of water draining out of the permeameter, what is the sample permeability?

15.7 A tank truck loaded with liquid ammonia is hit by a train, resulting in a steady and uncontrollable leak of ammonia gas at an estimated rate of 1 kg/s. You estimate the wind speed at 5 km/h, and it's a sunny afternoon. There is a small community about 2 km directly downwind. Realizing that the threshold limit of ammonia is about 100 ppm, should you recommend evacuation for the town?

15.8 In the 1970s we discovered many hazardous waste dumping sites in the United States. Some of these had been in operation for decades, and some were still active. Most of them, however, were abandoned, such as the notorious Valley of Drums in Tennessee. This was a rural valley where an almost unimaginable concoction of all manner of hazardous materials had been dumped, and the owner had simply abandoned the site after collecting hefty fees for getting rid of the hazardous materials for large chemical companies.

Why did we wait until the 1970s to discover the problem? Why did we not discover the problem of hazardous waste years ago? Why is the concern with hazardous waste so recent? Discuss your thoughts in a one-page paper.

15.9 Peter has been working with the Bigness Oil Company's local affiliate for several years, and has established a strong trusting relationship with Jesse, the manager of the local facility. The facility, on Peter's recommendation, has followed all of the environmental regulations to the letter, and has a solid reputation with the state regulatory agency. The local company receives various petrochemical products via pipeline and tank truck and blends them for resale to the private sector.

Jesse has been so pleased with Peter's work that he has recommended that Peter be retained as the corporate consulting engineer. This would be a significant advancement for Peter and his consulting firm, cementing Peter's steady and impressive rise in his firm. There is talk of a vice presidency in a few years.

One day, over a cup of coffee, Jesse starts telling Peter a wild story about a mysterious loss in one of the raw petrochemicals he receives by pipeline. Sometime during the 1950s, when operations were pretty lax, a loss of one of the process chemicals was discovered when the books were audited. There were apparently 40,000 L of the chemical missing. After running pressure tests on the pipelines, the plant manager found that one of the pipes had corroded and had been leaking the chemical into the ground. After stopping the leak, the company sank observation and sampling wells and found that the product was sitting in a vertical plume, slowly diffusing into a deep aquifer. Because there was no surface or groundwater pollution off the plant property, the plant manager decided to do nothing. Jesse thought that somewhere under the plant there still sits this plume, slowly diffusing into the aquifer, although the last tests from the sampling wells showed that the concentration of the chemical in the groundwater within 120 m of the surface was essentially zero. The wells were capped, and the story never appeared in the press.

Peter is taken aback by this apparently innocent revelation. He recognizes that the state law requires him to report all spills, but what about spills that occurred years ago, and where the effects of the spill seem to have dissipated? He frowns and says to Jesse. "We have to report this spill to the state, you know."

Jesse is incredulous. "But there *is* no spill. If the state made us look for it, we probably could not find it, and even if we did, it makes no sense whatever to pump it out or contain it in any way."

"But the law says that we have to report . . . ." argues Peter.

"Hey look. I told you this in confidence. Your own engineering code of ethics requires client confidentiality. And what would be the good of going to the state? There is nothing to be done. The only thing that would happen is that the company would get into trouble and have to spend useless dollars to correct a situation that cannot be corrected and does not need remediation."

"But . . ."

"Peter. Let me be frank. If you go to the state with this, you will not be doing anyone any good—not the company, not the environment, and certainly not your own career. I cannot have a consulting engineer who does not value client loyalty."

What should be Peter's response to the situation? What should he do? Who are the parties concerned? (Don't forget future generations!) What are all of his alternatives? List these in order of detrimental effect on Peter's career; then list them in order of detrimental effect on the environment. All things considered, what ought Peter to do?

15.10 Benzene (C_6H_6) is somewhat soluble in water (1.8 g/L). It is a known carcinogen. The MCL in drinking water is 5 ppb. Gasoline is being restricted to 0.62% benzene by volume. How much water will 3.78 L of gasoline contaminate to the MCL?

15.11 How long will it take for the radiation of a spill of Po-210 and Po-214 to decay by 80%?

15.12 Among other contamination, a site was found to have 4.5 Ci of tritium (H-3) on 15 January, 2000. Records indicate that the original release was likely in June 1963. The site is now owned by the city due to nonpayment of taxes.

    a. How many curies of H-3 were originally released?

    b. Who should be responsible for the cleanup of the site?

15.13 Suppose 1.0 g/d of dieldrin leaks into a 30,000-m^3 stormwater retention pond (i.e., it has no outfall). If the dieldrin has a half-life of 2 days and complete mixing occurs, what would be the steady state concentration in the pond?

15.14 Radon 222 is released from the earth and can become trapped in buildings. Suppose the concentration measured in a college classroom building is 1500 pCi/m^3, which is significantly above the recommended remedial action level of 4 to 8 pCi/m^3.

    a. If all leaks into the building are sealed (i.e., no more radon enters the building), how long will it take for the concentration of radon to reach the recommended level?

    b. Do you recommend some type of remedial action be taken instead (i.e., removal of the source)? Why or why not?

END NOTES

1. Geiser, Kenneth. 2001. *Materials Matter: Toward a Sustainable Materials Policy*. Cambridge, MA: MIT Press.

2. Anastas, P. T., and J. Warner. 1998. *Green Chemistry: Theory and Practice*. New York: Oxford University Press.

3. Grübler, A. 1998. *Technology and Global Change*. Cambridge, England: Cambridge University Press.

4. This example is based, in part, on a similar example in Davis, M. L., and D. A. Cornwell. 1991. *Introduction to Environmental Engineering*. New York: McGraw-Hill.

5. Richardson, S. 1990. Pollution prevention in textile wet processing: An approach and case studies. *Proceedings, Environmental Challenges of the 1990's*. USEPA 66/9–90/039.

6. Vesilind, P. Aarne, and Alastair Gunn. 1997. *Making a Difference: Engineering, Ethics and the Environment*. New York: Cambridge University Press.

Noise Pollution

Typical highway

Ha Ha Tonka State Park, Missouri

Environmental engineers consider the potential noise impacts of designs on humans and wildlife.

If a tree falls in the forest, does it make a sound?

You have heard this old question many times. Now you will have an answer for it. In fact, if a tree falls in the forest, it does indeed make a sound because sound is pressure waves in the air. But it might not make a *noise*, which is defined as unwanted sound—unwanted by humans. Thus, if there are no humans around, the tree falling in the forest does not make noise (However, the definition of noise is broadening to include sound that adversely impacts wildlife, so the answer then depends on if the wildlife find the sound of a falling tree objectionable, which may depend on whether or not they are in or near the tree when it falls.)

In this chapter we discuss first the basics of sound, defining some of the terms that acoustical engineers use to describe and control unwanted sound (noise). We then discuss some of the effects of noise, ending with a discussion on noise control.

16.1 SOUND

Pure sound is described by pressure waves traveling through a medium (air, in almost all cases). These pressure waves are described by their amplitude and their frequency. With reference to Figure 16.1, note that the pure sound wave can be described as a sinusoidal curve, having positive and negative pressures within one cycle. The number of these cycles per unit time is called the sound *frequency*, often expressed as cycles per second, or Hertz (Hz).[1] Typical sounds that healthy human ears hear range from about 15 Hz to about 20,000 Hz, a huge range. Low-frequency sound is deep (lower pitched) while high-frequency sound is high pitched. For example, a middle A on the piano is at a frequency of 440 Hz. Speech is usually in the range of 1000 to 4000 Hz.

The wide frequency spectrum is significantly reduced by age and environmental exposure to loud noise. The most significant sources of such damaging noise come from occupational sources and loud music, particularly rock concerts and earphones turned too high. High-quality sound reproduction equipment is designed to reproduce a full spectrum of frequencies, although advertising that the equipment is capable of less than 10 Hz reproduction is somewhat silly because most people can hear only down to 20 Hz.

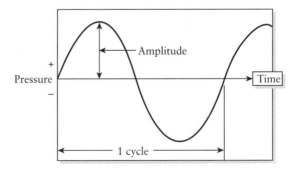

Figure 16.1   Pure sound travels as a perfect sinusoidal wave.

Young, healthy people (particularly young women) can hear very high frequencies, often including such signals as automatic door openers. With age, and unhappily with the damage so many young people do to their ears, the ability to detect a wide frequency range drops. Older people especially tend to lose the high end of the hearing spectrum and might start to complain that "everyone is mumbling."

The loudness of a noise is expressed by its *amplitude*. With reference again to Figure 16.1, the energy in a pressure wave is the total area under the curve. Because the first half is a positive pressure and the second is negative, adding these would produce zero net pressure. The trick is to use a *root mean square* analysis of a pressure wave, first multiplying the pressure by itself and then taking its square root. Since the product of two negative numbers is a positive number, the result is a positive pressure number. In the case of sound waves the pressure is expressed as Newtons per square meter (N/m^2), although sometimes sound pressure is also expressed in bars or atmospheres. In this text we use the modern designation of N/m^2.

The human ear is a phenomenal instrument, being able to hear sound at both a huge frequency range and an even more spectacular range of pressures. In fact, the audible range of hearing in humans ranges from 1 to 10^{18}—a range that makes it difficult to express sound pressure in some meaningful and useful way. What has developed over the years is a convention to use ratios to express sound amplitude.

Psychologists have known for many years that human responses to stimuli are not linear. In fact, the ability to detect an incremental change in any response, such as heat, cold, odor, taste, or sound, depends entirely on the original level of that stimulus. For example, suppose you are blindfolded and are holding a 10 kg weight in your hand, and 0.1 kg is added to it. You will probably not detect the change. On the other hand, if you are holding 0.1 kg in your hand and another 0.1 kg is added (a 100% increase), you would be able to detect the change. Thus, the ability to detect stimuli can be described on a logarithmic scale.

With respect to sound, it makes sense to consider not pressure but the energy level of the sound, so the ratio of two sound energies is

$$\log_{10}\left(\frac{W}{W_{\text{ref}}}\right)$$

where the W and W_{ref} represent the energy in watts of the sound wave and some reference energy, respectively. Because the energy is proportional to the pressure squared, this ratio can also be expressed as

$$\log_{10}\left(\frac{P^2}{P_{\text{ref}}{}^2}\right)$$

where $P =$ sound pressure, N/m^2
$P_{\text{ref}} =$ some reference pressure, N/m^2

By convention, this expression defines the *bel*.[2] Dividing this unit into 10 makes it easier to use and avoids fractions and is known as the *decibel* (dB).

But what reference pressure to use? It appears that the minimum pressure a human ear is able to detect is about 2×10^{-5} N/m^2 and this would make a convenient datum. Using

2×10^{-5} N/m^2 as a reference value, the *sound pressure level* (SPL) is then defined in terms of decibels as

$$\text{SPL} = 10 \log_{10} \left(\frac{P^2}{P_{\text{ref}}^2} \right) = 20 \ \log_{10} \left(\frac{P}{P_{\text{ref}}} \right)$$

where   SPL = sound pressure level, dB
P = the sound pressure as measured in N/m^2
P_{ref} = the reference sound pressure, 2×10^{-5} N/m^2

Typical sound pressure levels are shown in Table 16.1. Note that the highest possible SPL, at which point the air molecules can no longer carry pressure waves, is 194 dB while 0 dB is the threshold of hearing. The loudest sound recorded seems to be the Saturn rocket, at 134 dB, not far from the threshold of pain at 140 dB. Humans typically can detect changes in sound pressures of 3 dB and greater; changes of 1 to 2 dB are not noticeable. But remember that the scale is logarithmic. One dB difference from 40 to 41 is considerably less energy than a one-dB difference from 80 to 81. Every 10 dB increase produces a doubling of the energy level, and a doubling of the danger of damage from excessive noise.

Because sound pressure levels are logarithmic ratios, they cannot be added directly. If two sources of sound are combined, the ratio P^2/P_{ref}^2 has to be calculated from the SPL equation first. These ratios are then added, and a new SPL is calculated:

$$\text{SPL}_{\text{Total}} = 10 \log \left(\sum 10^{\text{SPL}_i/10} \right) \tag{16.1}$$

Table 16.1   Typical Sound Pressure Levels

Sound	SPL (dB)
Threshold of hearing	0
Inside of hearing test chamber	10
Remote area of Yellowstone	20
Library	40
Suburban subdivision not near major noise source	45
Typical classroom	50
Normal speech	60
Busy office	65
Ringing alarm clock next to head	80
Average street traffic	85
Lawnmower	90
Individual truck passby at 50 ft	90
Rock concert	110
Average newborn cry	110
F-16 fighter jet flyover at takeoff	120
Saturn rocket on takeoff	134
Threshold of pain	140
Maximum SPL in air	194

Two rules of thumb can be used when adding sounds. First, if two equal sounds are added, they result in approximately a 3-dB increase in overall sound level (which is just barely noticeable). Second, if the difference between two sounds is greater than 10 dB, the lesser of the two does not contribute much to the overall level of sound. (You can check these rules of thumb by following the procedure just outlined.)

EXAMPLE
16.1

Problem   A machine shop has two machines, one producing a sound pressure level of 70 dB and one producing 58 dB. A new machine producing 70 dB is brought into the room. What is the new sound pressure level in the room?

Solution   Using the 10-dB difference rule of thumb, $70 - 58 = 12$, which is greater than 10, so the effect of the 58 dB sound is negligible, and the room with the existing equipment would have a sound pressure level of 70 dB. (Using Equation 16.1, the answer is 70.2 dB, which rounds to 70 dB.)

If another 70-dB machine is brought in, the two sounds are equal, producing an increase of 3 dB. Thus, the sound pressure level in the room would be 73 dB. (Again, using Equation 16.1, the answer is 73.1 dB for the three machines, which rounds to 73 dB.)

Sound in the atmosphere travels uniformly in all directions, radiating out from its source. The sound intensity is reduced as the square of the distance away from the source of the sound according to the *inverse square law*. That is, the sound pressure level is proportional to $1/r^2$, where r is the radial distance from the source.

An approximate relationship can be developed if the sound power is expressed as a logarithmic ratio based on some standard reference power, such that

$$SPL_r \cong SPL_0 - 10 \log r^2 \tag{16.2}$$

where SPL_r = sound pressure level at some distance r from the sound source, dB
   SPL_0 = sound pressure level at the source, dB
   r = distance away from the sound, m

EXAMPLE
16.2

Problem   A sound source generates 80 dB. What would the SPL be 100 m from the source?

Solution   Using Equation 16.2, $SPL_r \cong 80 - 10 \log(100)^2 = 40$ dB

This is, of course, an approximation. We assume that the sound propagates in all directions evenly, but in the real world this is never true. If, for example, the sound occurs on a flat surface so that the area through which it propagates is a hemisphere instead of a sphere,

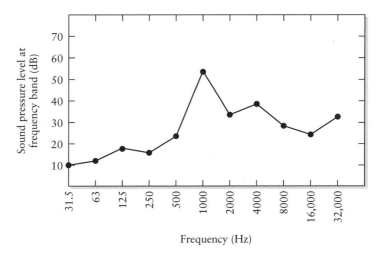

Figure 16.2 A typical frequency analysis of a "dirty" sound.

the approximate addition is 3 dB to the SPL calculated in Example 16.2. In enclosed spaces, reverberation can also greatly increase the sound pressure level because the energy is not dissipated. The most important point to remember is that the SPL is reduced approximately as the log of the square of the distance away from it.

The frequency in cycles per second (Hz) and the amplitude in decibels describe a pure sound at a specific frequency. All environmental sounds, however, are said to be quite "dirty," with many frequencies. A true picture of such sounds is obtained by a *frequency analysis*, in which the sound level at a number of different frequencies is measured and the results plotted as shown in Figure 16.2. Often a frequency analysis is useful in noise control because the frequency of highest sound pressure level can be identified and corrective measures taken. By convention, frequencies are designated as *octave bands* that represent a narrow range of frequencies, as shown in Table 16.2.

The average sound pressure can be calculated from a frequency diagram by adding the sound pressure levels at the individual frequencies. Again, because sound pressure level is

Table 16.2 Octave Bands

Octave Frequency Range (Hz)	Geometric Mean Frequency (Hz)
22–44	31.5
44–88	63
88–175	125
175–350	250
350–700	500
700–1400	1000
1400–2800	2000
2800–5600	4000
5600–11,200	8000
11,200–22,400	16,000
22,400–44,800	31,500

not an arithmetic quantity but rather a logarithmic ratio, such addition must be done by the following equation:

$$\text{SPL}_{\text{avg}} = 10 \log \left(\frac{1}{N} \right) \sum_{j=1}^{N} 10^{(\text{SPL}_j/10)} \tag{16.3}$$

where SPL_{avg} = average sound pressure level, dB
N = number of measurements taken
SPL_j = the jth sound pressure level, dB
$j = 1, 2, 3, \ldots, N$

E X A M P L E

16.3   **Problem**   What is the average sound pressure level of the sound pictured in Figure 16.2?

Solution   First, tabulate the sound pressure level at each octave band, then calculate $10^{(\text{SPL}_j/10)}$, sum these, and calculate the SPL_{avg} from Equation 16.3.

Octave Band (Hz)	SPL (dB)	$10^{(\text{SPL}_j/10)}$
31.5	10	10
63	12	15.8
125	17	50.1
250	15	31.6
500	24	251.2
1000	54	251,189
2000	34	2512
4000	39	7943
8000	28	631.0
16,000	25	316.2
32,000	33	1995
		264,945

$$\text{SPL}_{\text{avg}} = 10 \log \left[(1/11)(173974) \right] = 44 \text{ dB}$$

Note that Equation 16.3 is used to calculate an average SPL for a frequency diagram, but Equation 16.1 is used to calculate the new SPL when multiple distinct sources are combined.

16.2   MEASUREMENT OF SOUND

Sound is measured with an instrument that converts the energy in the pressure waves to an electrical signal. A microphone picks up the pressure waves, and a meter reads the sound pressure level, directly calibrated into decibels. Data thus obtained with a *sound pressure level meter* represent an accurate measurement of the energy level in the air.

But this sound pressure level is not necessarily what the human ear hears. While we can detect frequencies over a wide range, this detection is not equally effective at all frequencies. (Our ear does not have a *flat response* in audio terms.) If the meter is to simulate the efficiency of the human ear in detecting sound, the signal has to be filtered.

Using many thousands of experiments, researchers discovered that, on average, the human ear has an efficiency over the audible frequency range that resembles Figure 16.3. At very low frequencies our ears are less efficient than at middle frequencies, say between 1000 and 2000 Hz. At higher frequencies the ear becomes increasingly inefficient, finally petering out at some high frequency where no sound can be detected. This curve is called the *A-weighted* filtering curve (because there are other filtering curves for different purposes). Interestingly, the band of greatest efficiency for the human ear is very close to human speech range.

Using the curve shown in Figure 16.3, instrument designers have constructed a meter that filters out some of the very low-frequency sound and much of the very high-frequency sound so that the sound measured represents somewhat the hearing of a human ear. Such a measurement is called a *sound level* and is designated dB(A) because it represents a dB value modified with the A-weighted filter. The meter is often called a *sound level meter* (Figure 16.4) to distinguish it from a sound *pressure* level meter that measures sound as a flat response. Almost all sound measurements related to human hearing use the dB(A) scale because this approximates the efficiency of the human ear.

Frequency analyses are useful in measuring hearing ability. Using the hearing ability of an average young ear as the standard, an *audiometer* measures the hearing ability at

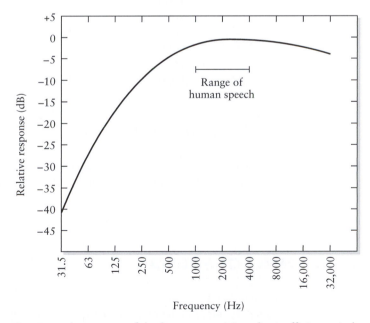

Figure 16.3 Typical response of the human ear. Note the inefficiency in hearing low frequencies, and note the high efficiencies, in the range of human speech.

© Susan Morgan

© Susan Morgan

Figure 16.4 (A) Typical sound level meter; and (B) frequency analyser.

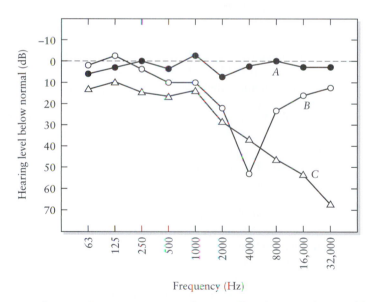

Figure 16.5 Three audiograms. Person A has excellent hearing, being able to hear at normal levels at all frequencies. Person B has lost considerable hearing in the 2000 to 8000-Hz range. Because this is the approximate speech range of frequencies, this person would have great difficulty hearing speech. Person C has a typical audiogram for either an older person who also has lost hearing in the higher frequencies or a younger person with damage to the ears from loud music or other insults.

various frequencies, producing an *audiogram*. Such audiograms are then used to identify the frequencies at which hearing aids must be able to boost the signal.

Figure 16.5 on the previous page, for example, shows three audiograms. Person *A* has excellent hearing with a basically flat response. Person *B* has lost hearing at a specific frequency range, in this case around 4000 Hz. Such hearing losses are frequently due to industrial noise, which often destroys the ear's ability to hear sounds at a specific frequency. Because speech is near that range, this person already has difficulty hearing. The third curve on Figure 16.5, Person *C*, shows a typical audiogram for either an older person who has lost much of the higher-frequency range and probably is a candidate for a hearing aid or a young person who has sustained severe damage to his/her ears by subjecting them to loud music.

16.3   EFFECT OF NOISE ON HUMAN HEALTH

Figure 16.6A shows a schematic of a human ear. The air pressure waves first hit the ear drum (*tympanic membrane*), causing it to vibrate. The cavity leading to the tympanic membrane and the membrane itself are often called the outer ear. The tympanic membrane is attached physically to three small bones in the middle ear that start to move when the membrane vibrates. The purpose of these bones, called colloquially the *hammer*, *anvil*, and *stirrup* due to their shapes, is to amplify the physical signal (to achieve some *gain* in audio terms). This air-filled cavity is called the *middle ear*. The amplified signal is then sent to the *inner ear* by first vibrating another membrane called the *round window membrane*, which is attached to a snail-shaped cavity called the *cochlea*.

Within this fluid-filled cochlea, a cross section of which is shown in Figure 16.6B, is another membrane, the *basilar membrane*, which is attached to the round window membrane. Attached to the basilar membrane are two sets of tiny *hair cells*, pointing in opposite directions. As the round window membrane vibrates, the fluid in the inner ear is set in motion, and the thousands of hair cells in the cochlea shear past each other, setting up electrical impulses that are then sent to the brain through the *auditory nerves*. The frequency of the sound determines which of the hair cells will move. The hair cells close to the round window membrane are sensitive to high frequencies, and those in the far end of the cochlea respond to low frequencies.

Damage to the human ear can occur in several ways. First, very loud impulse noise can burst the ear drum, causing mostly temporary loss of hearing, although frequently torn eardrums heal poorly and can result in permanent damage. The bones in the middle ear are not usually damaged by loud sounds, although they can be hurt by infections. Because our sense of balance depends very much on the middle ear, a middle ear infection can be debilitating. Finally, the most important and most permanent damage can occur to the hair cells in the inner ear. Very loud sounds will stun these hair cells and cause them to cease functioning. Most of the time this is a temporary condition, and time will heal the damage. Unfortunately, if the insult to the inner ear is prolonged, the damage can be permanent. This damage cannot be repaired by an operation or corrected by hearing aids. It is this permanent damage to young people, inflicted by loud music, that is the most frequent and insidious—and the saddest. Is it really worth it to spend your time in front of huge

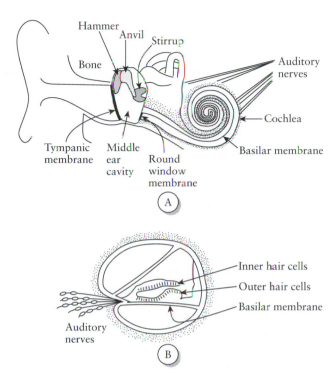

Figure 16.6 (A) A schematic of the human ear; (B) a cross section of the cochlea.

loudspeakers in concerts or to turn up the volume of the iPod when the result will be that you will not be able to hear *any* music by the time you are 40 years old?

But loud noise does more than cause permanent hearing damage. Noise is, in the Darwinian sense, synonymous with danger. Thus, the human body reacts to loud noise so as to protect itself from imminent danger. The bodily reactions are amazing—the eyes dilate, the adrenaline flows, the blood vessels dilate, the senses are alerted, the heartbeat is altered, blood thickens—all to get the person "up." Such an up state, if prolonged, is quite unhealthy. People who live and work in noisy environments have measurably greater general health problems, are grouchy and ill-tempered, and have trouble concentrating. Noise that deprives a person of sleep carries with it an additional array of health problems.

What is most important is that we cannot adapt to high noise levels. Industrial workers are often seen walking around with their ear protectors hanging around their necks as if this is some macho thing to do. They think they can "take it." Perhaps they can take it because they already are deaf at the frequency ranges prevalent in that environment.

16.4 NOISE ABATEMENT

Noise has been a part of urban life since the first cities. The noise of chariot wheels in Rome, for example, caused Julius Caesar to forbid chariot riding after sunset so he could sleep. In England, the birthplace of the Industrial Revolution, cities were incredibly noisy. As one American visitor described it:

The noise surged like a mighty heart-beat in the central districts of London's life. It was a thing beyond all imaginings. The streets of workaday London were uniformly paved in "granite" sets ... and the hammering of a multitude of iron-shod hairy heels, the deafening side-drum tattoo of tyred wheels jarring from the apex of one set (of cobblestones) to the next; the sticks dragging along a fence; the creaking and groaning and chirping and rattling of vehicles, light and heavy, thus maltreated, the jangling of chain harness; augmented by the shrieking and bellowings called for from those of God's creatures who desired to impart information or proffer a request vocally—raised a din that is beyond conception. It was not any such paltry thing as noise.[3]

In the United States, cities grew increasingly noisy, and all manner of ordinances were contrived to hold down the racket. In Dayton, Ohio, the city proclaimed that "it is illegal for hawkers or peddlers to disturb the peace and quiet by shouting or crying their wares." Most of these ordinances were, of course, of little use and soon disappeared into oblivion, as the noise increased.

The most effective legislation to pass Congress in the noise control area has been the Occupational Safety and Health Act of 1970, in which the Occupational Safety and Health Administration (OSHA) was given authority to control industrial noise levels. One of the most important regulations promulgated by OSHA is the limits of industrial noise. Recognizing that both the intensity and the duration are important in preventing damage to hearing, the OSHA regulations limit noise in both general industry and the construction industry, as shown in Table 16.3. Sound levels greater than 115 dB(A) are clearly detrimental and should not be permitted.

Industrial noise is mostly constant over an 8-hr working day. Community noise, however, is intermittent noise. If an airplane goes overhead only a few times a day, the sound level is quite low on average, but the irritation factor is high. For this reason a plethora of noise indexes have been developed, all of which are intended to estimate the psychological effect of noise. Most of these all begin by using the cumulative distribution technique, similar to the analysis of natural events, such as floods, discussed in Chapter 3.

The U.S. Department of Transportation has been concerned with traffic noise and has established maximum sound levels for different vehicles. Modern trucks and automobiles are considerably quieter than they were, but much of this has been forced not by governmental fiat but by public demand. The exception is renegade trucks that continue to be the major noisemakers on the highways and have little chance of being caught by police who

Table 16.3 OSHA Permissible Noise Exposures

Duration (hr)	Sound Levels [dB(A)]
8	90
6	92
4	95
3	97
2	100
1.5	102
1	105
0.5	110
≤ 0.25	115

FOCUS ON

The Valuation of Intangible Benefits

In Chapter 2, we described some of the tools for and challenges of placing value on options that range from easily quantifiable to highly subjective. Putting a monetary value on environmental and other intangible assets, including the value of human life, is challenging because such intangible assets typically do not have a market value. And it works only if a dollar has equal value to all people. It is particularly difficult when the potential impacts are irreversible, when there is a risk of a disastrous outcome, and when the environment's carrying capacities are being—or are close to being—overrun. In practice, most engineering decisions must be made without full knowledge of the benefits and costs involved. This is not an argument against trying to make as good an assessment as one can, but rather an acknowledgement of the importance of taking into account cost and benefit elements that cannot be expressed in monetary terms.

Sometimes, market information (and behavior related to traded goods) can be used to reveal preferences in the population-at-large regarding the values given to nonmarket goods. In this manner, the value associated with a particular intangible asset can be embedded into prices for marketed assets, and these values can be used to unbundle the values attached to different characteristics of the asset. It is possible to estimate how much people are will to pay for environmental benefit to support engineering decision making. For example, a house next to a busy street is exposed to high noise levels and would lose part of its value compared to similar houses farther from the highway. This difference can be used to estimate the cost of noise.

Researchers evaluated a large dataset of information on the housing market in Birmingham, England. They supplemented it with measures of noise exposure from road, rail, and aircraft traffic to investigate the public's demand for peace and quiet (Figure 16.7). They then determined mean monetary value for what people would pay for reduced transportation-related noise. For road noise, they determined that the public would pay £31.49 per year per household for a 1-dB reduction from a 56-dB baseline (a reduction that they wouldn't even be able to discern!). They would pay even more (£91.15 per year), for the same change from an 81-dB baseline. The equivalent values for rail noise were found to be even more valuable, ranging from £83.61 to £139.65 per year.[4]

In other situations, market-based schemes may be developed that force monetary values onto environmental and other assets. For example, market-based approaches have been developed to reduce various air pollutants, including CO_2. These

Figure 16.7   How does noise affect a residence's value?

approaches are based on pollution trading schemes, whereby each stakeholder is permitted so many pollution credits. Stakeholders who emit fewer pollutants than their credits can sell them to other stakeholders who are emitting more pollutants than their credits would allow. This provides an incentive (i.e., income) for cleaner operations and a disincentive (i.e., cost) for polluting operations.

What is important, however, is that a cap on the system is designed to lower the *total* pollution load. This scheme is sometimes called "cap-and-trade"

and takes advantage of market mechanisms. The price of the air pollutant being traded will depend on the supply, demand, and challenge of reaching the total cap. Some of problems associated with cap-and-trade schemes occur when the cap is not sufficiently ambitious or when the trading scheme allows for certain polluting industries to continue to pollute heavily and to create hot spots of high pollution near their locations.

To date we are unaware of any noise pollution trading schemes but the strategy could be applicable if deemed adequately important.

do not consider noise to be a serious problem compared to other highway safety concerns. Excellent work has also been done by acoustic engineers working to quiet commercial aircraft. The modern airliners are amazingly quite compared to the much louder models, such as the Boeing 727.

However, the control of noise in the urban environment or noise produced by machines and other devices has not received much governmental or public support. Although Congress passed the Noise Control Act of 1972, little progress has been made in controlling noise. In the 1980s the noise office of the USEPA was shut down in the near scuttling of the agency and has not been reestablished.

16.5 NOISE CONTROL

The level of noise can be reduced by using one of three strategies: protect the recipient, reduce the noise at the source, or control the path of sound.

16.5.1 Protect the Recipient

Protecting the recipient usually involves the use of earplugs or other ear protectors. Small earplugs, although easy and cheap, are not very effective for many frequencies of noise. The ear can detect sound not only coming through the ear canal but also from the vibration of the bones surrounding the ear. Thus, small earplugs are only partially effective. Far better are ear muffs that cover the entire ear and protect the wearer from most of the surrounding noise. The trouble with large ear protectors, however, is psychological. Too often people will shun what they consider clumsy and uncomfortable ear protectors and decide to take their chances, thus negating the effectiveness of the protection.

16.5.2 Reduce Source Noise

Reducing the source of the noise is, of course, often the most effective means of noise control. The redesign of commercial airplanes has already been mentioned as an example of the effectiveness of this control strategy. Changing the type of motors used in and around the home also often effectively reduces noise. For example, changing from a two-stroke gasoline motor for a lawn mower to an electric mower effectively eliminates a common and insidious source of neighborhood noise. The most offensive noisemakers now are the leaf blowers; often the noise level of leaf blowers is far above what OSHA would allow for industrial noise.

For traffic noise, reducing the source of the noise (vehicles) can be done through redesigning vehicles and pavements. In addition, traffic noise can be reduced to some extent by increasing the use of alternative transportation, including mass transit, walking, and bicycling. For residential areas and other areas where higher traffic noise levels are unacceptable, traffic noise can be reduced by reducing vehicle speed and by encouraging the use of alternate routes, through either speed control devices or road designs.

16.5.3   Control Path of Noise

Changing the path of the noise is a third alternative. The most visible evidence of this tactic is the growth of noise walls, or barriers, along our highways. A variety of materials and designs have been used for noise barriers (Figure 16.8 on the next page). While earth berms are the cheapest noise barrier to construct, they, of course, require large areas of land, something that is typically lacking in urban areas. The majority of walls in the United States have been built with concrete or concrete block.

Noise walls provide maximum noise reduction to only the first row of receivers (e.g., houses). The reduction is half as much for the second row and negligible for others. If a barrier is installed on only one side of a road, the receivers on the opposite side of the road actually receive more noise. These effects are due to the properties of sound and noise walls. Noise walls create a shadow zone (Figure 16.9 on the next page), in which the maximum noise reduction occurs. Unfortunately, noise is not like light. A noise shadow is not perfect, and noise can bend and bounce off the barrier and even off the air, depending on atmospheric conditions. Sometimes noises miles away can be heard as the pressure waves bounce off inversion layers.

In addition to providing protection only for the buildings immediately in the lee of the barrier, noise walls are quite costly. As Figure 16.10 on page 575 shows, the cost of building these walls has risen to more than $1 million per kilometer. Whether the benefits are sufficient reason to spend public funds is questionable.

Highway noise barriers are considered for new highway construction when noise impacts are found. Noise impacts are considered to occur when future noise levels are predicted (through computer modeling) to significantly exceed existing levels (defined by state departments of transportation) or when future levels are predicted to equal or exceed 66 dB(A). However, noise barriers are constructed only when it is reasonable and feasible. What does this mean? The state departments of transportation, again, get to define reasonable and feasible. Some type of benefit/cost analysis is usually included as a criterion.

Highway sound has also been abated by *heavy* natural growth. Trees by themselves are not very effective, but a dense growth will reduce the sound pressure level by several

© Courtesy of P. Aarne Vesilind

Figure 16.8 Highway noise barriers: (A) precast concrete, (B) Durisol,® and (C) brick.

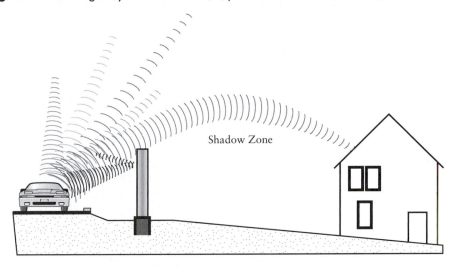

Shadow Zone

Figure 16.9 Noise barrier shadow zone.

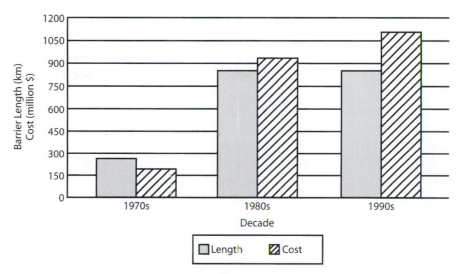

Figure 16.10 U.S. highway noise barrier construction and cost.

decibels per 31 m of dense forest. Cutting down the natural growth to widen a highway invariably will cause increased noise problems. In most urban settings, of course, there is not room for this type of dense growth, so plantings are more for aesthetics and psychological benefit than actual noise reduction.

PROBLEMS

16.1 A dewatering building in a wastewater treatment plant has a centrifuge that operates at 89 dB(A). Another similar machine is being installed in the same room. Will the operator be able to work in this room?

16.2 A lawn mower emits 80 dB(A). Suppose two other lawn mowers that produce 80 and 61 dB(A), respectively, join the mowing operation.

a. What will be the cacophony, in dB(A)?
b. What will be the sound pressure level 200 m away in a neighbor's yard?

16.3 A sludge pump emits the following sound as measured by a sound level meter at different octave bands:

Octave Band (Hz)	Sound Level [dB(A)]
31.5	39
63	42
125	40
250	48
500	22
1000	20
2000	20
4000	21
8000	10

a. What is the overall sound level, in dB(A)?
b. Does this meet OSHA criteria for an 8-hr working day?
c. What approaches could be taken to reduce the sound pressure level?

d. If the least expensive approach seems to be to use noise protectors (ear plugs), should this be used? What ethical considerations might there be involved in choosing this approach to noise control?

16.4 Using the frequency analysis in Problem 16.3, what would the sound *pressure* level be, in dB? [*Note*: You first have to correct the decibel readings to dB instead of dB(A), and then add them.]

16.5 A 20-year-old rock star has an audiogram done. The results are shown in Figure 16.11.

a. Is he able to hear normal conversational speech?
b. If the curve for Person *C* shown in Figure 16.5 represents a normal loss of hearing due to aging for a 65-year-old person, draw the audiogram for the rock star when he reaches 65 years old.
c. Will he be able to hear anything at all?

16.6 *Paternalism* is, as the name implies, a process of sheltering or caring for persons who cannot or will not care for themselves, as in the case of a parent and child. The name takes on a negative meaning, however, if the child is actually an adult and the parent is the government. The setting of OSHA industrial noise standards is a clear case of governmental paternalism.

Suppose an engineering friend of yours is working as an environmental control engineer for a corporation, and as part of her duties she is required to monitor the OSHA noise levels in her plant. She discovers that in one shop the noise level exceeds the allowable 8-hr OSHA level. There seems to be no means of reducing the source of the noise, nor can the path be altered. She suggests that the workers be required to wear ear protectors.

You are an OSHA compliance monitor and happen to be at the plant when your friend tells you about the shop that will require ear protection. But your friend tells you that the employees at this shop have asked her to please speak to you to see if you could allow them not to wear what they consider cumbersome and useless protectors. They insist that they know what is best for themselves, and

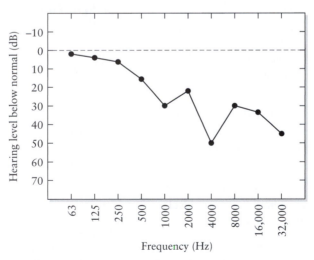

Figure 16.11   Audiogram for a rock star. See Problem 16.5.

that they do not want to be told how to protect themselves.

You meet with them and tell them and your friend, the plant engineer, that this is what the law says and that your hands are tied. But the workers insist that there must be reasonableness in the law, that it was passed so that workers who *want* protection can have the weight of the government on their side. But in this case the workers want the government to butt out, and leave them alone.

You have the power to overlook this small infraction in noise level in an otherwise safe industrial environment if you write a memo justifying the waiver of the regulations. Do you allow the waiver, or do you insist that the workers be protected from what appears to be excessive noise?

Write a one-page memorandum, addressed to your boss, recommending action.

16.7 A frequency analysis was conducted at an industrial site with the following results:

Octave Band (Hz)	Sound Pressure Level (dB)
31.5	30
63	40
125	10
250	30
500	128
1000	80
2000	10
4000	2
8000	0

a. What is the average sound pressure level?
b. What would the average sound pressure level be 100 m from this source?
c. If you had to reduce the sound pressure level, what would be your strategy for noise reduction?

16.8 A frequency analysis is conducted in an industrial plant with the following results:

Octave Band (Hz)	Sound Pressure Level (dB)
31.5	5
63	10
125	40
250	30
500	82
1000	48
2000	70
4000	125
8000	20
16,000	40
32,000	40

a. Suppose Person *B*, whose audiogram is shown in Figure 16.5, works at this plant. What can you say about the relationship between her audiogram and the frequency analysis conducted at the plant?
b. What is the average sound pressure level at the plant?
c. What is the average sound level [dB(A)] at the plant? (*Note*: Approximate by correcting the sound pressure levels, using Figure 16.3.)

16.9 A metal noise barrier was installed along an interstate. It experienced severe corrosion (Figure 16.12 on the next page). After a redesign, a subsequent installation of the barrier along a nearby stretch of interstate did not experience the corrosion. What contributed to the corrosion?

16.10 Noise effects on existing urban residences and institutions (such as schools, nursing homes, and hospitals) in an area are being studied.

a. Noise measurements of individual sources in one location were taken and found to be 67 dB, 58 dB, 57 dB, 59 dB, and 64 dB. What is the total noise level?
b. List three different methods to abate (i.e., actually reduce) the noise from increased traffic for the people living and working in the area.

Figure 16.12 Metal noise barrier: (A) original and (B) redesign. See Problem 16.9.

END NOTES

1. In honor of German physicist Heinrich Hertz.
2. In honor of Alexander Graham Bell.
3. Still, H. 1970. *In Quest of Quiet*. Harrisburg, PA: Stackpole Books.
4. Day, Brett, Ian Bateman, and Iain Lake. 2007. Beyond implicit prices: Recovering theoretically consistent and transferable values for noise avoidance from a hedonic property price model. *Environmental and Resource Economics* 37, No. 1: 211–32.

Ethics of Green Engineering[*]

© Keith and Susan Morgan

Fur seals at Admiral Arch, New Zealand

Environmental engineers work with others to do what is needed to protect human health and the environment and lead the way towards a sustainable standard of living.

[*]Contributions to this chapter by Jamie Hendry and Susan Hamill are gratefully acknowledged.[1]

It is one thing to do green engineering and quite another to explain why one *ought* to do green engineering. In this chapter, we consider why engineers choose to do green engineering and distinguish between green engineering that aids the survival and profitability of the firm and green engineering performed to benefit society-at-large and future generations.

We begin by describing the historical and political initiatives that have merged to create the concept of green engineering and then use examples to show how most green engineering is often simply self-serving actions that also happen to be beneficial in terms of global environmental pollution and resource use. Finally, we argue that true moral worth comes when engineers practice green engineering because doing so benefits others, not because it is necessarily profitable.

17.1 GREEN ENGINEERING

To attain sustainability, people need to adopt new and better means of using materials and energy. The operationalizing of the quest for sustainability is defined as *green engineering*, a term that recognizes that engineers are central to the practical application of sustainability principles to everyday life. The relationship between sustainable development, sustainability, and green engineering is as follows:

$$\text{Sustainable Development} \rightarrow \text{Green Engineering} \rightarrow \text{Sustainability}$$

Sustainable development is an ideal that can lead to sustainability, and one that needs green engineering to be successful. Green engineering is a growing field. Almost everyone is fairly confident that science and technology can support sustainability through the development of sustainable materials and processes, but we also recognize that technology is not a panacea for all global problems and that a higher level of technology will not guarantee sustainability. Technology can, however, provide innovative solutions when society challenges scientists and engineers to promote human and ecological well-being and to embrace systems concepts in their activities.

Green engineering is emerging as an umbrella field in which engineers from diverse disciplines pave the way for sustainable coexistence between humans and the rest of the natural world. The goal therefore is sustainability, and green engineering is the way to achieve this goal.

17.2 MOTIVATIONS FOR PRACTICING GREEN ENGINEERING

To understand why humans behave as they do, one has to identify the driving forces that lead to particular activities, as we do, for example, when determining that a concentration difference across an interface is causing gas transfer. In engineering and business alike, driving forces spur the adoption of new technologies or practices. The objective is to understand the motivational forces for adopting green engineering and business practices. The three driving forces that appear to support green engineering are legal considerations, financial considerations, and finally ethical considerations.

17.2.1   Legal Considerations

At the simplest level, green engineering is practiced to comply with the law. For example, a supermarket recycles corrugated cardboard because it is the law, either a state law such as in North Carolina or a local ordinance as in Pennsylvania. This behavior is, at best, "morality lite." Engineers and managers comply with the law because of the threat of punishment for noncompliance. The decision to comply with the law is thus largely an amoral decision. Complying with the law is neither morally good nor morally bad, although not complying may be considered morally bad. So in this situation, managers and engineers choose to do the right thing because they feel it is their only choice.

Occasionally, however, firms prioritize financial concerns over legal concerns, and the managers may determine that by adopting an illegal practice (or failing to adopt a practice codified in law) they can enhance profitability. In such cases they argue that either the chances of getting caught are low or the potential for profit is large enough to override the penalty if they do get caught. In other cases private firms can take advantage of loopholes in tax laws that inadvertently allow companies to pretend to be environmentally green while in reality doing nothing but gouging the taxpayer.

FOCUS ON

The EPA Versus the Electric Power Industry

In November 1999 the USEPA sued seven electric utility companies—American Electric Power, Cinergy, FirstEnergy, Illinois Power, Southern Indiana Gas & Electric Company, Southern Company, and Tampa Electric Company—for violating "the Clean Air Act by making major modifications to many of their coal burning plants without installing the equipment required to control smog, acid rain and soot."[2] On 7 August, 2003, "Judge Edmund Sargus of the U.S. District Court for the Southern District of Ohio found that Ohio Edison, an affiliate of FirstEnergy Corp., violated the Clean Air Act's [new source review] provisions by undertaking 11 construction projects at one of its coal-fired plants from 1984 to 1998 without obtaining necessary air pollution permits and installing modern pollution controls on the facility."[3] The companies had calculated that breaking the law and possibly getting caught was the lowest cost solution and that thus behaving illegally was "the right answer."

17.2.2   Financial Considerations

Decisions about the adoption of green practices are also driven by financial concerns. This level of involvement with greening is at the level promoted by the economist Milton Friedman, who stated famously, "The one and only social responsibility of business [is] to use its resources and engage in activities designed to increase its profits so long as it . . . engages in open and free competition, without deception or fraud."[4] In line with this stance, the firm calculates the financial costs and benefits of adopting a particular practice and makes its decision based on whether the benefits outweigh the costs or vice versa. Many companies seek out green engineering opportunities solely on the basis of their providing a means of lowering expenses, thereby increasing profitability.

FOCUS ON

The Synfuel Scam[2]

In the 1970s the U.S. Congress decided to promote the use of cleaner fuels to take advantage of both the huge coal reserves in the United States and the environmental benefits derived from burning a clean gaseous fossil fuel made from coal. Producing such synfuel from coal had already been successfully implemented in Canada, and the U.S. government wanted to encourage American power companies to get into the synfuel business. To promote the industry, Congress wrote in huge tax credits for companies that would produce synfuel and defined a synfuel as chemically altered coal, anticipating that the conversion would be to a combustible gas that could be used much as natural gas is used.

Unfortunately, the synfuel industry did not develop as expected because cheaper natural gas supplies became available. The synfuel tax credit idea remained dormant until the 1990s when a number of corporations (including some giants like the Marriott hotel chain) found the tax break and

went into the synfuel business. Because the only requirement was to change the chemical nature of the fuel, it became evident that even spraying the coal with diesel oil or pine tar would alter the fuel chemistry and that this fuel would then be legally classified as a synfuel. The product of these synfuel plants was still coal, and more expensive coal than raw coal at that, but the tax credits were enormous. Some companies were formed specifically to take advantage of the tax break, often with environmentally attractive names like Earthco, making huge profits by selling their tax credits to other corporations that needed them. The synfuels industry received a gift from the U.S. taxpayer of over $1 billion annually while doing nothing illegal but also while doing nothing to benefit the environment.

Source: Based on D. L. Barlett and J. B. Steele, "The Great Energy Scam," *TIME*, October 2003, Vol. 162, n. 15.

Some firms may believe that adopting a particular green engineering technology will provide them with public relations opportunities because green engineering is a useful tool for enhancing the company's reputation and community standing. If the result is likely to be an increase in sales for the business and if sales are projected to rise *more* than expenses so that profits rise, the firm is likely to adopt such a technology. The same is true if the public relations opportunities can be exploited to provide the firm with expense reductions, such as decreased enforcement penalties or tax liabilities. Similarly, green technologies that not only yield increased sales but at the same time decrease expenses are the perfect recipes for adoption by a company whose primary driving forces are financial concerns.

Cases in which engineers are simply trying to practice good business, seeking ways to increase the difference between revenues and expenses so that profits will rise, are not influenced by the desire to do the right thing for the environment. On occasion, though, managers are *forced* into considering the adoption of greener practices by the threat that not doing so will cause expenses to rise and/or revenues to fall.

One of the strongest forces in moving engineering into the green mode is standardization. For example, firms doing business in Europe are almost forced to adopt environmental management systems as defined under what are known as ISO 14000 standards. (ISO is the International Standards Organization, based in Switzerland.) Not doing so means that the firms would be locked out of the European market.

FOCUS ON

Green is Profitable

The following examples illustrate the adoption of green engineering for financial reasons:

- In one of its facilities at Deepwater, New Jersey, DuPont used phosgene, an extremely hazardous gas that was shipped to the plant. In an effort to reduce the chance of accidents, DuPont redesigned the plant to produce phosgene onsite and to use almost all of it in the manufacturing process, avoiding the costs associated with hazardous gas transport and disposal.

- Polaroid did a study of all of the materials it used in manufacturing and grouped them into five categories based on risk and toxicity. Managers were encouraged to alter product lines to reduce the amount of material in the most toxic groups. In the first five years, the program resulted in a reduction of 37% of the most toxic chemicals and saved over $19 million in avoided disposal costs.

- Dow Chemical challenged its subsidiaries in Louisiana to reduce energy use and sought ideas on how this should be done. Following up on the best ideas, Dow invested $1.7 million and received a 173% return on its investment.

Source: K. Gibney, "Combining environmental caretaking with sound economics, sustainable development is a new way of doing business," ASEE *Prism*. American Society of Engineering Education, January 2003.

FOCUS ON

Public Relations and Green Engineering

Positive public relations can be highly valuable to companies. For example, DuPont's well publicized decision to discontinue its $750-million-a-year business producing chlorofluorocarbons (CFCs) was a public relations bonanza. Not only did DuPont make it politically possible for the United States to become a signatory to the Montreal Protocol on ozone depletion, but it already had alternative refrigerants in the production stage and was able to smoothly transition to these. In 1990 the USEPA gave DuPont the Stratospheric Protection Award in recognition of the decision to get out of CFC manufacturing.[5] The fact that the decision also proved to be highly profitable for DuPont apparently did not matter to the judges.

Another example of the PR benefits of green engineering involves the seven electric companies sued by the USEPA in November 1999 for Clean Air Act violations. The companies heavily publicized their efforts to reduce greenhouse gas emissions. American Electric Power (AEP) issued news releases regarding emissions reduction efforts at various plants. Not coincidentally, the U.S. government, which had sued during the Clinton administration but which was now operating under the Bush administration, was in the process of revising the portions of the Clean Air Act that the company had violated. Presumably, regulators were favorably impressed with the company's hard work; in August 2003, the USEPA announced that it was dropping the suit and revamping that portion of the act.

FOCUS ON

Doing the Right Thing Because You Have To

In October 1998, ELF (Earth Liberation Front) targeted Vail Ski Resort, burning a $12 million expansion project to the ground.[6] In the wake of this damage, the National Ski Areas Association (NSAA) began developing its Environmental Charter in 1999 with "input from stakeholders, including . . . environmental groups,"[7] and officially adopted the charter in June 2000.[8] In accordance with the charter, NSAA produced a Sustainable Slopes Annual Report.[9] Apparently, the driving force behind the decision to adopt the Environmental Charter was largely a response to financial concerns rather than the desire to treat the environment responsibly: It was an amoral decision. That is, NSAA was spurred to create the Environmental Charter by concerns about member companies' bottom lines. Further ecoterrorist activity could occur, thereby causing expenses to rise. And the ELF action may have sufficiently highlighted the environmental consequences of resort development to the point that environmentally minded skiers might pause before deciding to patronize resorts where development was occurring, thereby causing revenues to fall.

There is ample evidence that one of the reasons engineers participate in the quest for sustainability is that it is good for business. The leaders of eight leading firms that adopted an environmentally proactive stance on sustainability were asked in one study to justify the firms' adoption of such a strategy.[10] All companies reported that they were motivated first by regulations, such as the control of air emissions, pretreatment of wastewater, and the disposal of hazardous materials. One engineer in the study admitted that "the [waste disposal] requirements became so onerous that many firms recognized the benefits of altering their production processes to generate less waste."

The second motivator identified in this study was competitive advantage. One director of a microprocessor company noted that "by reducing pollution, we can cut costs and improve our operating efficiencies." The company recognized the advantage of cutting costs by reducing its hazardous waste stream.[11]

Another study, conducted by Price Waterhouse Coopers, confirmed these findings.[12] When companies were asked to self-report on their stance on sustainable principles, the top two reasons for adopting sustainable development were found to be enhanced reputation (90%) and competitive advantage (cost savings) (75%). It is not clear whether the respondents were given the option of responding that they practiced sustainable operations because this was mandated by law. If it had, there is no doubt that all companies would have publicly stated that they are indeed law-abiding.

So it seems likely that the two primary driving forces behind the adoption of green engineering practices are legal concerns and financial concerns. Can we argue that such behavior is morally admirable simply on the basis that the outcomes (e.g., cleaner air and water) are morally preferable? No. Actions undertaken in response to legal and financial concerns are actually *obligatory*, in that society essentially demands that businesses make their decisions within legal and financial constraints.[13] For an action to be morally admirable, the motivating driving force has to be far different in character.

17.2.3 Ethical Considerations

Although most engineering decisions are made on the basis of legal or financial concerns, some companies believe that behaving more environmentally responsibly is simply the right thing to do. They believe that saving resources—and perhaps even the planet—for the generations that will follow is important. When making decisions, they are guided by the triple bottom line with the goal to balance the financial, social, and environmental impacts of each decision.

FOCUS ON

The Triple Bottom Line

The founder of Interface Carpet Company was Ray Anderson, now chairman of the board.[14] By the mid-1990s, Interface had grown to nearly $1.3 billion in sales, employed some 6600 people, manufactured on four continents, and sold its products in over 100 countries worldwide. In 1994, several members of Interface's research group asked Anderson to give a kickoff speech for a task force meeting on sustainability; they wanted him to provide Interface's environmental vision. Despite his reluctance to do so—Anderson had no environmental vision for the company except to comply with the law—he agreed. Fortuitously, as Anderson struggled to determine what to say, someone sent him a copy of Paul Hawken's *The Ecology of Commerce*[15]; Anderson read it, and it completely changed not only his view of the natural environment and not only his vision for Interface

Carpet Company, but his entire conception of business. In the coming years, he held meetings with employees throughout the Interface organization explaining to them his desire to see the company spearhead a sustainability revolution. No longer would they be content to keep pollutant emissions at or below regulatory levels. Instead, they were going to strive to be a company that created zero waste and did not emit any pollutants *at all*. The company began to employ The Natural Step[16] and notions of Natural Capitalism[17] as part of its efforts to become truly sustainable. The program continues today, and although the company has saved many millions of dollars as a result of adopting green engineering technologies and practices, the reason for adopting these principles was not to earn more money but rather to do the right thing.

Decisions to do green engineering proactively are morally admirable not simply because they enable a company to reduce toxic emissions (among many other positive outcomes for the environment); they are morally admirable when the *driving force* behind those decisions is the desire to stop harming the earth, to protect it so that future generations will be able to enjoy it as much as, or even more than, we do today. Conversely, in the cases where the *driving force* behind decisions to adopt green technologies is a desire to save the company money, the benefits to the earth were simply a fortunate byproduct of those decisions.

To understand some of these deeper motivational factors for adopting green engineering practices, a study asked a small set of engineers about their *real* reasons for participating in sustainability activities. When pressed for their personal drivers for adopting green engineering in their work, the respondents often said things like the following:

- My son challenged me to leave the earth a better and more livable place for my grandchildren.
- I hope to do my part to ensure that the planet remains a nice place for my children and grandchildren to explore.
- Waste is ugly and engineers in general can't stand things that are not designed well. I also now have a child and I want her to have a future.
- I'd rather be useful than famous.
- Some of my personal drivers come from a deep respect and love of the natural world. I think most human beings enjoy the fresh air and water that are found in the remote places in nature. Unfortunately, this freshness is limited. I would like to be in a world where I don't have to worry about breathing clean air or drinking uncontaminated water.
- I think we should at a minimum be aware of what we are doing (make it a conscious decision rather than side effect of our thoughtless consumer lifestyles), and in the ideal world we would be doing everything we could to stop the current destruction and restore at least some of what we have already damaged.
- I am interested in the environment in general. I am also a student of energy efficiency, but mostly I have a love for the outdoors. Sustainable design has a certain beauty in economy. But what is more intriguing as a designer is the focus is less narrow than traditional engineering. There is an appreciation for context and the way in which processes are integrated. It approaches artistic beauty, particularly when all the competing design goals are met. It is a real high.
- I have been interested in the environment most of my life. I suppose I express that through my involvement with the electric car project. It's fun and lets me be a "gear head." It allows my passion to intersect with my need to "do no harm."

Based on these interviews, it is evident that capitalist motives are inadequate to describe the driving forces that form the attitudes of all engineers and industrial leaders toward sustainability. Apparently some believe that green engineering is a way to protect the environment for future generations and it is an important part of their jobs. When *ethical considerations* are the motivational driving forces for corporate decisions to adopt green engineering, these decisions have moral value.

17.3 CONCLUSIONS

Green engineering will eventually help lead the world to sustainability, but green engineering today occurs most often when doing the right thing also results in adherence to laws and regulations and in achieving greater profitability for the organization. This kind of green engineering, although often beneficial to society, is not morally admirable.

The true heroes of green engineering are those who believe deeply in the principles behind green engineering and who try to work within these principles while still helping their corporations to be profitable. They enjoy working to promote sustainability and do so not for show or profit but because it gives them pleasure to *do the right thing*.

END NOTES

1. This article was published in *Sustainability Science and Engineering: Defining Principles*, Vol 1, M. Abraham, ed., Ethics of green engineering, by Vesilind, P. A., L. G. Heine, R. Hendry, and S. A. Hammill, Pages 33–46, Copyright Elsevier, 2006.

2. Lazaroff, C. U.S. government sues power plants to clear dirty air. *Environment News Service*. Accessed at http://ens.lycos.com/ens/nov99/1999L-11-03-06.html on 8 December, 2008.

3. Fowler, D. Bush administration, environmentalists battle over "new source review" air rules. Fall 2003. *Group Against Smog and Pollution Hotline*. Accessed at http://www.gasp-pgh.org/hotline/fall03_4.html on 3 January, 2004.

4. Friedman, M. 1962. *Capitalism and Freedom*. Chicago: University of Chicago Press, p. 133.

5. Billatos, S. B., and N. A. Basaly. 1997. *Green Technology and Design for the Environment*. Washington, DC: Taylor & Francis.

6. Associated Press. Two Formally Plead Guilty to Firebombing Ski Resort in Vail, Colorado. Dec. 14, 2006 Accessed at http://www.fxnews.com/story/0,2933,23657, 00.html on 9 January, 2009.

7. Ski Areas Association (NSAA). Environmental Charter. Accessed at http://www.nsaa.org/nsaa/environmentsustainableslopes.pdf

8. Jesitus, J. 2000. Charter promotes environmental responsibility in ski areas. *Hotel & Motel Management* 215, No. 19: 64.

9. National Ski Areas Association (NSAA). Sustainable slopes annual report. Accessed at http://www.nsaa.org/nsaa/environmentsustainable_slopes.

10. Lawrence, Anne T., and David Morell. 1995. Leading-edge environmental management: Motivation, opportunity, resources and processes. *Research in Corporate Social Performance and Policy*. J. s E. Post, D. Collins, and M. Starik (eds) Supplement 1, Greenwich, CT: JAI Press.

11. (ibid) Lawrence, Anne T., and David Morell. 1995. Leading-edge environmental management: Motivation, opportunity, resources and processes. *Research in Corporate Social Performance and Policy*. James Post, D. Collins, and M. Starik, eds. Supplement 1. Greenwich, CT: JAI Press.

12. Price WaterhouseCooper. 2002. *Sustainability Survey Report*. New York.

13. Sethi, S. P. 1975. Dimensions of corporate social performance: An analytical framework. *California Management Review* 17, No. 3 (Spring): 58–64.

14. Anderson, R. C. 1998. *Mid-course correction—Toward a sustainable enterprise: The interface model*. Atlanta GA: The Peregrinzilla Press.

15. Hawken, P. 1994. *The Ecology of Commerce: A Declaration of Sustainability*. New York: Harper Business.

16. Robert, K-H. 1991. Educating a nation: The natural step. *In Context* No. 28 (Spring).

17. Lovins, A. B., L. H. Lovins, and P. Hawken. 1999. A road map for natural capitalism. *Harvard Business Review* 7 No. 3: 145–58.

INDEX

Selected Elements

Element Name	Symbol	Atomic Weight (g/gmol)	Element Name	Symbol	Atomic Weight (g/gmol)
Aluminum	Al	26.981	Molybdenum	Mo	95.94
Antimony	Sb	121.75	Neon	Ne	20.183
Argon	Ar	39.948	Nickel	Ni	58.71
Arsenic	As	74.922	Niobium	Nb	92.906
Barium	Ba	137.34	Nitrogen	N	14.007
Beryllium	Be	9.012	Oxygen	O	15.999
Bismuth	Bi	208.98	Palladium	Pd	106.4
Boron	B	10.811	Phosphorus	P	30.974
Bromine	Br	79.909	Platinum	Pt	195.09
Cadmium	Cd	112.40	Potassium	K	39.102
Calcium	Ca	40.08	Radium	Ra	226
Carbon	C	12.01115	Radon	Rn	222
Cesium	Cs	132.90	Rhodium	Rh	102.91
Chlorine	Cl	35.453	Rubidium	Rb	85.47
Chromium	Cr	51.996	Ruthenium	Ru	101.07
Cobalt	Co	58.933	Scandium	Sc	44.956
Copper	Cu	63.54	Selenium	Se	78.96
Fluorine	F	18.998	Silicon	Si	28.086
Gallium	Ga	69.72	Silver	Ag	107.87
Germanium	Ge	72.59	Sodium	Na	22.990
Gold	Au	196.97	Strontium	Sr	87.62
Helium	He	4.003	Sulfur	S	32.064
Hydrogen	H	1.0080	Tellurium	Te	127.60
Indium	In	114.82	Tin	Sn	118.69
Iodine	I	126.90	Titanium	Ti	47.90
Iron	Fe	55.847	Tungsten	W	183.85
Krypton	Kr	83.80	Uranium	U	238.03
Lead	Pb	207.19	Vanadium	V	50.942
Lithium	Li	6.939	Xenon	Xe	131.30
Magnesium	Mg	24.312	Yttrium	Y	88.905
Manganese	Mn	54.938	Zinc	Zn	65.37
Mercury	Hg	200.59	Zirconium	Zr	91.22

Conversions

Multiply	by	to obtain
acres	0.404	ha
acres	43,560	ft^2
acres	4047	m^2
acres	4840	yd^2
acre ft	1233	m^3
atmospheres	14.7	lb/in^2
atmospheres	29.95	in mercury
atmospheres	33.9	ft of water
atmospheres	10,330	kg/m^2
Btu	252	cal
Btu	1.053	kJ
Btu	1,053	J
Btu/ft^3	8,905	cal/m^3
Btu/lb	2.32	kJ/kg
Btu/lb	0.555	cal/g
Btu/s	1.05	kW
Btu/ton	278	cal/tonne
Btu/ton	0.00116	kJ/kg
calories	4.18	J
calories	0.0039	Btu
calories/g	1.80	Btu/lb
calories/m^3	0.000112	Btu/ft^3
calories/tonne	0.00360	Btu/ton
centimeters	0.393	in
cubic ft	1728	in^3
cubic ft	7.48	gal
cubic ft	0.0283	m^3
cubic ft	28.3	L
cubic ft/lb	0.0623	m^3/kg
cubic ft/s	0.646	million gal/day
cubic ft/s	0.0283	m^3/s
cubic ft/s	449	gal/min
cubic ft of water	61.7	lb
cubic in of water	0.0361	lb
cubic m	35.3	ft^3
cubic m	264	gal
cubic m	1.31	yd^3
cubic m/day	264	gal/day
cubic m/hr	4.4	gal/min
cubic m/hr	0.00638	million gal/day
cubic m/s	1	cumec
cubic m/s	35.31	ft^3/s
cubic m/s	15,850	gal/min
cubic m/s	22.8	mil gal/day

Multiply	by	to obtain
cumec	1	m^3/s
cubic yards	0.765	m^3
cubic yards	202	gal
C-ration	100	rations
decacards	52	cards
feet	0.305	m
feet/min	0.00508	m/s
feet/s	0.305	m/s
feet/s	720	in/min
fish	10^{-6}	microfiche
foot lb (force)	1.357	J
foot lb (force)	1.357	Nm
gallon of water	8.34	lb
gallons	0.00378	m^3
gallons	3.78	L
gallons/day	43.8×10^{-6}	L/s
gallons/day/ft^2	0.0407	m^3/day/m^2
gallons/min	0.00223	ft^3/s
gallons/min	0.0631	L/s
gallons/min	0.227	m^3/hr
gallons/min	6.31×10^{-5}	m^3/s
gallons/min/ft^2	2.42	m^3/hr/m^2
gallons of water	8.34	lb water
ms	0.0022	lb
grams/cm^3	1,000	kg/m^3
hectares	2.47	acre
hectares	1.076×10^5	ft^2
horsepower	0.745	kW
horsepower	33,000	ft lb/min
inches	2.54	cm
inches/min	0.043	cm/s
inches of mercury	0.49	lb/in^2
inches of mercury	0.00338	N/m^2
inches of water	249	N/m^2
joules	0.239	cal
joules	9.48×10^{-4}	Bru
joules	0.738	ft lb
joules	2.78×10^{-7}	kWh
joules	1	Nm
joules/g	0.430	Btu/lb
joules/s	1	W

Conversions

Multiply	by	to obtain	Multiply	by	to obtain
kilocalories	3.968	Btu	million gal/day	1.547	ft^3/s
kilocalories/kg	1.80	Btu/lb	million gal/day	694	gal/min
kilograms	2.20	lb			
kilograms	0.0011	tons	newtons	0.225	lb(force)
kilograms/ha	0.893	lb/acre	newtons/m^2	2.94×10^{-4}	in mercury
kilograms/hr	2.2	lb/hr	newtons/m^2	$1.4 \times 10\ 10^{-4}$	lb/in^2
kilograms/m^3	0.0624	lb/ft^3	newtons/m^2	10	poise
kilograms/m^3	1.68	lb/yd^3	newton m	1	J
kilograms/m^3	1.69	lb/yd^3			
kilograms/tonne	2.0	lb/ton	pounds (mass)	0.454	kg
kilojules	9.49	Btu	pounds (mass)	454	g
kilojules/kg	0.431	Btu/lb	pounds (mass)/acre	1.12	kg/ha
kilometers	0.622	mile	pounds(mass)/ft^3	16.04	kg/m^3
kilometer/hr	0.622	miles/hr	pounds (mass)/ton	0.50	kg/tonne
kilowatts	1.341	horsepower	pounds (mass)/yd^3	0.593	kg/m^3
kilowatt-hr	3600	kJ	pounds (force)	4.45	N
kilowatt-hr	3.410	Btu	pounds (force)/in^2	0.068	atmospheres
			pounds (force)/in^2	2.04	in of mercury
lite year	365	days drinking low-calorie beer	pounds (force)/in^2	6895	N/m^2
			pounds (force)/in^2	6.89	kPa
liters	0.0353	ft^3	pound cake	454	Graham crackers
liters	0.264	gal			
liters/s	15.8	gal/min	pounds of water	0.01602	ft^3
liters/s	0.0228	million gal/day	pounds of water	27.68	in^3
			pounds of water	0.1198	gal
megaphone	10^{-12}	microphone			
meters	3.28	ft	square ft	0.0929	m^2
meters	1.094	yd	square m	10.74	ft^2
meters/s	3.28	ft/s	square m	1.196	yd^2
meters/s	196.8	ft/min	square m	2.471×10^{-4}	acres
microscope	10^{-6}	mouthwash	square miles	2.59	km^2
miles	1.61	km			
miles/hr	0.447	m/s	tons (2000 lb)	0.907	tonnes (1000 kg)
miles/hr	88	ft/min	tons	907	kg
miles/hr	1.609	km/hr	tons/acre	2.24	tonnes/ha
milligrams/L	0.001	kg/m^3	tonnes (1000 kg)	1.10	ton (2000 lb)
million gal	3,785	m^3	tonnes/ha	0.446	tons/acre
million gal/day	43.8	L/s	two kilomockingbirds	2000	mockingbird
million gal/day	3785	m^3/day			
million gal/day	157	m^3/hr	watts	1	J/s
million gal/day	0.0438	m^3/s			
million gal/day	0.0438	m^3/s	yard	0.914	m

PRINCIPAL UNITS USED IN MECHANICS

Quantity	International System (SI)			U.S. Customary System (USCS)		
	Unit	Symbol	Formula	Unit	Symbol	Formula
Acceleration (angular)	radian per second squared		rad/s^2	radian per second squared		rad/s^2
Acceleration (linear)	meter per second squared		m/s^2	foot per second squared		ft/s^2
Area	square meter		m^2	square foot		ft^2
Density (mass) (Specific mass)	kilogram per cubic meter		kg/m^3	slug per cubic foot		$slug/ft^3$
Density (weight) (Specific weight)	newton per cubic meter		N/m^3	pound per cubic foot	pcf	lb/ft^3
Energy; work	joule	J	$N \cdot m$	foot-pound		ft-lb
Force	newton	N	$kg \cdot m/s^2$	pound	lb	(base unit)
Force per unit length (Intensity of force)	newton per meter		N/m	pound per foot		lb/ft
Frequency	hertz	Hz	s^{-1}	hertz	Hz	s^{-1}
Length	meter	m	(base unit)	foot	ft	(base unit)
Mass	kilogram	kg	(base unit)	slug		$lb\text{-}s^2/ft$
Moment of a force; torque	newton meter		$N \cdot m$	pound-foot		lb-ft
Moment of inertia (area)	meter to fourth power		m^4	inch to fourth power		$in.^4$
Moment of inertia (mass)	kilogram meter squared		$kg \cdot m^2$	slug foot squared		$slug\text{-}ft^2$
Power	watt	W	J/s $(N \cdot m/s)$	foot-pound per second		ft-lb/s
Pressure	pascal	Pa	N/m^2	pound per square foot	psf	lb/ft^2
Section modulus	meter to third power		m^3	inch to third power		$in.^3$
Stress	pascal	Pa	N/m^2	pound per square inch	psi	$lb/in.^2$
Time	second	s	(base unit)	second	s	(base unit)
Velocity (angular)	radian per second		rad/s	radian per second		rad/s
Velocity (linear)	meter per second		m/s	foot per second	fps	ft/s
Volume (liquids)	liter	L	$10^{-3}\ m^3$	gallon	gal.	$231\ in.^3$
Volume (solids)	cubic meter		m^3	cubic foot	cf	ft^3